PHYSICS IN LOCAL LATTICE DISTORTIONS

Related Titles from AIP Conference Proceedings

To learn more about these titles, or the AIP Conference Proceedings Series, please visit the
webpage **http://www.aip.org/catalog/aboutconf.html**

PHYSICS IN LOCAL LATTICE DISTORTIONS

Fundamentals and Novel Concepts
LLD2K

Ibaraki, Japan   23–26 July 2000

EDITORS

Hiroyuki Oyanagi

Electrotechnical Laboratory, Ibaraki, Japan

Antonio Bianconi

University of Rome "La Sapienza", Rome, Italy

Melville, New York, 2001
AIP CONFERENCE PROCEEDINGS ■ VOLUME 554

Editors:

Hiroyuki Oyanagi
Electrotechnical Laboratory
1-1-4 Umezono, Tsukuba
Ibaraki 305-8568
JAPAN

E-mail: oyanagi@etl.go.jp

Antonio Bianconi
Department of Physics
INFN, University of Rome "La Sapienza"
Piazzale Aldo Moro 2
I-00185 Rome
ITALY

E-mail: antonio.bianconi@roma1.infn.it

L.C. Catalog Card No. 00-112076
ISBN 1-56396-984-X
ISSN 0094-243X
Printed in the United States of America

CONTENTS

1. HIGH TEMPERATURE SUPERCONDUCTIVITY

1.1 ELECTRONIC STRUCTURE

1.2 QUANTUM FLUCTUATION

1.3 QUANTUM ORDERING

1.4 QUANTUM ORDERING AND SUPERCONDUCTIVITY: THEORETICAL APPROACH

2. PHASE TRANSITIONS

2.1 LATTICE INSTABILITY AND STRUCTURAL PHASE TRANSITIONS

2.2 PHOTO-INDUCED PHASE TRANSITION

3. COLOSSAL MAGNETO RESISTANCE (CMR)

3.1 ELECTRONIC STRUCTURE AND CHARGE DYNAMICS

3.2 JAHN-TELLER DISTORTION AND MAGNETIC ORDERING

4. LATTICE EFFECTS ON ELECTRONIC STRUCTURE

4.1 ELECTRONIC STRUCTURE

4.2 LOCAL STRUCTURE

4.3 HETEROGENEITY AND DISORDER

Preface

Interest in local lattice distortions (LLD's) and physics is rapidly growing in relation with the recent dramatic developments in research of strongly correlated electron systems with a layered perovskite structure, such as cuprates and manganites. In these oxides, the interplays among spin, charge, and lattice (orbital) freedoms are believed to play a major role in their exotic transport and magnetic properties, such as high temperature super conductivity (HTSC) or colossal magnetoresistance (CMR). In the last decade, a number of experimental and theoretical works reported anomalous lattice effects which demonstrated that there exists a strong coupling between the local lattice and electronic structure, although, the lattice effects were neglected or at least underestimated for a long time. However, recent efforts to quantitatively evaluate the lattice effects by fast time-scale local probes strongly suggest that the LLD cannot be neglected for establishing the microscopic mechanism. The origin of complexity in the electronic structure of HTSC cuprates is ascribed to the intrinsic tendency of electronic phase separation due to competing interactions, *i.e.*, attractive magnetic and repulsive Coulomb forces which naturally involve LLD's. Generic features of these complex interactions are presently under extensive investigation. Reflecting the urgent need to focus on the LLD and related topics, the **International Symposium on Physics in Local Lattice Distortions 2000** (LLD2K) was organized. A three-day meeting was planned to review the recent developments in experiments and theories, to critically discuss the underlying physics, and to establish the role of LLD.

Since the lattice effects in HTSC compounds were discussed in the Santa Fe conference in 1992, reports on unusual local lattice effects have accumulated, demonstrating the importance of *electron-lattice interaction*. Above all, the recent discoveries of mesoscopic modulations (**quantum stripes**) involving spin, charge and lattice (orbital) freedoms have revealed that quantum ordering is a generic feature associated with electronic phase separation. As interest in stripe physics grew rapidly, the International Conference on Stripes and High T_c Superconductivity was held. The current understanding is that the LLD's are signatures of electronic phase separation and quantum ordering. In this book, the first chapter is devoted to electronic structure, quantum fluctuation and ordering phenomena, and theoretical approaches in HTSC. The fundamental electronic structure of cuprates and mechanism of HTSC based on LLD-induced spin correlation is given by H. Kamimura. Chapter 1 describes the experimental and theoretical studies on quantum fluctuation and ordering phenomena in cuprates. Local probes such as x-ray absorption spectroscopy (XAS) and neutron scattering revealed inhomogeneous charge distributions, fluctuations and orderings. Recent experiments further confirmed that the charge and lattice fluctuations are related to the electronic structure, such as pseudo gap opening or the Fermi surface geometries, which indicates that spin, charge and lattice freedoms are all vital ingredients in the microscopic mechanism of HTSC.

In Chapter 2, structural instabilities in ferroelectric perovskites are described from local lattice viewpoints. A modern interpretation of structural phase transition is the microscopic mechanism, *i.e.*, how the dynamic LLD coupled with a soft phonon mode gives rise to a structural phase transition. Recent efforts to model the dynamic LLD have successfully explained the transport properties, including superconductivity. A new feature of phase transition is the dynamic response of local lattice upon electronic excitation. Photo-induced lattice distortions and phase transitions are described in the last part of Chapter 2. Chapter 3 is dedicated to Jahn-Teller distortions in manganites. The crucial role of Jahn-Teller distortions in stabilizing magnetic ordering is a central topic. CMR effects depend on the charge-orbital correlation competing with ferromagnetic order. Chapter 3 describes, beyond the double- and super-exchange interactions, how antiferromagnetic ordering is stabilized. Chapter 4 describes how the static LLD associated with impurity atoms affects the electronic structure, which eventually leads to nano-scale heterogeneity.

The conference was held in Tsukuba, Ibaraki, Japan, during July 23rd to 26th, 2000. We wish to thank all invited speakers and other participants. The special session "Theory Forum" was planned to discuss some of the focused topics in theoretical achievements. A brief summary is given by D. Feinberg and F. Kusmartsev in Chapter 1. The conference was supported by Electrotechnical Laboratory, Physical Society of Japan, Japanese Applied Physics Society, and Japanese Society of Synchrotron Radiation. We would like to thank members of the international advisory committee for helpful suggestions. We wish to express our gratitude to LLD2K secretariat, M. Fujita, N. Fujinami and M. Tsutsui for their dedicated work in the preparation of as well as during the conference. We also wish to express our special thanks to M. Tsutsui for the technical editing of the electronic manuscripts. One of us (H. O.) wishes to express his special thanks to H. Kamimura, N. Saini, A. Bishop, T. Egami, A. Menushenkov, Y. Yacoby, K. Yamaji and M. Koyanagi for their support in organizing the conference.

In summary, we hope that LLD2K was successful in establishing a common understanding that the *local lattice effects are essential in describing complex exotic properties of strongly correlated electron systems.* The urgent need to systematically study the nature of LLD, *i.e.*, quantum fluctuation and ordering, was confirmed, which requires cross-linking of local probe experiments and theoretical approaches. We wish that this book will inspire researchers in this field. Lastly, we would like to pay our respects to the late Eugene Noda for his contributions. We hope that our joint efforts to understand the nature of LLD will deepen our knowledge of microscopic mechanisms of exotic properties.

Editors

HIROYUKI OYANAGI
Electrotechnical Laboratory, Japan
and
ANTONIO BIANCONI
INFM and Universit di Roma "La Sapienza", Italy.

International Symposium on Physics in Local Lattice Distortions
Tsukuba, Japan, July 23-26, 2000

LLD2K Supporting Organizations:

We gratefully acknowledge support by the following organizations:

Electrotechnical Laboratory
Japanese Society of Synchrotron Radiation
MAC Science Co., Ltd.
RIGAKU Corporation

Japanese Applied Physics Society
Physical Society of Japan
Niki Glass Co., Ltd.
SEIKO EG&G Co., Ltd.

1. HIGH TEMPERATURE SUPERCONDUCTIVITY

Further Understanding of the Electronic Structure of Cuprate Superconductors: Role of Local Distortions

H. Kamimura and T. Hamada

Institute of Physics, Graduate School of Science, Science University of Tokyo, 1-3 Kagurazaka, Shinjuku-ku, Tokyo 162-8601, Japan

Abstract. In the first part of this article (section 2 to 10) a review is given on the first-principles studies of the many-electron electronic structures of superconducting cuprates performed by Kamimura and his co-workers. In this review, unusual electronic states in the underdoped cuprates are clarified. This characteristic state is called the "Kamimura-Suwa model". In section 11 in the latter part we describe the very interesting results recently obtained by Kamimura, Hamada and Ushio, based on the Kamimura-Suwa model. According to them the microscopic origins of the "high energy" pseudogap and of the peculiar hole-concentration dependence of the electronic entropy correspond to a phase change between a phase consisting of the small Fermi surfaces and a phase with a large Fermi surface, although the phase change is not sharp due to the finite spin-correlation length for local antiferromagnetic ordering. Through this calculation we will show that the hole carriers in cuprates behave like heavy fermions.

1. INTRODUCTION

The discovery of high-temperature superconductivity in cuprates by Bednorz and Müller [1] has led to an intensive search for the mechanism of the superconductivity. Anderson [2] first pointed out the important role of the electron correlation in the mechanism of the superconductivity. Since then, a number of theoretical models have been proposed [3]. Owing to the interplay among the correlation effect, the local distortion and antiferromagnetic ordering, the electronic structures in the normal state as well as in the superconducting state of cuprates are still controversial. Especially the most of the models assume the electronic structures at the beginning.

From this standpoint Kamimura and Eto [4,5] performed the first principles calculations of the electronic structures of a CuO_6 cluster in cuprates, taking into account the effects of the electron correlation and local distortions explicitly. The computation method which they adopted is the Multi-Configuration Self-Consistent Field (MCSCF) method with Configuration Interaction (CI). In the MCSCF-CI method one-electron orbitals are determined in the presence of not only coulomb and exchange interactions but also the electron correlation [6-8]. By virtue of this method they obtained the clear-cut-view of the many-body states for the strongly correlated cluster systems such as a CuO_6 octahedron or a CuO_5 pyramid embedded in superconducting cuprates.

CP554, *Physics in Local Lattice Distortions*, edited by H. Oyanagi and A. Bianconi
© 2001 American Institute of Physics 1-56396-984-X/01/$18.00

3

In this article, we first review their first-principles studies of the many-electron electronic structures of underdoped cuprates. Then we discuss a microscopic origin of the pseudogap. The organization of the present article is the following. In section 2 we describe cluster models for hole-doped CuO_6 octahedron and CuO_5 pyramid embedded in cuprates. Then we describe the MCSCF-CI method in section 3, and in sections 4, 5 and 6 we summarize the calculated results applied to a CuO_6 octahedron embedded in $La_{2-x}Sr_xCuO_4$ compounds (abbreviated as LSCO), and a CuO_5 pyramid embedded in superconducting $YBaCu_3O_7$ compound (abbreviated as $YBCO_7$) with T_c = 90 K and also in superconducting $Bi_2Sr_2CaCu_2O_{8+\delta}$ compounds (abbreviated as Bi2212) with T_c = 80 K for δ = 0.25, respectively. In section 7 the Kamimura-Suwa model is explained, and experimental supports for the Kamimura-Suwa model are discussed in section 8. The effective Hamiltonian to describe the Kamimura-Suwa model approximately is presented in section 9. In section 10 the many-electron energy band structures and the Fermi surfaces are discussed, based on the Kamimura-Suwa model. In section 11 the microscopic origin of the pseudogap is discussed on the basis of the Kamimura-Suwa model.

2. CLUSTER MODELS FOR HOLE-DOPED CuO_6 OCTAHEDRON AND CuO_5 PYRAMID

2.1. Electronic structures of a single Cu^{2+} ion in a CuO_6 octahedron and of a hole-doped CuO_6 octahedron

The crystal structure of La_2CuO_4 is tetragonal at high temperatures and is of a layer-type. In this crystal structure a CuO_2 unit forms a square planar network in each layer (on x-y plane) perpendicular to the c-axis, as seen in Fig. 1, where each Cu^{2+} ion is surrounded by six O^{2-} ions nearly octahedrally. This CuO_6 octahedron is stretched along the c-axis, producing two long (2.41Å) and four short (1.89Å) Cu-O lengths [9] by Jahn-Teller effect [10]. The ground state of a Cu^{2+} ion ($3d^9$), placed in surroundings

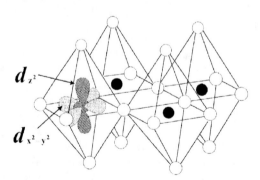

FIGURE 1. A CuO_2 plane which forms a planer network in each layer. The Cu $d_{x^2-y^2}$ and d_{z^2} orbitals are drawn together.

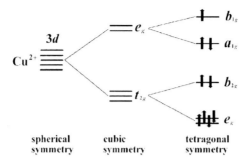

FIGURE 2. The energy splitting of Cu^{2+} orbitals in spherical, cubic and tetragonal symmetry.

that are of cubic symmetry, is ^2E_g, orbitary doubly- degenerate with the basis functions d_{z^2} and $d_{x^2-y^2}$. In a crystal field with tetragonal symmetry, ^2E_g state is further split into $^2A_{1g}$ and $^2B_{1g}$, where A_{1g} and B_{1g} are the irreducible representation of D_{4h} group. In La_2CuO_4 a Cu^{2+} ion in a CuO_6 octahedron is mainly subject to a crystal field with tetragonal symmetry, so that the five fold degenerate d orbitals of the Cu^{2+} ion with $3d^9$ electron configuration are split into b_{1g}, a_{1g}, b_{2g} and e_g orbitals as shown in Fig. 2, where the behavior of the orbital splitting by the cubic field is also shown in the figure. Thus a hole occupies an anitbonding b_{1g} orbital, denoted by b_{1g}^*.

When dopant holes are introduced in La_2CuO_4, there are two possibilities as to orbitals to accommodate a dopant hole in CuO_6. One case is that a dopant hole occupies antibonding a_{1g}^* orbital consisting of Cu d_{z^2} orbital and surrounding six oxygen p_σ orbitals, and its spin becomes parallel by Hund's coupling with localized spin of $S = 1/2$ around a Cu site. This many-electron state called "multiplet" is denoted by $^3B_{1g}$, as shown in Fig. 3(a). The other case is that a dopant hole occupies a bonding b_{1g} orbital consisting of in-plane oxygen p_σ orbitals with a small Cu $d_{x^2-y^2}$ component, and its spin becomes anti-parallel to the localized spin as shown in Fig.

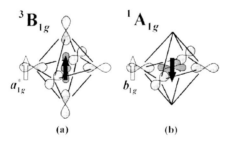

(a)   **(b)**

FIGURE 3. (a) Schematic view of $^3B_{1g}$ multiplet called the Hund's coupling triplet in a CuO_6 octahedron. A solid arrow represents a localized spin while an open arrow the spin of a hole carrier which occupies an antibonding a_{1g}^* orbital shown in the figure. (b) Schematic view of $^1A_{1g}$ multiplet, called the Zhang-Rice singlet, in which a hole occupies a bonding b_{1g} orbital shown in this figure.

3(b). This multiplet is denoted by $^1A_{1g}$. Since the $^1A_{1g}$ state is a key constituent state in the t-j model [11], we call the $^1A_{1g}$ multiplet "Zhang-Rice singlet".

2.2. Electronic structure of a hole-doped CuO_5 pyramid

Like LSCO, when holes are doped in superconducting YBCO$_7$ with T_c = 90 K or Bi2212 with δ =0.25 with T_c = 80 K, there are two possibilities as regards orbitals to accommodate a dopant hole in a CuO_5 pyramid. One case is that a dopant hole occupies an antibonding a_1^* orbital consisting of a Cu d_{z^2} orbital and five surrounding oxygen p_σ orbitals, and its spin becomes parallel to a localized spin of $S = 1/2$ around a Cu site, by Hund's coupling. This multiplet is the "Hund's coupling triplet" denoted by $^3B_1$, as shown in Fig. 4(a). The other case is that a dopant hole occupies a bonding b_1 orbital consisting of in-plane oxygen p_σ orbitals with a small Cu $d_{x^2-y^2}$ component, and its spin becomes anti-parallel to the localized spin as shown in Fig. 4(b). This multiplet is the "Zhang-Rice singlet" denoted by $^1A_1$.

2.3. Cluster models and the local distortion of a cluster by doping carriers

As a model for cluster calculations, we adopt a CuO_6 cluster in the LSCO compound, and a CuO_5 cluster in YBCO$_7$ or Bi2212 compounds. We label the oxygens in a CuO_2 plane as O(1), and the apical oxygens as O(2). We use the lattice constants reported in ref. [12] for LSCO, and in ref. [13] for YBCO$_7$. The number of electrons is determined so that the formal charge of copper is +2e and that of oxygen is –2e for an undoped case. Then we consider hole-doped systems for LSCO and YBCO by subtracting one electron.

To include the effect of the Madelung potential from the exterior ions outside the cluster under consideration, the point charges are placed at exterior ion sites (+2e for Cu and Ba, –2e for O, and +3e for La, Y and Bi). The number of point charges is 168 for CuO_6 and 300 for CuO_5. These point charges determine the Madelung potential at Cu, O(1) and O(2) sites within a cluster.

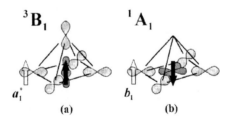

FIGURE 4. (a) Schematic view of $^3B_1$ multiplet called the Hund's coupling triplet in a CuO_5 pyramid. A solid arrow represents a localized spin while an open arrow the spin of a hole carrier which occupies an antibonding a_1^* orbital shown in the figure. (b) Schematic view of $^1A_1$ multiplet, called the Zhang-Rice singlet, in which a hole occupies a bonding b_1 orbital shown in this figure.

In superconducting cuprates the local distortions of a CuO_6 octahedron or CuO_5 pyramid play an important role. Thus the distortion effect of the apical oxygens which are located above (and below) the Cu atoms in the CuO_2 planes can be taken into account seriously in theoretical calculations. So far any model has not considered such distortion effect seriously, expect ours [4,5]. Recently a number of experimental results indicate that the distance between the apical O atom and the Cu atom is reduced when holes are doped in superconducting cuprates such as LSCO [9,15], YBCO [16,17] and Bi2212 [18,19] with $\delta = 0.25$. Theoretically, Shima et al. have predicted, by the LDA total energy calculation with the virtual crystal approximation, the doping concentration dependence of the apical O-Cu distance in LSCO [20].

In the case of a CuO_6 cluster in LSCO, Kamimura and Eto [4] have varied the Cu-O(2) distance c, according to the experimental results by Boyce et al. [9] and to the theoretical result by Shima et al. [20]. The distance c is taken as 2.41 Å, 2.35 Å, 2.30 Å and 2.24 Å, depending on the Sr concentration where 2.41 Å and 2.30 Å correspond to the value of c in the compounds of $x = 0$ (undoped) and of $x = 0.2$, respectively, in the $La_{2-x}Sr_xCuO_4$ formula. In the case of a CuO_5 cluster, on the other hand, the Cu(2)-O(2) distance is taken as 2.47 Å for insulating $YBCO_6$ and 2.29 Å for superconducting $YBCO_7$ with $T_c = 90$ K following the experimental results by neutron [16] and X-ray [17] diffraction measurements, where Cu(2) represents Cu ions in a CuO_2 plane while Cu(1) represents Cu(1) ions in a Cu-O chain. In the case of superconducting Bi2212 with $\delta = 0.25$ ($T_c = 80$ K), the distance between Cu and apical O in the CuO_5 pyramid is 2.15 Å [19]. It is very short compared with that of La_2CuO_4. On the other hand, the distance between Cu and O in a CuO_2 plane is 1.91 Å, which is nearly equal to that of La_2SrCuO_4.

2.4. Choice of basis sets in the MCSCF calculations

We express the one-electron orbitals by linear combinations of atomic orbitals, where Cu $1s$, $2s$, $3s$, $4s$, $2p$, $3p$, $3d$ and O $1s$, $2s$, $2p$ orbitals are taken into account as the atomic orbitals. Each atomic orbital is represented by a linear combination of several Gaussian functions. For Cu $3d$, $4s$ and O $2s$, $2p$ atomic orbitals we prepare two basis functions called "double zeta" for each orbital. Those are $(12s6p4d)/[5s2p2d]$ for Cu [21] and $(10s5p)/[3s2p]$ for O [22].

As to the oxygen ions, the diffuse components are usually used by researchers in the quantum chemistry. The diffuse components, however, cause problems with the point charge approximation outside of the cluster when a cluster is embedded in a crystal, because the diffuse components reach the nearest neighbor sites with considerable amplitudes. Instead of using the diffuse components for O^{2-}, Eto and Kamimura [4,5] used extended O $2p$ basis functions which have been originally prepared for a neutral atom, by introducing a scaling factor of 0.93. Then they multiplied all the Gaussian exponents in the double zeta base for the oxygen $2p$ orbitals by the same scaling factor of 0.93. This value of the scaling factor was determined so that the energy of an isolated O^{2-} ion should coincide with that obtained by the Hartree-Fock calculation.

3. THE MCSCF-CI METHOD

The MCSCF-CI method is the most suitable variational method applicable to a strongly correlated cluster system. With use of this method, Kamimura and Eto calculated the electronic structure of a hole-doped CuO_6 octahedron for the first time. Then, by applying this method to undoped Cu_2O_{11} dimer in La_2CuO_4, Eto and Kamimura [4,5] showed that the holes are localized around Cu sites and these localized holes form a spin-singlet state, consistent with the experimental results of Mott-Hubbard insulator for La_2CuO_4.

Later Kamimura and Sano [23] and Tobita and Kamimura [24] calculated the electronic structure of a CuO_5 pyramid embedded in $YBCO_7$ and $Bi2212$ with $\delta = 0.25$, respectively, by the MCSCF-CI method. In this section we give a brief review of how to use this method for the calculations of the lowest state energies of the $^1A_{1g}$ (or $^1A_1$) multiplet in the case of a CuO_6 octahedron (or CuO_5 pyramid) and $^3B_{1g}$ (or $^3B_1$) multiplet. First, the one-electron orbitals are determined by the MCSCF variational method [6-8]. The trial functions are taken for the Zhang-Rice singlet $^1A_{1g}$ (or $^1A_1$) as,

$$\Phi_S = C_0 |\psi_1 \alpha \psi_1 \beta \psi_2 \alpha \psi_2 \beta ... \psi_n \alpha \psi_n \beta |$$
$$+ \sum_i \sum_a C_{ii}^{aa} |...\psi_{i-1} \alpha \psi_{i-1} \beta \psi_{i+1} \alpha \psi_{i+1} \beta ... \psi_a \alpha \psi_a \beta |, \qquad (1)$$

and for the Hund's coupling triplet $^3B_{1g}$ (or $^3B_1$) as,

$$\Phi_T = C_0 |\psi_1 \alpha \psi_1 \beta ... \psi_{n-1} \alpha \psi_{n-1} \beta \psi_p \alpha \psi_q \alpha |$$
$$+ \sum_i \sum_a C_{ii}^{aa} |...\psi_{i-1} \alpha \psi_{i-1} \beta \psi_{i+1} \alpha \psi_{i+1} \beta ... \psi_a \alpha \psi_a \beta \psi_p \alpha \psi_q \alpha |, \qquad (2)$$

where $2n$ is the number of the electrons in the clusters, and $|......|$ represents a Slater determinant. Orbitals ψ_p and ψ_q are always singly occupied in Eq. (2). In Eqs. (1) and (2) all the two-electron configurations are taken into account in the summation over i and a so that the correlation effect is effectively included in this method. By varying ψ_i's and coefficients C_0 and C_{ii}^{aa}, the energy for each multiplet is minimized.

Next, the CI calculations are performed, by using the MCSCF one-electron orbitals ψ_i's determined above, as a basis set and the lowest energy of a multiplet is obtained. Since the main part of the correlation effect has already been included in determining the MCSCF one-electron orbitals, a small number of the Slater determinants are necessary in the CI calculations, and thus one can get a clear-cut-view of the many-body states by this MCSCF-CI method, even when the correlation effect is strong. Thus the MCSCF-CI method is the most suitable variational method for a strongly correlated cluster system [6].

In the MCSCF method all the orbitals consisting of the Cu $3d_{x^2-y^2}$, $3d_{z^2}$, $4s$ and O $2p_\sigma$ orbitals are taken into account in the summation over i and a in Eqs. (1) and (2). In the CI calculation, all the single-electron excitation configurations among these orbitals are taken into account.

4. CALCULATED RESULTS OF HOLE-DOPED CuO₆ OCTAHEDRA IN LSCO

In this section we discuss the calculated results of the CuO₆ cluster obtained by Kamimura and Eto [4]. They considered both the $^1A_{1g}$ state and the $^3B_{1g}$ state independently. Then they compared the respective energies of both states to determine which is the ground state, as a function of the Cu-O(2) distance.

4.1. The $^1A_{1g}$ state

The many-body wavefunctions of the $^1A_{1g}$ state are listed in Fig. 5, as a function of Cu-O(2) distance c. One-electron orbitals which are obtained by the MCSCF method are shown in Fig. 6. As seen in Fig. 5, the many-body wavefunction mainly consists

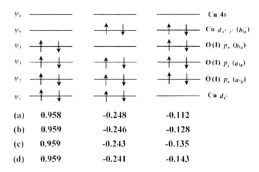

FIGURE 5. The many-body wavefunctions of $^1A_{1g}$ state in the hole-doped CuO₆ cluster. The Cu-O(2) distance, c, is (a) 2.41Å, (b) 2.35Å, (c) 2.30Å and (d) 2.24Å, respectively. The atomic orbital with the largest component in each MCSCF one-electron orbital is attached in the right side.

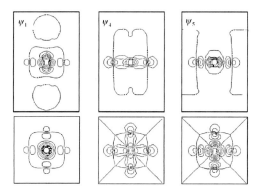

FIGURE 6. The MCSCF one-electron orbitals optimized for $^1A_{1g}$ state in the hole-doped CuO₆ cluster ($c = 2.41$Å). The upper row shows the wavefunctions perpendicular to the CuO₂ plane, while the lower row shows the wavefunctions in the CuO₂ plane. The contour lines are drawn every 0.05.

of three configurations. In the first configuration at the left column in the figure, which has the largest coefficient, the Cu $d_{x^2-y^2}$ -O(1) p_σ antibonding b_{1g} orbital, ψ_5, is unoccupied. In the second configuration at the center in the figure, the bonding orbital, ψ_4, is unoccupied while the antibonding orbital, ψ_5, is doubly occupied. Thus the mixing between the first and the second configurations indicates that the holes occupy both of the Cu $d_{x^2-y^2}$ and the O(1) p_σ orbitals of b_{1g} symmetry and that a dopant hole forms a spin-singlet pair with the localized hole which occupies an antibonding b_{1g} orbital, b^*_{1g}. This situation corresponds to the Zhang-Rice singlet state [11].

In the third configuration at the right column in Fig. 5, the a_{1g} orbital, ψ_1, is unoccupied while the b_{1g} orbitals, ψ_4 and ψ_5, are doubly occupied. The ψ_1, shown in Fig. 6, is almost localized at Cu d_{z^2}. This configuration appears by the following reason. When two holes are at a Cu site, the on-site Coulomb repulsion, the so-called Hubbard U, raises the energy. The Coulomb repulsion is smaller when the holes occupy both the d_{z^2} and the $d_{x^2-y^2}$ orbitals than when they remain only in the $d_{x^2-y^2}$ orbital. Thus the mixing of the $(d_{z^2})^2$ and the $(d_{x^2-y^2})^2$ configurations reduces the Hubbard U at the Cu site effectively, compared with the single configuration $(d_{x^2-y^2})^2$. This effect becomes larger as the Cu-O(2) distance decreases, as shown in Fig. 5.

4.2. The $^3B_{1g}$ state

The $^3B_{1g}$ many-body wavefunction is shown in Fig. 7. The a^*_{1g} orbital, ψ_4, and the b^*_{1g} orbital, ψ_5, are singly occupied and the two electrons couple to form the Hund's coupling triplet state. In ψ_5, $d_{x^2-y^2}$ is mixed with O(1) p_σ while ψ_4 consists almost entirely of d_{z^2}, as shown in Fig. 8. The strength of the on-site exchange energy, Hund's coupling, can be estimated from the energy difference between the $^3B_{1g}$ state and the excited $^1B_{1g}$ state. The estimated value is about 2.0 eV.

As the Cu-O(2) distance decreases and hence the CuO$_6$ cluster approaches the regular octahedron, the $^1B_{1g}$ state becomes more stable. This is because the energy difference between the b^*_{1g} orbital and the a^*_{1g} orbital becomes smaller, so that the Hund's coupling becomes more effective. The energy difference between the $^1A_{1g}$ and the $^3B_{1g}$ states is shown in Fig. 9, as a function of the Cu-O(2) distance. The figure indicates that the ground state of the CuO$_6$ cluster changes from the $^1A_{1g}$ state to the $^3B_{1g}$ state when the Cu-O(2) distance decreases. The distance at which the transition occurs corresponds to the doping concentration $x \sim 0.1$ in the La$_{2-x}$Sr$_x$CuO$_4$ formula.

5. CALCULATED RESULTS OF CuO$_5$ PYRAMID IN YBCO$_{7-\delta}$

In this section we discuss the calculated results by Kamimura and Sano [23] for a hole-doped CuO$_5$ pyramid in superconducting YBCO$_7$ with T_c = 90 K and insulating YBCO$_6$. The crystal structure of YBCO$_7$ is shown in Fig. 10(a). For comparison that of the insulating YBCO$_6$ is also shown in Fig. 10(b). A remarkable difference in the crystal structures of YBCO$_7$ and YBCO$_6$ is that there exists a Cu-O chain in YBCO$_7$.

Like LSCO, there are two orbitals, an antibonding a_1 orbital, a_1^*, and a bonding b_1 orbital, b_1^*, as possible orbital states to accommodate dopant holes, where a point

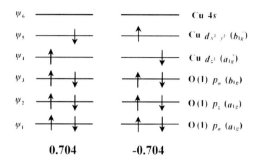

FIGURE 7. The many-body wavefunctions of $^3B_{1g}$ state in the hole-doped CuO_6 cluster ($c = 2.41$Å).

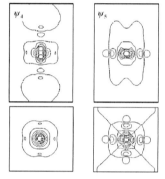

FIGURE 8. The MCSCF one-electron orbitals optimized for $^3B_{1g}$ state in the hole-doped CuO_6 cluster ($c = 2.41$Å). The upper row shows the wavefunctions perpendicular to the CuO_2 plane, while the lower row shows the wavefunctions in the CuO_2 plane. The contour lines are drawn every 0.05.

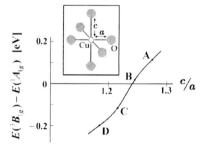

FIGURE 9. The energy difference between the $^3B_{1g}$ and the $^1A_{1g}$ multiplets, as a function of the Cu-O(2) distance, c, in the hole-doped CuO_6 cluster. The Cu-O(1) distance, a, is fixed at 1.889Å. c is (A) 2.41Å (undoped case), (B) 2.35Å, (C) 2.30Å ($La_{1.8}Sr_{0.2}CuO_4$) and (D) 2.24Å, respectively.

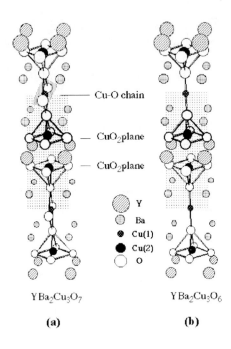

Cu-O chain

CuO$_2$plane

CuO$_2$plane

Y

Ba

Cu(1)

Cu(2)

O

YBa$_2$Cu$_3$O$_7$ YBa$_2$Cu$_3$O$_6$

(a)   (b)

FIGURE 10. The crystal structures of YBCO$_{7-\delta}$. (a) The orthorhombic structure of superconduting YBCO$_7$. (b) The tetragonal structure of insulating YBCO$_6$.

group of CuO$_5$ pyramid is C$_{4v}$. The sketch on the spatial extension of a_1^* and b_1 orbitals are shown in Fig. 4(a) and 4(b). As a result one has to deal with both the $^1A_1$ and the $^3B_1$ multiplets independently following the MCSCF-CI method. In doing so, we take into account the effect of Madelung potential from exterior ions outside the cluster by placing the point charge, +2 at Cu(2) in CuO$_2$ plane, +2 at Ba, +3 at Y, and -2 at O. As to the charge of Cu in Cu-O chain (Cu(1)), q, we have taken $q = +1$ for insulating YBCO$_6$ from experimental (NMR) result [25]. This value is consistent with a condition of charge neutrality. However, in superconducting YBCO$_7$, the value of q is not clear. Thus Kamimura and Sano [23] calculated the energy difference between the $^1A_1$ and $^3B_1$ multiplets in the case of YBCO$_7$ as a function of q and then investigated the effect of inhomogeneous hole distribution in Cu-O chain on the electronic state.

5.1. Energy difference between $^1A_1$ and $^3B_1$ multiplets

The calculated energy difference between the $^1A_1$ and the $^3B_1$ multiplets by Kamimura and Sano is shown in Fig. 11, as a function of the charge of Cu(1), q. The value of q and the existence of O^{2-} ions in a Cu-O chain play a crucial role in determining the Madelung energy at apical O site. There is the energy difference of

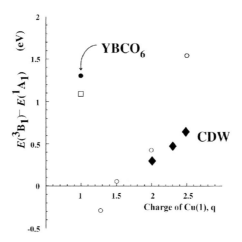

FIGURE 11. The energy difference between the $^3B_1$ and the $^1A_1$ multiplets, as a function of the charge of a Cu(1) ion in a Cu-O chain, q, in the hole-doped CuO_5 cluster embedded in $YBCO_6$ and $YBCO_7$. The closed circle represents the energy difference between the $^3B_1$ and the $^1A_1$ multiplets in insulating $YBCO_6$ [28,29]. The open circles represent the energy difference between the $^3B_1$ and the $^1A_1$ multiplets in superconducting $YBCO_7$ as a function of constant q for all Cu(1) ions, where c is fixed at 2.29 Å. Further the solid diamonds represent the calculated results in the case of CDW in a Cu-O chain.

1.3 eV bettwen $^3B_1$ and $^1A_1$ multiplets in insulating $YBCO_6$, as seen in Fig. 11 (closed circle), where the distance between Cu(2) and apical O, c, is fixed at 2.47 Å. To investigate the contraction effect of distance between Cu(2) and apical O ions in $YBCO_7$, Kamimura and Sano changed c from 2.47 Å to 2.29 Å and kept the other lattice constants fixed for $YBCO_6$. In this case the energy difference decreases to 1.1 eV (open square in Fig. 11) like LSCO. In Fig. 11 the open circles show the energy difference for superconducting $YBCO_7$ as a function of q, where c is fixed at 2.29 Å and oxygen atoms are introduced into a Cu-O chain. It is clear from this figure that, when the value of q decreases, the ground state of the CuO_5 pyramid in $YBCO_7$ changes from the $^1A_1$ to $^3B_1$ around the $q \approx 1.45$. This is because, as the value of q decreases and thus the Maderung potential at the apical oxygen site decreases, the energy difference between the a_1^* orbital which contains the p_z orbital at apical oxgen site and the b_1 orbital becomes smaller, so that the Hund's coupling becomes more effective.

5.2. Effect of charge density wave (CDW) in a Cu-O chain

In a previous subsection we have seen that, in superconducting YBCO$_7$ the calculated lowest state energy is very sensitive to the charge of Cu(1) in a Cu-O chain, q. In this subsection we discuss how the multiplets of a CuO$_5$ pyramid are affected by the inhomogeneous hole distribution in a Cu-O chain, that is the charge density wave (CDW), based on the calculated results by Kamimura and Sano [23]. Recently the existence of such CDW in a Cu-O chain in YBCO$_7$ has been reported by various experimental groups. For example, scanning tunneling microscopy (STM) experiment [26] and neutron inelastic scattering experiments [27] have reported on the existence of CDW in a Cu-O chain in YBCO$_7$. In this context Kamimura and Sano [23] tried to clarify theoretically how the CDW in the Cu-O chains affect the electronic structure of a hole-doped CuO$_5$ pyramid. This was the first theoretical study on the CDW effect in a Cu-O chain on the electronic structure of a CuO$_5$ pyramid in YBCO$_7$.

Following Kamimura and Sano [23], let us explain how the CDW in a Cu-O chains influences the electronic structure of a CuO$_5$ pyramid. Suppose that the charge of a Cu(1), q, is +2.5 and that the states of holes in a Cu-O chain are expressed by a one-dimensional energy band. This means that there are 1.5 holes in a Cu-O chain and that three quarters of the energy band for a Cu-O chain are filled by holes. In this case the Fermi wavenumber k_F is given by $\pi/4a$ approximately, where a is a Cu(1)-Cu(1) distance along the chain and it is 3.8 Å for YBCO$_7$. Thus the CDW modulation-wavelength becomes 15.2 Å, because the modulation wavelength λ_{CDW} is given by $\lambda_{CDW} = 2\pi/2k_F$ and it is nearly equal to $4a$. This value is consistent with the experimental results [26,27], since the observed modulation wavelength of CDW in a Cu-O chain takes a value between 13 ~ 16 Å.

In superconducting YBCO$_7$ an oxygen introduced in a Cu-O chain produces two holes in a unit cell consisting of a Cu-O chain and two CuO$_2$ planes. Considering the charge of +3e for Y, +2e for Ba, +2e for Cu(2) and –2e for O in CuO$_2$ plane, +1e for Cu(1) and –2e for O in a Cu-O chain, and further distributing the charge of dopant holes over both a Cu-O chain and two CuO$_2$ planes in a unit cell, the following equation holds for a relation between the number of holes in Cu-O chain, η, and that of a CuO$_2$ plane, ζ, in the unit cell from the condition of the charge neutrality;

$$\eta + 2\zeta = 2. \tag{3}$$

Then the charge of Cu in a Cu-O chain, q, is related to η by the relation $q=1 + \eta$. Since the values of η and ζ have not been determined experimentally so for, Kamimura and Sano calculated the lowest energies of the $^1A_1$ and $^3B_1$ multiplets by varying a value of η. In the case of the uniform charge distribution for the charge of Cu(1) in a Cu-O chain, for example, for $q = +2.5$, η becomes 1.5 and thus ζ is 0.25 from Eq. (3). This means that one hole exists per four CuO$_5$ pyramids. Since a CuO$_5$ pyramid is embedded in YBCO$_7$, one must take into account the effect of Madelung potential from exterior ions outside the pyramid by putting the point charge +2e at Cu(2) in CuO$_2$ plane, +2e at Ba, +3e at Y, –2e at O. As to the charge of Cu(1) in a Cu-O chain, one may place the point charges according to the CDW modulation-wavelength, as shown on the line A in Fig. 12. For example, $Cu(1)^{+1.75}$ ions are placed at the interval of every four Cu(1) sites along the line of the Cu-O chain. This corresponds to the case that the Cu(1) right above the CuO_5 pyramid under consideration has the charge of +1.75, while the charge of +2.75e is placed at remaining Cu(1) sites on the line A. Thus the averaged charge of Cu(1) atoms on the line A is +2.5e. In the same way one

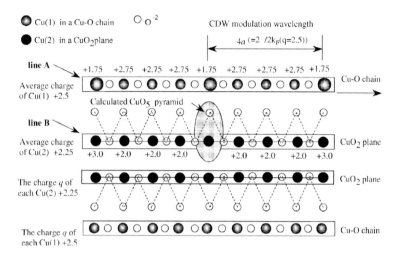

FIGURE 12. The charge distribution of Cu(1) ions in Cu-O chains and of Cu(2) ions in CuO_2 planes for the case in which the charge of Cu(1) is modulated by the CDW modulation wavelength and the average value of Cu(1)'s charge, $\bar{q}$, is equal to 2.5. The line A represents the Cu-O chain which includes the Cu(1) ion right above the hatched CuO_5 pyramid under consideration. The line B represents a CuO_2 plane which includes the calculating CuO_5 pyramid.

may put the charge of +3e at Cu(2) sites at the interval of four sites with the same modulation as that of the Cu-O chain and put the charge of +2e at the remaining Cu(2) sites on the line B as seen in Fig. 12. The line B includes the CuO_5 pyramid under consideration. Thus the averaged charge of the Cu(2) ions and the averaged hole concentration in a CuO_2 plane on the line B becomes 2.25e and 0.25, respectively. As to the charges of all the Cu(2) ions except those on the line B, one can take +2.25e as an averaged charge, while as regards the charges of all the Cu(1) ions except the Cu(1) ions on the line A, one can take +2.5e as an averaged charge, as shown in Fig. 12. In this way the CDW-like hole distribution is formed under the condition in which the charge neutrality is kept.

On the basis of the charge distribution shown in Fig. 12, Kamimura and Sano have calculated the lowest energies of the $^1A_1$ and $^3B_1$ multiplets by the MCSCF-CI method. The calculated results are shown by solid diamonds in Fig. 11, where the energy difference between the $^1A_1$ and $^3B_1$ multiplets is shown on the vertical axis and $\bar{q}$ on the horizontal line represents the averaged charge of Cu(1) ions in a Cu-O chain. For comparison, we also show by open circles the energy difference between the $^1A_1$ and

$^3B_1$ multiplets calculated for the case of the constant charge distribution in a Cu-O chain as a function of q [28,29].

As shown in Fig. 11, in the case of the constant charge distribution of the Cu(1) ions in Cu-O chain, the energy difference between $^1A_1$ and $^3B_1$ multiplets is larger than that in the case of CDW. For example, the former is 1.55 eV for $q = +2.5$. In the CDW case shown by solid diamonds in Fig. 11, the calculated energy difference between the $^1A_1$ and $^3B_1$ multiplets is significantly reduced. For example, in the case of $\bar{q} = +2.5$ it becomes 0.65 eV. Thus the electronic structure is strongly affected by the charge distribution in a Cu-O chain caused by CDW. The decrease of the energy difference between these two multiplets is reasonable because in this case the Madelung potential at the apical O in a CuO_5 pyramid becomes lower for hole carriers. However, since the holes in Cu-O chains occupy both Cu(1) and O sites, the charge of Cu(1) in a Cu-O chain becomes lower than $+2.5e$. This favors the $^3B_1$ multiplet energetically, because the Madelung potential at the apical O site becomes further lower. Thus we conclude that, when the averaged charge of Cu(1) ions takes a value between 2.0 and 2.3, the energy difference between the $^1A_1$ and $^3B_1$ multiplets becomes of the same order of magnitudes as transfer interaction between $^3B_1$ and $^1A_1$ multiplets at neighboring CuO_5 pyramids, 0.4 eV, by the existence of CDW in a Cu-O chain.

6. CALUCULATED RESULTS OF A CuO₅ PYRAMID IN Bi₂Sr₂CaCu₂O₈₊δ

6.1. Introduction

In this section we discuss the calculated results by Tobita and Kamimura [24] on the electronic structures of a CuO_5 pyramid embedded in $Bi_2Sr_2CaCu_2O_{8+\delta}$ with use of the MCSCF-CI method. The high T_c superconductors of the Bi-Sr-Ca-Cu-O system were discovered by Maeda et $al.$ in 1988 [18]. The composition of these materials is determined as $Bi_2Sr_2Ca_{n-1}Cu_nO_{4+2n+\delta}$ with n being 1, 2, and 3. To distinguish the values of different n, these compounds are distinguished as Bi2201 ($n = 1$), Bi2212 ($n = 2$) and Bi2223 ($n = 3$), where T_c of Bi2201, Bi2212 and Bi2223 are 20K and 80K, 110K, respectively. The number of the CuO_2 planes increases with increasing of n. These compounds have the Bi_2O_2 block layers. In the chemical formula of "$Bi_2Sr_2CaCu_2O_{8+\delta}$", δ represents the excess of oxygen. When the excess oxygen do not exist, $i.e.$ $\delta = 0$, this material is an insulator. When the excess oxygen are introduced, hole carriers are supplied into the CuO_2 planes, and this material shows superconductivity. With increasing the value of δ, T_c of Bi2212 raises. Thus many researcher regard the excess of oxygen as an origin of carriers which are responsible for superconductivity.

6.2. Cluster models

Figures 13(a) and (b) show the crystal structures of Bi2212 for $\delta = 0$ and $\delta = 0.25$, respectively. In these structures, the distance between Cu and apical O in the CuO_5 pyramid cluster is 2.15 Å for $\delta = 0.25$ [19]. It is very short, compared with the

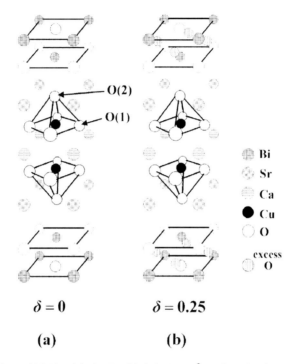

$\delta = 0$ $\delta = 0.25$

(a)   **(b)**

distance in insulating Bi2212 with $\delta = 0$ which is 2.47 Å. Thus the local distortion of CuO_5 pyramids is expected to play an important role in determining their electronic structure.

We consider that the case of $\delta = 0.25$ corresponds to the optimum doping in Bi2212. In this subsection, we pay attention to the electronic structures of a CuO_5 pyramid in **FIGURE 13.** The crystal structures of $Bi_2Sr_2CaCu_2O_{8+\delta}$. Here, (a) and (b) represent the structures for $\delta = 0$ [18] and 0.25, respectively.

the cases of $\delta = 0$ and $\delta = 0.25$. According to the observation by transmission electron microscope (TEM) [30], the Bi_2O_2 block layers are slightly distorted from the crystal structures shown in Figs. 13(a) and (b), and undulation appears along the b axis. Thus, a real crystal structure of Bi2212 is more complex than the structures shown in Figs. 13(a) and (b). The origin of this distorted structure may be considered due to following reasons: The excess oxygen enter into the middle of the Bi_2O_2 block layers.

17

Depending on whether the Bi_2O_2 block layers include the excess oxygen or not, the Bi_2O_2 block layers show a slight irregular structure. However, the hole carriers cannot recognize such slight change of the structure, because its mean free path is much longer than Cu-O-Cu distance. In this context Tobita and Kamimura [24] used the average structures shown in Figs. 13(a) and (b) for the calculation of electronic structures. Further, in the case of $\delta = 0.25$ the excess oxygen of charge $-0.5e$ are placed at all the four sites in every middle region between the Bi_2O_2 block layers, because the hole carriers are subject to the average Madelung potential from the excess oxygen of charge $-2e$ which are distributed randomly between the Bi_2O_2 block layers. We call the crystal structure shown in Fig. 13(b) a "virtual crystal structure" in this respect.

In their calculations, 742 and 846 ions outside the CuO_5 pyramid under consideration are treated as point charges to consider the effect of Madelung potential for the case of $\delta = 0$ and $\delta = 0.25$, respectively, following the melted by Kondo [31]. Then they calculated the electric structures of a single CuO_5 pyramid using the crystal structures shown in Figs. 13(a) and (b) for the case of $\delta = 0$ and $\delta = 0.25$, respectively.

6.3. Calculated results for the electronic structures of a CuO₅ pyramid in Bi2212

The calculated energy difference between the $^1A_1$ and $^3B_1$ states is about 2.15 eV for the case of $\delta = 0$. The energy of the $^1A_1$ state is lower than that of the $^3B_1$ state. Since the transfer interaction between neighboring pyramids is about 0.4 eV, a dopant hole is localized around a particular CuO_5 pyramid in the case of $\delta = 0$. As a result Bi2212 with $\delta = 0$ is an insulator, consistent with experimental results [18,19].

For the case $\delta = 0.25$, on the other hand, the energy difference between the $^1A_1$ and $^3B_1$ states is about 0.034 eV. The energy of the $^1A_1$ multiplet is still lower than that of the $^3B_1$ multiplet. Since this energy difference is very small, compared with the transfer interaction between the b_1 and a_1^* orbitals in the adjoined CuO_5 pyramids which is about 0.4 eV, two states are mixed by the transfer interaction between the neighboring CuO_5 pyramids, and a coherent state is expected to be composed in a superconducting Bi2212 material, when the localized spin form an antiferromagnetic ordering in a spin-correlated region, as will be described in the following section.

A reason why the difference between the $^1A_1$ and $^3B_1$ states decreases is the following: As the value of δ increase, the Madelung potential at an apical oxygen site decreases. As a result the energy difference between the energy of the a_1^* orbital which contains the p_z orbital at the apical oxygen site and that of the b_1 orbital becomes smaller, so that the Hund's coupling becomes more effective. Thus, the energy difference between the $^1A_1$ and $^3B_1$ states becomes smaller.

7. THE KAMIMURA-SUWA MODEL :
ELECTRONIC STRUCTURE OF UNDERDOPED CUPRATES

Now we construct the many-electron electronic structure of underdoped cuprates, based on the calculated results of a CuO_6 octahedron embedded in LSCO and of a CuO_5 pyramid in $YBCO_7$ and Bi2212. Before presenting results, we briefly describe the theoretical treatment made by Kamiumra and Suwa [14], which we call the Kamimura-Suwa model hereafter. As an example, we choose LSCO here. According to the Kamimura-Suwa model, there exist the areas in each CuO_2 layer in which the localized spins form the antiferromagnetic ordering. Here we call these areas "spin-correlated regions". The size of each spin-correlated region is characterized by the spin-correlation length. Then, following the results of Kamimura and Eto [4], a dopant hole with up spin in a spin-correlated region occupies an a_{1g}^* orbital, $\phi_{a_{1g}}$, at CuO_6's with localized up-spins, because of the energy gain of about 2 eV due to the intra-atomic exchange interaction between the spins of an a_{1g}^* hole and of a localized hole in an antibonding b_{1g} orbital (b_{1g}^*) (Hund's coupling) within the same CuO_6 octahedron, as shown in Fig. 14(c). As a result the spin-triplet $^3B_{1g}$ state is created. Since Hund's coupling prevents a hole with up spin from occupying an a_{1g}^* orbital in

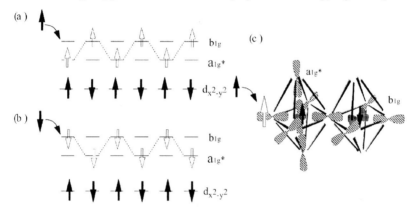

FIGURE 14. Schematic view for the coherent motion of a dopant hole from high-spin to low-spin states in the presence of the antiferromagnetic ordering of the localized spin system. Here (a) and (b) correspond to up-spin and down-spin bands of dopant holes, respectively. Figure (c) represents a coherent motion of an up-spin carrier from $^3B_{1g}$ to $^1A_{1g}$ multiplet. It should be noticed that the relative position of a_{1g}^* and b_{1g} levels changes according to the doping concentration. The energy levels in this figure are obtained from the results of Kamimura and Eto (ref. [4]).

a CuO_6 octahedron with a localized down-spin, a hole with up-spin can not hop into neighboring a_{1g}^* orbital. Instead, it can enter into a bonding b_{1g} orbital, $\phi_{b_{1g}}$, in a neighboring CuO_6 octahedron with localized down-spin without destroying the antiferromagnetic ordering. In this case there is the energy gain of about 4.0 eV due to the antiferromagnetic exchange interaction between holes in bonding and antibonding

b_{1g} orbitals, as shown in Fig. 14(c). This results in the Zhang-Rice singlet state. In this way the dopant holes can move resonantly from a CuO_6 to a neighboring CuO_6 in a CuO_2 layer by a transfer interaction of about 0.3 eV without destroying the local antiferromagnetic (AF) ordering, as shown in Figs. 14(a) and (b). Such coherent motion of the dopant holes is possible when the spin-correlation length is much larger than a distance between neighboring copper sites and the magnitudes of transfer interactions between neighboring CuO_6 octahedra are larger than the energy difference between the highest occupied orbital states in the Zhang-Rice singlet and the Hund's coupling triplet. As a result a metallic state is created, and it simultaneously causes d-wave superconductivity, as was shown by Kamimura *et al.* [32].

Kamimura and Suwa [14] have expressed the above coherent motion of dopant holes in a metallic state by the following forms of Bloch-type wave functions:

$$\Psi_{k\alpha}(r)\chi = \sum_R \exp(ik \cdot R)[A_k\phi_{a_{1g}^*}(r-R) + B_k\phi_{b_{1g}}(r-R-d)]\alpha\chi \quad (4)$$

and

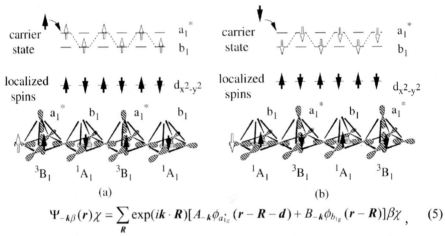

$$\Psi_{-k\beta}(r)\chi = \sum_R \exp(ik \cdot R)[A_{-k}\phi_{a_{1g}^*}(r-R-d) + B_{-k}\phi_{b_{1g}}(r-R)]\beta\chi, \quad (5)$$

FIGURE 15. Schematic view for coherent motion of a dopant hole from high-spin to low-spin states in the presence of the antiferromagnetic ordering of the localized spin system in the case of superconducting YBCO$_7$ and Bi2212. Here (a) and (b) correspond to up-spin and down-spin bands of dopant holes, respectively.

where α and β represent the up- and down-spin states of a dopant hole, respectively. The spin function χ represents the antiferromagnetic ordering state of the Cu localized spins in a CuO_2 layer, where the up and down spins are assigned at R and $R + d$ Cu sites, respectively. Furthermore, d is a vector representing a distance between Cu sites with localized up and down spins in an antiferromagnetic unit cell. The summation over R is taken for the antiferromagnetic unit cells. In both Eqs. (4) and (5), the first and the second terms in the square brackets represent the Hund's coupling and Zhang-

Rice multiplets, respectively. In the case of YBCO$_7$, the coherent motion of a dopant hole due to the alternate appearance of the ^{1}A$_1$ and ^{3}B$_1$ multiplets is also possible when the CDW exists in a Cu-O chain, as shown in Fig. 15, and in Bi2212 the coherent motion always occurs $\delta = 0.25$.

8. EXPERIMENTAL EVIDENCES FOR THE INCREASE OF THE SPIN-CORRELATION LENGTH WITH HOLE CONCENTRATION AND FOR THE COEXISTENCE OF THE ^{1}A$_1$ AND THE ^{3}B$_1$ STATE

In the Kamimura-Suwa model the spin-correlation length must increase in the underdoped region when the hole concentration increases, in order for every hole-carrier to move over a considerable distance without interacting with each other. As to the hole-concentration dependence of the spin correlation length, Mason et al. [33] and Yamada et al. [34] recently reported independently that the spin-correlation length in the underdoped region of La$_{2-x}$Sr$_x$CuO$_4$ increases from $x = 0.05$, the onset of superconductivity, with increase of hole concentration x and reaches a value of about 50Å for the optimum doping ($x = 0.15$). These experimental results support the Kamimura-Suwa model in which a metallic and superconducting state corresponds to a coherent state characterized by the coexistence of the local spin ordering and of the ordering with regard to the alternating appearance of the ^{3}B$_{1g}$ and ^{1}A$_{1g}$ multiplets in the carrier system.

In order to investigate the coexistence of the ^{3}B$_{1g}$ and ^{1}A$_{1g}$ states, Chen et al. [35] performed polarization-dependent X-ray absorption measurements for O K and Cu L edges in LSCO. For the Cu L edge, they observed the doping-induced satellite peak (L$_3$') for both polarizations of the electric vector of the X-rays E, parallel and perpendicular to the c-axis, in a shoulder area of the doping-independent Cu L$_3$ line, with the intensity ratio of about 1 to 9, where the main L$_3$ line corresponds to transitions from Cu 2p core level to the upper Hubbard Cu $d_{x^2-y^2}$ band, indicating the existence of the localized spins. Since the former ($E \parallel c$) and the latter ($E \perp c$) polarizations detect the Hund's coupling triplet (^{3}B$_{1g}$) and the Zhang-Rice singlet (^{1}A$_{1g}$), respectively, the appearance of the doping-induced satellite peak for both polarizations at the same energy suggested that the state of the dopant holes must be a single coherent state consisting of the Hund's coupling triplet state and the Zhang-Rice singlet state. For the compounds Tl$_2$Ba$_2$CaCu$_2$O$_8$ and Tl$_2$Ba$_2$Ca$_2$Cu$_3$O$_3$ as well as LSCO, Pellegrin et al. [36] have also found the polarization dependence similar to that found by Chen et al. [35].

In 1989 Bianconi et al. [37] also reported that the peak energy separation between transitions for polarizations parallel and perpendicular to the c-axis in LSCO decreases towards zero when the Sr concentration increases from a non-superconducting regime to a superconducting regime, consistent with the above experimental results. The existence of localized spins on Cu indicated by the observation of the Cu L$_3$ line is also supported by neutron scattering experiments. For example, Birgeneau et al. [38] showed the coexistence of the spin-correlation of localized spins and superconductivity in LSCO; that is, the spins of Cu $d_{x^2-y^2}$ holes form a two-dimensional local antiferromagnetic (AF) order even in the superconducting state.

Recently site-specific X-ray absorption spectroscopy of $YBa_2Cu_3O_{6.91}$ with $T_c = 90$ K by Merz *et al.* [39] has determined the hole distribution in a CuO_2 plane, at an apical O site and in a Cu-O chain. According to this result the experimental values of hole distribution in a CuO_2 plane, at an apical O site and in a Cu-O chain are 0.40, 0.27 and 0.24, respectively. These results are consistent with the theoretical values calculated by Kamimura and Sano [23]. In particular, this experimental result clarified an important role of an apical oxygen site in a CuO_5 pyramid in the electronic structure of superconducting $YBCO_{6.91}$. This is an important experimental evidence for the Kamimura-Suwa model.

9. EFFECTIVE HAMILTONIAN
FOR THE KAMIMURA-SUWA MODEL

Kamimura and Suwa [14] constructed the effective Hamiltonian to describe the Kamimura-Suwa model approximately. It consists of four parts: the effective one-electron Hamiltonian (H_{eff}) for a_{1g}^*- and b_{1g}-orbital states, the transfer interactions among a_{1g}^*- and b_{1g}-orbitals in the neighboring CuO_6 octahedra (H_{tr}), the superexchange interaction between the Cu $d_{x^2-y^2}$ localized spins (H_{AF}), and the exchange interactions between spins of dopant holes and $d_{x^2-y^2}$ holes within the same CuO_6 octahedron (H_{ex}). Thus the effective Hamiltonian is expressed as

$$H = H_{eff} + H_{tr} + H_{AF} + H_{ex}$$
$$= \sum_{i,m,\sigma} \varepsilon_m C_{im\sigma}^\dagger C_{im\sigma} + \sum_{\langle i,j\rangle,m,n,\sigma} t_{mn}(C_{im\sigma}^\dagger C_{jn\sigma} + \text{h.c.})$$
$$+ J\sum_{\langle i,j\rangle} S_i \cdot S_j + \sum_{\langle i,m\rangle} K_m s_{i,m} \cdot S_j \qquad (6)$$

where ε_m ($m = a_{1g}^*$ or b_{1g}) represents the effective one-electron energies of the a_{1g}^*- and b_{1g}-orbital states, $C_{im\sigma}^\dagger (C_{im\sigma})$ the creation (annihilation) operator of a dopant hole at the i-th CuO_6 octahedron, t_{mn} the effective transfer integrals of a dopant hole between m-type and n-type orbitals of neighboring CuO_6 octahedra, J the superexchange interaction between the spins S_i and S_j of $d_{x^2-y^2}$ localized holes in the antibonding b_{1g} orbital (b_{1g}^*) at the nearest neighbor Cu i and j sites ($J > 0$, antiferromagnetic), and K_m the exchange integrals between the spin of a dopant hole s_{im} and $d_{x^2-y^2}$ localized spin S_i in the i-th CuO_6 octahedron ($K_{a_{1g}} < 0$ for the Hund's coupling triplet and $K_{b_{1g}} > 0$ for the Zhang-Rice singlet).

10. ENERGY BANDS, FERMI SURFACE AND DENSITY OF STATES IN THE SUPERCONDUCTING STATE OF LSCO

Kamimura and Ushio [40] and later Ushio and Kamimura [41] calculated an effective one-electron-type energy band structure for the dopant hole-carriers from the Hamiltonian (6) by separating the localized spin system by adopting the molecular field approximation for S_i and S_j in the fourth term in Eq. (6). Since the spins of localized holes in b_{1g}^* orbitals form antiferromagnetic ordering, Kamimura and Ushio chose a unit cell so as to contain two neighboring CuO_6 octahedra with localized up and down spins, called A- and B-sites, and considered the 34×34 dimensional matrix ($\widetilde{H}$) of the Hamiltonian (6), where $2p_x$, $2p_y$ and $2p_z$ atomic orbitals for each of eight oxygen atoms and $3d_{yz}$, $3d_{xz}$, $3d_{xy}$, $3d_{x^2-y^2}$ and $3d_{z^2}$ atomic orbitals for each of two Cu atoms in the unit cell are taken as the basis functions. We call this unit cell the "antiferromagnetic unit cell". This Hamiltonian matrix $\widetilde{H}$ consists of two parts, the one-electron part $\widetilde{H}_0$ and the effective interaction part $\widetilde{H}_{int}$, the latter of which includes the exchange interactions between the carrier spins and the localized spins in the fourth term in Eq. (6) in a mean field sense. In the case of a dopant hole with up-spin, the energy of b_{1g}^* state in a CuO_6 octahedron with localized up-spin (A-site) is taken to be lower than that in a CuO_6 octahedron with localized down-spin (B-site) by Hubbard U interaction, which is taken as 10 eV in the present case. Further the energy of a_{1g}^* state at A-site is taken to be lower than that at B-site by Hund's coupling energy which is 2 eV, while the energy of b_{1g} state at B-site is taken to be lower than that at A-site by the antiferromagnetic exchange energy in the $^1A_{1g}$ state which is 4 eV. Thus we may take $K_{a_{1g}^*} = -2$ eV and $K_{b_{1g}} = 4$ eV in Eq. (6).

In this way the essential part of the many-body interaction terms in Hamiltonian (6) is taken into account as the effective one-electron terms in the the 34×34 dimensional effective interaction part $\widetilde{H}_{int}$. This kind of theoretical treatment is similar in its idea to the "LDA+U" method developed by Anisimov et al. [42] for copper oxides, but an essential point in the present treatment by Kamimura and Ushio [40,41] is to separate the localized spin system from the carrier system by expressing H_{ex} in Eq. (6) as $\sum_{i,m} K_m s_{i,m} \cdot \langle S_i \rangle$ by the molecular field approximation, where $\langle S_i \rangle$ is the average value of a localized spin at i-th Cu site. As a result the exchange interactions between spins of dopant and localized holes in the fourth term in Eq. (6) can be expressed as a form of an effective magnetic field acting to the carrier spins. Thus all the matrix elements in the 34×34 dimensional Hamiltonian matrix ($\widetilde{H}$) are expressed as the form of one-electron type, and then one electron type energy bands for the carrier system can be obtained by diagonalizing $\widetilde{H}$.

On doing so, all the matrix elements related to the transfer interactions which appear in the 34×34 dimensional Hamiltonian matrix ($\widetilde{H}_0$) were estimated from the

Slater-Koster parameters determined by DeWeert *et al.* [43] for La_2CuO_4. Further a value for the difference between $\varepsilon_{a_{1g}^*}$ and $\varepsilon_{b_{1g}}$ was taken so as to reproduce the energy difference between the ${}^3B_{1g}$ and ${}^1A_{1g}$ multiplet states calculated by Kamimura and Eto [4]. In this way Kamimura and Ushio [40,41] could separate the localized spin part from the carrier system, and they obtained a band structure including the many-body effects in a mean-field sense for the itinerant hole-carriers in LSCO. The band structure for up-spin dopant holes thus obtained is shown in Fig. 16, where the ordinary Brillouin zone which includes an ordinary unit cell corresponding to a CuO_6 octahedron is also shown in the upper part of the figure. The same shape of the band structure is also obtained for down spin dopant holes. In undoped La_2CuO_4, all the bands are fully occupied by electrons so that La_2CuO_4 is an insulator, consistent with the experimental result [44]. In Fig. 16 the highest occupied band is expressed by a solid line denoted by #1. The top of the highest band is located at Δ point in the Brillouin zone, where Δ corresponds to the edge of the antiferromagnetic Brillouin zone, $(\pi/2a, \pi/2a, 0)$, with a being the distance between neighboring Cu sites. In this respect the present concept of energy bands in Fig. 16 are completely different from the ordinary concept of energy bands in the one-electron picture such as those calculated by the LDA.

When Sr ions are doped, holes begin to occupy the top of the highest band (#1) at Δ. At the onset concentration at which the superconductivity takes place, the Fermi level in the #1 band is located at the energy much higher than that of G_1 state, where G_1 in the Brillouin zone corresponds to $(\pi/a, 0, 0)$ and also to a van-Hove singularity in a two-dimentional-like density of states. In Fig. 17 the wave functions at Δ and G_1 points for the #1 band are shown, where the right hand side of the figure corresponds to a CuO_6 octahedron with localized up-spin (A-site) and the left hand side to a CuO_6 octahedron with localized down-spin (B-site). One can see from this figure that, in the concentration below the onset of superconductivity the holes with up-spin are accommodated in b_{1g} orbital constructed mainly from oxygen p_σ orbitals in CuO_2 plane for the A-site, consistent with the result of the cluster calculation by Kamimura and Eto [4], while in the superconducting concentration regime the holes itinerate from a_{1g}^* orbital at the A-site to b_{1g} orbital at the B-site, consistent with the prediction

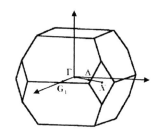

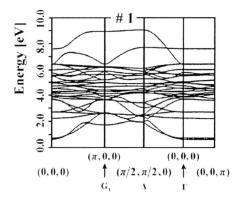

FIGURE 16. The many-body-effect-included-band-structure for up-spin dopant holes, obtained by solving the effective one-electron-type 34×34 dimensional Hamiltonian matrix $\tilde{H}$ for an antiferromagnetic unit cell, where the ordinary Brillouin zone corresponding to an ordinary unit cell consisting of a single CuO_6 octahedron is shown in the upper part of the figure. The highest occupied band is marked by the #1 band. The Λ-point corresponds to $(\pi/2a, \pi/2a, 0)$, while the G_1-point to $(\pi/a, 0, 0)$. In the figure the Cu-O-Cu distance, a, is taken to be unity.

by Kamimura and Suwa [14]. Here it should be remarked that the mixing ratio of the Zhang-Rice singlet is always dominant for the Hund's coupling triplet. Quantitatively the calculated ratio of the $^1A_{1g}$ to the $^3B_{1g}$ component is at most 8 to 2 in the underdoped region of LSCO, consistent with the experimental results by C. T. Chen *et al.* [35]

Kamimura and Ushio also calculated the Fermi surfaces and the density of states of the #1 band in LSCO. For example, the Fermi surfaces for $x = 0.15$ calculated by them are shown in Fig. 18, and the density of states for the #1 band is shown in Fig. 19. The Fermi surface for $x = 0.15$ consists of two pairs of extremely flat tube, which are directed along bisectors between k_x and k_y axis and are orthogonal to each other. The distance between two Fermi surfaces facing each other and its directions are close to the Q_1 and Q_2 vectors shown in Fig. 20, where the Q_1 and Q_2 vectors are equivalent to the spanning vectors of the 2D antiferromagnetic ordering; $(\pi/a, \pi/a, 0)$ and $(-\pi/a, \pi/a, 0)$, respectively. The cross section of each Fermi surface facing towards the Γ point, the center of the Brillouin zone, is very wide, and the dispersion of the highest

B-site   **A-site**

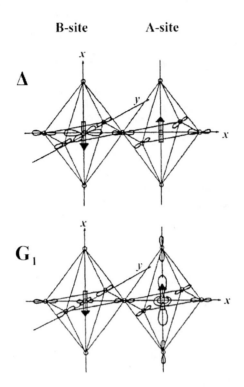

FIGURE 17. The wave functions at Δ and G_1 points. Here, the right hand side of the figure corresponds to a CuO_6 cluster with localized up-spin (A-site) and the left hand side to a CuO_6 cluster with localized down-spin (B-site).

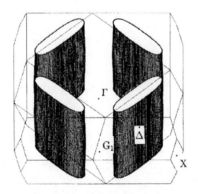

FIGURE 18. The Fermi surface for $x = 0.15$ calculated for the #1 band. Here two kinds of Brillouin zones are also shown. One at the outermost part is the ordinary Brillouin zone and the inner part is the folded Brillouin zone for the antiferromagnetic unit cell in LSCO. Here the k_x axis is taken along $\overline{\Gamma G_1}$, corresponding to the x-axis (Cu-O-Cu direction) in a real space.

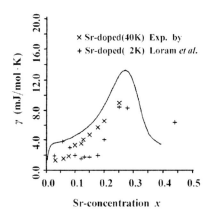

FIGURE 19. The electronic specific heat of LSCO as a function of hole concentration x. The solid lines are the calculated one for the #1 band [14] while the crosses are the experimental data by Loram *et al.* [52]. The energy is measured from the top of the band. Holes enter from the top.

band is relatively flat. These unique features of the energy dispersion of the #1 band and Fermi surfaces have been checked by angle resolved UPS experiments for $Bi_2Sr_2CuO_6$ (Bi2201), $Bi_2Sr_2CaCu_2O_{8+\lambda}$ (Bi2212) [45-47] and $Bi_2Sr_2Ca_{1-x}Dy_xCu_2O_{8+\delta}$. In particular, Aebi *et al.* [45,46] have observed a $\sqrt{2} \times \sqrt{2}$ superstructure, supporting the existence of the short-range antiferromagnetic correlations predicted by Kamimura and Suwa [14]. The feature of the calculated Fermi surface changes drastically in the overdoped region of around $x \sim 0.2$. In this overdoped region four flat sections of the Fermi surface merge into one "large Fermi surface", as shown in Fig. 21. Thus the Fermi surfaces change from small Fermi surfaces to a large Fermi surface when the hole concentration increases, as shown by the angle-resolved photoemission (ARPES) experiment by Norman *et al.* [49]. Such change of Fermi surfaces with the doping concentration is schematically shown in Fig. 22.

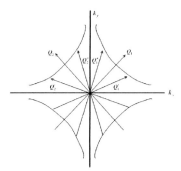

FIGURE 20. A way of expressing the location of the Fermi surfaces in the k_x–k_y plane in terms of the Q_1 and Q_2 vectors which are the bisectors between the k_x and k_y axes, and schematic view of the nesting vectors Q_1', Q_1'', Q_2' and Q_2'' between Fermi surfaces.

27

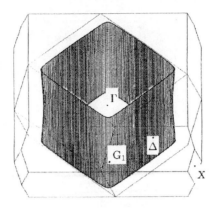

FIGURE 21. Schematic view for a large Fermi surface in the overdoped region.

In connection with the features of the Fermi surfaces shown in Fig. 18, we would like to point out a possibility of nesting between Fermi surfaces with different spins and between those with the same spin for the nesting vector of Q_1', Q_1'', Q_2' and Q_2'' which are deviated slightly from commensurate Q_1 and Q_2 vectors, as seen in Fig. 20. The former type of nesting may lead to an anomaly in spin excitation spectra while the latter is related to an anomaly in phonon spectra for the above nesting vectors. The appearance of incommensurate peak in spin excitation spectra of LSCO observed by neutron diffraction experiments [48] may be related to the above anomaly.

The density of states calculated by Ushio and Kamimura [40,41] for the #1 band is shown in Fig. 19. It exhibits a sharp peak at the vlue E_F corresponding to x ~ 0.3 in $La_{2-x}Sr_xCuO_4$. The appearance of this sharp peak is due to a two-dimensional-like singularity at the G_1 point, suppressed by the transfer interaction between adjacent CuO_2 layers.

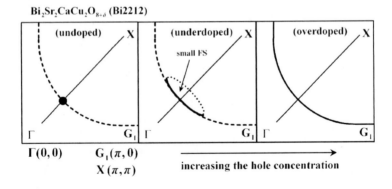

FIGURE 22. Schematic view on the change of Fermi surfaces with the carrier concentration, observed by Norman *et al.* [49] with the ARPES measurements.

11. MICROSCOPIC ORIGIN OF
THE "HIGH ENERGY" PSEUDOGAP

11.1. Introduction

Recently phenomena which can be attributed to the reduction of the density of states near the Fermi energy, called the *pseudogap*, have been observed [52]. In particular, the strange hole concentration dependence of the electronic entropy in LSCO observed by Loram *et al.* [52] has drawn considerable attention, related to the pseudogap. In this connection, Nakano *et al.* [50] have found from their magnetic susceptibility and resistivity measurements of LSCO that there are two crossover lines, $T_{max}(x)$ and $T^*(x)$ ($T_c < T^* < T_{max}$), in the T-x phase diagram of cuprates with a superconducting transition temperature T_c, both of which decrease monotonically with increasing the hole concentration x, as shown in Fig. 23. The upper crossover line T_{max} represents the temperature below which the magnetic susceptibility exhibits a broad peak, arising from the gradual development of antiferromagnetic spin correlation, while the lower crossover line T^* represents the temperature below which a spin gap may open up in the magnetic excitation spectrum around $q = (\pi, \pi)$. Now T_{max} and T^* are called the "high energy" and "low energy" pseudogaps, respectively.

Recently Kamimura, Hamada and Ushio [51] have shown on the basis of the Kamimura-Suwa model that the origins of both T_{max} and the strange hole concentration dependence of the electronic entropy are explained by a unified mechanism of phase change between the phase consisting of the small Fermi surface (abbreviated as "SF-phase") and that of the large Fermi surface (abbreviated as "LF-phase"). For this purpose they calculated the free energies of the SF- and LF-phases,

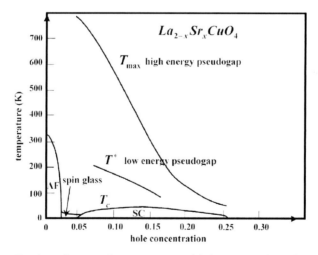

FIGURE 23. The phase diagram of temperature T and hole concentration x for LSCO, showing the "high-energy" pseudogap T_{max} and the "low-energy" pseudogap T^*.

29

and they showed that the hole carriers not only in the underdoped region but also in the overdoped region behave like "heavy fermions".

11.2. Calculation of free energies of the SF- and LF-phases

According to Kamimura, Hamada and Ushio [51], the electronic structure of LSCO changes from the SF-phase to the LF-phase at x_c in the overdoped region when x increases, as seen in Fig. 24, where the spin-correlation length is schematically plotted against the hole concentration. Similarly, when the temperature increases for a fixed hole concentration in the underdoped regime, the local AF order is destroyed so that the SF-phase is considered to change to the LF-phase at a certain temperature T_0. They considered that T_0 may correspond to T_{max} in the experiments by Nakano *et al.* [50].

When we write the difference between the free energies per Cu ion of the LF- and SF-phases,

$$\Delta F(T,x) = F_{LF}(T,x) - F_{SF}(T,x) , \qquad (7)$$

T_{max} and the critical concentration x_c are defined as $\Delta F(T_{max}, x) = 0$ and $\Delta F(T=0, x_c) = 0$, respectively. First the internal energy in the SF-phase is expressed in the following way;

$$E_{SF}(T,x) = E_{SF}(0,0) - E_{kin}^{(SF)}(T,x) - E_{kin}^{(SF)}(0,0) , \qquad (8)$$

where $E_{SF}(0,0)$ and $E_{kin}^{(SF)}(T,x)$ represent, respectively, the internal energy per Cu ion in a system of the local antiferromagnetic ordering with $T=0$ and $x=0$ and the kinetic

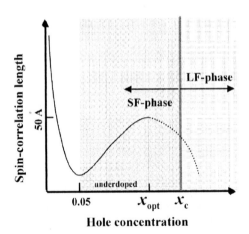

FIGURE 24. The x-dependence of spin-correlation length for LSCO observed by Mason *et al.* for its underdoped region and the classification of the SF- and LF-phases at x_c.

energy per Cu ion in a hole-carrier system with concentration x at T. Since $E_{SF}(0,0)$ includes the kinetic energy, we have to subtract $E_{kin}^{(SF)}(0,0)$ from $E_{kin}^{(SF)}(T,x)$ in order to avoid the double counting. By using the density of states per Cu ion for the highest (#) band calculated by Ushio and Kamimura [41], $\rho_{UK}(\varepsilon)$, and the Fermi distribution function $f(\varepsilon, \mu(x))$, $E_{kin}^{(SF)}(T,x)$ can be expressed as

$$E_{kin}^{(SF)}(T,x) = \int_{-\infty}^{\infty} \varepsilon \rho_{UK}(\varepsilon) f(\varepsilon, \mu(x)) d\varepsilon, \tag{9}$$

where they have considered explicitly the x-dependence of the chemical potential μ like $\mu(x)$. The entropy per Cu ion in the SF-phase is calculated by a well-known formula

$$S_{SF}(T,x) = -k_B \int_{-\infty}^{\infty} [f(\varepsilon, \mu(x)) \ln f(\varepsilon, \mu(x))$$
$$+ \{1 - f(\varepsilon, \mu(x))\} \ln\{1 - f(\varepsilon, \mu(x))\}] \rho_{UK}(\varepsilon) d\varepsilon. \tag{10}$$

Thus the free energy per Cu ion is calculated by the formula,

$$F_{SF}(T,x) = E_{SF}(T,x) - TS_{SF}(T,x) \quad . \tag{11}$$

Similarly we denote the internal energy per Cu ion in the LF-phase by $E_{LF}(T, x)$. In the LF-phase the local antiferromagnetic ordering does not exist so that they have first assumed that the electronic system may be treated by an ordinary band theory for the $(1+x)$ hole concentration. Thus $E_{LF}(T, x)$ can be expressed as

$$E_{LF}(T,x) = E_{LF}(0,0) + E_{kin}^{(LF)}(T,x) - E_{kin}^{(LF)}(0,0), \tag{12}$$

where a conduction band corresponds to a b_{1g}^{*} energy band with the main character of Cu $d_{x^2-y^2}$ orbital. $E_{LF}(0,0)$ represents the internal energy per Cu ion in the LF-phase with $T = 0$ K and $x = 0$, in which the b_{1g}^{*} energy band is half-filled. The entropy per Cu ion in the LF-phase is calculated by

$$S_{LF}(T,x) = -k_B \int_{-\infty}^{\infty} [f(\varepsilon, \mu(x)) \ln f(\varepsilon, \mu(x))$$
$$+ \{1 - f(\varepsilon, \mu(x))\} \ln\{1 - f(\varepsilon, \mu(x))\}] \rho_{b_{1g}^{*}}(\varepsilon) d\varepsilon \quad , \tag{13}$$

where $\rho_{b_{1g}^{*}}(\varepsilon)$ is the density of states for a system with $(1+x)$ hole concentration of the b_{1g}^{*} energy band. Then the free energy per Cu ion in the LF-phase is given by,

$$F_{LF}(T,x) = E_{LF}(T,x) - TS_{LF}(T,x) \quad . \tag{14}$$

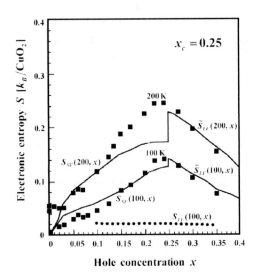

FIGURE 25. The calculated results for the electronic entropies of the SF- and LF-phases in LSCO for $T = 100$ K and $T = 200$ K, where the experimental results of Loram et al. [52] are also shown by solid squares. Further the electronic entropy calculated by the density of states of the LDA band is shown by dotted line. (After Ref. [51])

In Fig. 25, the x-dependence of the electronic entropies in the SF- and LF-phases calculated using Eqs. (10) and (13) is shown for LSCO for $T = 100$ K and 200 K, and the results are compared with those measured by Loram *et al.* [52] represented by solid squares. As regards $S_{SF}(T, x)$ for $T = 100$ K and 200 K, the agreement between theory and experiment for $T = 100$ K is fairly good for the concentration below $x = 0.25$, where the $S_{SF}(T = 100$ K, $x)$ and $S_{SF}(T = 200$ K, $x)$ have been calculated using $\rho_{UK}(\varepsilon)$ from the Kamimura-Suwa model without introducing adjustable parameters. Since the observed maxima in the x-dependence of entropy in LSCO appear around $x = 0.25$ both for $T = 100$ K and 200 K, Kamimura, Hamada and Ushio [51] have adopted $x_c=0.25$. For the overdoped region beyond $x_c = 0.25$, they have calculated the electronic entropy using Eq. (13) and the density of states obtained by the LDA band [20,43,53,54], $\rho_{b_{1g}^*}(\varepsilon)$. The entropy thus calculated for $T = 100$ K, $S_{LF}(100, x)$, is shown by dotted line in Fig. 25. As seen in Fig. 25, it is very small compared with experimental results shown by the solid squares. This means that the electronic state in the overdoped region beyond $x_c = 0.25$ can not be expressed by the ordinary LDA bands.

Therefore, they have modified the density of states of the LDA b_{1g}^* band so as to reproduce the experimental values of electronic entropy in the overdoped region above x_c. The density of states by Ushio and Kamimura in the SF-phase, $\rho_{UK}(\varepsilon)$, and the modified density of states in the LF-phase which we denote $\tilde{\rho}_{b_{1g}^*}(\varepsilon)$ are shown in Figs. 26(a) and (b), respectively. The modified density of states at a van Hove sigularity which lies at the center of the band is about 6 times larger than that of the LDA band. By using the modified density of states $\tilde{\rho}_{b_{1g}^*}(\varepsilon)$ for $\rho_{b_{1g}^*}(\varepsilon)$ in Eq. 13, we

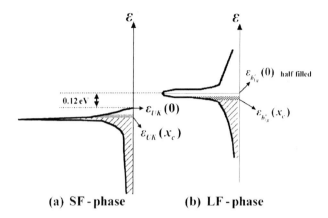

(a) SF - phase   (b) LF - phase

calculated

FIGURE 26. (a) Shematic view of the density of states calculated by Ushio-Kamimura $\rho_{UK}(\varepsilon)$ in the SF-phase. (b) Shematic view of the modified density of states for the b_{1g}^* band $\tilde{\rho}_{b_{1g}^*}(\varepsilon)$ in the LF-phase. In both cases the density of states is shown along the holizontal line.

the electronic entropies for $T = 100$ K and 200K, which are denoted by $\tilde{S}_{LF}(100, x)$ and $\tilde{S}_{LF}(200, x)$, respectively, and the results are also shown in Fig. 25. As seen in this figure, we can explain successfully the observed x-dependence of the electronic entropies in the underdoped and overdoped region.   Although the discontinuity appears at x_c in Fig. 25, this should be smeared out by taking account of the broadening effect of Fermi surfaces in the SF-phase. From this result we conclude that the appearance of the maximum in the observed entropy may be considered as an experimental evidence for a phase change from the SF-phase to the LF-phase at x_c. Further, from the modified density of states $\tilde{\rho}_{b_{1g}^*}(\varepsilon)$, we conclude that the effective mass of hole carriers are about 6 times heavier than the free electron mass in the overdoped region.   Since the effective mass of the hole carriers in the Ushio-Kamimura band [41] in the underdoped region is about ten times heavier than the free electron mass, we can say that the hole carriers in the superconducting cuprates behave like heavy fermions.

11.3. Calculated results of the x-dependence of the "high-energy" pseudogap

By using the density of states of the Ushio-Kamimura band, $\rho_{UK}(\varepsilon)$, for the underdoped to overdoped region below x_c and the modified density of states for the b_{1g}^* band, $\tilde{\rho}_{b_{1g}^*}(\varepsilon)$, for the overdoped region above x_c, Kamimura, Hamada and Ushio [51] have calculated T_{max} from the following equations;

$$\Delta F(T_{\max}, x) = 0 \quad , \qquad (15)$$

where

$$
\begin{aligned}
\Delta F(T,x) &= \widetilde{F}_{LF}(T,x) - F_{SF}(T,x) \\
&= \{\widetilde{E}_{LF}(T,x) - T\widetilde{S}_{LF}(T,x)\} - \{E_{SF}(T,x) - TS_{SF}(T,x)\} \\
&= \widetilde{E}_{LF}(0,0) - E_{SF}(0,0) + \{\widetilde{E}_{kin}^{(LF)}(T,x) - \widetilde{E}_{kin}^{(LF)}(0,0)\} \\
&\quad - \{E_{kin}^{(SF)}(T,x) - E_{kin}^{(SF)}(0,0)\} - T\{\widetilde{S}_{LF}(T,x) - S_{SF}(T,x)\} \qquad (16) \\
&= E_{AF} + \int_{-\infty}^{\infty} \varepsilon \widetilde{\rho}_{b_{1g}^*}(\varepsilon) f(\varepsilon,\mu(x)) d\varepsilon - \int_{-\infty}^{\varepsilon_{b_{1g}^*}(0)} \varepsilon \widetilde{\rho}_{b_{1g}^*}(\varepsilon) d\varepsilon \\
&\quad - \left\{ \int_{-\infty}^{\infty} \varepsilon \rho_{UK}(\varepsilon) f(\varepsilon,\mu(x)) d\varepsilon - \int_{-\infty}^{\varepsilon_{UK}(0)} \varepsilon \rho_{UK}(\varepsilon) d\varepsilon \right\} \\
&\quad - T\{\widetilde{S}_{LF}(T,x) - S_{SF}(T,x)\} \quad .
\end{aligned}
$$

Here $F_{AF} \equiv \widetilde{E}_{LF}(0,0) - E_{SF}(0,0)$. Further $\varepsilon_{UK}(0)$ and $\varepsilon_{b_{1g}^*}(0)$ represent the band energies of the Ushio-Kamimura band and of the "modified b_{1g}^* band" for $x=0$, respectively, as shown in Figs. 26(a) and (b). T_{max} is calculated from Eqs. (15) and (16) as a function of x. In doing so, the value of E_{AF} was found to be 0.043 eV by inserting $T_{max} = 1000$ K at $x = 0$ to Eqs. (15) and (16), where $T_{max} = 1000$ K at $x = 0$ was obtained by extrapolating the experimental data. E_{AF} thus determined corresponds to the energy change from the half-filled band in the LF-phase to a Mott-Hubbard insulator in the SF-phase. Further information of the energy difference, $\varepsilon_{b_{1g}^*}(0) - \varepsilon_{UK}(0)$, is necessary.

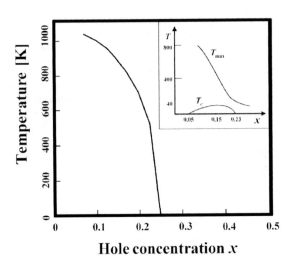

FIGURE 27. The calculated results for the hole concentration dependence of T_{max} in LSCO. The experimental data for T_{max} by Nakano *et al.* [50] are also shown, together with the sketch of the x dependence of superconducting transition temperature T_c in the inset.

Kamimura, and Ushio determined it at $T = 0$ K by requiring that $\Delta F(T = 0$ K, $x)$ vanishes at x_c. As to x_c they chose 0.25 from the concentration at which the entropy shows a maximum in the experimental data by Loram *et al.* [52]. Then they determined $\varepsilon_{b'_g}(0) - \varepsilon_{UK}(0)$ to be 0.12 eV. Using $E_{AF} = 0.043$ eV and $\varepsilon_{b'_g}(0) - \varepsilon_{UK}(0) = 0.12$ eV, they calculated the x-dependence of T_{max} from Eqs. (15) and (16) quantitatively. The calculated result is shown in Fig. 27. When the calculated x-dependence of T_{max} is compared with that observed for LSCO by Nakano *et al.* [50] shown in the inset of Fig. 27, we find that the agreement is fairly good. Thus one may say that the phase change between the SF- and LF-phases corresponds to the "high-energy" pseudogap, T_{max}.

In this section we have seen from the calculations of the electronic entropies in the SF- and LF-phases and also from the calculated free energy difference between the SF- and LF-phases that both the observed peculiar x-dependence of entropy by Loram *et al.* and the observed T_{max} by Nakano *et al.* can be explained by a unified mechanism of a phase change between the SF- and LF-phases. Further, by fitting the calculated x-dependence of the electronic entropy to the one observed in the overdoped region beyond $x_c = 0.25$, we have known that the hole carriers in cuprates behave like heavy fermions. One may say that the origin of the heavy effective mass may be due to both the electron correlation and the local lattice distortion of CuO_6 octahedra.

11.4. A new method to test the validity of theoretical models

Concerning theoretical models for high temperature superconductivity in cuprates, there exist an enormous number of models. In order to test the validity of these models, we would like to suggest the authors of these models to calculate the hole concentration dependence of electronic entropies in the underdoped and overdoped region. If the calculated concentration and temperature dependences of electronic entropy as well as its magnitude would agree with experimental results by Loram *et al.*, one can say that these models are valid quantitatively. If not, these models are not appropriate in explaining the superconducting as well as normal-state properties of real cuprates quantitatively.

ACKNOWLEDGEMENTS

It is a pleasure to acknowledge the collaboration of Hideki Ushio, Mikio Eto, Yuji Suwa and Shunichi Matsuno on the subjects reviewed in this article. One of the authors (H. K.) would like to thank Dr. John Loram and Prof. Yao Wei Liang for valuable discussions on various topics mentioned in this article when H. K. stayed in the Interdisciplinary Research Centre in Superconductivity, University of Cambridge.

REFERENCES

1. Bednorz, J.G., and Müller, K.A., *Z. Phys. B* **64**, 189 (1986).
2. Anderson, P.W., *Science* **235**, 1196 (1987).
3. See, for example, Dagotto, E., *Reviews of Modern Physics* **66**, 763-840 (1994).
4. Kamimura, H., and Eto, M., *J. Phys. Soc. Jpn.* **59**, 3053-3056 (1990).
5. Eto, M., and Kamimura, H., *J. Phys. Soc. Jpn.* **60**, 2311-2323 (1991).
6. Eto, M., and Kamimura, H., *Phys. Rev. Lett.* **61**, 2790-2793 (1988).
7. Wohl, A.C., and Das, G., "Vol. 3, Methods of Electronic Structure Theory" in *Modern Theoretical Chemistry*, edited by Schaefer, H.F., III, New York, Plenum, 1977, pp. 51.
8. Kato, S., and Morokuma, K., *Chem. Phys. Lett.* **65**, 19 (1979).
9. Boyce, J.B., Bridges, F., Claeson, T., Geballe, T.H., Chu, C.W., and Tarascon, J.M., *Phys. Rev. B* **35**, 7203-7206 (1987).
10. Jahn, H.A., and Teller, E., *Proc. Roy. Soc. London A* **161**, 220 (1937).
11. Zhang, F.C., and Rice, T.M., *Phys. Rev. B* **37**, 3759-3761 (1988).
12. Cava, R.J., Santoro, A., Johnson D.W., Jr., and Rhodes, W.W., *Phys. Rev. B* **35**, 6716-6720 (1987). We use the lattice constants at 10 K reported there.
13. Beno, M.A., Soderholm, L., Capone, D.W., II, Hinks, D.G., Jorgensen, J.D., Grace, J.D., Schuller, Ivan K., Segre, C.U., and Zhang, K., *Appl. Phys. Lett.* **51**, 57 (1987).
14. Kamimura, H., and Suwa, Y., *J. Phys. Soc. Jpn.,* **62**, 3368-3371 (1993).
15. Egami, T. *et al.*, in *High Temperature Superconductivity*, edited by J. Ashkenazi and Vezzoli, G., Plenum Press, New York, 1992 pp. 389, and related references therein.
16. Cava, R.J. *et al.*, *Physica C* **153-155**, 560 (1988).
17. Shmahl, W.W. *et al.*, *Phil. Mag. Lett.* **58**, 181 (1988).
18. Maeda, H. *et al.*, *Jpn. J. Appl. Phys.* **27**, L209 (1988).
19. Tarascon, J.M. *et al.*, *Phys. Rev. B* **37**, 9382-9389 (1988).
20. Shima, N., Shiraishi, K., Nakayama, T., Oshiyama, A., and Kamimura, H., in Proc. the 1st Int. Conf. on Electronic Materials, edited by Sugano, T., Chang, R.P.H., Kamimura, H., Hayashi, I., and Kamiya, T., MRS, Pittsburgh, 1988, pp. 50-53.
21. Roos, B., Veillard, A., and Vinot, G., *Theoret. Chim. Acta.* **20**, 1, (1971).
22. Dunning, T.H., Jr., and Hay, P.J., "Vol. 3, Methods of Electronic Structure Theory" in *Modern Theoretical Chemistry* edited by Schaefer, H.F., III, Plenum, New York, 1977, pp. 1.
23. Kamimura, H., and Sano, A., *Solid State Commun.* **109**, 543-547 (1999).
24. Tobita, Y., and Kamimura, H., *J. Phys. Soc. Jpn.* **68**, 2715-2720 (1999).
25. Shimizu, T., Yasuoka, H., Imai, T., Tsuda, T., Takabatake, T., Nakazawa, Y., and Ishikawa, M., *J. Phys. Soc. Jpn.* **37**, 2494 (1988).
26. Edward, H.L., Barr, A.L., Markert, J.T., and de Lozanne, A.L., *Phys. Rev. Lett.* **73**, 1154-1157 (1994).
27. Mook, M.A., Dai, P., Salama, K., Lee, D., Dögan, F., Aepli, G., Boothroyd, A.T., and Mostoller, M.E., *Phys. Rev. Lett.* **77**, 370-373 (1996).
28. Sano, A., Eto, M., and Kamimura, H., *J. Mod. Phys. B* **11**, 3733-3750 (1997).
29. Kamimura, H., and Sano, A., *J. Superconductivity* **10**, 279-284 (1997).
30. Matsui, Y. *et al.*, *Jpn. J. Appl. Phys.* **27**, L372 (1988).
31. Kondo, J., *Springer Series in Solid-State Sciences* **89**, Springer, 1989, pp. 57.

32. Kamimura, H., Matsuno, S., Suwa, Y., and Ushio, H., *Phys. Rev. Lett.* **77**, 723-726 (1996).

33. Mason, T.E. *et al.*, *Phys. Rev. Lett.* **77**, 1604-1607 (1996).

34. Yamada., K. *et al.*, *J. Superconductivity* **10**, 343 (1997).

35. Chen, C.T. *et al.*, *Phys. Rev. Lett.* **68**, 2543-2546 (1992).

36. Pellegrin, E., Nücker, N., and Fink, J. *et al.*, *Phys. Rev. B* **47**, 3354-3367 (1993).

37. Bianconi, A. *et al.*, *Physica C* **162-164**, 207 (1989).

38. Birgeneau, R.J. *et al.*, "Mechanism of High Temperature Superconductivity" in *Springer Series in Materials Science 11*, edited by Kamimura, H., and Oshiyama, A., Springer, Heidelberg, 1989, pp. 120. See also Birgeneau *et al.*, *Phys. Rev. B* **38**, 6614-6623 (1988).

39. Merz, M. *et al.*, *Phys. Rev. Lett.* **80**, 5192-5195 (1998).

40. Kamimura, H., and Ushio, H., *Solid State Commun.* **91**, 97-100 (1994).

41. Ushio, H., and Kamimura, H., *Int. J. Mod. Phys. B* **11**, 3759-3796 (1997), related references therein.

42. Anisimov, V.I., Korotin, M.A., Zaanen, J., and Andersen, O.K., *Phys. Rev. Lett.* **68**, 345-348 (1992).

43. Deweert, M.J., Papaconstantopoulos, D.A., and Pickett, W.E., *Phys. Rev. B* **39**, 4235-4248 (1989), and related references therin.

44. Takagi, H., Ido, T., Ishibashi, S., Uota, M., Uchida, S., and Tokura, Y., *Phys. Rev. B* **40**, 2254-2261 (1989).

45. Aebi, P. *et al.*, *Phys. Rev. Lett.* **72**, 2757-2670 (1994).

46. Aebi, P. *et al.*, *Physica C* **235-240**, 949 (1994).

47. Marshall, D.S. *et al.*, *Phys. Rev. Lett.* **76**, 4841-4844 (1996).

48. Shirane, G. *et al.*, *Phys. Rev. Lett.* **63**, 330-333 (1989).

49. Norman, M.R. *et al.*, *Nature* **392**, 157-160 (1998).

50. Nakano, T. *et al.*, *J. Phys. Soc. Jpn.* **67**, 2622-2625 (1998).

51. Kamimura, H., Hamada, T., and Ushio, H., submitted to PRL.

52. Loram, J.W. *et al.*, *J. Phys. Chem. Solids.* **59**, 2091-2094 (1998).

53. Mattheiss, L.F., *Phys. Rev. Lett.* **58**, 1028-1030 (1987).

54. Yu, J., Freeman, A.J., and Xu, J.H., *Phys. Rev. Lett.* **58**, 1035-1037 (1987).

Competing Interactions in Complex Oxides: Polaron Stability in Manganites and Cuprates

T. Egami

*Department of Materials Science and Engineering and
Laboratory for Research on the Structure of Matter,
University of Pennsylvania, Philadelphia, PA 19104 USA*

Abstract. Near the metal-insulator transition competing forces to localize and delocalize charges are in delicate balance, and small perturbations can produce drastic effects. For the manganites that show colossal magnetoresistivity (CMR) the electron-lattice interaction plays a key role in producing and altering the balance through formation of polarons. Our recent measurements of high-energy LO phonons by inelastic neutron scattering demonstrate that polaronic charge localization tendency exists also in the cuprates that exhibit high-temperature superconductivity (HTSC). It is suggested that the lattice is an integral part of the HTSC mechanism, and that *marginal* stability of polaron is important for the CMR effect in manganites as ell as for the HTSC in cuprates.

INTRODUCTION

The lattice has been considered by many to be irrelevant, or even harmful, to high-temperature superconductivity (HTSC) of cuprates [1]. This view was formed mainly by the observations that the isotope effect is so small, and there is no evidence of strong electron-phonon coupling in the transport properties of the normal state. Furthermore, since the most popular magnetic mechanism of the d-wave superconductivity [1,2] is based upon the repulsive mechanism, s-wave phonon mechanism, which is attractive, cancels the spin contribution, and is therefore considered to be harmful. However, it is well known that the isotope effect can be small in the case of strong coupling [2], and the normal state conductivity can be explained by the two-component model, having one component with weak electron-phonon coupling that provides high conductivity and the other with strong coupling that contribute to pairing [3,4]. In addition the phonon mechanism can positively contribute to the d-state superconductivity in the presence of antiferromagnetic (AFM) spin correlations [5].

In the meantime evidences of lattice anomalies and charge inhomogeneity associated with superconductivity are increasing [6]. The observation of the charge/spin stripe structure in non-superconducting $(La,Nd)_{1.875}Sr_{0.125}CuO_4$ [7] provided a strong evidence that the charge and spin in cuprates want to phase-segregate. In this paper we review our recent measurements of phonon dispersion in cuprates that strongly indicate charge inhomogeneity. The phonon results are not easily reconciled with the stripe state observed for the non-superconducting samples,

CP554, *Physics in Local Lattice Distortions*, edited by H. Oyanagi and A. Bianconi
© 2001 American Institute of Physics 1-56396-984-X/01/$18.00

and suggest a different kind of dynamic polaronic or incipient charge density wave (CDW) state. We suggest, while stable polarons produce an insulating stripe state, marginal stability of polaronic state is a required ingredient of the HTSC phenomenon. This situation has a strong parallelism in the colossal magnetoresistivity (CMR) of manganites. In manganites stable polarons result in an insulating state, but marginal polarons that can be destroyed by applied fields produce the CMR effect. The stability of polaron depends upon the balance between the localizing forces (electron-phonon coupling and spin disorder) and delocalizing forces (electron kinetic energy and elasticity). We discuss similarities and differences in the role of these factors between manganites and cuprates, and suggest the lattice has an important role in the HTSC mechanism.

NEUTRON INELASTIC SCATTERING

Recent neutron inelastic scattering measurements of dispersion of longitudinal-optical (LO) phonons in cuprates provide strong evidence for the existence of charge inhomogeneity in cuprates. The phonon branch of interest is the LO phonons along [100] direction in the tetragonal indices, along the Cu-O bond in the CuO_2 plane. The phonon mode in this branch changes the Cu-O distance and thus induces charge transfer between Cu and O, as shown in Fig. 1 [8,9]. In particular the zone-boundary mode at (1/2, 0, 0), or (π, 0, 0), is most effective in causing charge transfer between Cu ions, and results in very strong electron-phonon coupling [10,11]. Note that the zone-center (0, 0, 0) mode is polar, and related to ferroelectricity. The zone-edge mode is half-breathing, and induces charge transfer between alternating Cu ions at the opposite phonon phases.

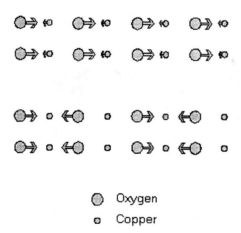

FIGURE 1. The high-energy longitudinal-optical (LO) phonon Cu-O bond-stretching mode in the CuO_2 plane. At the zone center (above) the mode is polar, while at the zone-edge (below) it is half-breathing, causing charge transfer from one Cu ion to the other.

The early studies of this mode revealed strong mode-softening at the zone-boundary due to doping, and diffuse behavior at the middle of the zone around (1/4, 0, 0) [12,13]. A series of measurements made recently on $La_{1.85}Sr_{0.15}CuO_4$ (LSCO) and $YBa_2Cu_3O_{6+x}$ (YBCO) single crystals using a triple-axis spectrometer at the High-Flux Isotope Reactor (HFIR) of the Oak Ridge National Laboratory uncovered further important details [14-16]. When YBCO is doped (x is increased from 0 to 0.95), one might expect gradual softening of the zone-boundary phonon. Instead, as shown in Fig. 2 [16], the dispersion is basically unchanged, with a strong split in the middle of the zone into two branches, and only the spectral weight changes with doping. The high-energy branch (70 – 75 meV) becomes weaker as doping is increased, while the low-energy branch (50 – 55 meV) increases its intensity. The result strongly indicates that the system is not homogeneous, and there are two microscopic states, with and without charges. The high-energy branch must be associated with the undoped state, and the low-energy branch with the doped state. Doping only increases the *volume fraction* of the doped state, and the local charge density in these states must remain unchanged. These two states, however, are not separate macroscopic phases. If there is phase separation into doped and undoped phases, we should observe two phonon branches with similar Q dependences in intensity. But the high-energy branch has high intensity around the zone-center ($Q_x = 3.0$, in the unit of the reciprocal lattice vector $= 1.6$ Å^{-1}), while the low-energy branch has strong intensity at the zone-boundary ($Q_x = 3.5$). In the simple ball and spring model the structure factor decreases from 3.0 to 3.5, since at $Q_x = 3.0$ the Cu amplitude and O amplitude add up to the neutron scattering intensity, while at 4.0 they subtract from each other. Thus the strong intensity at the zone-boundary is quite unusual. A part of this intensity is due to the Cu-O bond-bending mode, but it does not diminish the importance of this peculiar nature.

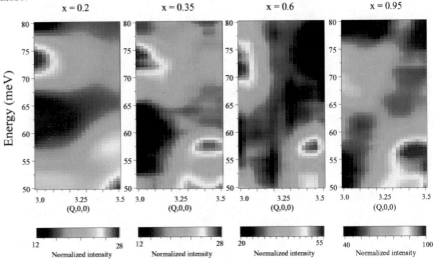

FIGURE 2. The composition dependence of the LO phonon dispersion for YBCO x = 0.2, 0.35, 0.6, and 0.95.

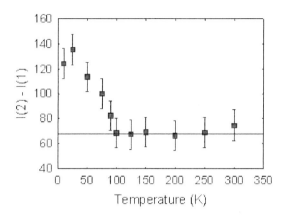

FIGURE 3. Temperature dependence of the intensity difference between the two energy zones at (1/4, 0, 0). See text for definition.

Moreover the neutron scattering intensity distribution was found to depend on temperature. As shown in Fig. 3 [16], the difference in the average intensities at $Q_x =$ 3.25 between the energy ranges 56 – 68 meV and 51 – 55 meV for the optimally doped YBCO (x = 0.95) varies with temperature just as the superconducting order parameter. At lower doping levels the temperature dependence becomes more gradual, with the difference parameter starting to rise around the pseudo-gap temperature and reaches half-way to saturation around T_c [17]. It resembles the temperature dependence of the hole-pair density deduced from the specific heat [18].

NATURE OF CHARGE INHOMOGENEITY

The results shown above provide a strong evidence of microscopic charge inhomogeniety in HTSC cuprates. However, they do not necessarily point to the presence of the stripe state. Current opinion on charge/spin stripes appears to be divided into two camps; one view is that they are harmful to HTSC since they appear to compete against superconductivity [19], and the other is to regard them central to the mechanism of superconductivity [20]. In the second view it is assumed that stripes exist even in the superconducting state but they are dynamic and short-range. In this view static stripes compete against HTSC, but dynamic stripes help it. The energy of stripe fluctuations must be of the order of a few meV, since otherwise much stronger phonon broadening will be observed. It is then difficult to understand how such slow fluctuations could affect HTSC so strongly.

Actually so far there is no hard evidence to support the presence of stripes in the superconducting cuprates. The continuity of the incommensurate periodicity of the dynamic magnetic scattering [21] and the discontinuity in the phonon dispersion [22] are often cited as the evidence of dynamic stripes, but these data allow alternative interpretations. Since the charge stripe corresponds to the AFM domain wall [23], the charge stripe can move without disturbing the long-range magnetic order. Conversely well-defined magnetic correlation does not necessarily imply well-defined charge correlation. It is possible, and likely, that the charge correlation is much more short-range than the magnetic correlation.

The phonon dispersion data by Mook and Dogan [22] show discontinuity and broadening at the wave vector corresponding to the stripe periodicity, $q_d = 2\pi/d$. At an optimum doping $d = 4a$, where a is the unit cell size, or the Cu-Cu distance in the CuO_2 plane. Thus the discontinuity appears at $q_d = 0.25$, in the unit of a^*. This result agrees with our observation shown above, except that the gap is larger, about 20 meV for the LO phonon. However, the results are not necessarily consistent with the stripe phase. If there are static stripes elastic superlattice peaks are observed at $Q = K \pm q_d$, where K is the reciprocal lattice vector. Then phonons are expected to soften at these wave vectors. The discontinuous gap observed in the phonon dispersion reminds the phonon gap at the Brillouin zone boundary, and in that case the underlying charge periodicity is $2a$, not $4a$ [14]. Note that the charge inhomogeneity becomes enhanced below T_C, as shown in Fig. 3.

These phonon results shown in Figs. 2 and 3 strongly suggest that the charge state in the superconducting phase is inhomogeneous, but it is not consistent with the picture of dynamic stripes as the AFM domain boundary. Actually it is more reasonable to assume that there are *two* kinds of charge inhomogeneities in the cuprates near 1/8 or optimum doping. One is the well-known static stripes with the periodicity of $4a$ (for the optimum doping) that compete against HTSC. The other is dynamic, characterized with the short-range periodicity of $2a$, and facilitates HTSC. The latter state may be related to the Peierles instability of the half-filled band. An exactly half-filled band with the strong on-site coulomb repulsion results in the Mott insulator, but if the Mott state is disturbed by doping and the long-range AFM state is lost, the electron-lattice coupling may induce the Peierles (CDW) instability with the periodicity of $2a$. So let us call this charge inhomogeneity as the CDW instability state, to differentiate from the $4a$ stripes. The stripe state is dominated by the magnetic correlation, while the CDW state is dominated by the coulomb repulsion among electrons and electron-phonon coupling.

In the stripe state it may not be easy to tunnel from one charge stripe to the other, since charges are separated by $4a$. In fact the results of the Hall effect measurement [24] and the photoemission study [25] of $(La,Nd)_{1.875}Sr_{0.125}CuO_4$ suggest that the conductivity across the stripes is very small. On the other hand if the periodicity is $2a$ as in the CDW state it should be very easy for charge to hop across, maintaining high conductivity. The CDW instability may lead to HTSC via the electron-phonon coupling enhanced by the coulomb repulsion and spin fluctuation [26].

POLARON STABILITY IN MANGANITES

We now turn to manganites that show the CMR behavior, to illustrate the similarities and differences between the manganites and the cuprates. The most crucial point with the CMR phenomenon is the marginal instability of polarons. If polarons are too stable the system is insulating, since polarons are easily Anderson-localized by random electrostatic potential of the dopants, while if polarons are not stable the system is just an ordinary metal. When polarons are marginally stable they can easily be destroyed by small external perturbations such as applied field [27,28]. This is most clearly demonstrated in the ionic-size effect [29]. When the ionic-size of the A-site ion is small the Mn-O-Mn bond is bent, and the system is insulating even at a high doping level. When the A-site ion is large the system is a regular metal, and at the crossover region one observes strong CMR effect.

This effect has been interpreted in terms of the band narrowing due to the Mn-O-Mn bond-bending, but this effect is too small to account for the observed effect, amounting only to a few percent [30]. We proposed, instead, the change in the local elastic compliance explains the observed size effect. In other words the phonon modes involved in polaron formation changes as the A-site ionic size is changed [31]. The undoped manganites, such as $LaMnO_3$, are Jahn-Teller (JT) distorted [32]. In these compounds a Mn^{3+} ion has six oxygen neighbors, two with a long Mn-O bond (~2.2 Å) and four with a short bond (~ 1.95 Å). In order to create a polaron the JT distortion has to be locally suppressed by making the long Mn-O bonds shorter. If a MnO_6 octahedral cluster is isolated this can be done easily. However, MnO_6 cluster is just a part of the solid, and this local deformation has to be accommodated by the surrounding. This is where the structure come to play a role. If the Mn-O-Mn bond is linear, in order to reduce the length of one of the two Mn-O bonds the other has to be stretched. This action, longitudinal accommodation, is energetically expensive, since the longitudinal elastic constant is high. On the other hand if the Mn-O-Mn bond is sufficiently bent, the shortening of one of the Mn-O bonds can be accommodated by unbending the bond. This transverse accommodation is energetically much more favorable. Thus one could assume that when the longitudinal accommodation is required the elastic energy cost is too high and polarons are unstable, while when the transverse accommodation can be invoked polarons are stable.

This mechanism was strongly supported by a recent PDF study [33]. Also the phase diagram calculated with this mechanism agrees well with the observation. It is important to note that in this mechanism the crossover ionic size for polaron stability is easily determined without any fitting parameter. In the band-narrowing scenario, on the other hand the critical value of the ionic size depends upon many variables, and there is no physical reason why a particular ionic size is important.

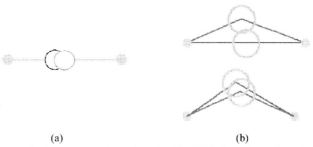

(a)   (b)

FIGURE 4. Longitudinal accommodation when the Mn-O-Mn bond is straight (a), and transverse accommodation when the Mn-O-Mn bond is bent [31].

Recently much studies have been made on the effect of charge ordering, and it is sometimes argued that charge ordering produces the CMR effect. It is important to recognize, however, that the *polaronic phonon dressing is a critical prerequisite for charge ordering*. Without such heavy dressing Wigner crystal would form only at a very low charge concentration and a very low temperature. Charge ordering should be considered as polaron lattice formation or polaron crystallization.

POLARON STABILITY AND SUPERCONDUCTIVITY

It is tempting to extend the same argument to the case of cuprates. There are strong evidences of polaron formation in cuprates at low charge densities [34,35]. Also the charge/spin stripe phase can be considered as the formation of a one-dimensional polaron phase. Since the stripe phase competes against HTSC [36], polaron formation is damaging to HTSC, just as the charge density wave. However, if polarons are marginally stable and fluctuating, they might help HTSC [26,37]. Thus it is useful to consider the characteristics of the polarons in cuprates.

Just as the manganites the ionic size appears to be one of the critical factors here. By substituting smaller rare-earths such as Nd [38,39] for larger La HTSC becomes unstable and at the right charge density ($x = 0.125$) the stripe phase is observed. Indeed the tilt angle of the CuO_6 octahedron appears to be the critical parameter [39]. If the tilt angle exceeds a certain value HTSC is not observed. Thus the polaron stability in cuprates could be argued in just the same way as in the manganites, in terms of the bending of the Cu-O-Cu bond. If the Cu-O-Cu bond is sufficiently bent polarons become stable, killing HTSC. On the other hand if the Cu-O-Cu bond is straight the regular metallic behavior is observed, as in the strongly overdoped phase. According to the neutron PDF studies the HTSC phase is composed of microscopic mixture of tilted and untilted phases, the former representing the insulating phase and the latter the metallic phase [40,41]. This observation is completely consistent with the phonon results above, and strongly suggest that the spatial charge distribution in cuprates is microscopically inhomogeneous, possibly due to unstable polarons, and this inhomogeneity is crucially important to HTSC.

This result, however, does not necessarily imply that HTSC is caused by electron-phonon interaction alone, while it must be involved somehow in the mechanism. It definitely does not implicate the bipolaron mechanism, since formation of polaron or bipolaron in the form of stripes competes against HTSC. Indeed our recent numerical calculation of the one-dimensional Holstein-Hubbard Hamiltonian with dynamic phonons using the density matrix renormalization group method suggests that the main force of pair formation comes from the on-site coulomb repulsion, and not from the electron-phonon coupling [26].

In this study we started with the Mott-Hubbard insulator, and increased the strength of the electron-phonon coupling. The Hamiltonian is given by

$$H = H_e + H_p + H_{e-p} \qquad (1)$$

$$H_e = \sum_{ij} t_{ij} c_i^+ c_j + U \sum_i n_{i\uparrow} n_{i\downarrow} \qquad (2)$$

$$H_p = \hbar\omega \sum_i a_i^+ a_i, \quad H_{e-p} = g \sum_i \left(a_i^+ + a_i\right)\left(c_i^+ c_i - 1\right) \;, \qquad (3)$$

where c_i^+, c_i and a_i^+, a_i are electron and phonon creation and annihilation operators, and to save computer time the phonons were assumed to couple to a hole, rather than electron. This, however, will not introduce qualitative difference. In this system as the strength of the electron-phonon coupling, g, is increased until the ground state changes from the Mott-Hubbard insulator to the charge-density wave (CDW) insulator. At the crossover, however, it was found that the system becomes metallic, indicated by the marked decrease in the electron kinetic energy, due to the disorder created by phonons scrambling the Mott-Hubbard gap [42]. Near the metal-insulator transition (MIT), very interesting pairing behavior was discovered. Before we get into the details some comments are in order, to make a connection between this model and the real cuprates. Firstly in this model for $g = 0$ the gap in the density of states is the Mott-Hubbard gap, $U - W$, where W is the bandwidth ($= 2t$), but in the real cuprates it is the charge transfer gap, $U - \Delta$, where Δ is the difference between the Cu-d level and the O-p level. Since the magnitude of the charge transfer gap is about 2 eV [43] the value of U has to be set smaller than it really is. Secondly as the cuprates are doped the doped hole occupies a d-state strongly hybridized with the O-p level. The effective U becomes smaller, and thus the gap narrows due to in-gap states, eventually driving the system to a metallic state. Therefore in our model system the increase in g is supposed to represent the increased doping. The MIT brought about by increasing g should be viewed as representing the MIT due to increased doping, revealed by applying high magnetic fields that suppress HTSC [44].

When the system was doped with holes they were found to form pairs, possibly suggesting superconductivity, although in small *1-d* systems proving the existence of superconductivity is difficult. As it turned out that the pairing energy does not come from the electron-phonon coupling as anticipated for bipolaron formation, but originates from the reduction in the coulomb repulsion energy. As the Mott-Hubbard state is scrambled by phonons, many sites are doubly occupied, increasing the expectation value of the Hubbard U term. Having hole pairs reduces the double occupation, thus decreases the Hubbard on-site repulsion energy. In this mechanism phonons are the **drivers**, but the **engine** is the Hubbard U. Phonons produce charge inhomogeneity by setting the phases of the electrons, but the real binding energy is provided by the on-site repulsion. This is just as the driver of a car steers the wheels and pushes the pedal, but the real force of propulsion is provided by the engine.

CONCLUSIONS

While the majority opinion in the field of HTSC is that phonons are either irrelevant or even harmful to HTSC, many experimental observations have been documented indicating the importance of lattice contributions [6]. The recent research on the manganites that show the CMR effect provided parts of the missing links that connect the lattice effects and HTSC more directly. In particular the inhomogeneous charge distribution in the metallic state and marginal stability of polarons are the key issues [37]. In CMR manganites the insulating state is produced by the formation of anti-Jahn-Teller polarons that become Anderson localized by the random electrostatic potential. When the polarons are marginally stable so that they can easily be destroyed by small applied fields the CMR effect is observed. The stability of polarons is determined by a delicate balance between the competing forces to localize and delocalize charges. Our recent measurements of high-energy LO phonons by inelastic neutron scattering demonstrate that charge localization tendency exists also in the cuprates that exhibit high-temperature superconductivity (HTSC). While the real polaron formation, as in the stripe phase, is detrimental to HTSC just as to the CMR effect, the tendency for polaron formation and *marginal* stability of polaron could be important for the HTSC in cuprates as well as for the CMR effect in manganites. In particular in the system of strongly interacting electrons with dynamic phonons the on-site coulomb repulsion, not the electron-phonon interaction, can provide the binding energy for holes. These results indicate that the lattice contribution to the HTSC, long overlooked by the majority, may be the most important factor in the mechanism of HTSC.

ACKNOWLEDGMENTS

The author is grateful to the collaborators of the phonon and CMR projects, R. J. McQueeney, Y. Petrov, D. Louca, J. F. Mitchell, M. Yethiraj, H. A. Mook, G. Shirane and Y. Endoh. He is also thankful to M. Arai, S. J. L. Billinge, A. R. Bishop, A. Bussmann-Holder, V. J. Emery, J. B. Goodenough, L. P. Gor'kov, S. Kohmoto, V. Kresin, K. A. Müller, J. C. Phillips, M. Tachiki and S. Uchida for useful discussions. Research at the University of Pennsylvania was supported by the National Science Foundation through DMR96-28136.

REFERENCES

1. Anderson, P. W., *Theory of Superconductivity in the High-T_C Cuprates*, Princeton University Press, Princeton, 1997.
2. Kresin, V., Morawitz, H., and Wolf, S. A., *Mechanism of Conventional and High T_C Superconductivity*, Oxford Press, Oxford, 1993.
3. Micnas, R., Ranninger, J., and Robaszkiewicz, S., *Rev. Mod. Phys.* **62**, 113 (1990).
4. Bar-Yam, Y., *Phys. Rev. B* **43**, 359 (1991), *ibid.* **43**, 2601 (1991).
5. Pao, C. H., and Schüttler, H. B., *J. Superconductivity* **8**, 633 (1995).
6. Egami, T., and Billinge, S. J. L., in *Physical Properties of High Temperature Superconductors V*, edited by D. Ginsberg:, World Scientific, Singapore, 1996, pp. 265.
7. Tranquada, J. M., *et al.*, *Nature* **375**, 561 (1995).
8. Egami, T., Ishihara, S., and Tachiki, M., *Science* **55**, 3163 (1993).

9. Egami, T., Dmowski, W., McQueeney, R. J., Arai, M., Seiji, N., and Yamauchi, H., *J. Superconductivity*, **8**, 587 (1995).

10. Ishihara, S., Egami, T., and Tachiki, M., *Phys. Rev. B* **55**, 3163 (1997).

11. Petrov, Y., and Egami, T., *Phys. Rev. B* **58**, 9485 (1998).

12. Pintschovius, L., *et al., Physica C* **185-189**, 156 (1991).

13. Pintschovius, L., and Reichardt, W., in *Physical Properties of High Temperature SuperconductorsIV*, edited by D. Ginsberg, World Scientific, Singapore, 1994, pp. 295.

14. McQueeney, R. J., Petrov, Y., Egami, T., Yethiraj, M., Shirane, G., and Endoh, Y., *Phys. Rev. Lett.* **82**, 628 (1999).

15. Egami, T., McQueeney, R. J., Petrov, Y., Yethiraj, M., Shirane, G., and Endoh, Y., *AIP Conf. Proc.* **483**, 231 (1999).

16. Petrov, Y., Egami, T., McQueeney, R. J., Yethiraj, M., Mook, H. A., and Dogan, F., *Cond-mat/0003414*.

17. Egami, T., Chung, J. H., McQueeney, R. J., Yethiraj, M., Mook, H. A., and Dogan, F., unpublished.

18. Lorum, J. W., *et al., J. Superconductivity* **7**, 243 (1994).

19. Tranquada, J. M., *et al., Phys. Rev. B* **54**, 7489 (1997).

20. Kivelson, S. A., Fradkin, E., and Emery, V. J., *Nature (London)* **393**, 550 (1998).

21. Yamada, K., *et al., Phys. Rev. Lett.* **75**, 1626 (1995).

22. Mook, H. A., and Dogan, F., *Nature (London)* **401**, 145 (1999).

23. Zaanen, J., and Gunnarson, O., *Phys. Rev. B* **40**, 7391 (1989).

24. Noda, T., Eisaki, H., Uchida, S. *Science* **286**, 265 (1999).

25. Zhou, X. J., *et al., Science* **286**, 269 (1999).

26. Mills, A. J., Littlewood, P. B., and Shraiman, B. I., *Phys. Rev. Lett.* **74**, 5144 (1995).

27. Röder, H., Zang, J., and Bishop, A. R., *Phys. Rev. Lett.* **76**, 1356 (1996).

28. Petrov, Y., and Egami, T., *cond-mat / 9912449*.

29. Hwang, H. Y., Cheong, S. W. Radaelli, P. G., Marezio, M., Batlogg, B., *Phys. Rev. Lett.* **75**, 914 (1995).

30. Radaelli, P. G., Iannone, G., Marezio, M., Hwang, H. Y., Cheong, S. W., Jorgensen, J. D., and Argyriou, D. N., *Phys. Rev. B* **56**, 8265 (1997).

31. Egami, T., and Louca, D., *J. Superconductivity* **12**, 23 (1999).

32. Goodenough, J. B., *Phys. Rev.* **100,** 564 (1955).

33. Louca, D., Egami, T., and Mitchell, J. F., unpublished.

34. Reagor, D. W., *et al., Phys. Rev. B* **38**, 5106 (1988).

35. Chen, C. Y., *et al., Phys. Rev. B* **51**, 3671 (1995).

36. Tranquada, J. M., *et al., Phys. Rev. B* **54**, 7489 (1997).

37. Egami, T., *J. Low Temp. Phys.* **105**, 791 (1996).

38. Crawford, M. K., *et al., Phys. Rev. B* **47**, 11623 (1993).

39. Büchner, B., *et al., Phys. Rev. Lett.* **73**, 1841 (1994).

40. Billinge, S. J. L., and Egami, T., *Phys. Rev. B* **47**, 14386 (1993).

41. Bozin, E. S., Kwei, G. H., Takagi, H., and Billinge, S. J. L., *Cond-mat / 9907017*.

42. Denteneer, P. J. H., Scalettar, R. T., and Trivedi, N., *Phys. Rev. Lett.* **83**, 4610 (1999).

43. Uchida, S., *et al., Phys. Rev. B* **43**, 7942 (1991).

44. Boebinger, G. S., *et al., Phys. Rev. Lett.* **77,** 5417 (1996).

Local Lattice Distortions Probed by X-Ray Absorption Spectroscopy

H. Oyanagi

Electrotechnical Laboratory, 1-1-4 Umezono, Tsukuba, Ibaraki 305-8568, Japan

Abstract. In the last decade, we have witnessed a number of cases where a small deviation from crystallographic or average structure strongly modifies electron states, physical and chemical properties and functions. The crucial role of breaking local symmetry, stabilizing a certain electronic (spin-state) configuration, has been recognized recently, by x-ray absorption spectroscopy (XAS). In biological systems such as haeme proteins, having $3d$ transition metal ions at active sites in a large molecule, XAS studies clarified the intimate relation between the local structure and the spin state. A number of high T_c superconducting oxides have been investigated and a generic behavior of local lattice as an intrinsic tendency of highly correlated electrons is revealed. We find anomalous local lattice distortions in $YBa_2Cu_3O_y$ *emerge at characteristic temperatures T^* and T_0 associated with changes of electron states, indicating a strong electron-lattice coupling.* We demonstrate that *in-situ* XAS can probe the local lattice distortions during photo-excitation. XAS showed that photo-induced nonthermal melting occurs in chalcogenide glasses and the frozen-in local lattice distortions cause Coulomb repulsion observed as photo-structural change.

1. INTRODUCTION

The role of local lattice distortion (LLD) in modern condensed matter physics arising from the interplay between highly correlated electrons and lattice has been recently recognized. Local lattice distortions competing with charge dynamics play an essential role in exotic physical properties of these materials, such as high T_c superconductivity (HTSC) in cuprates or colossal magnetoresistance (CMR) in manganites. Such a revival of interests is a natural consequence of the fact that the lattice distortions in those materials are a signature of frustrating interactions among charge, spin and orbitals. The competition between short-range ferromagnetic and long-range antiferromagnetic Coulomb interactions results in an electronic phase separation [1-3]. Such a phase separation gives rise to spin and charge ordering observed as a one-dimensioal modulations (stripes) [4,5]. Microscopically, stripe ordering is usually associated with the anomalous LLD below a specific crossover temperature T^*, indicating the importance of electron-lattice interaction [6]. X-ray absorption fine structure (XAS) is one of the powerful techniques for probing the local lattice with a fast time scale (10^{-15} sec), high sensitivity (10^{19} atoms/cm^3) and a high spatial resolution (0.1 Å). Since synchrotron radiation (SR) facilities as a stable and high-intensity x-ray source became available in early 1980's, XAS has been successfully utilized to investigate the local lattices for numerous materials [7]. In

CP554, *Physics in Local Lattice Distortions*, edited by H. Oyanagi and A. Bianconi
© 2001 American Institute of Physics 1-56396-984-X/01/$18.00

Sec. 3, the advantage of XAS over other techniques and recent progress are briefly reviewed. The effect of LLD on the electron states is familiar in chemistry as Jahn-Teller distortions often observed in inorganic complexes stabilizing the crystal [8]. In solids where electrons are localized due to the electron correlation, LLD is often observed as a signature of frustrated magnetic and Coulomb interactions. In proteins having transition metal ions at active centers, the distortion of coordination is strongly related with the electron state of central ions and their biological functions. In the next section, we describe the fundamental aspects of LLD and in Sec. 3, principle of XAS is briefly described. The intimate relation between the spin state and coordination of central atom in metal protein (myoglobin) [9] is discussed in Sec. 4. The local lattice anomalies in $YBa_2Cu_3O_y$ [10] is discussed in Sec. 5. In Sec. 6, we report a photo-induced nonthermal melting and photo-induced phenomena such as photo-darkening by illumination in chalcogenide glasses [11].

2. LOCAL LATTICE DISTORTIONS

The octahedral coordination of oxygen O around a central metal ion M is illustrated in Fig. 1a. Let us consider a single d electron in an octahedral field where five-fold degenerate $3d$ orbitals split into the e_g orbitals and t_{2g} orbitals as shown in Fig. 1b. Because of the electrostatic interactions between these orbitals and negative ligands (oxygen atoms), the t_{2g} orbitals are more stable as illustrated in Fig. 2b. In most of solids, however, the crystal structure is slightly different from the ideal octahedral

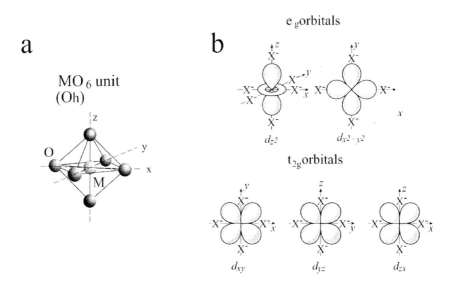

FIGURE 1. (a) Local structure of MO_6 unit in an octahedral coordination where M denotes $3d$ metal ions, (b) electron density of five $3d$ orbitals. Because of crystal field, the five-fold degeneracy is removed and $3d$ orbitals split into two eg orbitals and three $t2g$ orbitals

(Oh) symmetry. Thus the observed M-O distances are not equal. Instead, the local distortion such as the shortening of M-O distance (apical oxygens) and elongation of in-plane M-O distances occurs. This distortion (Q_3 mode) shown in Fig. 1a is known as a typical Jahn-Teller (JT) distortion [12] which effectively splits the orbital energies of both e_g and t_{2g} orbitals as illustrated in Fig. 2b. The split in energy increases with the degree of crystal field (CF) which is a function of lattice distortion. In an extreme case MO_4 (square planar geometry) without apical oxygen atoms, the lowered d_{z^2} orbital energy becomes lower than d_{xy} orbital. According to a point charge model of distorted MO_6 cluster, the degeneracy in $^3T_{1g}$ symmetry is removed by the self-distortion and the total energy of d-electrons is lowered when they occupy the lower orbitals, stabilizing the system [13]. In addition to the static JT distortion, a dynamic JT in which the two coordinations are resonating each other may occur. As XAS is a fast time scale probe, both static and dynamic JT distortions are observed. Polarized XAS can probe the ab-plane and c-axis M-O distance separately, one can evaluate the magnitude and direction of JT distortions.

The split energy levels due to CF and JT are important as a dominant factor of d-electron configuration. For Fe^{3+} ion with five $3d$ electrons in a distorted octahedral

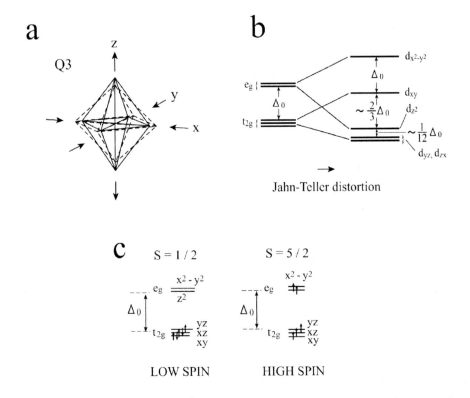

FIGURE 2. (a) Displacement of atom positions due to Jahn-Teller distortion, (b) energy diagram of 3d orbitals as a function of local lattice distortion, (c) spin states of iron (+2) ions in haemprotein.

symmetry, if the magnitude of distortion (not necessarily JT) is large enough, S = 5/2 state (high spin state) is formed while S = 1/2 (low spin state) is the case if the distortion is small. The electron configuration is therefore intimately related to the local lattice, as illustrated in Fig. 2c. When the local structure is sterically limited or deformed by the strain, the distortion sometimes controls the electron configuration. The consequence of nine $3d$ electron configuration (Cu^{2+}) with the presence of distorted octahedron will be given elsewhere in this book by Kamimura and Hamada [14]. Here, we note that the effect of LLD is essentially important in the ground-state and excited-state electron states. In order to properly evaluate the magnitude and types of LLD, one should study the radial distribution function for M-O pair in detail. How we probe the local structure is described in the following section.

3. X-RAY ABSORPTION SPECTROSCOPY

Fundamental characteristics of XAS are briefly described in this section. In an x-ray absorption spectrum above a core excitation threshold, extended x-ray absorption fine structure (EXAFS) and x-ray absorption near edge structure (XANES) are observed. Fig. 3 shows a schematic of principle of XAS including a new technique to

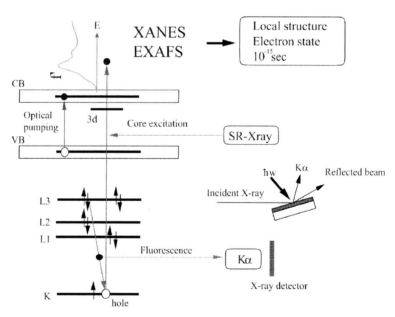

FIGURE 3. Principle of x-ray absorption spectroscopy (XAS). X-ray absorption near-edge structure (XANES) and extended x-ray absorption fine structure (EXAFS) occur due to the interference of photoelectrons scattered by adjacent atoms around an excited atom. Polarization dependence can be measured by orienting a single crystal to the electrical field vector E monitoring fluorescence x-ray.

observe the local structure of excited states. In general, EXAFS refers to the sinusoidal oscillations observed over a wide energy range, extending ~ 1000 eV above the edge while XANES refers to sharp fine features around the edge extending typically ~ 100 eV. In contrast to diffraction techniques which gives atomic coordinates as a *macroscopic average* of equilibrium positions, XAS provides the information on a *local structure*, *i.e.*, radial distribution and unoccupied electron states around a particular species of atom. The uniqueness of XAS is attributed to its physical origin, the interference of out-going and scattered photoelectron waves at the excited central atom [15]. The interference modulates the matrix element of a dipole transition which is experimentally given as the fine structures in absorption coefficient. Since a photoelectron interference effect does not require a long range order, a "partial" radial distribution within several Å is therefore measured because of a rather short mean free path of ejected photoelectrons. The information is thus intrinsically atom-specific since the interference is observed for a specific pair correlation at the origin (central atom) as a modulation of absorption coefficient. In Fig. 4, polarized Cu K-XANES spectra for La_2CuO_4 and Pr_2CuO_4 single crystals are shown [16,17]. Comparing the two geometries, *i.e.*, the electrical field E parallel with *ab*-plane (E//*ab*) and E parallel with the *c*-axis (E//*c*), one can notice that the two orientations give a large anisotropy in the near-edge features and secondly the E//*c* orientation data is highly sensitive to the apical oxygen coordination. The latter is due to the fact that the features A_1 and A_2 are the transitions from the deep core level (K-shell) to the empty $4p_z$ states affected by multiple scattering and shape resonance among nearby atoms [18,19]. Thus a spectral evolution around features A_1 and A_2 for E//*c* on going from T to T' structure can be used as a measure of LLD, *i.e.*, to what extent the apical oxygen atoms are displaced away the central atom along the *c*-axis as illustrated in Fig. 2a. As the E//*ab* features (B_1 and B_2) are reflecting the square planar lattice and therefore insensitive to the apical ligand, a relative ratio $A_1(A_2)/B_1(B_2)$ is used as a practical measure of the out-of-plane distortion [20].

Another important feature of XAS is the fact that the time scale is the order of 10^{-15} sec, being much faster than phonon frequencies. This makes the technique a powerful "dynamical" tool which provides a "snapshot" of the short-range lattice distortions both in static and dynamic aspects. Since a typical lattice relaxation occurs with a time scale of 10^{-12} sec, we can obtain the direct information on excited states also. EXAFS is an interference phenomena of photoelectrons between the out-going waves and the scattered waves. Since the dominant interference term involves the path lengths between the absorbing (central) atom and a scatterer atom $2r$, a $2kr$ oscillations are observed in k-space, in contrast to kr oscillations in a well-known formula of scattering. Within a single scattering formalism, the interference among the three components, *i.e.*, the out-going wave, the single scattering path and the double scattering path of the lowest order is dominant. Real space radial distribution functions are obtained by a Fourier transform (FT) over a measured k-range typically upto 18 Å^{-1}. Using theoretically calculated phase-shift functions [21], we can correct the phase-shift of scattered photoelectrons and obtain a reliable information on the local structure.

Here we describe the experimental details for measuring the local structure of photo-excited and/or trapped (metastable) states. Time-resolved coincidence experiments using two excitation sources in optical spectroscopy is called "pump and probe" technique. This technique is widely used to investigate the dynamics or intermediate state of photoexcitation. Recently, synchrotron radiation x-ray diffraction is used with laser excitation in a time-resolved mode where a probe x-ray is used to study the optically pumped state [22]. In a similar manner, pump & probe XAS experiments in principle can provide the local structure of optically excited atoms. However, since a single-shot experiment can not provide a enough statistics (10^6 photons), the experiment must be repetitive. Therefore a time resolution in XAS is strictly limited in the application to condensed matter physics. Previous optical pump and x-ray probe experiments have been performed for a relatively slow process, such as the photo-dissociation of ligands in haemeproteins [23,24]. Structural study on CO-binding in myoglobin has been reported [25]. Here we discuss the application

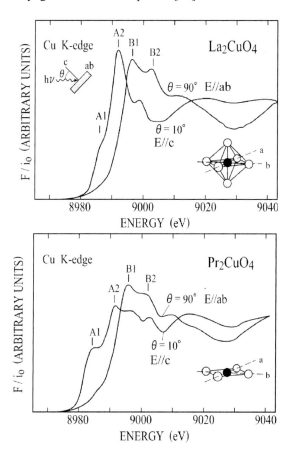

FIGURE 4. Polarized Cu K-XANES spectra for La_2CuO_4 (upper column) and Pr_2CuO_4 (lower column). The out-of-plane (E//c) orientation, the apical oxygen atoms strongly affect the near-edge features (A_1, A_2) due to the transition from K-shell to $4p_z$ unoccupied orbitals, making these features a sensitive probe of displacement of apical oxygens.

of this technique to solids, attempting "snapshot" measurements of lattice distortion, lattice relaxation and possible phase transition by optical excitation of valence electrons. Fig. 3 illustrates the schematic principle of the experiment [26]. For a steady-state experiment, a mixture of ground-state and excited-state is observed. If the fraction of excited states is given, the local structure of excited state (or metastable states) can be reproduced from a difference spectra between light-on and light-off states.

The difficult problem in pump & probe XAS applied to condensed matters is the mismatch in the extinction lengths between x-ray and visible photons. Penetrating power for solids in a hard x-ray region is in the order of a few microns while that for visible photons is smaller by two orders of magnitude. Thus x-ray thick samples are difficult to optically excite considering poor transmission. Recently, it was shown that this problem can be reconciled by introducing a surface-sensitive fluorescence excitation [27]. Using a grazing-incidence geometry for controlling x-ray penetration depth, use of a high-brilliance photon source and a dense-packed solid-state detector array [28], one can study the local structure of optically excited states [26,29].

4. LOCAL STRUCTURE AND SPIN STATES IN HAEM PROTEINS

4.1 Thermal Spin Equilibrium

Metal proteins contain a small amount of metal ions at the biologically active sites and the electron (spin) states and local structure are strongly related each other and control biological activity. For instance, some ferric haemoproteins such as hydroxide complexes of ferric haemoglobin, and myoglobin (Mb) show values of spin susceptibilities intermediate between those characteristic of 5- and 1-unpaired electrons. Oxygen binding affinity is controlled by the spin state. The spin-states of these complexes have been interpreted in terms of thermal spin equilibrium between two magnetic isomers, one in a high-spin and the other in a low-spin state (S = 5/2 and 1/2 states). These results have been obtained from the analysis of the temperature dependence of the magnetic susceptibilities and light-absorption spectra [30,31]. Spin states are strongly related to the coordination geometry and species of ligand. While fluoride and cyanide complexes were found respectively to be in purely-high, and purely-low spin states, hydroxide, azide, imidazole and cyanate complexes exhibited intermediate magnetic susceptibilities and the optical spectra characteristic of an intermediate spin state.

Although the thermal spin-equilibrium has been studied extensively for a variety of haemoproteins and their model haeme-compounds, from a thermodynamic view-point [30–32], the relationship between the local structure of haeme-iron and the spin states has not yet been established. As illustrated in Fig. 5, haeme-iron is expected to be out of the haeme plane in high spin states, which is favored from the view-point of metal-d – ligand-p interaction. The magnitude of the displacement of haeme-iron from the haeme plane in deoxyhaemoglobin (Hb(II)) has been of considerable interest in

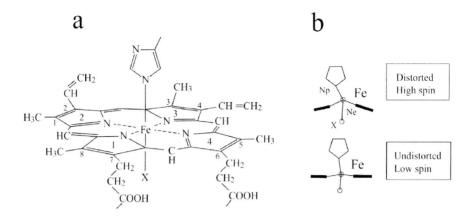

FIGURE 5. (a) Local structure of haem-iron in Fe-protopophyrin IX. Haem-iron is coordinated by four nitrogen atoms (N_p), another nitrogen atom (N_e) of proximal histidine and the 6th ligand (X). Oxygen atoms are bound to the 6th ligand. (b) displacement of haem-plane as the spin state of iron changes from the low-spin to high-spin state.

relation to the affinity of oxygen binding [33–35]. The movement of haeme-iron into the haeme plane in an oxygenated form has been proposed as the mechanism for transfer of the information on oxygen-binding from one subunit to the other, causing the transition from the low-affinity T state to the high-affinity R state [33].

4.2 XANES Spectra

The local structure of haeme-iron has been studied for a variety of haemoproteins by means of EXAFS [34-37], which determined the distance between haeme-iron (Fe) and the nitrogen atoms of porphyrin (N_p) and that between haeme-iron and the nitrogen atoms of proximal histidine (N_e), together with the distance between haeme-iron and the sixth ligand, such as oxygen, within 0.01 Å accuracy [35,36]. However, the magnitude of displacement of haeme-iron from the haeme plane is difficult to determine by EXAFS because it is insensitive to the three-dimensional arrangement of the atoms, *i.e.* the bond angle [35,38]. On the other hand, XANES is suitable for probing the coordination geometry in haemoproteins because of its sensitivity to the three-dimensional atomic arrangement [39]. Fig. 6 shows the temperature dependent Fe K-XANES for dilute (2 mM) Mb(III)OH$^-$ [9]. At 80 K (solid line), where Mb(III)OH$^-$ is purely in the low-spin state, several characteristic features (P, A-C and C_2) are observed [38]. On going from the predominantly high-spin state indicated by dashed line (300 K) to purely low-spin state (80 K), the spectrum changes systematically. Feature A indicates the position of a shoulder structure, observed for low-spin Mb(III)OH$^-$ at 7125 eV, which appears as an inflection point in the first derivative. This shoulder gradually disappears with increase of the high-spin content; at 300 K no shoulder structure is observed. A broad bump structure C_2 observed at

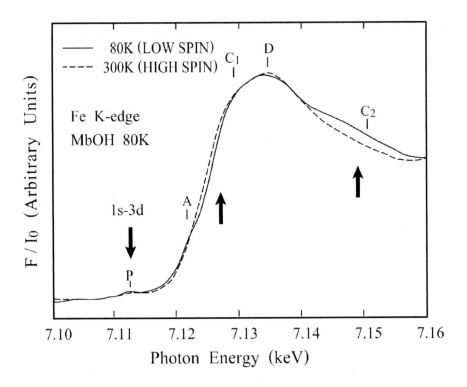

FIGURE 6. Fe K-XANES spectra for 2 mM Mb(III)OH- taken at 300 K and 80 K. Near-edge features A-C_1 and C_2 change as the spin state varies the high spin (300 K) to low spin (80 K) state. A weak 1s-3d dipole-forbidden transition peak P also changes reflecting the change of empty 3d orbitals corresponding to the spin states.

7148 eV for low-spin MbOH⁻ reduces in its intensity at higher temperature and is not observed at 300 K. Peak P, observed at 7112 eV, is a quadrupole-allowed transition from Fe 1s to empty 3d states. This dipole-forbidden transition is observed as a weak broad feature at 300 K but it becomes a sharp feature at 80 K. It is found that there are three features, P, A and C_2 within 40 eV from the edge, which are spin-state sensitive. In particular, feature A is strongly correlated with a decrease in broad peak intensity at C_2 in the high-spin state.

   A difference between a sample measured at T (> 80 K) and a reference (T = 80 K) can detect and analyze *quantitatively* the local structure and spin states. The inset in Fig. 7 indicates the difference spectra around feature A as a function of temperature. The intensity of this peak decreases with the decrease in temperature consistent with the high-spin fraction calculated from thermodynamic data [30,31]. The overall behavior is well predicted by thermodynamic data suggesting that the *difference is proportional to the high-spin content*. The magnitudes of C_2 and A in a difference spectrum were found to be inversely correlated with one another. The change of feature "A" is due to the increased multiple-scattering of photoelectrons within the haeme plane [38] which is sensitive to the motion of haeme-iron out of the haeme plane. The near-edge features 40 - 50 eV above the edge are sensitive to the *haeme-iron displacement from the haeme plane.*

On the other hand, recent polarized XANES studies [40] on square-planar Cu complexes have shown that sharp features ascribed to a $1s$-$4p_z$ transition appear 4 - 5 eV above the $1s$-continuum transition. The former transition is polarized normal to the plane while the latter is polarized along the plane. The interaction between Fe $4p_z$ and ligand $4p^*(\pi)$ states are sensitive to the distance between metal and axial ligand [40]. The features C_1 and C_2 observed for single-crystalline Mb(II)CO are strongly polarized in the direction normal to the haeme plane [41]. The interaction between the $4p_z$ and ligand π^* orbital reflects the symmetry around a metal atom. If the metal-ligand distance is not changed appreciably between the high- and low-spin states, the observed change in spin-state sensitive structures arises from the *change in the metal-ligand interaction which is caused by the displacement of haeme-iron out of the haeme plane.*

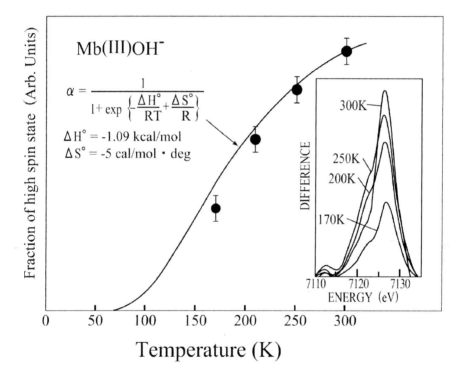

FIGURE 7. Temperature dependence of the magnitude of difference peak at 7122 eV, corresponding the features A-C$_1$. High spin content α calculated using thermodynamical data given in this figure is normalized at 300 K.

4.3 Spin State and Local Structure

The haeme-iron of Mb(III)OH⁻ is coordinated with four nitrogen atoms of pyrrole rings and with another nitrogen atom of proximal histidine, as indicated in Fig. 5. The sixth ligand of Mb(III)OH⁻, denoted by X, is a hydroxyl ion. In Mb(II), haeme-iron has no sixth ligand and is displaced from the haeme plane by ~ 0.40 Å and from the nitrogen plane by ~ 0.27 Å, according to the structural analysis [42]. Because of this doming of the pyrrole ring, the iron-nitrogen distance is slightly shorter than the iron-haeme plane distance. Such a displacement of haeme-iron from the mean haeme plane has been reported for various high-spin porphyrin compounds [43]. On the other hand, haeme-iron in low-spin myoglobin derivatives such as Mb(III)CN⁻, Mb(II)CO, and Mb(II)O_2 is expected to be within the haeme plane or only slightly displaced from it. The displacement of haeme-iron out of the haeme plane is expected to weaken the octahedral ligand field (O_h) by lowering the symmetry. High-spin states are stabilized by this distortion of the square-planar geometry because the weaker CF reduces the e_g-t_{2g} splitting energy, which is favorable for high-spin states. In haemoprotein, the spin states of haeme-iron are strongly affected by the chemical character of the sixth ligand. Low-spin states are stable for strong ligands such as CN⁻ while high-spin states are commonly found in haemoproteins with a weak ligand, such as H_2O. The haeme plane-normal component of CF due to the axial distortion also allows p-d mixing and lowers the d_{z^2} orbital energy, which contributes to the stabilization of high-spin states.

The change in distance between haeme-iron (Fe) and the center of the haeme plane (C_t) on going from a low-spin to high-spin state is not associated with an appreciable change in Fe-N_p proximal, Fe-N_e or Fe-O(OH_2) bond lengths, since the difference in the nearest-neighbor distance would appear as an energy shift. These results are consistent with recent EXAFS studies [34,35] on oxy- and deoxy-haemoglobin which found that the Fe-N_p distance is 2.05 Å for both forms. If the movement of haeme-iron is not associated with the change in Fe-N_p distance, the four nitrogen atoms (N_p) are pulled toward the center of a square. The strain energy caused by the Fe-N_p bond-bending can be relaxed if the porphyrin ring is further deformed. In this respect, the XANES results are consistent with the recent interpretation of EXAFS data for oxy- and deoxy-haemoglobin by Perutz et al. [34].

The Fe-C_t and Fe-N_p distances in carp azide haemoglobin, associated with the T-R transition are less than 0.1 Å and 0.01 Å, respectively [44]. The specific spectral regions change during the T-R transition indicates that the T-R transition involves the movement of haeme-iron out of the haeme plane. Variations of the Fe-N_p distance in carp haemoglobin associated with the T-R transition are less than the detectable limit (0.01 Å) from EXAFS experiments [45]. It has been demonstrated that there are particular spectral regions in the near-edge spectra for Mb(III)OH⁻ within 40 eV from the threshold are spin-state sensitive. The spin-state sensitive near-edge features are primarily due to the change of the haeme-iron and ligand-orbital interaction as a result of the axial movement of haeme-iron and following deformation of the porphyrin ring. These features are generally observed for other myoglobin derivatives. It is expected that the variation in the Fe-N_e distance for Mb(III)OH⁻ should not be large on going from a low-spin to high-spin state, suggesting that the porphyrin ring should be deformed to enhance the doming.

5. LOCAL LATTICE DISTORTIONS AND HIGH T_C SUPERCONDUCTIVITY

5.1 Local Lattice Anomalies in HTSC Materials

Since the discovery of HTSC, numerous efforts have been taken to understand the mechanism of pairing. Up to now, however, attempts to *consistently and quantitatively* explain the mechanism of HTSC and normal-state properties have failed. This fact simply cast doubts on some of the fundamental assumptions. For instance, theories based on magnetic interactions (spin fluctuations) assume that HTSC occurs in a *homogeneous* two-dimensional CuO_2 planes [46]. However, a number of experiments recently demonstrated that the CuO_2 plane is rather *inhomogeneous* in microscopic viewpoints. Common understanding is that the basal plane is characterized by two domains having different local structures and transport properties, *i.e.*, carrier-rich and carrier-poor regions. The two domains have distorted and undistorted local structures and form stripe-order under certain conditions [4]. Such a mesoscopic-scale inhomogeniety is believed to be a consequence of the two competing interactions in highly correlated electron systems, *i.e.*, attractive magnetic and repulsive Coulomb interactions [47]. Structural evidence for the two-component charge carriers was also provided by neutron pair distribution function (PDF) [48]. Spin susceptibility obtained from the relaxation rate $(T_1)^{-1}$ of NMR [49] or inelastic neutron scattering [50] have shown anomalies well above T_c. Recently, unusual (non-Fermi liquid-like) properties of underdoped samples observed near the pseudogap temperature T^* is often discussed in relation to a quantum critical point (QCP) but quantum criticality is sill a subject of discussion and not yet established. On the other hand, XAS experiments have shown that anomalous LLD indicating stripe ordering as a signature of frustrated phase separation is observed below T^*. In this paper, we report another pronounced local lattice anomalies observed near the characteristic temperature T_0 below whichh broad spin susceptibility peak shows a maximum and a possible electronic phase transition from a Fermi liquid to non-Fermi liquid is proposed to take place [14]. We believe that the local lattice behavior around these cross-over temperatures (T^*, T_0) is a key feature to understand the nature of *electron-lattice interaction* in normal state.

The fact that only local and fast techniques can detect the LLD in HTSC materials suggests that the distortion is local and dynamical in nature. Therefore XAS with a time scale of 10^{-15} sec. is an ideal probe of the radial distribution. Recent EXAFS studies have established that the temperature-dependent lattice anomalies above T_c are one of the common features of HTSC materials: $Tl_2Ba_2CuO_y$ [51], $Bi_2Sr_2CaCu_2O_8$ [7], $La_{1.85}Sr_{0.15}CuO_4$ [4] and $La_2CuO_{4+\delta}$ [52]. The detailed analysis of LLD and superstructure indicated the stripe ordering of distorted low temperature tetragonal (LTT) domains and undistorted low temperature orthorhombic (LTO) domains [4].

Current understanding is that stripe ordering is a consequence of frustrated phase separation due to the competition between short-range magnetism and long-range Coulomb interactions. More detailed Information on LLD is now needed to characterize the role of electron-lattice interactions in HTSC materials.

5. 2 Polarized XAS of YBa$_2$Cu$_3$O$_y$

Taking YBa$_2$Cu$_3$O$_y$ as an example, we will discuss how the anisotropic local lattice is probed by XAS. For YBa$_2$Cu$_3$O$_y$, much attention has been paid to *out-of-plane* oxygen motions in an early XAS studies [53]. On the other hand, recent polarized XAS study for YBa$_2$Cu$_3$O$_y$ showed that the *in-plane* local lattice anomaly exists below a characteristic temperature T^* well above T_c, close to the characteristic temperature of spin gap opening [54]. Whether the in-plane lattice anomaly is related to the charge and lattice stripe or simply due to the spin correlation, *i.e.*, the lattice sees the spin singlet formation through a spin-lattice interaction is an interesting problem. Spin gap observed in various experiments such as NMR [49], neutron scattering [50] and transport properties [55] have been related to the short-range ordering of Zhang-Rice singlets [56]. In case of optimally doped YBa$_2$Cu$_3$O$_y$, T^* is close to T_c, which made it difficult to relate the two phenomena, *i.e.*, LLD (as a signature of stripe ordering) and pseudogap opening. Later, we reported the doping dependence of in-plane lattice anomalies in YBa$_2$Cu$_3$O$_y$ over a wide range in y (6.3 - 6.8) to understand the critical condition of LLD in relation to pseudogap opening. These experiments have been performed for the in-plane geometry. In this paper, we report the results for out-of-plane geometry ans show that the c-axis lattice anomaly starts at much higher temperature close to T_1 and T_0. T_1 is the upper characteristic temperature for lattice anomalies while T_0 is originally referred as the temperature at which the maximum in spin susceptibility is observed. T_0 is recently proposed as the electronic phase transition from Fermi liquid to non-Fermi liquid state [14].

Highly oriented YBa$_2$Cu$_3$O$_y$ thin film samples were prepared by a pulsed laser ablation technique using a KrF excimer laser ($\lambda = 248$ nm) [57] onto single crystal SrTiO$_3$ substrates at 750°C. Oxygen stoichiometry y was carefully controlled by an oxygen partial pressure after the growth; samples were slowly cooled down from the growth temperature under oxygen pressure of 0.1 - 200 mbar yielding samples with T_c ranging from 31.3 K to 90 K. Temperature dependence of resistivity was measured prior to the experiment for all samples. As-grown samples (100 nm thick) showed a sharp superconducting transition with a typical transition width of about 5 K indicating a high degree of oxygen ordering along the Cu1-O1 chain and homogeniety. Three samples with a sharp transition width have been chosen to study doping dependence of local distortions in YBa$_2$Cu$_3$O$_y$ over a wide range in T_c (31.3 - 78.5 K). In addition to oriented thin films, bulk single crystals grown by traveling seed floating zone (TSFZ) and other growth techniques were used.

Polarized EXAFS experiments were performed in a fluorescence mode at the 27-pole wiggler station BL13B1 of the Photon Factory using synchrotron radiation from a 2.5 GeV storage ring. For polarized fluorescence detection at low temperature down to 15 K, a closed-cycle He refrigerator was mounted on a high precision two-axis Huber 422 goniometer [58]. For an in-plane polarization geometry, a sample was mounted on the horizontal ω axis and a grazing-incidence angle was optimized for monitoring a thin layer of 100 nm (1.6 degree). The stability of temperature achieved by a precision temperature controller was within ± 0.1 K. The Cu K_α fluorescence signal was collected by a solid-state x-ray detector array consisting of nineteen pure Ge elements [28]. The output of each detector was energy-analyzed in order to remove channels affected by strong diffractions from substrates. EXAFS scans were made by a sagittaly focusing Si(111) double crystal monochromator. A typical energy resolution was 2 eV at 9 keV with a photon flux of about 10^{11} photons/sec when the storage ring was operated at 350 mA.

Normalized EXAFS oscillations $\chi(k)$ were Fourier-transformed after multiplying k over the range $2.5 < k < 18$ Å^{-1}. In the left column of Fig. 8, the magnitude and imaginary part of the FT results for YBa$_2$Cu$_3$O$_y$ (T_c=78.5 K) taken with the electrical field vector E parallel to the ab-plane (E//ab) are shown. The E//c FT results for YBa$_2$Cu$_3$O$_y$ (T_c=90 K) are shown in the right column of Fig. 8. A prominent peak observed at around 2.0 Å, after the phase shift correction, consists of square-planar oxygen atoms (O2, O3) forming the CuO$_5$ pyramid and a small contribution of Cu1-O1 pair. As seen from the FT magnitude between the two temperatures, for both geometries, the Cu-O radial distribution becomes *asymmetric* at low temperature. The asymmetry toward the longer radial distribution indicates that the appearance of distorted domain with an elongated Cu-O distance. This is a direct evidence for the manifestation of LLD essentially of Q_2-type distortion where half of the in-plane Cu-O bonds are stretched while the apical oxygen atom is displaced toward the central atom (Cu). In case of YBa$_2$Cu$_3$O$_y$, the in-plane Cu-O distance is the average over Cu2-O2, O3 and Cu1-O1 pairs, while E//c orientation the Cu-O peak around 1.5 Å is the weighted average of Cu2-O4 and Cu1-O4. Note that the peak positions are shifted

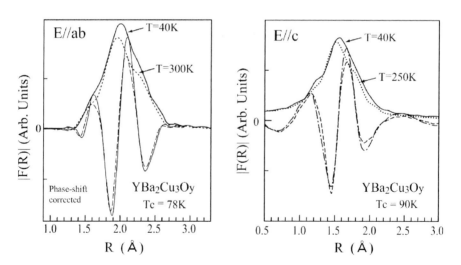

FIGURE 8. Magnitude and imaginary part of Fourier transform of the E//ab polarized Cu K-EXAFS oscillations for YBa$_2$Cu$_3$O$_y$ with T_c=78 K (left), and those for E//c polarized data for the sample with T_c=90 K (right). Low temperature data show asymmetric radial distribution of oxygen atoms.

toward a smaller value due to the phase shift effect. In order to correct the phase shift for the Cu-O pair, theoretical phase shift functions obtained by FEFF6 [21] were used.

In the unit cell of optimally doped $YBa_2Cu_3O_y$ (y = 7), copper atoms (Cu2) in the CuO_2 plane are coordinated by square-planar oxygen atoms (O2 and O3) and an apical oxygen atom (O4) while Cu1 atoms form a linear chain (Cu1-O1) along the b-axis. The in-plane (E//ab) polarized Cu K-EXAFS probes the Cu-O pair correlation averaged over Cu2-O2, O3 and Cu1-O1. The contribution of Cu1-O1 to the average coordination (2y-4)/3 in a twinned sample is ~ 14% for T_c = 78.5 K sample (y = 6.79) and ~ 12% for T_c = 55 K sample (y = 6.57). Inspecting the imaginary part which grows only at the large R side, one can find that the asymmetry is caused by an additional component with a slightly longer Cu-O bond.

5.3 Anomalous Temperature-dependence of Cu-O Pair Correlation

In Fig. 9, the magnitude of FT for E//c geometry is shown as a function of temperature for the Cu-Ba and Cu-O pairs. The magnitude for the Cu-Ba pair is a smooth function of temperature and except a small jump at T_c, there is no anomalous change. Here, temperatures T_1 and T_2 indicate the two characteristic temperatures where the anomalous atomic displacement was found in ion channeling experiment [59]. However, the same plot for the Cu-O pair indicates an anomalous behavior. A pronounced deviation from a normal phonon contribution begins at T_1 followed by an upturn below T_2 or T^*. The shaded region indicates the temperature range where the FT magnitude shows asymmetric pattern due to the distorted domain. T^* and T_2 are quite close and we believe that the distortion or stripe

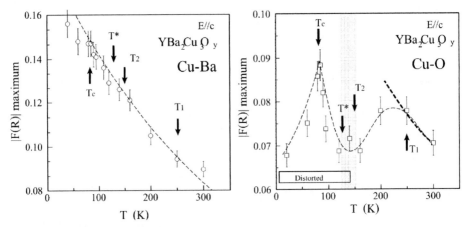

FIGURE 9. Magnitude of Fourier transform of the E//c polarized Cu K-EXAFS oscillations corresponding to the Cu-Ba correlation in $YBa_2Cu_3O_y$ with T_c = 90 K (left), the same plot for the Cu-O correlation (right). T_c and T^* indicate superconducting and pseudogap opening temperatures while T_1 and T_2 are characteristic temperatures where lattice anomalies are observed. Temperature range where the radial distribution shows an asymmetry is indicated. The shaded region indicates the temperature range between T^* and T_2.

ordering occurs at the crossover temperature $T^* \sim T_2$. This suggests that the phase separation and subsequent stripe-ordering take place as the pseudogap is opened. Secondly, we note that the first anomaly occurs at T_1 where the elastic anomalies such as internal friction Q^{-1} [60] and ion channeling [59] were reported. In fact, both experiments reported similar values for the two temperatures (T_1, T_2) characterizing lattice anomalies above T_c. In summary, the fact that only the Cu-O pair peak in FT has an unusual temperature dependence shows that *the unusual lattice response is due to the displacement of oxygen atoms*, in agreement with the recent report of LO phonon anomalies observed in neutron scattering involving half breezing mode of oxygen atoms [61]. However, even more striking point is the fact that the magnitude increases after the two domains are formed below T^*. This may look inconsistent with a simple expectation that the FT magnitude would decrease due to the interference between the closely separated two shells. The decrease of FT is indeed found in phase separated systems such as Spinodal decompositions of compound solid solution. If the FT magnitude increases in spite of the phase separation, it means that the mean square relative displacement (MSRD) decreases. MSRD includes both static and phonon term. We believe that the phonon contribution (less fluctuation) overcomes the distortion effect.

In Fig. 10, the FT magnitudes for E//ab with $T_c = 78$ K and $T_c = 55$ K are compared. Although the number of data points is limited between T^* and 300 K, the same trend with E//c data, *i.e.*, the increase below T^* and the decrease below Tc, is observed for the data with $T_c=78$ K. The decrease of FT magnitude is less clear in case of the underdoped sample with $T_c=55$ K. If the temperature dependence of FT magnitude above T^* is related to the lattice fluctuation, the overall trend after subtracting the normal phonon contribution can be illustrated schematically in Fig. 11.

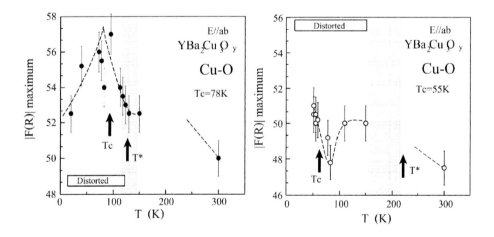

FIGURE 10. Magnitude of Fourier transform of the E//ab polarized Cu K-EXAFS oscillations corresponding to the Cu-O correlation in YBa$_2$Cu$_3$O$_y$ with $T_c = 78$ K (left), the same plot for the sample with $T_c = 55$ K (right). T_c and T^* indicate superconducting and pseudogap opening temperatures. Temperature range where the radial distribution shows an asymmetry is indicated. The shaded region indicates the temperature range where the radial distribution becomes asymmetric.

Decreasing temperature from 300 K to T_1 (upper anomaly temperature), the FT magnitude varies in a normal behavior predicted by Einstein formula and thus the lattice fluctuation contribution is expected to be constant. However, below T_1, the FT magnitude sharply deviates from a normal phonon behavior, indicating the *increase of lattice fluctuation*. At T^* ($T^* \sim T_2$, lower anomaly temperature), the negative slope indicating the *decrease of fluctuation* is observed, as the LLD or stripe-ordering is indicated in the asymmetric RDF. The decrease of fluctuation changes its sign at T_c and below T_c, the fluctuation becomes large. As T_1 is close to T_0 where the electronic phase transition from large Fermi surface (Fermi liquid) to small Fermi surface (non-Fermi liquid) is proposed [14], this implies that *as the electron states become unusual (non-Fermi liquid-like) metal, the lattice fluctuations are pronounced*. We conjecture that when the "quantum critical fluctuation (QCF)" is reached, the stripe-ordering takes place which reduces the fluctuation. In Fig. 11, dashed line indicates the magnitude of lattice distortion.

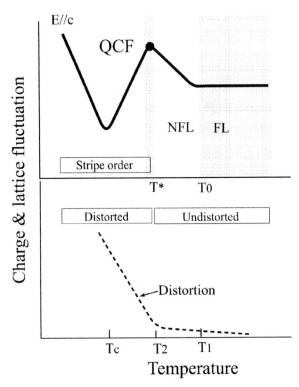

FIGURE 11. Schematic diagram of the magnitude of lattice fluctuations (solid line) and lattice distortion (dashed line) as a function of temperature. T_1 and T_2 are characteristic temperatures where lattice anomalies are observed as the decrease and increase of the magnitude of the Fourier transform of the Cu-O pair EXAFS oscillations, respectively. T_0 is the temperature below which the magnetic susceptibility exhibits a broad peak, which is proposed to be due to the electronic phase transition from a non-Fermi liquid to Fermi liquid (Ref. 7).

5.4 Oxygen Displacement

One can imagine various models of local lattice distortions involving in-plane Cu-O bonds. One of such models is a LTT-like tilting which causes a Q_2 mode Jahn-Teller distortion where the two in-plane Cu-O bonds are elongated while the apical bonds are shortened. This model is consistent with the fact that the in-plane RDF becomes asymmetric at lower temperature below T^*. The fact that the out-of-plane RDF becomes also asymmetric suggests that the distorted CuO$_5$ pyramid becomes more rigid and the apical oxygen displacement toward the copper atom affects the energy levels of e_g orbitals. Such a displacement would decrease the energy difference between $^3B_{1g}$ and $^1A_{1g}$ multiplets in the hole-doped CuO$_6$ cluster, favoring $^3B_{1g}$ configuration [13]. Detailed mechanism of HTSC based on Kamimura-Suwa model is in ref. 14. Here, we note that the displacement of apical oxygen toward the Cu atom is essential in this model. Our observation of LLD enhanced below T_1 is larger in magnitude in the c-axis direction than that within the basal plane. The apical Cu-O peak in Fig. 8 (right column) is the average of Cu2-O4 and Cu1-O4 pairs where Cu1 and Cu2 denote copper atoms in the Cu-O linear chain and CuO$_2$ planes, respectively. The Q_2 type distortion shifts the two pairs located at 2.32 Å and 1.85 Å approach each other. The asymmetry of the apical Cu-O peak below T^* is likely due to this distortion. The important implication is that the apical Cu-O displacement occurs at higher temperature than T^*, T_1, indicating a relation to the lattice anomalies observed in elasticity and T_0. Further, if this "high energy" pseudogap temperature corresponds to the electronic transition involving the spin correlation, the present results imply that *the apical Cu-O displacement initiates such a transition*. It should be noted that the antiferromagnetic bilayer coupling may reduce the fluctuation. The suppression of the fluctuation below T^* in this scenario is due to the bilayer coupling. Further investigations are required to clarify the nature of magnetic interaction at high

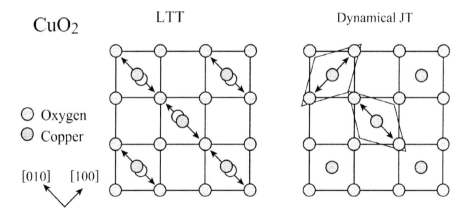

FIGURE 12. Displacement of atom positions in the CuO$_2$ plane after the low temperature tetragonal (LTT) type tilting of octahedron (left) and dynamic Jahn-Teller distortion in which two spins at copper sites and one spin on oxygen sites are stabilized (Ref. 70).

temperatures above T_c. However, we believe that the apical oxygen displacement near the crossover temperatures T^* and T_1 (T_0) is crucial to trigger the anomalous non-Fermi liquid like normal state properties.

Here let us consider what is expected from the increase of fluctuation below T_1. In Fig. 12, two cases of in-plane oxygen atom displacements, *i.e.*, Q_2 type distortion as a result of LTT-like tilting (left) and dynamical Jahn-Teller distortion (right) are illustrated. In the former distortion, the displacement of oxygen atoms are along the axis 45 degrees from the crystallographic axis. The in-plane azimuthal angle dependence of Cu K-XANES for untwinned single crystal was observed. This supports the LTT-like displacement below stripe ordering temperatures [62]. Unlike LTT-case in the latter displacement model, the direction of the distortion alternates so that the time-average is isotropic. Recently, three-spin magnetic polaron (two copper spin and one doped hole on bridging oxygen atom) stabilized by the dynamical Jahn-Teller distortion was proposed. In both cases, the in-plane and apical Cu-O displacements are correlated, LLD can be observed in E//*ab* and E//*c* RDF.

5.5 Implications of Local Lattice Distortions

In summary, we report the temperature-dependent LLD in $YBa_2Cu_3O_y$ single crystals using *ab*-plane and *c*-axis polarized EXAFS. The results indicate that the anomalous displacement occurs only at oxygen sites and sublattices formed by Cu and Ba atoms are normal from 300 K down to 20 K. The apical Cu-O and in-plane Cu-O RDF shows the evolution of asymmetry indicating the oxygen atom displacement at two characteristic temperatures T^* (~ 150 K) and T_1 (~ 250 K). The asymmetric RDF appears below T^* in both directions, indicating the stripe ordering occurs below the pseudogap opening temperature. Another unusual temperature dependence of LLD is the increase of mean square relative displacement below T_1 followed by a suppression after the stripe-order is introduced. We conclude that there could be a quantum critical fluctuation around the crossover temperature T^* leading to the quantum critical point at $T = 0$. When the fluctuation exceeds this limit, the pseudogap opening and electronic phase separation seem to occur. LLD discussed in this paper coincides with the elastic anomalies and lattice anomalies observed by ion channeling above T_c. Future experiments on an untwinned single crystal would clarify the nature of LLD above T_c and thereby the anomalous normal state properties. Such information would be helpful to critically evaluate the role of spin correlation in HTSC.

Arai *et al.* reported that T-dependence of $S(Q, E)$ obtained by an inelastic neutron scattering technique for optimally doped $YBa_2Cu_3O_7$ (T_c ~ 93 K) shows an anomalous increase around 120 K close to T^* (123 K) for $YBa_2Cu_3O_y$ (T_c ~ 78.5 K). A sharp increase of $S(Q, E)$ maximizing at T_c is ascribed to an expansion of dynamical correlation length associated with a local structural distortion [50]. A sharp increase in the FT magnitude below T^* is explained by the increased fluctuations reflects in the instantaneous Cu-O distance. Similar anomalies have been reported in internal friction Q^{-1} at T^* (~ 150 K) and T_1 (~ 250 K) [59]. The present observation is consistent with Raman experiments which observed anomalies in the electronic peak at $T \sim 1.6 T_c$ [63]. It should be noted that the temperature-range for enhanced fluctuation coincides with the anomalous $S(Q, E)$ variation around T_c [50].

The fact that the onset of LLD and T^* coincide is interpreted as a signature of stripe ordering occurs as the pseudogap opens up [55]. We conjecture that *stripe ordering stabilizes the charge fluctuation.* Below T^*, in-plane optical conductivity changes from a single-component to two-component carrier regime [64], which is consistent with a charge and lattice stripe of a distorted localized and undistorted itinerant domains [50]. In the Bose-Einstein Condensation picture, T^* is interpreted as the onset of pairing [64]. On the other hand, T^* observed in NMR $(T_1 T)^{-1}$ and Raman frequency shift is related to the onset of short range order of Zhang-Rice singlets [56]. It is still a long way from distinguishing the mechanism of HTSC but the LLD or stripe ordering below T^* and suppression of fluctuation enhanced below T_0 (T_1) seem to be a universal feature of HTSC materials.

6. PHOTO-INDUCED LOCAL LATTICE DISTORTIONS

6.1 In-situ XAS Studies of Chalcogenide Glasses

A new kind of melting phenomenon which is not based on thermal excitation has been observed. X-ray absorption spectroscopy (XAS) experiments under optical pumping provide a "snap-shot" information on the local structure under excitation. We have studied the local structure of chalcogenide glasses such as vitreous selenium and As_2Se_3 under optical excitation and confirmed the local melting phenomenon under light illumination at low temperature. The photo-induced nonthermal melting (PNM) in chalcogenide glasses is interpreted as the result of pairing of excited lone pair electrons during the illumination. Trapped states in this photo-assisted metastable phase result in a local structural disorder which is partially quenched at room temperature. The increased short-range disorder causing Coulomb repulsion is likely to be the origin of red shift of the absorption coefficient known as photodarkening effect. We found that the bond alternation of chalcogens occur during the photo-excitation indicating melting induced by photoexcitation.

The EXAFS oscillations for *a*-Se and *c*-Se extracted from the Se K_α fluorescence yield spectra taken at 30 K were analyzed. The magnitude of Fourier transform (FT) provides the radial distribution of scatterer atoms as a function of the radial distance R from the central atom. For *a*-Se, the nearest neighbor peak shifts to a shorter distance together with an increased magnitude (confirmed to be due to decreased mean-square relative displacement, MSRD) and decreased intensity of inter-chain second-nearest peak [29,65]. Such a change of FT magnitude indicates that the increased intra-chain and weakened inter-chain correlations. The FT-filtered first shell (nearest neighbor) contribution has been further analyzed. The first-shell EXAFS oscillations $k\chi(k)$ extending from 4.5 to 15 Å^{-1} were curve-fitted by a single-scattering formula and the structural parameters such as bond length, coordination number and MSRD were obtained after analyzing the correlation between these parameters. In Fig. 13, variations in the normalized coordination number and MSRD associated with sequences of sample treatment are summarized [29,65].

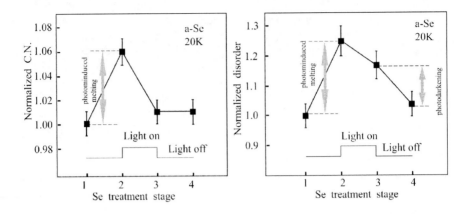

FIGURE 13. Normalized coordination number (left and the relative disorder parameter (MSRD) for *a*-Se (right). (1: as-prepared, 2: under illumination, 3: after illumination and 4: after heat treatment at 300 K). The increased coordination number and disorder upon illumination is due to the photo-induced melting. A part of disorder is frozen which causes the optical absorption edge shifts to lower energy (Refs. 65 and 66).

We find that the coordination number increases reversibly in the sample kept under irradiation by about 4% after a long exposure time at 30 K. This clearly demonstrates that the *photoexcitation induces the average over-fold coordinated states and rules out the model which assumes only the change in the bond angle* (constant average coordination number). The MSRD also increases upon light irradiation. After illumination, the original coordination number is restored while the light-enhanced structural disorder remains. The observed structural change, averaged over all sites, indicates a locally increased coordination number such as three-fold coordinated sites. Annealing of the irradiated sample at 300 K (near T_g) resulted in recovery of the initial values for both the coordination number and the structural disorder. These changes were not observed in the data of *c*-Se, confirming that they are not related to thermal effects upon irradiation.

6.2 Photo-induced Bond-alternation

The observation of a larger coordination number than two in *a*-Se is interpreted as the experimental evidence of the three-fold coordinated sites: $2C^*_2 => (C^0_3 - C^0_3)$ where C^*_2 denotes two-fold excited lone pair states [29,65]. Photo-induced effects in chalcogenides have been intensively discussed based on the valence-alternation-pairs (VAP) model which assumed that dangling bonds are transformed pairs of positively charged three-fold coordinated and negatively charged singly coordinated sites: $2C^0_2 => (C^+_3 - C^-_1)$. Although such a configuration is energetically favored because of Coulomb interaction, the formation of $C^+_3 - C^-_1$ pair is unlikely since the average coordination for such a pair would be constant while the change of average coordination was experimentally observed. Although the overcoordinated defect pairs are also formed in As_2Se_3 [67], the neutral defect pair is not the case as discussed in the later section.

Recent molecular dynamics calculations for a-Se predicted that the average coordination number exceeds 2.0 (2.17 at 350 K), *i.e.*, the three-fold coordinated sites dominate over one-fold coordinated sites at moderately high temperature [68]. This implies that neutral three-fold defect pairs serving as inter-chain links are formed. More recent calculations showed that upon elevating the temperature, the fraction of three-fold coordination increases and saturates while that of the one-fold coordinated site keeps increasing. The total average coordination number thus maximizes around 700 K [68]. The experimentally observed fraction at 30 K is about 5%, which is close to the maximum value at 700 K. The fact that the same local structure can be achieved either by thermal excitation or photoexcitation suggest that PNM is realized for chalcogenide glasses.

6.3 Photo-induced Nonthermal Melting

Fig. 14 shows a schematic electron states for different configurations [29,65]. Under steady-state photoexcitation, lone pair electrons in the top of the valence band are excited into the conduction band leaving one electron in the former lone-pair orbital. Because of Coulomb repulsion of extended excited electron and lone pair electrons of the adjacent chain, the bond alternation would stabilizes the cross-linking coordination. The three-fold coordinated sites act as a "switching board" and may result in the regular two-fold coordinated chains or the two one-fold coordinated sites, disrupting the chains. The latter is enhanced at elevated temperature according to the molecular dynamics calculation. We believe that this neutral defect pair is a metastable state where the relaxed electron repulsion by bond formation compensates the stress energy. The formation of these new inter-chain bonds (photo-induced nonthermal melting) introduce a local distortion observed as an increase in the MSRD inspite of a free rotation of the dihedral angle. Dynamical cross linking disappears after switching off the light restoring the ground-state coordination, yet the stress is not completely relieved with a low mobility at low temperature.

Switching off the light, the defect pairs immediately split away because of the repulsive force between lone pair electrons. However, the atoms trapped in metastable potential pockets can not move at low temperature and the structural disorder is quenched even after switching off the light since the energy barrier between the potential minima corresponding to different configurations is higher than kT. Thus the displacement is not relaxed and the disorder is quenched to room temperature. The effect of disorder naturally decreases the optical gap and additional tailing. This frozen deviation from the minimum energy configuration from Coulomb repulsion of lone pair electrons is the physical origin of photodarkening. Since MSRD directly measures the relative displacement between the first nearest neighbors, the nature of disorder is likely to be the bond-stretching and not bond-bending. Fig. 15 illustrates the coordination diagram for the ground state (g), excited state (e), metastable state (m) and quasi-ground state (g'). The g'-state is associated with frozen disorder and enhanced Coulomb repulsion U.

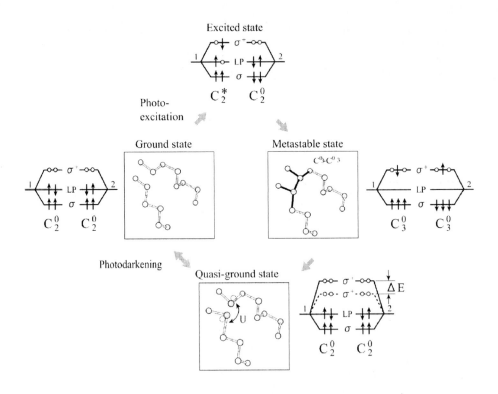

FIGURE 14. Electron states of selenium for various configurations. Excited lone pair electrons form neutral defect pairs (C^0_3-C^0_3) which is the same configuration of a liquid state (photo-induced melting). This metastable state is short-lived and stable only during the light illumination. After switching off these defects disappears but the disorder partly remains. Coulomb repulsion between the lone pairs causes the energy shift of unoccupied antibonding p-states.

Finally we consider the role of three-fold defect pairs in the photo-induced effects, *i.e.* photo-induced fluidity [69]. The quenched structural disorder observed as broadening in bond length distribution decreases the conduction band tail leading to the photo-darkening. During the photo-excitation, inter-chain inks are sequentially formed competing with the relaxation process just like a liquid state. We can expect naturally that fluidity increases during the photoexcitation as observed. The relation between the photoinduced local structure change and optical gap is illustrated in Fig. 10. Upon illumination, the cross link is formed as a metastable state (leading photofluidity) and the incomplete relaxation gives a repulsion-enhanced state as a quasi-ground state, which is the origin of photodarkening.

Photo-induced melting

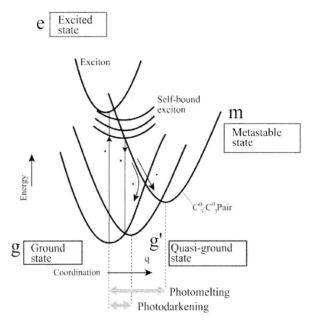

FIGURE 15. Schematic configuration diagram. Ground state (g), excited state (e), metastable state (m) and quasi-ground state (q) are indicated. Radiative and nonradiative decay paths are shown by arrows.

7. CONCLUDING REMARKS

It was demonstrated that XAS is an ideal tool probing the LLD in highly correlated electron systems, metal proteins and photo-excited states, providing the magnitude and types of distortion. The spin state and LLD in myoglobin was investigated by XANES and biological function, *i.e.*, affinity for oxygen binding , was shown to be controlled by the LLD. HTSC materials often exhibit LLD as a signature of electron-lattice interaction and XAS studies have shown that *a strong electron-lattice coupling emerges below a characteristic temperature $T*$ and T_0* associated with the change in electron states. The LLD during photo-excitation in chalcogenide glasses was studied by in-situ XAS and it was found that optical excitation induces nonthermal local melting.

ACKNOWLEDGMENTS

This work has been conducted as a collaboration with N. Saini, A. Bianconi, T. Haage, J. Zegenhagen, K. Oka, A.M. Moe, T. Ito, S. Tajima, T. Masui, T. Iizuka to whom the author expresses his greatest thanks. The author also would like to thank K. Yamaji, H. Kamimura, J. Ranninger, D. Mihalovic for helpful discussions.

REFERENCES

1. Emery, V.J., and Kivelson, S.A., *Phys. Rev. Lett.* **64,** 475 (1990).
2. *Phase separation in cuprate superconductors,* edited by Müller, K.A., and Bendek, G., World Scientific, 1992.
3. *Phase separation in cuprate superconductors*, edited by Sigmund, E., and Müller, K.A., World Scientific, 1994.
4. Bianconi, A,. Saini, N.L., Lanzara, A., Missori, M., Oyanagi, H., Yamaguchi, H., Oka, K., and Ito, T., *Phys. Rev. Lett.* **76,** 3412 (1996).
5. Tranquada, J.M., Butley, D.J., Sachan, V., and Lorenzo, J.E., *Phys. Rev. Lett.* **73,** 1003 (1994).
6. Di Castro, C., Feiner, L.F., Grilli, M., *Phys. Rev. Lett.* **66,** 3209 (1991).
7. Saini, N.L., Lanzara, A., Oyanagi, H., Yamaguchi, H., Oka, K., Ito, T., and Bianconi, A., *Phys. Rev.* B **55,** 12759 (1997); *ibid* Bianconi, A., Saini, N.L., Roseetti, T., Lanzara, A., Perali, A., Misori, M., Oyanagi, H., Yamaguchi, H., Nishihara, Y., and Ha, D.H., *Phys. Rev. B* **54,** 12018 (1996).
8. Jahn, H.A., and Teller, E., *Proc. Roy. Soc. (London)*, A**161,** 220 (1937).
9. Oyanagi, H., Iizuka, T., Matsushita, T., Saigo, S., Makino, R., Ishimura, Y., and Ishiguro, T., *J. Phys. Soc. Jpn.* **56,** 3381 (1987).
10. Oyanagi, H., Masui, T., and Tajima, S., in preparation.
11. Keneman, S.A., *Appl. Phys. Lett.* **19,** 205 (1971).
12. See, for instance, *Introduction to ligand field theory,* edited by Ballhausen, C.J., McGraw-Hill, 1962.
13. Kamimura, H., and Eto, M., *J. Phys. Soc. Jpn.* **59,** 3053 (1990).
14. Kamimura, H., Hamada, T., Nishimura, M., and Ushio, H., to be published in *Physica C.*
15. Stern, E.A., Sayers, D.E., and Lytle, F.W., *Phys. Rev.*, **B11,** 4836 (1975).
16. Oyanagi, H., Oka, K., Unoki, H., Nishihara,Y., Murata, K., Yamaguchi, H., Matsushita, T., Tokumoto, M., and Kimura, Y., *J. Phys. Soc. Jpn.* **58,** 2896 (1989).
17. Oyanagi, H., Yokoyama, Y., Yamaguchi, H., Kuwahara, Y., Katayama, T., and Nishihara, Y., *Phys. Rev. B* **42,** 10136 (1990).
18. Li, C., Pompa, M., Castellano, A.C., Della Longa, S., and Bianconi, A., *Physica C* **175,** 369-380 (1991).
19. Kosugi, N., Yokoyama, T., Asakura, K., and Kuroda, H., *Chem. Phys.* **91,** 249(1984).
20. Lanzara, A., Zhao, G.-m., Saini, N.L., Bianconi, A., Conder, K., Keller, H., and Müller, K.A., *J. Phys.:Condens. Matter* **11,** L541 (1999).
21. Rehr, J.J., Zabinsky, S.I., and Albers, R.C., *Phys. Rev. Lett.* **69,** 3397 (1992).
22. Lindenberg, A.M. *et al.*, *Phys. Rev. Lett.* **84,** 111-114 (2000).
23. Chance, B., Fischetti, R., and Powers, L., *Biochemistry* **22,** 3820-3829 (1983).

24. Powers, L., Chance, B., Chance, M., Campbell, B., Khalid, J., Kumar, C., Naqui, A., Reddy, K.S., and Zhou, Y., *Biochemistry* **26**, 4785-4796 (1987).

25. Mills, D.M., Lewis, A., Harootunian, A., Huang, J., and Smith, B., *Science* **223**, 811-813 (1984).

26. Oyanagi, H., Kolobov, A., and Tanaka, K., *J. Synchrotron Rad.* **5**, 1001-1003 (1998).

27. Becker, R.S., Golvchenko, J.A., and Patel, J.R., *Phys. Rev. Lett.* **50**, 153 (1983); Heald, S.M., Keller, E., and Stern, E.A., *Phys. Lett. A* **103**, 155 (1984).

28. Oyanagi, H., Martini, M., and Saito, M., *Nucl. Instrum. & Methods* **A403**, 58-64 (1998).

29. Kolobov, A., Oyanagi, H., Tanaka, K., and Tanaka, K., *Phys. Rev. B* **55**, 726-734 (1997).

30. George, P., Beetlestone, J., and Griffith, J.S., *Rev. Mod. Phys.* **36**, 441 (1964).

31. Iizuka, T., and Yonetani, T., *Adv. Biophys.* **1**, 157 (1970).

32. Neya, S., Hada, S., Funasaki, N., Umemura, J., and Takenaka, T., *Biochim. Biophys. Acta* **827**, 157 (1985).

33. Perutz, M.F., *Nature (London)*, **228**, 726 (1970).

34. Eisenberger, P., Shulman, R.G., Kincaid, B.M., Brown, G.S., and Ogawa, S., *Nature (London)* **274**, 30 (1978).

35. Perutz, M.F., Hasnain, S.S., Duke, P.J., Sessler, J.L., and Hahn, J.E., *Nature (London)* **295**, 535 (1982).

36. Eisenberger, P., Shulman, R.G., Brown, G.S., and Ogawa, S., *Proc. Natl. Acad. Sci. USA* **73**, 491 (1976).

37. Chance, B., Fischetti, R., and Powers, L., *Biochemistry* **22**, 3820 (1983).

38. Bianconi, A., Congiu-Castellano, A., Dell'Ariccia, M., Giovannelli, A., Durham, P.J., Burattini, E., and Barteri, M., *FEBS Lett.* **178**, 165 (1984).

39. Bianconi, A., Alema, S., Castellani, L., Fasella, P., Giovannelli, A., Mobilio, S., and Oesh, B., *J. Mol. Biol.* **165**, 125 (1983).

40. Smith, T.A., Penner-Hahn, J.E., Berding, M.A., Doniach, S., and Hodgson, K.O., *J. Am. Chem. Soc.* **107**, 5945 (1985).

41. Bianconi, A., Congiu-Castellano, A., Durham, P.J., Hasnain, S.S., and Phillips, S., *Nature* (London) **318**, 685 (1985).

42. Takano, T., *J. Mol. Biol.* **110**, 537 (1977).

43. Hoard, J.L., Hamor, M.J., Hamor, T.A., and Caughey, W.S., *J. Am. Chem. Soc.* **87**, 2312 (1965).

44. Bianconi, A., Congiu-Castellano, A., Dell'Ariccia, M., Giovannelli, A., and Morante, S., *FEBS Lett.* **191**, 241 (1985).

45. Chance, M.R., Parkhurst, L.J., Powers, L.S., and Chance, B., *J. Biol. Chem.* **261**, 5689 (1986).

46. Anderson, P.W., Baskaran, G., Zou, Z., and Hsu, T., *Phys. Rev. Lett.* **58**, 2790 (1987).

47. Low, U., Emery, V.J., Fabricius, K., and Kivelson, S.A., *Phys. Rev. Lett.* **72**, 1918 (1994).

48. Egami, T., and Billinge, S.J.L., *Prog. Mater. Sci.* **38**, 359 (1994).

49. Takigawa, M., Reyes, A.P., Hammel, P.C., Thompson, J.D., Heffner, R.H., Fisk Z., and Ott, K.C., *Phys. Rev. B* **43**, 247 (1991).

50. Arai, M. Yamada, K., Hidaka, Y. Itoh, S., Bowden, Z.A., Taylor, A.D., and Endoh, Y., *Phys. Rev. Lett.* **69**, 359 (1992).

51. Yamaguchi, H., Nakajima, S., Kuwahara, Y., Oyanagi, H., and Shono, Y., *Physica C* **213**, 375 (1993).

52. Lanzara, A., Saini, N.L., Bianconi, A., Soldo, Y., Chou, F.C., Johnson, D.C., *Phys. Rev. B* (in press, 1997).

53. Mustre de Leon, J., Conradson, S.D., Batistic, I., Bishop, A.R., Raistrick, I.D., Aronson, M.C., and Garzon, F.H., *Phys. Rev. B* **45**, 2447 (1992).

54. Oyanagi, H., and Zegenhagen, J., *J. Superconductivity* **10**, 415 (1997).

55. Ito, T., Takenaka, K., and Uchida, S., *Phys. Rev. Lett.* **70**, 3995 (1993).

56. Tanamoto, T., Kohno, K., and Fukuyama, H., *J. Phys. Soc. Jpn.* **61**, 1886 (1992); Tanamoto, T., Kohno, K., and Fukuyama, H., *J. Phys. Soc. Jpn.*, **62**, 717(1993).

57. Ludwig, C., Jiang, Q., Kuhl, J., Zegenhagen, J., *Physica C* **269**, 249 (1996).

58. Oyanagi, H., Shioda, R., Kuwahara, Y., Haga, K., *J. Synchrotron Rad.* **2**, 99 (1995); Oyanagi, H., *J. Synchrotron Rad.* **5**, 48 (1998).

59. Sharma, R.P., Ogale, S.B., Zhang, Z.H., Liu, J.R., Chu, W.K., Veal, B., Paulikas, A., Zheng, H., and Venkatesan, T., *Nature (London)* **404** 736 (2000).

60. Cannelli, G., Cantelli, R., Cordero, F., Costa, G.A., Ferretti, M., and Olcese, G.L., *Europhys. Lett.,* **6**, 271 (1988).

61. McQueeney, R.J., Petrov, Y., Egami, T., Yethiraj, M., Shirane, G., and Endoh, Y., *Phys. Rev. Lett.* **82**, 628 (1999).

62. Saini, N.L. *et al.*, unpublished.

63. Ruani, G., and Ricci, P., *Phys. Rev. B* **55**, 93 (1997).

64. Mihailovic, D., Mertelj, T., and Müller, K.A., *Phys. Rev. B* **57**, 6116 (1998).

65. Kolobov, A., Oyanagi, H., Roy, A., and Tanaka, K., *J. Non-Cryst. Solids* **277-230**, 710-714 (1998).

66. Kolobov, A., Oyanagi, H., and Tanaka, K., *MRS Buletin* **24**, 32-35 (1999).

67. Hohl, D., and Jones, R.O., *Phys. Rev. B* **43**, 3856-3870 (1991).

68. Zhang, X., and Drabold, D.A., *Phys. Rev. Lett.* **83**, 5042-5045 (1999).

69. Hisakuni, H., and. Tanaka, K., *Science* **270**, 974-975 (1995).

70. Kochelaev, B.I., Sichelschmidt, J., Elschner, B., Lemor, W., and Loidl, A., *Phys. Rev. Lett.* **24**, 4274 (1997).

Evidence for Electronic Phase Separation in High T_c Cuprates and CMR Manganites

R. P. Sharma

Center for Superconductivity Research, Physics Department, University of Maryland,
College Park, Maryland-20742, USA

Abstract. By using ion channeling, an ultrafast real space probe of small (sub picometer range) incoherent atomic displacements, direct experimental evidence for electronic phase separation in $YBa_2Cu_3O_{7-x}$ as a function of oxygen doping has been obtained. The stripe dynamics in YBCO has been predicted to undergo three phase crossovers/transitions as a function of temperature, at the stripe phase (T_1), spingap formation temperature (T_2) and the global superconducting transition temperature T_c. Three similar temperatures are seen in ion channeling investigations, in the form of non analycities in the incoherent lattice fluctuations. Their observed weak doping dependence conforms to the suggested behavior of the pseudo-gap formation temperature T_2, while T_1 has shown a clear difference in the spin and charge dynamics. Similar incoherent atomic displacements have been observed in CMR manganites at the phase transition from paramagnetic insulator to ferromagnetic metal. Replcing ^{16}O by ^{18}O in high quality single crystalline $Nd_{0.67}Sr_{0.33}MnO_3$ films (pulse laser deposited), a clear change in the atomic displacement is seen in the two cases, close to the respective phase transitions

INTRODUCTION

The growing evidence for the existence of complex textures of charges and spins in high temperature superconductors has drawn attention to the stripe phase scenario [1-4] as a possible basis for the mechanism of superconductivity in these materials. The new ideas refer to a state that is radically different from a Fermi liquid. Large number of experimental observations until now have been restricted to systems like LSCO [5,6] where the texture dynamics are slow or suppressed altogether, and do not include the important case of YBCO. The corresponding evidence is less direct in other high T_c materials such as YBCO and BSCCO. Neutron scattering data [7,8] on underdoped YBCO and ARPES data on BSCCO [9] suggest that there are stripe-like but more disordered local charge fluctuations in these superconductors. Hauff *et al.* [10] studied Zn doped YBCO by optical reflectivity and concluded that a coupling of phonons to spin fluctuations accompanies the opening of the spin gap. It appears that the fluctuation dynamics in these systems is fast and needs ultra fast ($< 10^{-12}$ sec) probes for investigation. One such technique is ion channeling which is an ultra fast, unique method, providing a direct real space probe of extremely small ($<$ picometer) incoherent displacements (static and dynamic) of atoms in single crystalline materials.

In manganites, shot into prominence on account of the observation of the colossal magneto-resistance, the phenomenon of electronic phase separation has been gaining

CP554, *Physics in Local Lattice Distortions*, edited by H. Oyanagi and A. Bianconi
© 2001 American Institute of Physics 1-56396-984-X/01/$18.00

ground with significant experimental support. Direct evidence of static stripes has been seen by Cheong *et al.* in the charge ordered (CO) region of the $La_{1-x}Ca_xMnO_3$ phase diagram [11]. Further, within the ferromagnetic (FM) regime, a phase separation scenario has been proposed by Uehara *et al.* [12], wherein clusters of FM regions surrounded by CO regions with charge defects are supposed to form a percolating network. The phase boundaries are clearly the surfaces where significant lattice distortions can be expected to occur. Other phase separation scenarios involving small polaron to large polaron crossover across the metal-insulator transition [13,14] have also been proposed.

The CMR and oxide superconductors have a number of similarities. Both of them belong to perovskite class of crystals, and have strong Jahn-Teller ions (Cu^{2+} in cuprates and Mn^{3+} in manganites). They strongly favor lattice distortions to reduce the electronic degeneracy of valence electrons and play an intimate role in the transport process, both embody the physics of doped holes in anti-ferromagnetic background, and both exhibit several magneto-electronic phase transitions and crossovers.

We present here direct experimental evidence of incoherent lattice fluctuations in $YBa_2Cu_3O_{7-\delta}$ single crystals at optimum doping and at two other lower doping levels, as obtained by ion channeling measurements.

In a similar way in CMR manganites the oxygen-isotope effects on the local structure distortions are shown.

ION CHANNELING

Ion channeling [15] is a unique method, providing a direct real space probe of extremely small (< picometer) uncorrelated displacements (static and dynamic) of atoms in single crystalline materials. Ion channeling occurs when energetic ions, incident along a major crystallographic direction, are steered by a series of glancing collisions with the atoms of close packed atomic rows or planes. The critical angle for channeling to occur or the full width at half maximum (FWHM) of the channeling angular scan depends on the incident ion energy, the atomic numbers of the projectile and target, the interatomic spacing, electron screening potential, and most importantly, any displacements (dynamic or static) of the atoms from their regular lattice sites. The FWHM is affected dramatically by incoherent atomic displacements (both static and dynamic), as opposed to an equivalent coherent change in lattice parameter (*e.g.* thermal expansion). Lindhard [16] has given the basic continuum model of channeling, while Barrett [17] has incorporated corrections to the same based on Monte Carlo simulations involving binary ion-atom collisions. We have extracted the magnitude of the atomic displacement u from the measured FWHM using the continuum model, with corrections based upon the simulation of Barrett [17].

SAMPLE PREPARATION

Single crystals of $YBa_2Cu_3O_{7-\delta}$ were grown by a partially nonstoichiometric melting technique at Argonne National Laboratory. These were annealed in flowing oxygen at 450 to 500 C for ~ 72 h and showed a sharp (ΔT_c < 1K) superconducting transition at 92.5 K. Different sets of such crystals were annealed a second time for about 180 hours in the requisite O_2/N_2 ambient to reduce the O doping to get crystals with T_c at 65 and 45 K respectively, and one non-superconducting.

The epitaxial thin films of $Nd_{0.67}Sr_{0.33}MnO_3$ were grown on <100> $LaAlO_3$ single crystal substrates by pulsed laser deposition using a KrF excimer laser. The substrate temperature was 820°C and the oxygen pressure was kept constant at 400 mTorr during the deposition. The film thickness was about 190 nm. Two halves were cut from the same piece of film for oxygen-isotope diffusion. The diffusion for sample pair I was carried out for 10 h at about 900°C and oxygen pressure of about 1 bar. The diffusion for sample pair II was performed for 10 h at 940°C. The cooling rate was 300°C/h. The ^{18}O-isotope gas is enriched with 95% ^{18}O, which can ensure 95% ^{18}O in the ^{18}O thin films.

EXPERIMENTAL DETAILS

The ion channeling measurements were carried using a well-collimated beam (0.5 mm diameter and < 0.01° divergence) of 1.5 MeV He^+ ions obtained from a 1.7 MV tandem pelletron accelerator. The crystal was suitably mounted on a precision five axis goniometer having an angular resolution < 0.01° and could be cooled down to 33 K with the help of a closed-cycle refrigeration unit. Necessary precautions were taken to keep the effective pressure of condensable gases to a negligible level (~ 3 × 10^{-11} Torr) at the target surface; the background pressure in the target chamber was maintained at ~ 2 × 10^{-8} Torr. The specimen temperature could be varied between 30 and 300 K by a small heater mounted on the back of the target holder.

The back-scattered He^+ particles were analyzed using an annular silicon surface barrier detector of 300 mm^2 active area with a 4 mm diameter central hole which was mounted along the beam axis at a distance of ~ 5 cm from the target. This arrangement provided good statistics (~ 10^4 counts in the gate at each angular setting in random direction) for back scattered (170°) particles at a dose of only 3 nC of the incident He-ions in a channeling angular scan. The [001] single crystals of $YBa_2Cu_3O_{7-\delta}$ with suitable oxygen doping were first aligned precisely parallel to the incident beam direction. Angular channeling scans were made by measuring the Rutherford back scattering (RBS) yield as a function of the tilt angle about the [001] axis of the sample at sixteen different temperatures in the range 300 to 30K in each case respectively. In each case two to three sets of measurements are made using 8 to 10 high quality YBCO single crystals of respective T_c values. A standard deviation of < 1% is determined in these measurements. The atomic displacement u was determined from the measured full width at half maximum (FWHM) of the channeling angular scans.

RESULTS AND DISCUSSION

Measurements in $YBa_2Cu_3O_{7-\delta}$

The u vs. temperature data for the optimally oxygen doped and underdoped non superconducting YBCO samples are shown in Fig. 1. It is interesting to note that the data for the optimally doped sample shows well-defined structures in the temperature dependence of the u value, which are considerably suppressed in the case of non-superconducting sample. At least three crossover temperatures can be visually identified. To bring out these crossover/transition temperatures more clearly it is desirable to subtract the thermal vibration component, which is expected to exhibit a monotonic temperature dependence. The calculated Debye curve, with the constraint that the excess distortion measured above the thermal background can not be negative, is also shown in Fig. 1. We followed this procedure also for the T_c = 65 K and 45 K samples. By taking the difference of the measured curve with the Debye calculation we get the "excess" lattice distortions shown in Fig. 2. Let us first consider the optimally doped case. From room temperature to a temperature of about 230K there is a rapid drop in the amplitude. At about 230K (T_1) the nature of the atomic displacement u is indicative of some kind of phase transition in the system which enhances these fluctuation effects (dynamic or static). One more cusp is seen close to 150K (T_2). Near the electrically measured superconducting transition a sudden drop in the u value is seen at T_3. Remarkably below T_3, the u value increases with decreasing temperature. When we look at the oxygen depleted YBCO (T_c = 65 K and 45 K respectively), these features are reproduced fairly consistently, with the drop at T_3 scaling with the T_c, and below T_c an increasing excess fluctuation with decreasing temperature. The data about the characteristic transition/crossover temperatures as a function of oxygen doping are summarized in Fig. 3. It can be seen that while T_1 shows a small increase with increased underdoping, T_2 shows little dependence on doping.

The observed transitions at three temperatures (T_1, T_2, and T_3) seem to be consistent with the electronic and magnetic phase evolution as predicted by the stripe phase model. Specifically, transitions are seen at temperatures corresponding to a) stripe formation b) spin-gap and pair formation and c) the superconducting transition. Comparison of our data on lattice (charge) fluctuations with single particle tunneling and photoemission data has suggested the role of spin-charge separation in the normal state properties of cuprate superconductors. Further, the observed increasing excess lattice fluctuations below T_c in our measurements provides a clear indication of slowing boundery fluctuations which would enable the lattice to follow the fluctuations more easily.

The existence of non-analyticities in the normal state of the high T_c materials are found to occur in the incoherent lattice displacements and not in the structure of the material. Fluctuation of both spin and charge are seen by Mook *et al.* [8] by neutron scattering in highly under doped *i.e.* in $YBa_2Cu_3O_{6.6}$ system, supporting the existence of dynamic stripe phase in this material. The anomalous lattice fluctuations could

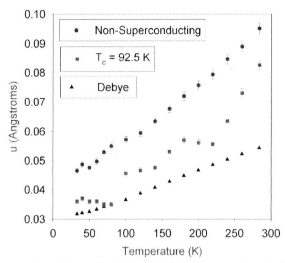

FIGURE 1. Temperature dependence of incoherent atomic displacement in optimal doped $YB_2Cu_3O_{7-x}$ and non-superconducting YBCO crystal (x = 0.065). The normalized Debye behavior is also shown.

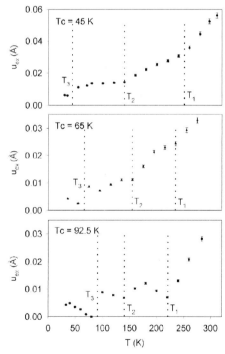

FIGURE 2. The excess atomic displacement u_{ex} in $YBa_2Cu_3O_{7-x}$ as obtained after subtracting the Debye component from the u values for 45K, 65K and 92.5K samples respectively as a function of temperature.

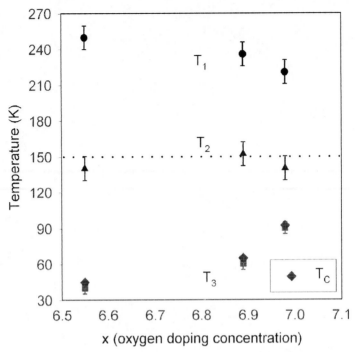

FIGURE 3. The phase transitions or crossover temperatures as shown in Fig. 2 are plotted as a function of oxygen doping.

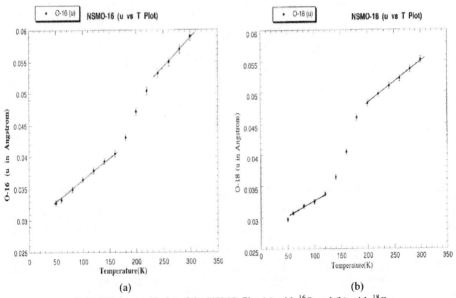

(a)   (b)

FIGURE 4. u vs T plot of the NSMO film (a) with ^{16}O and (b) with ^{18}O.

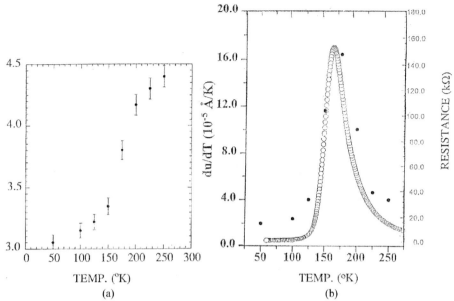

FIGURE 5. (a) A plot of the atomic displacement u versus temperature for NSMO film. (b) The du/dT vs T is shown by solid squares. Superimposed on it is the variation of resistance with temperature.

arise due to the influence of the electron system on the lattice via the known coupling between the lattice fluctuations and the low energy excitations of the electron system. These results suggest that the metallic state in the cuprates can not be regarded as a normal Fermi liquid and is not easily understood in terms of conventional theory of metals. According to Zaanen [18], superconductivity is associated with dynamic stripes. In this state the system of stripes is disordered by long-wavelength quantum fluctuations. More details are given in Ref. [19].

Isotope Effect Measurements in $Nd_{0.67}Sr_{0.33}MnO_3$ Films

Detailed measurement of incoherent lattice fluctuations in oxygen isotope substituted $Nd_{0.67}Sr_{0.33}MnO_3$ thin films are made by ion channeling in a similar way as described above. These studies have clearly shown (Fig. 4) an increase in the incoherent displacement of the Mn atoms, which is much larger than the phonon background, as the sample goes from the feromagnetic to the paramagnetic state. The observed anomaly is about 15% larger for [18]O than for [16]O samples. The peak in resistivity, which is close to the magnetic phase transition, is shifted to lower temperature by about 25 K in the former case as compared to the later one [20]. Similar shift is seen in the ion channeling measurements. The transport measurements have also shown that in the paramagnetic state the transport activation energy of the [18]O samples is higher than the [16]O samples by about 10%. Also the [18]O samples have a sharper resistivity drop below T_c than the [16]O samples.

In the case of manganites, we have established earlier that the magnetic and electronic phase changes in these systems are strongly reflected in the incoherent lattice fluctuations as revealed by ion channeling experiments [21], in view of the Jahn Teller character of Mn^{3+} ions. Such fluctuations are seen in manganite films made by laser ablation, particularly when they are cooled through the ferromagnetic transition. In (Fig. 5a) the atomic displacement u as obtained by ion channeling measurements in $Nd_{0.7}Sr_{0.3}MnO_3$ films is plotted as a function of temperature. The large fluctuation in the u value is clearly seen in the vicinity of ferromagnetic transition where a peak in resistivity is observed. The distortions have shown a clear correlation with the transport properties of these materials as seen in Fig. 5b, where du/dt as obtained from Fig. 5a, is plotted as a function of temperature along with the resistance measurements in the same film. It is clear that the temperature range where large incoherent lattice fluctuations are observed overlaps with the peak magneto-resistance regions of paramagnetic semiconductor to ferromagnetic metal phase transitions in the system.

The observed effect can not be explained simply by double exchange mechanism. Even after taking into account the Jahn Teller distortion of Mn-O octahedra the transport properties can not be explained completely. Dynamic distortions seem to be very important.

CONCLUSION

In $Yba_2Cu_3O_{7-x}$ system the incoherent lattice fluctuations which reflect charge dynamics show a non analytic dependence on temperature. Three distinct temperatures (T_1, T_2, and T_3) are seen reflecting phase crossovers in the system. T_1 is very close to the AF Neel temperature and exhibits a weak increase with under doping. Behavior of T_1 is quite unlike the variation of T^* as seen in the neutron scattering measurements reflecting spin dynamics. Thus the spin dynamics and charge dynamics are different in the normal state of YBCO. T_2 at 150 K shows very little oxygen doping dependence. This behavior is similar to that of the spin gap temperature in stripe phase scenario. Temperature T_3 closely corresponds to T_c showing increased coherence of the lattice at the superconducting transition.

ACKNOWLEDGMENTS

This work was supported by the NSF – MRSEC under Grant No. DMR – 9632521. The following are the collaborators in the work described above:

S. B. Ogale, R. Marcia, D. J. Kang, M. Rajeswari, H. D. Drew, R. L. Greene and T. Venkatesan
Center for Superconductivity Research, Department of Physics, University of Maryland, College Park MD 20742.

Z. H. Zhang, J. R. Liu, and W. K. Chu
Texas Center for Superconductivity, University of Houston, Houston, TX 77204.

Boyed Veal, A. Paulikas, and H. Zheng
Materials Science Division, Argonne National Laboratory, Argonne, IL 60439.

Guo-meng Zhao, and H. Keller
Physik-Institut, Universität Zürich-Irchel, Winterthurerstr 190 CH-8057 Zürich Switzerland.

REFERENCES

1. Zaanen, J., and Gunnarsson, O., *Phys. Rev. B*. **40**, 7391-7394 (1989).
2. Emery, V.J., Kivelson, S.A., and Lin, H.Q., *Phys. Rev. Lett.* **64**, 475-478 (1990).
3. Schulz, H.J., *Phys. Rev. Lett.* **64**, 1445-1448 (1990).
4. White, S.R., and Scalapino, D.J., *Phys. Rev. Lett.* **80**, 1272-1275 (1998).
5. Hunt, A.W., Singer, P.M., Thurber, K.R., and Imai, T., *Phys. Rev. Lett.* **82**, 4300 (1999).
6. Bianconi, A., Saini, N.L., Lanzara, A., Missori, M., Rossetti, T., Oyanagi, H., Yamaguchi, H., Oka, K., and Ito, T., *Phys. Rev. Lett.* **76**, 3412-3415 (1996).
7. Mook, H.A., and Dogan, F., *Nature* **401**, 145-147 (1999).
8. Mook, H.A., Dal, P., Dogan, F., and Hunt, R.D., *Nature* **404**, 729 (2000).
9. Tranquada, J.M., Sternlieb, B.J., Axe, J.D., Nakamura, Y., and Uchida, S., *Nature* **375**, 561-563 (1995).
10. Hauff, R., Tajima, S., Jang, W.J., and Rykov, A.I., *Phys. Rev. Lett.* **77**, 4620-4623 (1996).
11. Mori, S., Chen, C.H., and Cheong, S.W., *Nature* **392**, 473-476 (1998).
12. Uehara, M., Mori, S., Chen, C., and Cheong, S.W., *Nature* **399**, 560 (1999).
13. Millis, A.J., Littlewood, P.B., and Shraiman, B.I., *Phys. Rev. Lett.*, **74**, 5144 (1995).
14. Alexandrov, A.S., and Bratkovsky, A.M., *Phys. Rev. Lett.* **82**, 141 (1999).
15. Gemmell, D.S., *Rev. Mod. Phys.* **46**, 129 (1974).
16. Lindhard, J., and Kalckar, J., *Dan. Vidensk. Selsk. Mat. Phys. Medd.* **34**, No. 14 (1965).
17. Barrett, J.H., *Phys. Rev. B* **3**, 1527 (1971).
18. Zaanen, J., *Nature* **404**, 714 (2000).
19. Sharma, R.P., Ogale, S.B., Zhang, Z.H., Liu, J.R., Chu, W.K., Veal, B., Paulikas, A., Zheng, H., and Venkatesan, T., *Nature* **404**, 736 (2000).
20. Zhao, G.-M., Kang, D.J., Prellier, W., Rajeswari, M., Keller, H., Venkatesan, T., and Greene, R.L., to be published.
21. Sharma, R.P., Xiong, G.C., Kwon, C., Ramesh, R., Greene, R.L., and Venkatesan, T., *Phys. Rev. B* **54**, 10014 (1996).

Dynamical Properties of the Spin and Charge Stripes in La$_{2-x}$Sr$_x$CuO$_4$

S. Sugai and N. Hayamizu

Department of Physics, Faculty of Science, Nagoya University,
Chikusa-ku, Nagoya 464 8602, Japan

Abstract. The dynamical stripes of separated spin and charge densities were investigated in La$_{2-x}$Sr$_x$CuO$_4$ utilizing the high frequency response of Raman scattering. The two-magnon peak splits into double peaks below about 300 - 350 K at $0.035 \leq x \leq 0.06$ and $x = 0.115$ due to the formation of the stripe structure. The spectra recover metallic electronic response below about 1100 cm^{-1}. It suggests that one observes the instantaneous picture of magnetic excitations above this characteristic frequency of the stripe motion and the averaged picture of the fluctuating charge stripes below it. The threshold frequency is assigned to the two-phonon energy of the breathing mode in which a hole is migrating synchronously with the vibration between the anti-phase CuO$_6$ pair. At high temperature or at the carrier concentration less relating to the stripe structure, the influence of this phonon mode is reserved in the B_{2g} spectra. On the other hand the B_{1g} spectra include the influence of only magnetic excitations. The B_{1g} and B_{2g} spectra correspond to the electronic excitations along the $(0, 0)$ - $(\pi, 0)$ and the $(0, 0)$ - (π, π) directions in the k-space, respectively. The anisotropic spectra in the k-space can be interpreted as the anisotropic charge propagation, that is, the charge transfer along the Cu-O-Cu direction excites magnons, while the charge transfer along the diagonal direction excites phonons. This anisotropic charge transfer mechanism gives fundamental properties to the high T_c superconductors.

INTRODUCTION

The stripe structure was observed first by Tranquada *et al.* in La$_{1.48}$Nd$_{0.4}$Sr$_{0.12}$CuO$_4$ [1]. This phase is related to the suppression of the superconductivity at the carrier concentration of 1/8 holes / Cu atom and the low temperature tetragonal phase (LTT, P42/ncm) [2-4]. In this phase carriers are sited on the one-dimensional domain walls of antiferromagnetic spin stripes. In the case of La$_{2-x}$Sr$_x$CuO$_4$ (LSCO) the decrease of T_c is only a few degree and the static LTT phase is not observed. The magnetic order has been observed below 32 K at $x = 0.115$ by nuclear magnetic resonance (NMR) and nuclear quadrupole resonance (NQR) [5-7]. The spatially modulated dynamical spin correlation has been observed at $x = 0.05$ - 0.25 as incommensurate magnetic peaks by neutron scattering [8-13]. The direction of the stripe changes from the diagonal direction on the CuO$_2$ lattice in the spin glass composition ($x = 0.02$ - 0.05) to the vertical direction in the superconducting composition ($x \geq 0.055$). The inter-domain wall distance increases as $l = a/2x$ at $x < 1/8$ and becomes constant above it, where a is the Cu-Cu atomic distance. The domain walls are half-occupied by holes at $x < 1/8$.

CP554, *Physics in Local Lattice Distortions*, edited by H. Oyanagi and A. Bianconi
© 2001 American Institute of Physics 1-56396-984-X/01/$18.00

The present Raman scattering experiment disclosed the clear split of the two-magnon peak into double peaks at about $2J$ and $3J$, where J is the exchange interaction energy in the undoped antiferromagnet La_2CuO_4 (LCO). Furthermore the metallic spectra are recovered below about 1100 cm^{-1}. It indicates that one observes the instantaneous picture of the magnetic stripe above 1100 cm^{-1} and the averaged metallic picture of fluctuating charge stripes below it. The threshold energy is assigned to twice the energy of the breathing mode coupled with the hole migration between two anti-phase CuO_6 clusters. At high temperatures and/or at the carrier concentrations where the stripe structure disappears, the B_{1g} spectra loose the influence of this phonon mode, but the B_{2g} spectra preserve the influence of this mode. These different electronic excitation spectra indicate the anisotropic conducting mechanism that the charge transfer along the diagonal direction in the CuO_2 plane is accompanied by the phonon excitations, and the charge transfer along the Cu-O-Cu direction is accompanied by the magnetic excitations.

EXPERIMENTAL PROCEDURE

The single crystals of LSCO were synthesized by the traveling solvent floating zone method utilizing an infrared radiation furnace with four elliptic mirrors (Crystal system, FZ-T-4000). The Néel temperature T_N is 293 K at $x = 0$. The superconducting transition temperatures T_c's are 12 K, 33 K, 33 K, 42 K, 32 K, and 13 K for $x = 0.06$, 0.1, 0.115, 0.15, 0.2, and 0.25, respectively. Raman spectra were measured on fresh cleaved surfaces in a quasi-back scattering configuration utilizing a triple monochromator (JASCO, NR-1810), a liquid nitrogen cooled CCD detector (Princeton, 1100PB), a 5145 Å Ar-ion laser (Spectra Physics, stabilite 2017). The laser beam of 10 mW was focused on the area of 50 μm × 500 μm.

RAMAN SPECTRA IN THE STRIPE PHASE

Fig. 1 shows the temperature dependence of the B_{1g} and B_{2g} Raman spectra. The 3213 cm^{-1} peak in the B_{1g} spectrum at $x = 0$ and 5 K is the two-magnon scattering peak. The sharp peaks below 700 cm^{-1} are due to one-phonon scattering, the peaks from 850 to 1450 cm^{-1} are due to the resonant two-phonon scattering. When carriers are doped, the two-magnon peak shifts to low energy monotonically at 300 K, but at low temperatures the two-magnon peaks split into double peaks at $x = 0.035$, 0.05, 0.06, and 0.115. The B_{2g} two-magnon peak is observed at 4128 cm^{-1} at 300 K in LCO, which is about 4/3 times the B_{1g} two-magnon peak energy. Both spectra becomes similar in the stripe phase at $x = 0.05$, 0.06, and 0.115 at low temperatures. It indicates that the symmetry is relaxed in the stripe structure. The B_{1g} and B_{2g} spectra are obtained in the $(E_i, E_s) = (x, y)$ and (a, b) polarization configurations, respectively, where E_i and E_s are the electric fields of the incident and scattered light. a and b are parallel to the Cu-O-Cu directions and x and y are the intersecting directions with them at an angle of $1/4\pi$.

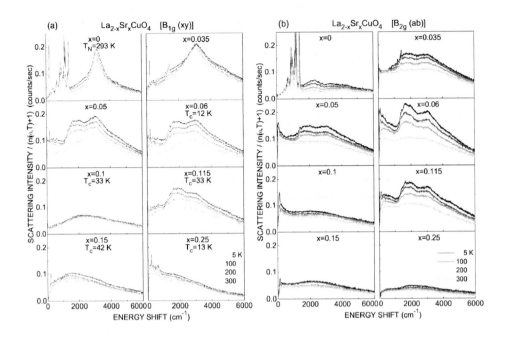

FIGURE 1. The (a) B_{1g} and (b) B_{2g} Raman spectra in $La_{2-x}Sr_xCuO_4$.

The split of the two-magnon peak into double peaks at about 3050 cm^{-1} and 1900 cm^{-1} can be explained as follows. The two-magnon Raman scattering in the $S = 1/2$ antiferromagnet is caused by the exchange of nearest neighbor spins. When the two-magnon scattering process occurs within the spin stripe, the exchange energy increases at six bonds, which corresponds to the increase of energy $3J$. When the process occurs in contact with the charge stripe, the number of bonds with increased exchange energy is four on the assumption that the magnetic moment on the charge stripes is zero. The increased energy is $2J$ in consistent with experimental observation that the energy of the lower energy peak is 2/3 times of the higher energy peak. The constant energies of the split two-magnon peaks at $0.035 \le x \le 0.06$ and $x = 0.115$ indicates that the carriers are not included in the spin stripes.

The small components of the split two-magnon peaks are observed at all other carrier concentrations at low temperatures. These onset temperature is about 300 K for $x = 0.035$, 0.05, 0.06, 350 K for $x = 0.115$, 100 K for $x = 0.1$ and 0.15. At $x = 0.2$ and 0.25 the small sign of the higher energy peak is noticeable at 300 K, but it does not increase at low temperatures. The onset temperatures at $0.035 \le x \le 0.06$ and $x = 0.115$ are much higher than those obtained from low frequency probes such as NMR, NQR, and neutron scattering [6, 7, 10, 12, 13]. These results indicate that the high frequency fluctuation of stripes starts at much higher temperatures than reported. The frequency of the fluctuation decreases as temperature decreases and the quasi-static component appears at the temperatures reported so far.

POHONON-ASSISTED STRIPE MOTION

The Raman spectra in the stripe phase at $0.035 \leq x \leq 0.06$ and $x = 0.115$ at low temperatures has dual properties; the split two-magnon spectra above 1100 cm^{-1} and the metallic electronic spectra below it. This duality can be interpreted by the model that the charge stripes are fluctuating at the frequency corresponding to 1100 cm^{-1}. One observes the instantaneous picture of magnetic scattering above this frequency and averaged picture of fluctuating charge stripes below it. This energy is found to be just one of the two-phonon peak energy.

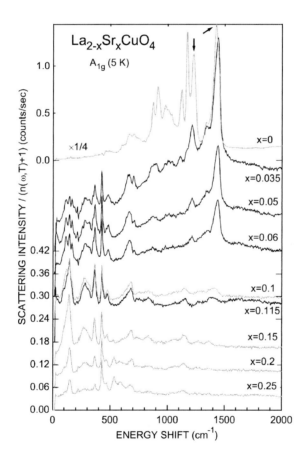

FIGURE 2. The A_{1g} spectra in $La_{2-x}Sr_xCuO_4$ at 5 K. The spectra in the stripe phase are shown by black curves and the others by gray curves. The arrows indicate the breathing modes with and without a hole.

Fig. 2 shows the carrier concentration dependence of the A_{1g} spectra at 5 K. The two-phonon scattering from 700 cm^{-1} to 1400 cm^{-1} is strongly enhanced by the resonant Raman effect at $x = 0$. The intensity decreases rapidly as the carrier concentration increases. The 1146 cm^{-1} and the 1363 cm^{-1} peaks are dominant at $x = 0.15$. The relative intensity of the lower energy peak increases as the carrier concentration increases. This suggests that the lower energy mode is related to carriers.

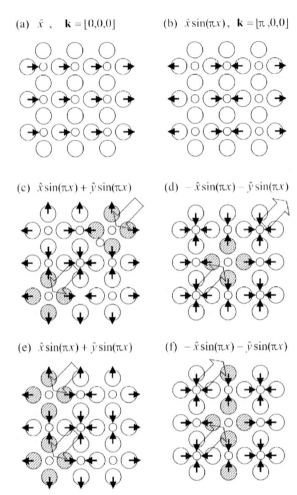

(a) $\hat{x}$, $\mathbf{k} = [0,0,0]$ (b) $\hat{x}\sin(\pi x)$, $\mathbf{k} = [\pi,0,0]$

(c) $\hat{x}\sin(\pi x) + \hat{y}\sin(\pi x)$ (d) $-\hat{x}\sin(\pi x) - \hat{y}\sin(\pi x)$

(e) $\hat{x}\sin(\pi x) + \hat{y}\sin(\pi x)$ (f) $-\hat{x}\sin(\pi x) - \hat{y}\sin(\pi x)$

FIGURE 3. (a) Displacement of the LO phonon mode at (0, 0, 0), and (b) at (π, 0, 0), (c)-(f) combined LO phonon modes at (0, 0, 0) and (π, 0, 0). The small circles represent copper atoms and the large circles oxygen atoms in the CuO$_2$ plane. The CuO$_6$ clusters vibrating in the anti-phase breathing mode are located along the diagonal directions. Four hatched oxygen atoms include one hole. The open arrow indicates the motion of a hole. The paired CuO$_6$ clusters joined by the open arrow are vibrating at about 550 cm^{-1}.

The energies of these modes, 1146 cm^{-1} and 1363 cm^{-1}, coincide exactly with twice the energies of the anomalous LO phonon branches observed by neutron scattering at 10 K by McQueeney *et al.* [14]. This LO phonon branch exchanges the dominant intensity at (0.5π, 0, 0) from the almost dispersionless branch of 85 meV at (0, 0, 0) - (0.5π, 0, 0) to the almost dispersionless 70 meV branch at (0.5π, 0, 0) - (0.8π, 0, 0).

It is known in the bimolecular crystals that the energy of the anti-phase vibration of the breathing mode is reduced, if a carrier migrate between the two molecules synchronously with the vibration [15, 16]. Fig. 3(a) and (b) shows the atomic displacement for the LO modes at (0, 0, 0) and (π, 0, 0). Figs. (c)-(f) are the displacement for the combined LO modes at (π, 0, 0) and (0, π, 0). The displacement of four oxygen atoms around a copper atom is the breathing mode type (four oxygen atoms simultaneously go away from a copper atom or approach a copper atom) or the quadrupole type (two oxygen atoms go away and two approach a copper atom). The direction connecting two anti-phase breathing vibrations is the diagonal direction. The diagonal charge stripe is shown in (c) and (d) and the vertical charge stripe is show in (e) and (f). The four hatched oxygen atoms around a copper atom has one hole. The hole moves from the shrunk CuO_6 cluster to the inflated CuO_6 cluster as shown by the open arrow. The hole goes back after the phase π. Holes are located at every two CuO_6 units for both diagonal and vertical charge stripes in consistent with the neutron scattering experiments at $x \leq 1/8$ [11, 13]. The breathing mode is not formed by the combination of the (0, 0, 0) LO phonon modes with *a*- and *b*-directional displacements. Therefore the low energy mode coupled with charge transfer disappears around (0, 0, 0) in consistent with the neutron scattering experiment [14].

The energy of the local lattice vibration in the CuO_6 clusters associated with a moving hole is assigned to 600 cm^{-1} (the two-phonon energy is 1200 cm^{-1}) and the energy without a hole is assigned to 700 cm^{-1} (the two-phonon energy is 1400 cm^{-1}). The 600 cm^{-1} mode increases in number as the carrier concentration increases in consistent with the increase of the relative intensity of the lower energy peak between 1146 cm^{-1} and 1363 cm^{-1} peaks as shown in Fig. 2. The metallic conductivity is induced by the movement of the hole to a different CuO_6 cluster from the paired CuO_6 clusters.

ANISOTROPIC CHARGE TRANSFER MECHANISM
IN THE *K*-SPACE

The Raman spectra in the stripe phase are characterized by twice the energy of the breathing mode coupled with a migrating hole. The influence of this mode to the spectra is not limited to the stripe phase. Even if charged pairs of CuO_6 clusters are distributed randomly in the CuO_2 plane, the influence is expected to appear in the spectra only when charges are coupled with the phonon mode. In practice the decrease of the scattering intensity is observed below around 1100 cm^{-1} at 300 K, where the stripe structure almost disappears, in the B_{2g} spectra from $x = 0.05$ to 0.25.

On the other hand the B_{1g} spectra at 300 K have no effect of this mode and the scattering intensity is depleted below the two-magnon peak energy whose energy decreases as the carrier concentration increases.

The B_{1g} spectra correspond to the excitation along the (0, 0) - (π, 0) and the B_{2g} along the (0, 0) - (π, π) [17]. Thus the excitation spectra along (0, 0) - (π, 0) are depleted below the two-magnon peak and those along (0, 0) - (π, π) are depleted below the two-phonon energy of the breathing mode. This directional dependence in the k-space is consistent with the experiments of angle resolved photoemission spectroscopy [18-22]. These anisotropic excitation spectra can be interpreted as follows. When a hole moves into the nearest neighbor CuO_6 site along the Cu-O-Cu direction, the magnetic energy increases, because the spins at copper atomic sites are opposite. On the other hand when a hole modes into the diagonal CuO_6 site, the magnetic energy does not increase, but the hole excites the breathing phonon mode. This phonon mode is excited only by the diagonal charge transfer, because the pairs of anti-phase CuO_6 clusters are sited along the diagonal directions as shown in Fig. 3.

The same kinds of anisotropic charge transfer spectra are also observed in the Raman spectra of $Bi_2Sr_2Ca_{1-x}Y_xCu_2O_{8+\delta}$ [23], although the depletion in the B_{2g} spectra is much smaller in BSCYCO then in LSCO. It indicates that the electron-phonon interaction is larger in LSCO then in BSCYCO. It is consistent with the much larger two-phonon scattering intensity in LSCO than in BSCYCO.

CONCLUSIONS

The dynamical properties of the stripe structure were investigated by virtue of the high frequency response of Raman scattering. The high frequency fluctuation of stripes starts at 300 - 350 K which is much higher than the reported onset temperatures. The stripe phase is observed at $x = 0.115$ (~ 1/8) and also $0.035 \leq x \leq 0.06$ which is the transition region from insulator to metal. The charges on the stripes are vibrating in cooperation with the pairs of anti-phase breathing modes. The electronic conductivity is induced by changing the partner CuO_6 cluster. This electron-phonon interaction is retained even if the stripe structure disappears at high temperature and/or carrier concentrations less relevant to the stripe structure. In such a case the influence of the phonon mode appears in the B_{2g} spectra. On the other hand the B_{1g} spectra include the influence of only magnetic excitations. This anisotropic Raman spectra indicate that the charge transfer along the diagonal direction in the CuO_2 plane excites the breathing phonon mode and the charge transfer along the Cu-O-Cu direction induces the magnetic excitations. This anisotropic charge transfer mechanism gives the fundamental properties of electronic excitations in high T_c superconductors. The anomalous ARPES spectra in LSCO may be related to this problem [24].

ACKNOWLEDGMENTS

The authors thank K. Takenaka for the characterization of single crystals. This work was supported by CREST of the Japan Science and Technology Corporation.

REFERENCES

1. Tranquada, J.M. *et al.*, *Nature* **375**, 561 (1995).
2. Kumagai, K. *et al.*, *J. Mag. Mag. Materials* **76 & 77**, 601 (1988).
3. Moodenbaugh, A.R. *et al.*, *Phys. Rev.* B **38**, 4596 (1988).
4. Axe, J.D. *et al.*, *Phys. Rev. Lett.* **62**, 2751 (1989).
5. Ohsugi, S. *et al.*, *J. Phys. Soc. Jpn.* **63**, 2057 (1994).
6. Goto, T. *et al.*, *J. Phys. Sco. Jpn.* **63**, 3494 (1994).
7. Hunt, A.W. *et al.*, *Phys. Rev. Lett.* **82**, 4300 (1999).
8. Cheong, S-W. *et al.*, *Phys. Rev. Lett.* **67**, 1791 (1991).
9. Thurston, T.R. *et al.*, *Phys. Rev.* B **46**, 9128 (1992).
10. Suzuki, T. *et al.*, *Phys. Rev.* B **57**, R3229 (1998).
11. Yamada, K. *et al.*, *Phys. Rev.* B **57**, 6165 (1998).
12. Kimura, H. *et al.*, *Phys. Rev.* B **59**, 6517 (1999).
13. Wakimoto, S. *et al.*, *Phys. Rev.* B **60**, R769 (1999).
14. McQueeney, R.J. *et al.*, *Phys. Rev. Lett.* **82**, 628 (1999).
15. Rice, M.J. *et al.*, *Phys. Rev. Lett.* **39**, 1359 (1977).
16. Rice, M.J., *Solid State Commun.* **31**, 93 (1979).
17. Devereaux, T.P., and Kampf, A.P., *Phys. Rev.* B **59**, 6411 (1999).
18. Marshall, D.S. *et al.*, *Phys. Rev. Lett.* **76**, 4841 (1996).
19. Loeser, A.G. *et al.*, *Science* **273**, 325 (1996).
20. Shen, Z.-X., and Schrieffer, J.R., *Phys. Rev. Lett.* **78**, 1771 (1997).
21. Ino, A. *et al.*, *J. Phys. Soc. Jpn.* **68**, 1496 (1999).
22. Ino, A. *et al.*, *Phys. Rev.* B **62**, 4137 (2000).
23. Sugai, S., and Hosokawa, T., *Phys. Rev. Lett.* **85**, 1112 (2000).
24. Zhou, X.J. *et al.*, *Science* **286**, 268 (1999).

Resistivity Anomalies in Polaronic Systems

M. Zoli

Istituto Nazionale di Fisica della Materia - Dipartimento di Matematica e Fisica,
Università di Camerino, 62032 Camerino, Italy. e-mail: zoli@campus.unicam.it

Abstract. Different classes of physical systems with sizeable electron-phonon coupling and lattice distortions present anomalous resistivity behaviors versus temperature. We study a molecular lattice Hamiltonian in which polaronic charge carriers interact with non linear potentials provided by local atomic fluctuations between two equilibrium sites. A path integral model is developed to select the class of atomic oscillations which mainly contributes to the partition function and the electrical resistivity is computed in a number of representative cases. We argue that the common origin of the observed resistivity anomalies lies in the time retarded nature of the polaronic interactions in the local structural instabilities.

Theoretical investigation on the physics of polarons has become intense after that a polaronic mechanism has been proposed [1] as a viable possibility to explain high T_c superconductivity with its peculiar transport properties. Evidence has been recently provided [2] for the polaronic nature of carriers also in perovskite manganites with colossal magnetoresistance (CMR) and a strong coupling of carriers to Jahn-Teller lattice distortions has been suggested [3] to explain the resistivity peak located at the Curie temperature. Although polarons are invoked to account for different resistivity behaviors in systems with some lattice distortions [4], the common origin, if any does exist, of such behaviors remains unexplained and a unifying theory is lacking. In general, a strong electron-phonon coupling implies violation of the Migdal theorem and polaron collapse of the electron band with the appearance of time retarded interactions in the system. The path integral method has proved successful in dealing with this problem since a retarded potential naturally emerges in the exact integral action. Both the Fröhlich and the Holstein polaron properties have been studied by using path integrals techniques.

Here I propose a path integral approach to the problem of a polaron scattered by a local lattice instability arising from an anharmonic phonon mode. Small polarons are assumed to exist by virtue of the strong *overall* electron-phonon coupling and independently of the *single* anharmonic mode. As an example, the latter could be associated with the motion of apical oxygen atoms in $YB_2Cu_3O_7$ superconductor where a c-axis polarized high frequency phonon couples to in-plane charge carriers [5]. On the other hand, the highly correlated motion of in-plane holes and c-axis lattice vibrations could in *itself* give origin to a small polaron [6]. We model the local instability by a double well potential in its two state configuration, a Two Level System, and the one dimensional atomic path between the two equilibrium minima is taken as time dependent. Retardation effects are thus naturally introduced in the full partition function and the path integral method permits to derive the effective time (temperature) dependent coupling strengths which control the resistivity.

Our starting point is contained in the following Hamiltonian

$$H_0(\tau) = \bar{\varepsilon}(g)\tilde{c}^\dagger(\tau)\tilde{c}(\tau) + \sum_q \omega_q a_q^\dagger(\tau)a_q(\tau) + H_{TLS}(\tau)$$

$$(H_{TLS}(\tau)) = \begin{pmatrix} 0 & \lambda Q(\tau) \\ \lambda Q(\tau) & 0 \end{pmatrix}$$

$$H_{int}(\tau) = -2\lambda Q(\tau)\tilde{c}^\dagger(\tau)\tilde{c}(\tau); \quad Q(\tau) = -Q_0 + \frac{2Q_0}{\tau_0}(\tau - t_i). \qquad (1)$$

τ is the time which scales as an inverse temperature according to the Matsubara Green's function formalism. $H_0(\tau)$ is the free Hamiltonian made of: a) a polaron created (destroyed) by $\tilde{c}^\dagger(\tau)(\tilde{c}(\tau))$ in an energy band $\bar{\varepsilon}(g)$ whose width decreases exponentially by increasing the strength of the overall electron-phonon coupling constant g, $\bar{\varepsilon}(g) = Dexp(-g^2)$; b) a diatomic molecular lattice whose phonon frequencies ω_q are obtained analytically through a force constant approach; c) a Two Level System (TLS) in its symmetric ground state configuration due to an anharmonic phonon mode.

The interaction is described by $H_{int}(\tau)$ with λ being the coupling strength between TLS and polaron, τ_o is the bare hopping time between the minima of the TLS which are located $\pm Q_o$ and t_i is the instant at which the ith-hop takes place. One atomic path is characterized by the number $2n$ of hops, by the set of t_i ($0 < i \leq 2n$) and by τ_o. The closure condition on the path is given by: $(2n - 1)\tau_s + 2n\tau_0 = \beta$, where β is the inverse temperature and τ_s is the time one atom is sitting in a well. $\lambda Q(\tau)$ is the renormalized (versus time) tunneling energy which allows one to introduce the τ dependence in the interacting Hamiltonian. The full partition function of the system is obtained by integrating over the times t_i and summing over all possible even number of hops:

$$Z_T = Z_0 \sum_{n=0}^{\infty} \int_0^\beta \frac{dt_{2n}}{\tau_0} \cdots \int_0^{t_2 - \tau_0} \frac{dt_1}{\tau_0} \exp[-\beta E(n, t_i, \tau_0)]$$

$$\beta E(n, t_i, \tau_0) = L - (K^A + K^R) \sum_{i>j}^{2n} \left(\frac{t_i - t_j}{\tau_0} \right)^2 \qquad (2)$$

with $E(n, t_i, \tau_0)$ being the one path atomic energy. L, which is a function of the input parameters, is not essential here while the second addendum in Eq. (2) is *not local in time* as a result of the retarded polaronic interactions between successive atomic hops in the double well potential. K^A and K^R are the one path coupling strengths containing the physics of the interacting system. K^A (negative) describes the polaron-polaron attraction mediated by the local instability and K^R (positive) is related to the repulsive scattering of the polaron by the TLS. Computation of $E(n, t_i, \tau_0)$ and its derivative with respect to τ_0 shows that the largest contribution to the partition function is given by the atomic path with $\tau_s = 0$. The atom moving back and forth in the double well minimizes its energy if it takes the path with the highest τ_0 value allowed by the boundary condition, that is with $(\tau_0)_{max} = (2nK_BT)^{-1}$. This result, which is general, provides a criterion to determine the set of dominant paths for the atom at any temperature. Then, the effective interaction strengths $< K^A >$ and $< K^R >$ can be obtained as a function of T by summing over n the dominant paths contributions.

The lattice Hamiltonian is made of diatomic sites whose intramolecular vibrations can favor trapping of the charge carriers. The *intra*molecular frequency ω_0 largely influences the size of the lattice distortion associated with polaron formation while the dispersive features of the phonon spectrum controlled by the *inter*molecular couplings are essential to compute the polaron properties both in the ground state and at finite temperatures [7]. We take a simple cubic lattice model with first and second neighbors molecular sites interacting via a force constants pair potential. γ and δ are the intermolecular first and second neighbors force constants, respectively. Second neighbors couplings remove the phonon modes degeneracy (with respect to dimensionality) at the corners of the Brillouin zone thus permitting to estimate with accuracy the relevant contributions of high symmetry points to the momentum space summations. Let's define $\omega_1^2 = \gamma / M$ and $\omega_2^2 = \delta / M$, M being the reduced molecular mass.

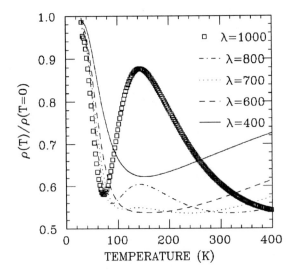

FIGURE 1. Electrical Resistivity normalized to the residual resistivity for five values of the polaron-TLS coupling λ. $g = 1$. The force constants which control the phonon spectrum are: $\omega_0 = 100meV$, $\omega_1 = 20meV$, $\omega_2 = 10meV$.

We turn now to compute the electrical resistivity due to the polaronic charge carrier scattering by the impurity potential with internal degree of freedom provided by the TLS. The input parameters of the model are six that is, the three molecular force constants, g, D and the bare energy λQ_0. Q_0 can be chosen as ~ 0.05 $\mathring{A}$ consistently with reported values in the literature on non magnetic TLS's which are known to exist in glassy systems, amorphous metals, A15 compounds and likely in some cuprate superconductors. The bare electronic band D is fixed at 0.1eV. In Fig. 1, we take a rather large phonon spectrum setting the system in moderately adiabatic conditions, an intermediate ($g = 1$) electron-phonon coupling regime which ensures polaron mobility [8] and tune the TLS-polaron coupling λ. The broad resistivity maximum developing at low temperatures clearly signals that the TLS's are at work here while, by increasing T, the hopping time shortens and the incoming polarons cannot distinguish any more the TLS internal degree of freedom, then diagonal scattering prevails and a *quasi linear* resistivity behavior emerges at $T > 200K$ at low λ values. A resistivity peak located at $T \sim 150K$ arises at $\lambda \geq 700$ $meV\,\mathring{A}^{-1}$ with height and width of the peak being strongly dependent on λ hence, on the TLS energy. The low T resistivity still displays the maximum at the unitary limit while the high T ($T > 300K$) behavior can be metallic like ($\lambda < 800meV\,\mathring{A}^{-1}$) or semiconducting like ($\lambda > 800meV\,\mathring{A}^{-1}$). $\lambda \sim 700 - 800meV\,\mathring{A}^{-1}$ corresponds to a TLS energy of $\sim 35 - 40meV$ which is comparable to the value of the bare polaron energy band $\bar{\varepsilon}$ ($g = 1$) $\sim 37meV$. In this picture, the resistivity peak has a structural origin and it can be ascribed to resonant TLS-polaron scattering with effective attractive and repulsive interaction strengths becoming of the same order of

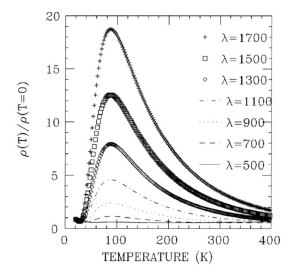

FIGURE 2. Electrical Resistivity normalized to the residual resisitivity for nine values of the polaron-TLS coupling λ. $g = 1$. $\omega_0 = 50meV$, $\omega_1 = 20meV$, $\omega_2 = 10meV$.

magnitude at $T \simeq 150K$. At larger λ values the resonance peak is higher and broader since an increasing number of incoming polarons can be off diagonally scattered by the TLS. However, the appearance of this many body effect mediated by the local potential does not change the position of the peak which, instead, can be shifted towards lower temperatures (see Fig. 2) by reducing the phonon frequencies. This effect is accompanied by a huge increase in the height of the resistivity peak due to the fact that the polaron effective mass is heavier when lower energy phonons build up the quasiparticle. Although our model does not account for the superconducting transition, the non metallic behavior at T larger than $\simeq 90K$ reminds of the anomalous c-axis resistivity observed in underdoped high T_c superconductors. $\overline{\omega} \simeq 75meV$ (as we take in Fig. 2) is the characteristic energy of c-axis polarized phonons due to apical oxygen vibrations which couple to the holes in the Cu-O planes in $YBa_2Cu_3O_{7-\delta}$. Anharmonic features of the oxygen modes have been recognized to be larger in underdoped samples and doping dependent polaron formation has been correlated to distortions of the oxygen environment. We believe that the semiconducting like ρ_c in underdoped high T_c superconductors can be ascribed to anharmonic potentials due to oxygen displacements strongly coupled via λ to polaronic carriers. To attempt a comparison with experiments we need to fix the residual resistivity. By extrapolating to $T = 0$ the normal state data on $YBa_2Cu_3O_{7-\delta}$ one derives $\rho\ (T = 0) \simeq 3m\Omega\text{-}cm$, hence the experimental peak value $\rho \simeq 20m\Omega\text{-}cm$ observed in $YBa_2Cu_3O_{6.87}$ can be reproduced in our model by $\lambda \simeq 1300meV\ \mathring{A}^{-1}$ which corresponds to a local mode energy of $\simeq 65meV$, in fair agreement with the measured energies of phonons strongly coupled to the charge carriers. A resistivity peak at $T \simeq 250K$ is observed in CMR materials and a splitting of the Mn states due to a local lattice distortion has been proposed as a possible mechanism. In our description, only extended polarons dragging a cloud of high frequency optical phonons can produce a peak at $T \simeq 200\text{-}250K$ which is however broader than the experimental one. High phonon frequencies imply reduced mass renormalization and, consistently, weak overall electron-phonon coupling values. Extended polarons move therefore in wide energy bands and the resonance effect can take place only if the TLS splitting energy is large enough. In fact, by taking $D = 0.5eV$ and $\lambda \simeq 5000meV\ \mathring{A}^{-1}$ we get the peak at $g \leq 1$ whereas, in a small polaron regime ($g > 1$) the resonance is lost and the peak is suppressed. Although more specific models accounting for magnetic field effects are required for the CMR systems, we argue that the strong coupling of polarons to structural distortions is likely relevant to their transport properties.

In conclusion, the path integral method provides a powerful tool to describe *non local (in time)* scattering of the charge carriers by *local (in space)* potentials due to structural distortions. Lattice polarons can either form in anharmonic potentials or exist by virtue of a sizeable overall electron-phonon coupling. We suggest that the retarded nature of the interactions is the *key* ingredient to explain some anomalous transport properties observed in real materials. A resistivity peak is obtained as a consequence of resonant scattering of polarons strongly coupled to local double well potentials.

REFERENCES

1. Alexandrov, A.S., and Mott, N.F., *Rep. Prog. Phys.* **57**, 1197 (1994).
2. Zhao, G.-m. *et al.*, *Nature* (London) **381**, 676 (1996).
3. Millis, A.J. *et al.*, *Phys. Rev. B* **74**, 5144 (1995).
4. Proceedings of the Conference on *Stripes, Lattice Instabilities and High T_c Superconductivity* edited by Bianconi, A., and Saini, N.L., *J. Supercond.* **10** (1997).
5. Timusk, T., Homes, C.C., and Reichardt, W., in *Anharmonic properties of High T_c Cuprates* edited by Mihailovi'c D. *et al.*, World Scientific, Singapore, 1995, pp.171.
6. Mustre de Leon, J. *et al.*, *Phys. Rev. B* **65**, 1675 (1990).
7. Zoli, M., *Phys. Rev. B.* **57**, 10555 (1998).
8. Zoli, M., *Phys. Rev. B* **61**, 14523 (2000).

Instability Criteria for Local Lattice Distortions in Cuprate Superconductors

N. Kristoffel[*] and M. Klopov[†]

[*]Institute of Physics, University of Tartu, Riia 142, 51014 Tartu, Estonia
[†]Tallinn Technical University, Ehitajate tee 5,19086 Tallinn, Estonia

Abstract. Green's function method is used to calculate destabilization criteria for local perturbations causing structural distortions in cuprate superconductors. The build-up of soft low-frequency dynamical resonances is traced. Chain O(1) a-axis and apex O(4) <110>-instabilities in YBa$_2$Cu$_3$O$_7$ have been investigated. The Cu out-of-plane c-axis destabilizations have been considered also for La$_2$CuO$_4$.

INTRODUCTION

Various structural distortions on bulk, mesoscopic and local level are known in cuprate superconductors. Macroscopic averaged crystal structures does not reflect the richness of events present on microscopic scale in these compounds. Most of these structural events are driven by the superconductivity stimulating doping.

The knowledge of the necessary strength of the perturbation which leads to a local lattice destabilization is useful in various aspects. One solves this traditional problem of local lattice dynamics by the usual Green's function method. The physical nature of localized perturbations acting in cuprates can be of various origin. Localization of extra hole carriers introduced by doping leads to essential redistribution of the electronic density and change the dynamical force constants. E.g. if there is a hole on an oxygen, its radius diminishes to 0.066 nm from 0.136 nm of O$_2^-$. Further, structural distortion in one subsystem can cause "structural pressure" on other structural elemens. In the case of impurities the "chemical pressure" is present. It must be held in mind that the Cu^{2+} is a Jahn-Teller ion and vibronic contributions to lattice dynamics can lead to essential structural consequences.

The Green's functions are defined as

$$G_{\alpha\beta}^{st}(\omega) = \frac{1}{N}\sum_{\vec{q},j}\frac{e_{\alpha}^{s}(\vec{q}j)e_{\beta}^{t}(\vec{q}j)}{\omega^2 - \omega^2(\vec{q}j)}e^{i\vec{q}\vec{r}_a}$$

with the phonon eigenvectors $e_{\alpha}^{s}(\vec{q}j)$ and frequencies $\omega(\vec{q}j)$ of the unperturbed crystal. The local dynamics in the perturbed region (force constant change matrix $\Delta\hat{D} \neq 0$) is given by the equation for the displacement field

CP554, *Physics in Local Lattice Distortions*, edited by H. Oyanagi and A. Bianconi
© 2001 American Institute of Physics 1-56396-984-X/01/$18.00

$$\hat{U} = \hat{G}\Delta\hat{D}\hat{U}.$$

The perturbed Green's function is expressed as

$$\hat{\psi} = (I - \hat{C}\Delta\hat{D})^{-1}\hat{G}.$$

The localized mode (if present) and quasilocal resonance frequencies follow from the condition

$$\mathrm{Re}\det\left|\hat{I} - \hat{G}(\omega_l^2)\Delta\hat{D}\right| = 0.$$

The case $\omega_l^2 \leq 0$ corresponds to dynamical instability and the displacement field of the resonance mode must be frozen into the lattice to restore its stability. The situation $\omega_l = 0$ determines the critical perturbation (force constants changes). The soft dynamical resonance is described by $\mathrm{Im}\hat{\psi}(\omega^2)$ which gives the spectral density of the squared mode displacement under interest.

The unperturbed lattice dynamics data we used are taken for $YBa_2Cu_3O_7$ from [1] and for La_2CuO_4 [2]. Effective force constants have been used. We denote the central ones by A and the noncentral ones by B. The used values (in N/m) are given for various bonds in Table 1.

TABLE 1.

Y	O(1)-Cu(1)	O(1)-O(4)	Cu(1)-O(4)	Cu(2)-O(4)	Cu(2)-O(2)	Cu(2)-O(3)
A				40.1	94.0	78.6
B	19.8	9.2	59.6	16.0	18.5	18.5
La	Cu-O(1)	Cu-O(2)				
A	47.0	25.7				
B	26.7	18.9				

First, we mention the result on chain oxygen O(1) a-axis off-center distortion in $YBa_2Cu_3O_7$ by weakening of its noncentral forces (in the same ratio) to two nearest Cu(1) and four O(4) apex ions [3]. Only O(1) and Cu(1) are allowed to move. The critical force constant decrease is moderate: 29.2%, and at this the sole O(1) a-axis shift forms the condensing mode. This result agrees with the observed [4,5] O(1)-distortion leading to zigzag chain fragments and points to O(1) position as the most unstable in the Y-system. Correspondingly the spectral distribution of O(1) a-axis displacement is most low-frequency concentrated one in this system.

The CuO_2 plane buckling in the Y-system creates compressed c-axis bonds of apex O(4) with its associated <110> off-centre displacement [6-11]. This destabilization has been investigated by us in [12]. The perturbation shortens the Cu(2)-O(4) bond by 0.1 Å [7], however according to our estimation the shortening of the Cu(1)-O(4) bond by 0.03 Å must be also taken into account. Noncentral force weakening by the O(4) c axis bonds compression creates a well expressed soft resonance which condenses at the total 40% perturbation for the <110>-displacement. The <100> direction instability has a slightly stronger criterion. The force constants change corresponding to the observed O(4) c-axis shortening can be calculated using the O^{2-}-Cu^{2+} short range potential from [13]. We have found the value –30.6 N/m which is very close to the necessary 40% mentioned joining in this manner the theory and experiment. This apex position distortion starts at the same doping level where T_c and the plane buckling become maximal [16].

The most essential structural events in cuprates are connected with the superconductivity playground CuO_2 planes [9, 10, 14, 15]. Doped carriers induce here striped phase separation. In the distorted regions of the material the plane becomes corrugated, the Cu and O ions are no longer in the same plane. The largest distortive displacements are connected with the out-of plane c-axis shifts of the ions.

In La_2CuO_4 the spectral density of the Cu c-axis shift is concentrated in the low-frequency region and this ion is expected to be easily destabilized out-of plane. The defect region incorporates the building-block CuO_6 octahedra. The distortion mode basis includes c-axis displacements of the central Cu, plane O(1) and apex O(2) ions. Only Cu-O(1) bonds are supposed to be perturbed, e.g. by the oxygens radii reduction through a shared (trapped) hole. Among the three A_{2u} (D_{4h} symmetry) modes a soft low-frequency resonance is present (see Fig. 1). It condenses for 14% noncentral force weakening.

In this mode the plane Cu and O(1) move in the same direction and O(4) oppositely. However with the force softening the displacement of the Cu dominates even more, so that at the instability practically only the Cu is displaced. In the case of the sole Cu

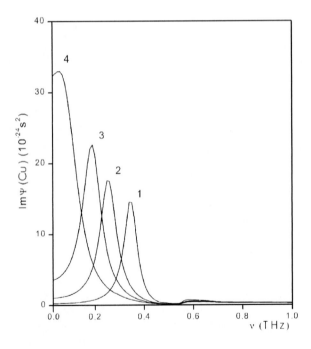

FIGURE 1. The A_{2u} low-frequency resonance softening for Cu-O(1) noncentral force constant weakening in La_2CuO_4 by 1 -- 12.5; 2 -- 13.2; 3 -- 13.6; 4 -- 14 (%).

displacement with perturbed bonds to all the 6 oxygens (note that $A_2 \approx B_1$) the instability is reached at only 7% force softening. These results illustrate the relatively soft c-axis dynamics of Cu in La_2CuO_4 and the light deformability of the CuO_6 octahedra by vibronic coupling between the electronic states of this complex.

An analogous calculation has been made also for the Cu(2) dynamics in the CuO_2 plane of the Y-system. For the Cu c-axis destabilization a 52% softening of noncentral forces to nearest neighbors (practically equal for O(2) and O(3)) is necessary. The movement of these ions in the resonance mode is opposite and at its condensation only the Cu-displacement survives. The rigidity of the O-ions network will add to the instability criterion 20%. One concludes that the plane Cu c-axis dynamics in the Y-system is much more stabile as compared with the La-system.

In the case if one diminishes the central and noncentral force constants (in the same ratio) for the Cu a-axis displacement in the plane one finds a relatively moderate destabilization criterion of 32% force softening.

One must have in mind that for the local symmetry breaking configurations considered multiple equivalent ones exist according to the symmetry of the nondistorted structure. This circumstance introduces the possibility of "two-center behaviour", etc.

This work has been supported by the Estonian Science Foundation Grant No 3591.

REFERENCES

1. Yim, K.K., Oitmaa, J., and Elcombe, H.M., *Australian J. Phys.* **45**, 221 (1992).

2. Choudhuri, N., Rao, K.R., and Shaplot, S.L., *Physica C* **171**, 567 (1990).

3. Kristoffel, N., and Klopov, M., *Phys. Rev. B* **54**, 9074 (1996).

4. Müller, V. et al. *Solid State Commun.* **72**, 997 (1989).

5. Cannelli, G. et al., *Phys. Rev. B* **45**, 931 (1992).

6. Arai, M. et al., *Phys. Rev. Lett.* **69**, 359 (1992).

7. Sullivan, J.D. et al., *Phys. Rev.* **48**, 10638 (1993).

8. Schweiss, P. et al., *Phys. Rev. B* **49**, 1387 (1994).

9. Bianconi, A. et al., *Phys. Rev. B* **54**, 12018 (1996).

10. Egami, T., *J. Low Temp. Phys.* **105**, 791 (1996).

11. Sharma, R.P. et al., *Phys. Rev. Lett.* **77**, 4624 (1996).

12. Kristoffel, N., and Klopov, M., *Physica C* **307**, 261 (1998).

13. Iguchi, E. and Yonezawa, Y., *J. Phys. Chem. Solids* **51**, 313 (1990).

14. Jorgensen, J.P. et al., *Physica C* **185**, 184 (1991).

15. Tranquada, J., *Phys. Rev. B* **54**, 7489 (1996).

16. Kristoffel, N., *phys. stat. solidi b* **213**, R9 (1999).

Dynamic Lattice Distortions
in High-T_c Cuprates

T. Sakaï[*] and D. Poilblanc[†]

[*]Faculty of Science, Himeji Institute of Technology, Hyogo 678-1297, Japan
[†]Laboratoire de Physique Quantique, Université Paul Sabatier, 31062 Toulouse, France

Abstract. The effect of the dynamic vibrations of the in-plane oxygen atoms on the hole pairing and charge ordering in the high-T_c cuprates is investigated based on generalized 2D t-J Holstein models. In-plane (breathing) and out-of-plane (buckling) vibrational modes are considered. The exact diagonalization of the finite clusters revealed that the buckling vibrations stabilize the d-wave hole pairing, while the breathing ones suppress it. It is also found that the breathing modes possibly play an important role in some static order formations, while the buckling ones yield no effect. In addition we discuss on the polaron states with some phonon-dressed hole operators.

INTRODUCTION

The local lattice distortion gives rise to various interesting effects on the strongly correlated electron systems like high-T_c cuprate superconductors. One of the most important effects is the polaronic self localization [1] inducing CDW and SDW instabilities [2], which might lead to the charge stripe order. In fact the superlattice due to polaron ordering was observed in the electron diffraction experiments of $La_{2-x}Sr_xNiO_{4+y}$ [3]. On the other hand, the dynamical lattice distortion also plays the key role on the pair formation in the conventional BCS superconductivity. Although the hole-hole attraction is supposed to be mainly mediated by the strong antiferromagnetic correlation in the high-T_c superconductors, it is still possible that the phonon assists the hole pairing. The observed oxygen isotope effect in the high-T_c cuprates [4] indicates some contribution of the electron-phonon interaction to the superconductivity. The interplay between the strong electronic correlation and dynamical lattice distortion still remains an open question.

We describe the electronic properties of the high-T_c cuprates by the t-J model and consider the two types of the vibrations of the in-plane oxygen atoms of the CuO_2 plane; (a) in-plane breathing modes and (b) buckling modes, as shown schematically in Fig. 1. For the latter one the electron-phonon interaction becomes linear, when the equilibrium position of the oxygen atom lies away from the CuO_2 plane by u_0 in Fig. 1(b). Such a buckling structure is realized in $YBa_2Cu_3O_{7-\delta}$. The two modes of interaction yield an effective nearest-neighbor (NN) hole-hole repulsion (a) and attraction (b) respectively, in the antiadiabatic limit. Some mean-field-level approximations revealed that the breathing mode suppresses the d-wave superconductivity [5], while the buckling one can enhance it [6,7]. In the present

CP554, *Physics in Local Lattice Distortions*, edited by H. Oyanagi and A. Bianconi
© 2001 American Institute of Physics 1-56396-984-X/01/$18.00

paper, we investigate the polaronic effects of the two modes on some static orders and hole pairing in the high-T_c cuprates, using the exact diagonalization of small t-J phonon clusters [8]. We also discuss on the properties of the polaron state with some phonon-dressed hole operators.

MODELS AND METHODS

We consider a generalization of the t-J-Holstein Hamiltonian

$$H = -t \sum_{<i,j>,\sigma}(\tilde{c}^\dagger_{j,\sigma}\tilde{c}_{i,\sigma} + \tilde{c}^\dagger_{i,\sigma}\tilde{c}_{j,\sigma}) + J \sum_{<i,j>}(S_i \cdot S_j - \frac{1}{4}n_i n_j)$$
$$+ \sum_{i,\delta}(\frac{p^2_{i,\delta}}{2m} + \frac{1}{2}m\Omega^2 u^2_{i,\delta}) + g\sum_{i,\delta}u_{i,\delta}(n^h_i \mp n^h_{i+\delta}) \tag{1}$$

where $\tilde{c}^\dagger_{i,\sigma}$ is the usual hole creation operator, n_i and n^h_i are the electron and hole local densities respectively, m is the oxygen ion mass, Ω is the phonon frequency and $\delta = $ **x, y** differentiates the bonds along the x- and y-direction respectively. The sign $-(+)$ in the last term corresponds to the breathing (buckling) mode.
Throughout, energies are measured in unit of the hopping integral t. The electron-phonon g-term involves the coupling of each copper hole with the displacements of the four neighboring oxygens $u_{i,\delta}$ and $u_{i-\delta,\delta}$. We re-write the electron-phonon interaction in the boson representation of the phonons,

$$H_{e-ph} = \Omega \sum_{i,\delta}(b^\dagger_{i,\delta}b_{i,\delta} + \frac{1}{2}) + \lambda_0 \sum_{i,\delta}(b_{i,\delta} + b^\dagger_{i,\delta})(n^h_i \mp n^h_{i+\delta}) \tag{2}$$

where $\lambda_0 = \sqrt{1/2m\Omega}$. Since the phononic Hilbert space has an infinite dimension, we truncate it to a finite number of bosonic states i.e. $b^\dagger_{i,\delta}b_{i,\delta} \leq n_{ph}$ at each oxygen site. We restrict ourselves to $n_{ph} = 1$. Calculation of the one-hole ground state (GS) energy

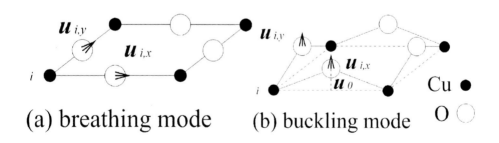

(a) breathing mode   (b) buckling mode   O

FIGURE 1. Schematic lattice displacements of breathing (a) and buckling (b) modes in the CuO$_2$ plane.

of the 2×2 cluster for several values of n_{ph} [8] suggested that the one-phonon calculation is a good approximation in the weak-coupling region ($\lambda_0 \leq 0.3$) for $J = 0.3$ and $\Omega = 0.2$. This truncation procedure enables us to study a $\sqrt{8} \times \sqrt{8}$-unit cells cluster with all the phonon modes (16 modes). We investigate the one and two-hole GS of hamiltonian (1) in a regime ($0.3 \leq J \leq 0.4$) where, in the absence of phonons, the two-hole pairing state is stabilized by the antiferromagnetic correlation, and we take a realistic phonon frequency $\Omega = 0.2$. Since the $\sqrt{8} \times \sqrt{8}$ cluster with periodic boundary conditions has the C_{4v} symmetry, we concentrate on the lowest state with the $d_{x^2-y^2}$ symmetry as the two-hole GS.

HOLE MOTION

The lattice distortions make the effective mass of the hole larger and lead to self-localization of the hole. The breathing lattice deformations around the hole raise the potential at the NN Cu sites, while the buckling ones lower it. Thus the mass renormalization due to the buckling mode is expected to be smaller. The absolute value of the kinetic energy in the one-hole GS is shown as a function of λ_0 in Fig. 2(a).

$$E_{kin} = \langle -t \sum_{<i,j>,\sigma} (\tilde{c}^{\dagger}_{j,\sigma} \tilde{c}_{i,\sigma} + \tilde{c}^{\dagger}_{i,\sigma} \tilde{c}_{j,\sigma}) \rangle \qquad (3)$$

$|E_{kin}|$ decreases significantly with increasing λ_0 for the breathing mode, while it does not change significantly for the buckling mode. It implies that only the breathing mode leads to a polaronic self-trapping process. The difference between the two modes is also clear from the behavior of the spin structure factor in the one-hole GS

$$S_S(\pi, \pi) = \langle (\sum_i (-1)^{(ix+iy)} S^z_i)^2 \rangle \qquad (4)$$

shown in Fig. 2(b). A significant increase of $S_s(\pi, \pi)$ occurs around $\lambda_0 = 0.1$ almost independently of J for the breathing mode. The agreement between the behaviors of $S_s(\pi, \pi)$ and E_{kin} vs λ_0 suggests that the increase of the effective mass of the hole due to a polaronic self-localization effect leads, for the breathing mode, to an enhancement of the antiferromagnetic spin correlation towards a static order.

HOLE PAIRING

The two-hole binding energy is a good probe to test the formation of pair of holes. It is defined as

$$\Delta_2 = E_0^{(2)} + E_0^{(0)} - 2E_0^{(1)} \qquad (5)$$

where $E_0^{(p)}$ is the GS energy for a system with $N_h \equiv \sum_i n_i^h = p$. $E_0^{(0)}$ corresponds to the energy of the antiferromagnetic background. A negative value of Δ_2 indicates the stability of a two-hole bound state, as was established for the pure t-J model [9]. Fig. 2(c), where Δ_2 is displayed as a function of λ_0, clearly shows that the buckling mode stabilizes the two-hole bound state while the breathing mode suppresses it. The effect of the electron-phonon interaction is to shift the boundary of the pairing phase of the t-J model: the buckling mode enlarges the phase toward small J while the breathing mode reduces it. The behavior of Δ_2 suggests the possibility that the buckling mode assists superconductivity in the high-T_c cuprates, while the breathing mode suppresses it. We note that, for the buckling mode, no self-trapping process occurs even in the two-hole state, since there is no significant decrease of the kinetic energy in Fig. 2(a). Thus the hole pair is not localized and can contribute to superconductivity.

POLARONIC QUASIPARTICLES

We propose phonon-dressed operators describing hole-polaron and bi-polaron states and discuss the quasiparticle weight associated to these operators. The larger weight means the better description of the quasiparticle.

First, we shall consider a single polaron operator describing the one-hole state with a momentum k. Since the size of the polaron is small, phonons at the nearest oxygen sites of the hole play the most important role and, hence, we only take them into

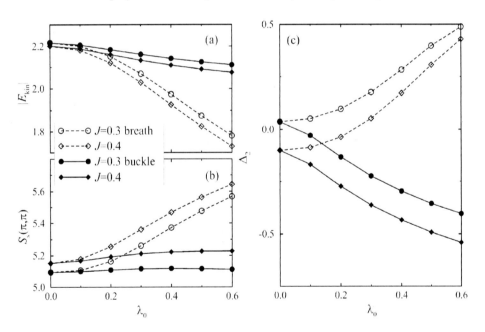

FIGURE 2. (a) Absolute value of the kinetic energy $|E_{kin}|$ and (b) spin structure factor $S_s (\pi, \pi)$ in the one-hole GS. (c) Two-hole binding energy Δ_2.

account. Thus, in our framework, up to 4 phonons could be included. The quasiparticle operator should have the same symmetry as the one-hole ground state. Under these conditions a general form of a composite polaron operator with momentum k is

$$\bar{c}_{k,\sigma} = \sum_i e^{ik \cdot i} \tilde{c}^\dagger_{i,\sigma} \left[\gamma^{(0)} + \sum_\alpha \gamma^{(1)}_\alpha \hat{\phi}^{(1)}_\alpha + \sum_\alpha \gamma^{(2)}_\alpha \hat{\phi}^{(2)}_\alpha + \cdots \right]_i \qquad (6)$$

where $\tilde{c}^\dagger_{i,\sigma}$ is the usual hole creation operator of the t-J model. $\hat{\phi}^{(n)}_\alpha$ is the symmetrized operator creating n phonons with the same local point group symmetry as the one-hole ground state and Σ_α corresponds to the sum over all such independent n-phonon configurations. [...]$_i$ means that the boson part is centered around site $\mathbf{i}$. The coefficients $\gamma^{(n)}_\alpha$ (independent of $\mathbf{i}$) are determined so as to optimize the quasiparticle weight

$$Z_{1h} = \frac{|\langle \Psi^{N-1}_0 | \bar{c}_{k,\sigma} | \Psi^N_0 \rangle|^2}{\langle \Psi^N_0 | \bar{c}^\dagger_{k,\sigma} \bar{c}_{k,\sigma} | \Psi^N_0 \rangle} \qquad (7)$$

where Ψ^N_0 and Ψ^{N-1}_0 are the Néel and one-hole ground states, respectively on the N-site cluster. We also define the operator $\bar{c}^{(n)}_{k,\sigma,\alpha} \equiv \sum_i e^{ik \cdot i} \tilde{c}^\dagger_{i,\sigma} [\hat{\phi}^{(n)}_\alpha]_i$. Since all the states $\bar{c}^{(n)}_{k,\sigma,\alpha} | \Psi^N_0 \rangle$ are orthogonal to each other, the coefficients can be obtained as

$$\gamma^{(n)}_\alpha = \frac{\langle \Psi^{N-1}_0 | \bar{c}^{(n)}_{k,\sigma,\alpha} | \Psi^N_o \rangle}{\langle \Psi^N_0 | \bar{c}^{(n)\dagger}_{k,\sigma,\alpha} \bar{c}^{(n)}_{k,\sigma,\alpha} | \Psi^N_0 \rangle} \qquad (8)$$

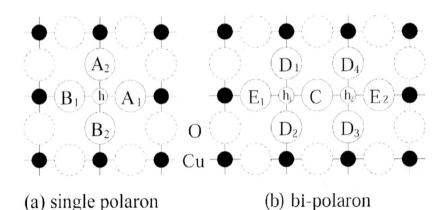

(a) single polaron   (b) bi-polaron

FIGURE 3. Schematic oxygen sites for the phonon dressing to construct the single polaron (a) and bi-polaron (b) operators.

106

The one-hole ground state of the $\sqrt{8}\times\sqrt{8}$ cluster has the momentum $\mathbf{k} = (\pi/2, \pi/2)$ and it is symmetric under the reflection with respect to $y = x$. Thus the independent sites for one-phonon dressing are A_1 and B_1 in Fig. 3(a). The two independent symmetrized one-phonon operators are written as $\phi_1^{(1)} = b_{A_1}^\dagger + b_{A_2}^\dagger$ and $\phi_2^{(1)} = b_{B_1}^\dagger + b_{B_2}^\dagger$. The four two-phonon operators can be given in the same way; $\phi_1^{(2)} = b_{A_1}^\dagger b_{A_2}^\dagger$, $\phi_2^{(2)} = b_{A_1}^\dagger b_{B_1}^\dagger + b_{A_2}^\dagger b_{B_2}^\dagger$, $\phi_3^{(2)} = b_{A_1}^\dagger b_{B_2}^\dagger + b_{A_2}^\dagger b_{B_1}^\dagger$ and $\phi_4^{(2)} = b_{B_1}^\dagger b_{B_2}^\dagger$. In a similar way we also define the three and four phonon-dressed hole operators. The quasiparticle weights up to n-phonon dressing ($n = 0, 1, 2, 3$, and 4) are shown in Fig. 4(a) and (b) for the breathing and buckling modes, respectively. It suggests that larger-n dressed operators describe the polaronic quasiparticle better for both modes. The polaron effect is revealed to be larger for the breathing deformation.

A bi-polaron operator can be constructed in a similar way. We start from the conventional $d_{x^2-y^2}$ BCS spin singlet operator for nearest-neighbor sites $\bar{\Delta}_d = \sum_i \sum_{l=0}^3 (-1)^l (\hat{R}_{\pi/2}(\mathbf{i}))^l \{\tilde{c}_{\mathbf{i},\uparrow}^\dagger \tilde{c}_{\mathbf{i+x},\downarrow}^\dagger\}$, where $\hat{R}_{\pi/2}(\mathbf{i})$ is the $\pi/2$-angle rotation of all the coordinates around the site $\mathbf{i}$. The bi-polaron operator is given by

$$\hat{\Delta}_d = \sum_\mathbf{i} \sum_{l=0}^3 (-1)^l (\hat{R}_{\pi/2}(\mathbf{i}))^l$$
$$\left\{ \tilde{c}_{\mathbf{i},\uparrow}^\dagger \tilde{c}_{\mathbf{i+x},\downarrow}^\dagger \left[\Gamma^{(0)} + \sum_\alpha \Gamma_\alpha^{(1)} \hat{\Phi}_\alpha^{(1)} + \sum_\alpha \Gamma_\alpha^{(2)} \hat{\Phi}_\alpha^{(2)} + \cdots \right] \right\}_\mathbf{i} \tag{9}$$

where each $\Phi_\alpha^{(n)}$ corresponds to an independent n-phonon creation operator. As previously, the coefficients $\Gamma_\alpha^{(n)}$ are determined by the optimization of the quasiparticle weight

$$Z_{2h} = \frac{|\langle \Psi_0^{N-2} | \bar{\Delta}_d | \Psi_0^N \rangle|^2}{\langle \Psi_0^N | \bar{\Delta}_d^\dagger \bar{\Delta}_d | \Psi_0^N \rangle} \tag{10}$$

where Ψ_0^{N-2} is the two-hole ground state. Thus, each coefficient $\Gamma_\alpha^{(n)}$ can be obtained from a formula similar to (8). Independent one-phonon dressing operators for $\bar{\Delta}_d$ are $\hat{\Phi}_1^{(1)} = b_C^\dagger$, $\hat{\Phi}_2^{(1)} = b_{D_1}^\dagger + b_{D_2}^\dagger + b_{D_3}^\dagger + b_{D_4}^\dagger$ and $\hat{\Phi}_3^{(1)} = b_{E_1}^\dagger + b_{E_2}^\dagger$, where the points C, D_1 and E_1 are defined in Fig. 3(b). There are eight independent two-phonon dressing operators;

$$\hat{\Phi}_1^{(2)} = b_C^\dagger b_{D_1}^\dagger + b_C^\dagger b_{D_2}^\dagger + b_C^\dagger b_{D_3}^\dagger + b_C^\dagger b_{D_4}^\dagger, \quad \hat{\Phi}_2^{(2)} = b_C^\dagger b_{E_1}^\dagger + b_C^\dagger b_{E_2}^\dagger,$$

$$\hat{\Phi}_3^{(2)} = b_{D_1}^\dagger b_{E_1}^\dagger + b_{D_2}^\dagger b_{E_1}^\dagger + b_{D_3}^\dagger b_{E_2}^\dagger + b_{D_4}^\dagger b_{E_2}^\dagger, \quad \hat{\Phi}_4^{(2)} = b_{D_1}^\dagger b_{E_2}^\dagger + b_{D_3}^\dagger b_{E_1}^\dagger + b_{D_4}^\dagger b_{E_1}^\dagger + b_{D_2}^\dagger b_{E_2}^\dagger,$$

$$\hat{\Phi}_5^{(2)} = b_{D_1}^\dagger b_{D_2}^\dagger + b_{D_3}^\dagger b_{D_4}^\dagger, \qquad\qquad \hat{\Phi}_6^{(2)} = b_{D_1}^\dagger b_{D_3}^\dagger + b_{D_2}^\dagger b_{D_4}^\dagger,$$

$$\hat{\Phi}_7^{(2)} = b_{D_1}^\dagger b_{D_4}^\dagger + b_{D_2}^\dagger b_{D_3}^\dagger, \qquad\qquad \hat{\Phi}_8^{(2)} = b_{E_1}^\dagger b_{E_2}^\dagger.$$

We also consider the dressed operators up to four phonons. The quasiparticle weights of thus-defined bi-polaron operators exhibit almost the same feature as that of single polarons; the breathing vibrations induce a larger polaronic effect. However, the renormalized quasiparticle weight Z_{2h}/Z_{1h}^2 in Fig. 5 is larger for the buckling modes

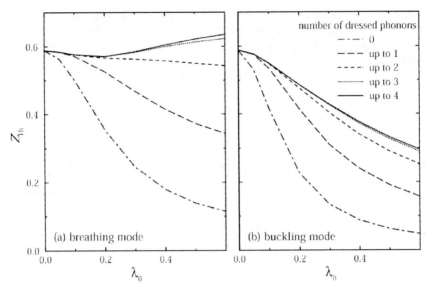

FIGURE 4. Quasiparticle weight of n-phonon dresses single polarons (n is up to 0, 1, 2, 3 and 4) for (a) breathing and (b) buckling modes.

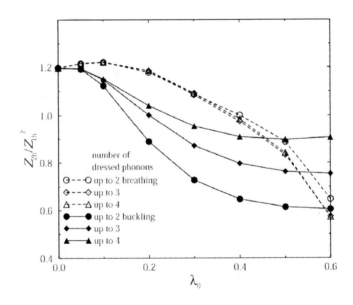

FIGURE 5. Normalized quasipparticle weights for the breathing and buckling modes.

for sufficiently large electron-phonon couplings. It implies that the bi-polaron is more stable for the buckling modes. It is consistent with the result of the hole binding energy discussed above. Thus we confirm the conclusion that the dynamical buckling distortions enhance the hole pairing again.

REFERENCES

1. Zhon J., and Schüttler, H.-B., *Phys. Rev. B* **69**, 1600 (1992).
2. Röder, H., Fehske, H., and Silver, R.N., *Europhys. Lett.* **28**, 257 (1994).
3. Chen, C.H., Cheong, S.-W., and Cooper, A.S., *Phys. Rev. Lett.* **71**, 2461 (1993).
4. Frank, J., in *Physical Properties of High Temperature Superconductors IV*, edited by Ginsberg, D.M., World Scientific, Singapore, 1994.
5. Song J., and Annet, J.F., *Phys. Rev. B* **51**, 3840 (1995).
6. Bulut, N., and Scalapino, D.J., *Phys. Rev. B* **54**, 14971 (1996).
7. Nazarenko, A., and Dagotto, E., *Phys. Rev. B* **53**, R2987 (1996).
8. Sakai, T., Poilblanc, D., and Scalapino, D.J., *Phys. Rev. B* **55**, 8445 (1997).
9. Kaxiras, E., and Manousakis, E., *Phys. Rev. B* **38**, 566 (1988); Bonča, J., Prelovšek, P., and Sega, I., *Phys. Rev. B* **39**, 7074 (1989); Riera, J., *Phys. Rev. B* **40**, 833 (1989); Hasegawa, Y., and Poilblanc, D., *Phys. Rev. B* **40**, 9035 (1989); Poilblanc, D., *Phys. Rev. B* **48**, 3368 (1993).

Possible Effect of Periodic Tilt Order

S. Shamoto

Department of Applied Physics, Faculty of Engineering,
Tohoku University, Sendai 980-8579, Japan

Abstract. In a strong covalent bond network system, tilt order of MO_6 octahedra (M; Transition metals), produced by twin order or microscopic phase mixture, is proposed as an origin of charge order or stripe formation in the vicinity of insulator-metal transition.

INTRODUCTION

In the vicinity of insulator-metal (I-M) transition, carriers could induce various electronic instabilities, such as Wigner crystallization, phase separation or charge order accompanied by lattice distortion. Here, we will focus on the effect of periodic lattice distortion there. In a strong covalent bond network system such as cuprates, however, it is difficult to make a periodic lattice distortion, such as charge density wave (CDW) by ions, since the displacement of ions along covalent bonds costs much energy. On the other hand, they exhibit structural phase transitions accompanied by the change of Cu-O-Cu bond angles, *i.e.*, the uniform tilting of CuO_6 octahedra. Local twin order or phase mixture could result in nonuniform tilting of CuO_6 octahedra, where transfer integrals between the nearest neighbor transition metal atoms are modified locally, although the change is expected to be small. In the vicinity of I-M transition, however, such a small change could affect the electronic ground state seriously, if there was any periodic tilt order which is discovered as a local tilt disorder by recent PDF analysis [1,2]. Significant interdomain correlations are suggested to exist in about 1 nm scale length as the shortest dimension of the domain [2,3]. This tilt correlation, *i.e.*, periodic tilt order, could produce a wave of local transfer integral by which the electronic state could result in stripe or charge order (CO). Polarized X-ray scattering experiments in the $La_{2-x}Sr_xCuO_4$ have also identified the presence of a variety of phases such as the low-temperature orthorhombic (LTO) and low-temperature tetragonal (LTT) phases [4]. From a theoretical point of view, band Jahn-Teller effect is discussed in the relation to uniform structural phase transition. However, the tilt angle used in the LDA calculation [5], where unrealistic peak of electronic density of states play an important role, is known to be much larger than the real angle. Furthermore, in a simple *t-J* model the striped states are not ground states [6]. Therefore, the coupling of the electrons with periodic lattice distortions should be taken into account to stabilize stripe phases besides long-range Coulomb interaction.

The broadening of mosaic spread has been observed by applying magnetic field more than 8 T on $La_{2-x}Sr_xCuO_4$ ($x \sim 0.115$) single crystal, where an indication of

CP554, *Physics in Local Lattice Distortions*, edited by H. Oyanagi and A. Bianconi
© 2001 American Institute of Physics 1-56396-984-X/01/$18.00

structural phase transition from LTO phase to LTT phase is also observed [7]. This structural phase transition can be explained thermodynamically because of larger magnetization of LTT phase than that of LTO phase. It should be noted that the structural phase transition on $La_{1.885}Sr_{0.115}CuO_4$ competes with its superconductivity resulting to average structure of nearly LTO phase at low temperature. The experimental result suggests that LTT phase has strong lattice strain in $La_{1.885}Sr_{0.115}CuO_4$ due to strong coupling between electronic state and lattice. This strong coupling has also been revealed by many kind of experiments [8,9]. Here, we propose a simple twin order model (Fig. 1) [10] as a possible LTT local structure of $La_{2-x-y}Nd_ySr_xCuO_4$ (LNSCO).

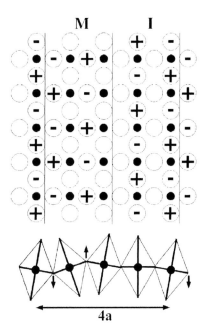

Figure 1. An example of the periodic twin order of LTT phase. Top: The closed circles denote Cu atoms, while the open large circles denote oxygen atoms, where signs mean the displacement direction normal to CuO_2 basal plane. Solid lines are twin boundaries. In the second grain M from the left end, the vertical 1D array of copper and oxygen atoms does not suffer any displacement, whereas in the third grain I the vertical 1D Cu-O array has a zigzag structure. Bottom: Projection of CuO_6 octahedra along $[010]_t$. Oxygen displacements in CuO_2 basal plane are shown by arrows. a is the lattice parameter in CuO_2 basal plane of HTT.

DISCUSSION

A critical value of the static stripe phase has been reported for the tilt angle of CuO6 octahedra [9,11], accompanied by rhombus distortion of CuO4 rectangle. In addition to the rhombus distortion, the tilt reduces the transfer integral between the nearest neighbor Cu $3d_{x^2-y^2}$ orbitals. As demonstrated in the study of (La,Y)TiO3 [12], the tilt angle of octahedra modulates the transfer integral between the nearest neighbor Ti $3d$ orbitals. The tilt angle increases by decreasing the radii of the rare earth ion. The critical tilt angle for the static stripe phase of LNSCO is 3.6 degrees [9]. On the other hand, rhombus distortion of the CuO4 basal plane estimated as $2(d-e)/(d+e)$ is about 0.002 in La$_{1.88}$Ba$_{0.12}$CuO4 [11], where d is the longer diagonal length of CuO4 basal plane and e is the shorter one. The transfer integral, t, between Cu $3d_{x^2-y^2}$ and O $2p$ modified from the original value, t_0, by both of the tilt and the rhombus distortion of CuO6 octahedra can be roughly evaluated based on simple estimation [13] in the small limits of Δr and θ as

$$t \approx t_0 (1 \pm 3.5 \Delta r)(1 - 0.5 \theta^2) , \tag{1}$$

where t_0 is the original value of transfer integral, Δr is the deviation from the original Cu-O distance r_0 that is normalized by r_0, $+(-)$ is for $e(d)$-direction, and θ is the tilt angle of CuO6 octahedra in radian. Above values obtained from the average structure ($\Delta r \sim 0.001$ and $\theta \sim 0.06$ rad.) gave us 0.6%, i.e., about 8 meV, reduction of transfer integral. In La$_{2-x}$Sr$_x$CuO4 system, however, the critical hole concentration of I-M transition is $x = 0.06$, which is close to the x-region of the static stripe phase. Therefore, the 0.6% increase of U/W could be responsible for the electronic transition around 60 K ($\sim$ 5 meV) observed as the anomaly in the temperature dependence of Seebeck coefficient at $x = 0.10 - 0.15$ [14], if the local transfer integral was reduced periodically. The periodic modulation can be produced by twin order or microscopic mixture of more than two phases, leading to the modulation of tilt angles along the conduction pass. An example of the possible patterns for the periodic twin order of LTT phase along a_t-axis is shown in Fig. 1. Recent angle resolved photoemission study has revealed $2k_F$ Fermi surface nesting condition corresponding to the present 4×1 domain wall unit cell around $x \sim 0.1$ [15]. On the other hand, 2 dimensional (2D) Fermi temperature of cuprates, i.e., about 2000K, deduced from superconducting carrier density by μSR measurement [16] is high enough to explain the present Fermi surface instability.

Once we assume that a hole stripe would be formed in the smooth 1 dimensional (1D) Cu-O bond array without any displacements, M region in Fig. 1 is the most suitable for the hole stripe. On the other hand, the smooth 1D Cu-O networks in I region of Fig. 1 are blocked by twin boundaries, indicating carrier localization. It would lead to antiferromagnetic stripes, which can be regarded as 2-leg ladder stripes with spin gap. This pattern of charge distribution is identical to a calculated result [6]. If one recognize that the stripe structure has 4×1 domain wall unit cell, the highest Cu-O σ* band of the stripe can be regarded as quarter-filling. If one introduces interatomic Coulomb repulsion V, the carriers in the hole stripe tends to localize. This would be one of the reasons of 1/8 anomaly.

Inelastic neutron scattering experiment provides evidence about the existence of dynamical stripes from high energy phonon dispersion [17], where strong correlation between T_c and inelastic neutron scattering intensity has been observed [18,19]. In $YBa_2Cu_3O_{6+x}$ phonon modes of 42 - 44 meV B_{1g} [19] or 24 meV B_{2u} [20] oxygen atoms in CuO_2 plane vibrate along c-axis (normal to the conducting plane). This kind of phonon at a certain Q could affect the electronic state in the similar manner as stated above, *e.g.*, periodic modulation of transfer integral t and super-exchange interaction J in CuO_2 plane, resulting to the formation or destruction of dynamical stripes. The appearance of magnetic coherent peaks in $YBa_2Cu_3O_{6+x}$ below T_c might relate to the formation of dynamical stripes

CONCLUSION

Periodic twin order is proposed as an origin of stripe phase or charge order, where local periodic transfer integral modulation by tilting of MO_6 octahedra may reduce the total energy in the vicinity of insulator-metal transition, *i.e.*, $U/W \sim 1$. Such a coupling between electronic state and the displacement of oxygen atoms normal to the conducting plane may play an important role not only in the static stripes but also in dynamics of stripes.

ACKNOWLEDGMENTS

The author would like to thank Prof. T. Tohyama, Institute for Material Research, Tohoku University, for invaluable discussion. This work was supported by a Grant-in-Aid for Scientific Research on Priority Areas "Novel Quantum Phenomena in Transition Metal Oxides" (12086211) of the Ministry of Education, Science, Sports and Culture of Japan.

REFERENCES

1. Bozin, E. S. *et al.*, *Phys. Rev. B* **59**, 4445 (1999).
2. Billinge, S. J. L. *et al.*, *Phys. Rev. Lett.* **72**, 2282 (1994).
3. Haskel, D., Stern, E. A., Dogan, F., and Moodenbaugh, A. R., *Phys. Rev. B* **61**, 7055 (2000).
4. Bianconi, A. *et al.*, *Phys. Rev. Lett.* **76**, 3412 (1996).
5. Pickett, W. E., Cohen, R. E., and Krakauer, H., *Phys. Rev. Lett.* **67**, 228 (1991).
6. Hellberg, C. S., and Manousakis, E., *Phys. Rev. Lett.* **83**, 132 (1999).
7. Yamada , K., and Fukase, T. *et al.*, unpublished
8. Tranquada, J. M. *et al.*, *Nature* (London), **375**, 561 (1995).
9. Buechner, B. *et al.*, *Phys. Rev. Lett.* **73**, 1841 (1994).
10. Shamoto, S., *Physica C* in press.
11. Kobayashi, A. *et al.*, *Physica B* **194-196**, 1945 (1994).
12. Okimoto, Y. *et al.*, *Phys Rev B* **51,** 9581 (1995).

13. Harrison, W. A., *Electronic Structure and the Properties of Solids: The Physics and the Chemical Bond*, W. H. Freeman and Company (1980)

14. Sera, M. *et al.*, *Solid State Commun.* **69,** 851 (1989).

15. Ino, A., Kim, C., Nakamura, M., Yoshida, T., Mizokawa, T, Shen, Z, -X, Fujimori, A, Kakeshita, T, Eisaki, H, and Uchida, S. *et al.*, cond-mat/0005370.

16. Uemura, Y. J. *et al.*, *Physica B* in press.

17. McQueeney, R. J. *et al.*, *Phys Rev. B* **82,** R628 (1999), Mook, H. A. et al., *Physica C* in press.

18. Arai, M. *et al.*, *Physica C* **181,** 45 (1991).

19. Arai, M. *et al.*, *Phys. Rev. Lett.* **69,** 359 (1992).

20. Harashina, H. *et al.*, *J. Phys. Soc. Jpn.* **67,** 3216 (1998).

Fluctuation Effects by Boson Dispersion in Fermion-boson Coupled Systems

Y. Motome[*] and G. Kotliar[†]

[*] Institute of Materials Science, University of Tsukuba, Tsukuba, Ibaraki, JAPAN 305-0006
[†] Department of Physics, Rutgers University, Piscataway, New Jersey, USA 08854-8019

Abstract. Nonlinear feedback effects due to dispersive bosons are studied in fermion-boson coupled systems by using an extension of dynamical mean-field theory. In the nonperturbative regime, strong coupling and nonadiabaticity reduce effective interaction between fermions as the width of boson dispersion increases. The boson field is completely softened at the crossover to this regime. The softening is significantly enhanced by the boson dispersion. We elucidate the controlling parameters of this nonperturbative regime where fluctuations of the dispersive bosons accelerate the delocalization of fermions.

INTRODUCTION

Control of bandwidth or dispersion of fermions relative to the interaction strength has been the subject of a great deal of theoretical and experimental works. Tuning of the fermion dispersion has been extensively studied, for instance, in terms of localization-delocalization problems such as the Mott transition [1]. Compared with this situation, effects of boson dispersion have been paid less attention in this field. Bosons have been considered in the limit of zero dispersion (Einstein bosons) in many studies of fermion-boson coupled systems [2, 3].

Dispersion of bosons is, however, relevant in many realistic systems. For instance, phonon modes in various oxide materials are dispersive due to intersite coupling through oxygens shared by adjacent unit cells. This should be relevant to variety of cooperative phenomena of lattice and charge degrees of freedom in those materials.

In this contribution, we study dispersive bosons interacting with fermions using dynamical mean-field (DMF) theory. The DMF theory provides a local view of a many-body problem in terms of an impurity problem obeying a self-consistency condition [4]. The DMF method is extended to include feedback effects through the fermion-boson interaction [5]. We control the width of the boson dispersion to examine fluctuation effects of boson fields.

FORMALISM AND MODEL

For the fermion-boson coupled systems, the action for a local view by the DMF theory has the general form [4, 5]

CP554, *Physics in Local Lattice Distortions*, edited by H. Oyanagi and A. Bianconi
© 2001 American Institute of Physics 1-56396-984-X/01/$18.00

$$S_{\text{eff}} = \int d\tau\, d\tau' [\sum_\alpha c_\alpha^\dagger(\tau) g_{0\alpha}^{-1}(\tau-\tau') c_\alpha(\tau') + \sum_\nu x_\nu(\tau) D_{0\nu}^{-1}(\tau-\tau') x_\nu(\tau')]$$
$$+ \int d\tau \sum_{\alpha_1 \alpha_2 \nu} \lambda_{\alpha_1 \alpha_2 \nu} c_{\alpha_1}^\dagger(\tau) c_{\alpha_2}(\tau) x_\nu(\tau), \tag{1}$$

where g_0 and D_0 are the Weiss fields for fermion and boson, respectively. Note that the action (1) is quite general which contains also fermion interactions such as the Coulomb repulsion through the Hubbard-Stratonovich transformation [6, 7].

The full Green's functions g and D are related to the Weiss fields by $g_\alpha^{-1}(i\omega_n) = g_{0\alpha}^{-1}(i\omega_n) - \Sigma_\alpha(i\omega_n)$ and $D_\nu^{-1}(i\omega_n) = D_{0\nu}^{-1}(i\omega_n) - \Pi_\nu(i\omega_n)$, where Σ and Π are the self-energy for fermion and boson, respectively. The DMF framework is extended to determine both g and D simultaneously by the self-consistency conditions

$$g_\alpha = \sum_q [i\omega_n + \mu - \varepsilon_{q_\alpha} - \Sigma_\alpha]^{-1}, \quad D_\nu = \sum_q [(i\omega_n)^2 - \omega_{q_\nu}^2 - \Pi_\nu]^{-1} \tag{2}$$

where ε_{q_α} and ω_{q_ν} give the dispersion relations for fermions and bosons, respectively. Here the bosons are described as harmonic oscillators.

In the following sections, the above DMF equations are examined for a simple model with fermions interacting with one branch of bosons, which is a straightforward extension of the so-called Holstein model [3] to include the boson dispersion:

$$H = \sum_{\alpha=1,2} \sum_{ij} t_{ij} c_{i\alpha}^\dagger c_{j\alpha} + \frac{M}{2}(\sum_i \dot{x}_i^2 + \sum_{ij} \omega_{ij}^2 x_i x_j) + \lambda \sum_{\alpha,i} (c_{i\alpha}^\dagger c_{i\alpha} - \frac{1}{2}) x_i. \tag{3}$$

We focus on the half-filled case in the DMF solutions without any symmetry breaking at low temperatures.

The DMF theory has been tested to give useful insights into three-dimensional systems [4]. We therefore assume semicircular density of states for both fermions and bosons as $D_F(\varepsilon) = 2\sqrt{W^2 - \varepsilon^2}/\pi W^2$ and $D_B(\varepsilon) = 2\sqrt{\omega_1^2 - (\varepsilon - \omega_0)^2}/\pi \omega_1^2$ where W is the half-bandwidth of the fermion density of states; ω_0 and ω_1 are the center and the half-bandwidth of the boson density of states, respectively.

The action (1) is solved by quantum Monte Carlo method. Details of the method and conditions of calculations are reported elsewhere [5].

RESULTS

In this section, effects of the boson dispersion are examined in the boson Weiss field D_0 and the imaginary part of the fermion self-energy $\text{Im}\Sigma$. Both quantities are closely connected with the effective interaction between fermions. In particular, D_0 is directly related to the retarded effective interaction between fermions as $U_{\text{eff}} = \lambda^2 D_0$ when the bosons are integrated out. We take three independent paramaters, ω_0, $U \equiv U_{\text{eff}}(\omega_1 = 0) = \lambda^2 / M\omega_0^2$, and ω_1 in the energy unit $W = 1$.

In the perturbative regimes, i.e., in the weak-coupling limit ($W \gg \omega_0$, U) and in the atomic limit ($W \ll \omega_0$, U), both $|D_0|$ and $|\text{Im}\Sigma|$ increase as ω_1 increases in our DMF solutions. This indicates that the effective interaction between fermions is enhanced by the width of the boson dispersion. The results are consistent with the perturbative arguments [8, 3, 2].

In contrast to this, both $|D_0|$ and $|\text{Im}\Sigma|$ decrease as ω_1 increases in the nonperturbative regime, i.e., in the strong coupling region away from the anti-adiabatic limit ($U > W$ and $\omega_0 \sim W$). Surprisingly, the effective interaction between fermions are weakened by the dispersive bosons.

Fig. 1(a) shows the boson Weiss field $|D_0|$ as a function of ω_1 for various values of U. The fermion self-energy $|\text{Im}\Sigma|$ changes in a similar way. For small U, $|D_0|$ increases by ω_1, as in the perturbative regime. When the value of U increases, a kink appears and $|D_0|$ decreases for larger values of ω_1. This kink corresponds to the crossover from the perturbative to the nonperturbative regime.

The crossover is correlated well with the complete softening of the boson field. Fig. 1(b) exhibits the effective frequency of the boson field defined by a pole of the boson Green's function as $\omega^* = \sqrt{(\omega_0 - \omega_1)^2 + \Pi(i\omega_n = 0)}$, where Π is the self-energy for boson. ω^* goes to zero at the value of ω_1 where $|D_0|$ shows a kink. The nonperturbative regime is characterized by the weakening of the effective interaction due to the strong boson fluctuations and by the softening of the boson field.

DISCUSSIONS

A systematic study of the crossover for various ω_0 and U elucidates two important parameters in the present system [5,9]. One is the pair formation energy divided by the fermion kinetic energy u, and the other is the fermion-boson coupling divided by

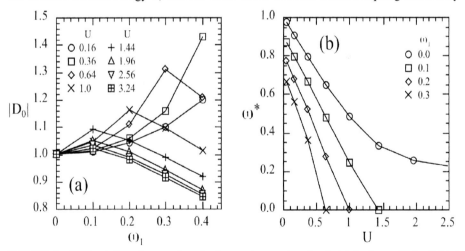

FIGURE 1. DMF solutions for $\omega_0 = 1$ at the inverse temperature $\beta = 8$; (a) the boson Weiss field at zero Matsubara frequency and (b) the effective frequency of the boson field.

the stored energy in the boson field η. The former describes the competition between the itinerant character and the localization of fermions. The latter distinguishes the single-boson and the multiboson regimes. The both conditions $u > 1$ and $\eta > 1$ are necessary to cause the crossover accompanied by the complete softening of bosons.

Our DMF solutions in the regime with $u > 1$ and $\eta > 1$ indicate a reduction of the effective interaction by the boson dispersion. Strong boson fluctuations tend to accelerate the delocalization of fermions and to make polarons unstable. This striking result is revealed for the first time by our extended DMF method which successfully includes the mutual feedback in many-body systems. Note that the criteria of $u > 1$ and $\eta > 1$ have been discussed for the formation of a small polaron in a system with a single fermion interacting with bosons [10, 11]. In the single-fermion problem, however, there is no feedback to bosons from changes of the fermion state in the thermodynamic limit.

The softening of the boson field is significantly affected by the boson dispersion. Small width of the dispersion reduces the crossover value of U markedly. The reduction is notable especially for $\omega_0 > W$. For instance, in the case of $\omega_0 = 4$, the boson field is completely softened at $U = 4$ when $\omega_1 = 0.2$ ($\omega_1 / \omega_0 = 5\%$) although ω^* remains finite even at $U = 16$ for the dispersionless case. This is a consequence of the strongly nonlinear feedbacks by dispersive bosons.

The softening may be related with charge freezing accompanied by a lattice ordering in electron-phonon problems if we allow symmetry breakings in an unfrustrated model. Therefore, our results strongly suggest the importance of the boson dispersion in describing cooperative phenomena of lattice and charge degrees of freedom. More realistic models are under investigation which include orbital degrees of freedom of electrons, several modes of phonons, and Coulomb interactions.

REFERENCES

1. Imada, M., Fujimori, A., and Tokura, Y., *Rev. Mod. Phys.* **70**, 1039 (1998).
2. Mahan, G.D., *Many Particle Physics*, Plenum Publishing, New York, 1981.
3. Holstein, T., *Ann. Phys.* **8**, 325 (1959); ibid. **8**, 343 (1959).
4. Georges, A. *et al.*, *Rev. Mod. Phys.* **68**, 13 (1996).
5. Motome, Y., and Kotliar, G., *cond-mat*/0005395, to be published in *Phys. Rev. B*.
6. Hubbard, J., *Phys. Rev. Lett.* **3**, 77 (1959).
7. Stratonovich, R.L., *Sov. Phys. Dokl.* **2**, 416 (1957).
8. Migdal, A.B., *Zh. Eksp. Teor. Fiz.* **34**, 1438 (1958); *Sov. Phys. JETP* **7**, 996 (1958).
9. Feinberg, D., Ciuchi, S., and de Pasquale, F., *Int. J. Mod. Phys. B* **4**, 1317 (1990).
10. Capone, M., Stephan, W., and Grilli, M., *Phys. Rev. B* **56**, 4484 (1997).
11. Ciuchi, S. *et al.*, *Phys. Rev. B* **56**, 4494 (1997).

Strongly Correlated Lattice States and Thermal Hysteresis of Elastic, Thermal and Optical Properties of HTSC Compounds

V. Gusakov and A. Saiko

Institute of Solid State & Semiconductor Physics National Academy of Science of Belarus,
P. Brovki str. 17, Minsk, 220072, BELARUS

Abstract. It is shown that in crystal lattices with a basis the cooperative behavior of a certain type of atoms performing optical long-wavelength vibrations in a double-well potential of the field of the matrix lattice may lead to the formation of a bistable states (sublattice). As a result of the interaction of the metastable states of such a sublattice with the vibrational states of the matrix lattice, the elastic, optic, and thermal properties of the crystal acquire anomalous, hysteresis-like, temperature curves. The concepts developed in the paper make it possible to obtain a qualitative interpretation, which agrees with the experimental data, of the hysteresis-like temperature dependence of the speed and absorption of ultrasonic waves, the specific heat, optic, and the thermal conductivity in high temperature superconductors.

INTRODUCTION

The precision measurements have revealed thermal hysteretic behavior of elastic [1], thermal [2] and optic [3] properties of HTSC crystals. The hysteretic behavior has been observed not only for the copper oxides but for non-copper oxide superconductors as well. So recently a thermal hysteresis of ultrasonic wave velocity and attenuation [4] and the hysteretic behavior of Raman line intensity [5] was observed in $Ba_{1-x}K_xBiO$ crystals. What is remarkable is the large interval of temperature hysteresis and at the same time the absence of relaxation in the measured parameters in the hysteresis region (some samples were kept at a fixed temperature for several hours). Although there is still no universal opinion concerning the nature of the observed anomalies, it is obvious that they are related in one way or another to the metastable states of the crystal lattice. The effect of hysteretic behavior of some physical properties of HTSC compounds is extraordinary and interesting not only by itself but also because of its occurrence specifics unusual dynamics (changes) of a HTSC crystalline lattice.

RESULTS

In the present work we give a common theoretical description of unusual thermal hysteretic behavior of HTSC compounds. Our theoretical model is based on the

CP554, *Physics in Local Lattice Distortions*, edited by H. Oyanagi and A. Bianconi

following ideas. The hysteretic behavior of thermal properties about all indicates that there are some metastable states of atoms in a crystal. The fact that the hysteretic loops are collapsed at law temperatures and there is no relaxation in a metastable region indicates that the heat fluctuations are not of crucial importance in the transition between metastable states. Otherwise the local fluctuations of energy will destroy metastable states and the relaxation processes will be observed. The HTSC crystals are characterized by strong electron-electron correlations. A correlation between atom displacements is universally present for crystals with strong electron-electron correlations. The correlation of atoms can be so strong that in the transition between metastable states is involved a group of atoms. In other words metastable states are formed for a cooperative motion of atoms. In this case the transition between metastable states is described as a cooperative transition. Relative fluctuations of the total energy of the correlated state are decreased by the factor $\sqrt{N}$ (N the number of atoms in the correlated state) in comparison with that for a single atom in the correlated state. The number of atoms involved in the correlated state are determined by the interactions in the crystal by even though we suppose that the correlated state is formed from some atoms of the nearest elementary cells we have $\sqrt{N} \geq 3$ and the influence of heat fluctuations is decreased considerably (by the factor $\propto \exp(\sqrt{N})$).

To thoroughly study the dynamics of such strongly correlated system we need to write down the Hamiltonian. As atoms on the correlation length move coherently we can write Hamiltonian as

$$H_{coh} = N \cdot H_{anh} , \qquad (1)$$

where N the number of particles in the correlative volume; H_{anh} is reduced (to a single particle) Hamiltonian of strongly correlated system and can be represented in the form of an anharmonic oscillator in a double-well potential with asymmetric wells $H_{anh} = p^2/2m + U(q) = p^2/2m + (\alpha/2)q^2 - (\beta/3)q^3 + (\gamma/4)q^4$, q and p are the configuration coordinate and canonically conjugate momentum of the particles. The thermodynamics of the system described by the following expression for the free energy F

$$F \leq k_B T \ln\left(2sh\left(\frac{\hbar\Omega}{2k_B T}\right)\right) + \left(\frac{\alpha}{2}<q>^2 - \frac{\beta}{3}<q>^3 + \frac{\gamma}{4}<q>^4\right) +$$
$$\left(\frac{\alpha}{2} - \beta<q> + \frac{3}{2}<q>^2 \gamma - \frac{m\Omega^2}{2}\right)\sigma + \frac{3}{4}\gamma\sigma^2 \qquad (2)$$

where $\sigma = <(q-<q>)^2>_0 = \frac{\hbar}{2m\Omega} cth\left(\frac{\hbar\Omega}{2k_B T}\right)$ and statistical-mean displacement $<q>$

and effective frequency Ω are determined from the system of equations

$$-(3<q>\gamma-\beta)\frac{\hbar}{2m\Omega}cth\left(\frac{\hbar\Omega}{2k_BT}\right)=\alpha<q>-\beta<q>^2+\gamma<q>^3$$

$$m\Omega=\alpha-2\beta<q>+3\gamma\left[<q>^2+\frac{\hbar}{2m\Omega}cth\left(\frac{\hbar\Omega}{2k_BT}\right)\right]$$

(3)

(for more details of the calculation see [4]).

Figure 1 depicts the temperature dependence of the mean displacement $<q>$, effective frequency Ω and free energy as function of temperature. The solution has the shape of a hysteresis curve, describing the transition of system from the global minimum quantum states to passing quantum states (above- barrier oscillation). In this case we can say that in crystal there is a bistable system. The interaction of the bistable system with the phonon excitations of the matrix lattice may lead to experimentally observed effects. For instance, the scattering of a traveling acoustic mode of the matrix lattice by perturbations caused by thermodynamics of bistable system give rise to singularities in the real and imaginary parts of these modes, which must be observable in experiments, in particular, in the anomalous behavior of the elastic and thermal characteristic of the crystals, such as the speed (the elastic modulus) and decay of ultrasound and the thermal conductivity. The renormalized frequencies of the crystal harmonic modes can be calculated using the equations of motion for the retarded two-time Green functions [6].

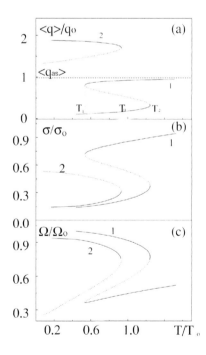

FIGURE 1. Temperature dependence of the statistical-mean displacement $<q>$ (a), the variance σ (b), and the effective frequency Ω(c) of a strongly correlated bistable system. The curves 1 and 2 describe the motion of the lattice in the global and local minima of the potential, respectively. The dotted curves represent unstable solutions.

In Figure 2 one can see an example of such calculations. The theoretical curves reflect fairly well the experimentally observed behavior of the speed of ultrasound, attenuation and thermal conductivity. Thus, at reasonable values of the model parameters, not only do the size and temperature interval of the hysteresis loop agree with the experimental data but so does the sense of tracing of the hysteresis loop in the cooling-heating cycle. It is interesting to note that Raman spectra also have hysteretic behavior because the integral intensity proportional to the number of atoms moving in double well potential.

The question on the nature of strongly correlated states in high-temperature superconductors is most interesting. It is difficult to answer this question now but we want to pay attention on the following interesting fact. The intriguing feature of the hysteretic behavior is that the temperature of opening of a hysteretic loop coincides with the pseudogap temperature. As the heat transport is rather insensitive to the spin order we attribute this hysteretic behavior to the formation of stripe order. Charge inhomogeneities can be strongly coupled to the lattice in high temperature superconductors and can lead to the formation of strongly correlated states.

This work was support in part by INTAS organization under grant No.97-1371.

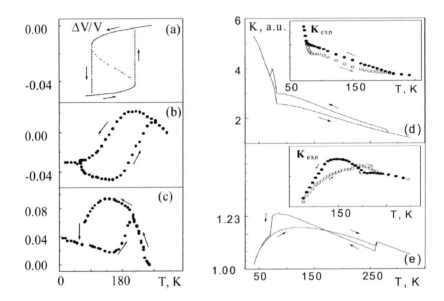

FIGURE 2. Temperature dependence of the relative variation of the speed of an ultrasonic wave (a)-(c) and thermal conductivity (d)-(e). (a) represent the results of calculations for a crystal with bistable strongly correlated system of atoms. The experimental data for $YBa_2Cu_3O_7$ in the direction of the crystallographic axis c at 12 MHz [7] are depicted in (b), and those at frequency $1.25 \times 10^5 s^{-1}$ in (c) [8]. (d)-calculated thermal conductivity dominates the cubic interaction of bistable system with matrix lattice; (e)-competition between cubic and quartet interactions. The insets schematically depict the experimental result [9] for $RBa_2Cu_4O_8$, $YBa_2Cu_3O_7$ and $RBa_2Cu_4O_6$.

REFERENCES

1. Pal'-Val', P.P., Pal'-Val', L.N., Demirsky, V.V. *et al.*, *J de Phys.* **6**, C8 489-492 (1996).
2. Gusakov, V., Jezowski, A., Barilo, S. *et al.*, *M2S-HTSC-VI*, 2000, pp. 297.
3. Gomez, P., Jimenez, J., Martin, P. *et al.*, *J. Appl. Phys.* **74** 6289 (1993).
4. Zerlitsyn, S., Luthi, B., Gusakov, V. *et al.*, *Eur. Phys. J. B* **16**, 59-66 (2000).
5. Pashkevich, Yu., Gnezdilov, V. *et al.*, *M2S-HTSC-VI*, 2000, pp. 451.
6. Saiko, A.P., Gusakov, V.E., *J. Experim and Theor. Phys.* **89**(1), 92-106 (1999); *Zh, Eksp. Teor. Fiz.* **116**, 168-193 (July 1999).
7. Xu, M.-F., Schenstrom, A., Hong, Y. *et. al.*, *IEEE Trans Magn.* **25**, 2414-2416 (1989).
8. Natsik, V.D., Pal'-Val', P.P., Engert, J., and Kaufmann, H.-J., *Fiz Nizk. Temp.* **15**, 836-840 (1989); *Sov. J. Low Temp.* **15**. 463-467 (1989).
9. Jezovski, A., Klamut, J., and Dabrowski, B., *Phys. Rev. B* **52**, R7030-R7033 (1995).

The Strain Quantum Critical Point for Superstripes

A. Bianconi[*], D. Di Castro[*], G. Bianconi[§] and N. L. Saini[*]

[*]Unitá INFM and Dipartimento di Fisica, Università di Roma "La Sapienza", 00185 Roma, Italy
[§]Department of Physics, Notre Dame University,46566 Indiana, USA

Abstract. The experimental determination of the quantum critical point (QCP) that controls the self-organization of charged striped domains in cuprate perovskites is reported. The phase diagram of doped cuprate superconductors is determined by the hole doping δ and by a second variable: the micro-strain ε of the Cu-O bond length. We have determined the $T_c(\varepsilon,\delta)$ phase diagram by measuring the micro-strain ε by Cu K-edge EXAFS. The superconducting phase occurs around the QCP at the critical strain where T_c reaches its maximum. The critical charge, lattice and spin fluctuations near this strain QCP provide the singular interaction for the pairing and show self-organization of local lattice distortions into a metallic superlattice of quantum wires "superstripes" that favors the amplification of the critical temperature.

INTRODUCTION

While the electron-phonon interaction provides the mechanism for pairing in the low temperature superconductors, collective charge fluctuations are expected to be involved in the pairing of high temperature superconductors [1,2].

The critical temperature increases by increasing the particle-particle attraction i.e., by decreasing the size of the pairs, i.e., of the coherence length, ξ_0, of the superconducting phase. The critical temperature T_c reaches a maximum in a optimum intermediate coupling regime where $k_F\xi_0 \sim 2\pi$ [2] and in the normal phase we have a complex phase of the condensed matter: a non Fermi liquid formed by a mixture of fermions and bosons (preformed pairs). The focus of the research in this field is the experimental determination of the relevant interactions in the normal phase driving the superconducting state from the range $0 < T_c < 23K$ of metals and alloys to the high temperature range $20 < T_c < 150$ K of doped cuprate oxide perovskites [3] where $k_F\xi_0 \sim 2\pi$ [4] i.e., enhancing the critical temperature by a factor ~ 10.

The normal phase shows a non Fermi-liquid behavior that is indicated by the temperature dependence of transport properties. The non-Fermi-liquid behavior and quantum lattice, charge and spin fluctuations are typical of a system close to a quantum critical point (QCP) [5-7]. In fact quantum fluctuations near a QCP drive an electron gas toward a regime of strong interactions well beyond the Fermi liquid approximations [8]. In this regime the pairing is described by a singular interaction [9-13].

CP554, *Physics in Local Lattice Distortions*, edited by H. Oyanagi and A. Bianconi
© 2001 American Institute of Physics 1-56396-984-X/01/$18.00

A quantum phase transition (QPT) is a zero temperature generically continuous transition tuned by a parameter in the Hamiltonian. Near this transition quantum fluctuations take the system between two distinct ground states. Examples of QPT include the metal-to-insulator transition in disordered alloys, the integer and fractional Quantum-Hall effect, magnetic transitions in heavy Fermion alloys, the superconducting-to-insulator transition in granular superconductors. The unconventional normal state can be described in terms of fluctuations of the local order parameter and the superconducting phase is in the regime of quantum fluctuations.

It has been proposed that the critical point is due to the doping of the 2D AF Mott Hubbard insulator, in fact a phenomenological model for the low energy spin dynamics in the normal state shows that these systems are close to a QPT [9]. However in this case the predicted maximum T_c is expected at the critical point for disappearing of AF order in the doping range $0.02 < \delta < 0.06$ in disagreement with the experiments. Other authors have considered a case of QPT transition in the weak electron-lattice coupling limit as a function of the hole doping from an insulating CDW phase at low doping to a normal metal at high doping [13].

Here we provide an alternative solution to this problem showing that the parameter in the Hamiltonian which controls the QCP is the electron lattice interaction. This parameter can be tuned to the critical value for the onset of local lattice distortions by the micro-strain of the Cu-O (planar) bonds, due to the stress parallel to the CuO_2 plane. Introducing this new axis we obtain:

1) the phase diagram of the normal phase;
2) the location of the quantum critical point;
3) the nature of the quantum fluctuations that drive the pairing in cuprate superconductors to reach $T_c \sim 150$ K.

The cuprate perovskites are heterogeneous materials [25,26] made of three different portions:

1) *metallic bcc CuO_2 layers*, intercalated between
2) *insulating rock-salt fcc AO_{1-x} layers* (A = Ba, Sr, La, Nd, Ca, Y.) and
3) *(CR) charge reservoir layers* BO_{1-x} where the chemical dopants are stored.

The bcc CuO_2 layer has a tetragonal structure with crystallographic axis $a_t = b_t$. The Cu ion form a square pyramid or bi-pyramid with planar Cu-O(P) distance and a long axial Cu-O(A) distance due to cooperative Jahn Teller effect for the Cu^{2+} $3d^9$ ion that removes the degeneracy of the $Cu(3d_{x^2-y^2})$ $m\ell = 2$, and $Cu(3d_{3z^2-r^2})$ $m\ell = 0$, orbital.

The CuO_2 layers are intercalated between insulating rock-salt fcc AO layers, that its orthorhombic axis $a_0 = b_0$ is rotated by 45°. The inter-atomic distance in this layer is $R(AO) = R(Cu-O(P))\sqrt{2}$ for a good crystalline matching. The metallic phase in the CuO_2 plane is obtained by two separate steps in the design of the material: *first*, chemical dopants are introduced in the charge reservoir blocks. These localized charges play the role of acceptors and pump electrons from the CuO_2 plane. *Second*, multiple substitutions of metallic ions A (A = Ba, Sr, La, Nd, Ca, Y...) in the oxygen deficient rock-salt layers are made in such a way to change the average ionic radius $<r_A>$ of the rock-salt layers.

Doping introduces holes in the O $2p$ orbital (L) and a single hole remains in the Cu site. However the symmetry of the molecular orbital for the added hole can be either of pure $m\ell = 2$ symmetry or of mixed symmetry with a component of local a_1 symmetry $L(a_1) = (1/2)(p_{x_1} + p_{y_2} - p_{x_3} - p_{y_4})$ mixed with $d_z = Cu(3d_{3z^2-r^2})$, $m\ell = 0$, [14] and the doped charges are associated with a local lattice distortion (LLD) of the CuO_4 square plane mixing states with different orbital momentum such as dimpling of the oxygen ions or Cu displacements, tilting of the pyramids and rhombic distortion of the square plane. These LLD distortions indicate a pseudo Jahn Teller electron lattice interaction of the doped holes. We have found in 1992 that in Bi2212 the pseudo JT-LLD get self organized in linear arrays, i.e., stripes [15-17]. The co-existing itinerant particles form "rivers" of charges and at the Erice workshop in 1992 [16] the scenario of superconducting stripes, where *"the free charges move mainly in one direction, like the water running in the grooves of a corrugated iron foil"*, was introduced for the first time in the field of high T_c superconductors.

The complex band structure of the CuO_2 plane of cuprate perovskites is due to its deformation that increases with the elastic strain field, due to the mismatch. The strain induces local lattice deformations such as the dimpling of Cu ions relative to the oxygen ions, the buckling of the plane, the tilting of the CuO_4 square planes and their rhombic distortions. The states near the Fermi level [18] can be described by an effective tight band model with near neighbor hopping integral t and next-nearest neighbor hopping integral t'. The large on site Coulomb repulsion U gives a correlated electron gas described by the Hubbard term. Moreover there is a relevant electron lattice interaction of the type of cooperative pseudo Jahn-Teller coupling of charges with Q_2-type local modes can be described by the Holstein-Hubbard Hamiltonian:

$$H = H_{el} + H_U + H_{ph} + H_I =$$
$$= -t \sum_{<i,j>\sigma} \left(c_{i\sigma}^+ c_{j\sigma}\right) + t' \sum_{<<i,j>>,\sigma} \left(c_{i\sigma}^+ c_{j\sigma}\right) + U \sum_i n_{i\uparrow} n_{i\downarrow}$$
$$+ \omega_0 \sum_q a_q^+ a_q + g\omega_0 \sum_i c_i^+ c_j [a_i^+ + a_i] - \mu_0 \sum_{i\sigma} n_{i\sigma}$$

The first two terms describe the itinerant charges in a 2D square lattice simulating the CuO_2 plane where t is the electron transfer integral between nearest-neighbor sites $<i,j>$ and t' is the electron transfer integral between next-nearest-neighbor sites $<<i,j>>$, $n_{i\sigma} = c_{i\sigma}^+ c_{j\sigma}$ is the local electron density, $c_{i\sigma}^+$ denotes the electron creator operator at site i.

The third term is the Hubbard Hamiltonian describing the electronic correlation in the CuO_2 plane. The Hubbard term induces a renormalization of the electron effective mass that for realistic values of the parameters give $m^*/m_0 \sim 2$ near the Fermi surface in the (π, π) direction in agreement with experiments.

The coupling of the charges with local lattice distortions (LLD) of the CuO_4 unit can be described by the Holstein Hamiltonian ($H_{ph} + H_I$). The position of the lattice site is indicated by R_i and a_q^+ represents the creation operator for phonon with wavevector q, ω_0 is the frequency of the optical local phonon mode and g indicates the coupling of the charge with this local lattice mode. The Holstein coupling is known to describe both the weak coupling regime and the strong coupling regime [19]. In the weak coupling regime the free charges are not bound to local lattice distortions (LLD), and can be called "large polarons" in the low density limit. In the strong coupling limit the charges are trapped in on a single site and associated with a LLD and they can be called "small polarons" in the low density limit. There is a relevant regime in the intermediate coupling regime where the each charge is trapped in a domain of local lattice distortions of about 4 lattice sites in the low density limit [20]. At finite charge density a finite number M of charges are trapped in strings of cooperative pseudo Jahn Teller lattice deformations. There is a critical value g^* for phase separation, where strings of trapped charges coexist with free carriers (a generalization of the idea of large polarons). For particular case of cuprate superconductors the finite value of t' shift the peak of the density of states due to the van Hove singularity (VHS) to finite doping. In the case of Bi2212 it is shifted at $\delta=0.25$. Moreover it is possible to identify two electron lattice coupling

constants $\lambda_1 = \dfrac{g^2 \omega_0}{2td}$, $\lambda_2 = \dfrac{g^2 \omega_0}{2t'd}$ where $d = 2$ for a 2D electron gas. In the Bi2212

cuprate perovskites, $\omega_0 \sim 55$ meV and $t \sim 200$ meV and $t' \sim 50$ meV as determined by angle resolved photoemission [18]. Therefore while the first coupling constant is in the weak coupling limit, $\lambda_1 < 1$, the second one (that is associated with pseudo JT vibronic modes) is in the intermediate-strong coupling limit $1 < \lambda_2 < 2.5$ and it gives local lattice distortions, via modulation of t'. In this situation we are in an intermediate regime where charges trapped into LLD can coexist with itinerant charges [20]. The critical value g^* for the charge lattice phase separation is given by:

$$g^* = \sqrt{\frac{1}{D_0 \omega_0} + 4t\varepsilon_F \frac{D_0}{\omega_0}}$$

The search for the QCP failed in previous investigation since a single variable, the charge density was considered. Here we show that the relevant electron lattice interaction (the pseudo JT cooperative vibronic coupling) is triggered to the critical value for co-operative local lattice distortions by the micro-strain of the in plane Cu-O bond length. The micro-strain provides a new relevant axis and the local lattice fluctuations in the shape of a superconducting superlattice of quantum wires appears at a critical micro-strain.

THE MICRO STRAIN QUANTUM CRITICAL POINT

The electron-lattice interaction g of the pseudo JT type in the cuprates is controlled by the static distortions of the CuO_4 square plane and the Cu-O(apical) distance. In fact $g = \Psi(Q)\phi(\Delta_{JT})\gamma(\beta)$ where Q is the conformational parameter for the distortions of the CuO_4 square, like the rhombic distortion of CuO_4 square; β is the dimpling angle given by the displacement of the Cu ion from the plane of oxygen ions; and Δ_{JT} is the JT energy splitting that is controlled by the Cu-O(apical) bond length [23-25].

There is an external field acting on the CuO_2 plane of the cuprates that controls $g(\varepsilon)$ via the micro-strain of the CuO_2 lattice due to the compressive stress generated by the lattice mismatch between the metallic bcc CuO_2 layers and the insulating rock-salt fcc AO layers [21]. The bond-length mismatch across a block-layer interface is given by the Goldschmidt tolerance factor $t = [r(A-O)]/\sqrt{2}[r(Cu-O)]$ where $[r(A-O)]$ and $[r(Cu-O)] = d_0$ are the respective equilibrium bond lengths in homogeneous isolated parent materials A-O and CuO_2 [22]. The hole doped cuprate perovskite heterostructure is stable in the range $0 < t < 0.9$ that corresponds to a mismatch $\eta = 1-t$ of $0 < \eta < 10$ %.

We have focused our attention to 3 main cuprate perovskite systems 1) $Hg1212$, 2) $Bi2212$ and 3) La_2CuO_{4+y}. In these materials the hole concentration in the CuO_2 plane is controlled by the oxygen doping in the charge reservoir blocks. The stress due to the mismatch or the chemical pressure acting on the CuO_2 plane is controlled by the average ionic radius in the rocksalt layers. The stress increases going from Ba to Sr to La. We have measured the average Cu-O bond lengths by Cu K-edge EXAFS, a local structural probe, and shown in Fig. 1 as a function of average ionic radius of metallic ions in the rocksalt layers $<r>$. Decreasing $<r>$ is equivalent to a stress in the CuO_2 plane direction, or anisotropic chemical pressure, acting on the CuO_2 plane. We define the local or micro strain of the CuO in plane bond length $\varepsilon = 2(1-<Cu-O>/d_0)$, where $d_0 = 1.97$ Å is the equilibrium Cu-O distance at doping $\delta = 0.16$ in many different systems [26-30]. In the cuprate perovskites the micro-strain ε drives the system to a quantum critical point $g_c(\varepsilon_c)$ for the formation of a superlattice of quantum stripes. The stripes of self organized one dimensional cooperative ordering of pseudo-Jahn Teller local lattice distortions are detected by x-ray diffraction above a critical micro-strain $\varepsilon > \varepsilon_c \sim 4$ %. We have plotted in Fig. 1 the variation of the critical

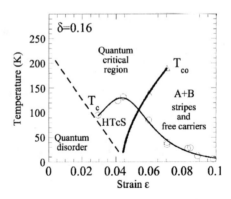

FIGURE 1. The critical temperature for charge ordering T_{co} and the superconducting critical temperature T_c as function of the micro strain ε at optimum doping $\delta=0.16$.

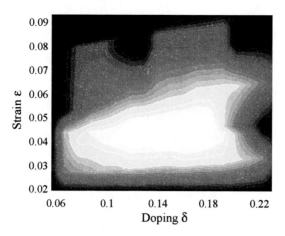

FIGURE 2. The superconducting critical temperature T_c plotted in a color scale (from $T_c = 0K$, black, to $T_c \sim 135K$, white) as a function of the micro-strain ε and doping δ. The maximum T_c occurs at the critical point $\delta_c = 0.16$, and $\varepsilon_c = 0.04$.

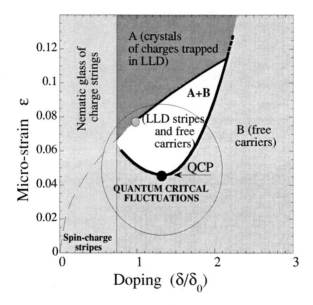

FIGURE 3. The phase diagram of the normal phase of doped cuprate perovskites. The high T_c superconductivity occurs in the region with the center at the quantum critical point QCP. The doping is measured in units of the critical doping of 1/8 holes per Cu site.

129

temperature for charge ordering T_{co} and the superconducting critical temperature T_c as function of the micro-strain at constant optimum doping $\delta = 0.16$.

In Fig. 2 we report the critical temperature T_c in a color plot (the critical temperature increases from black, $T_c = 0K$, to white, the maximum $T_c \sim 135K$) as a function of the micro-strain ε and doping δ for all superconducting cuprate families. The figure shows that the maximum T_c occurs at the critical point P ($\delta_c = 0.16$; $\varepsilon_c \sim 0.04$).

From these data we can derive a qualitative phase diagram for the normal metallic phase of all cuprate perovskites that give high T_c superconductivity that is shown in Fig. 3. This phase diagram solves the long standing puzzle of the phase diagram of the normal phase of the cuprates. There was a hidden physical parameter, the micro-strain, that triggers the electron-lattice interaction at a critical value for the onset of charges trapped into pseudo JT-LLD. The doping of the strained antiferromagnetic lattice forms both free carriers and charges trapped into the JT-LLD above the critical micro-strain ε_c. For $\varepsilon > \varepsilon_c$, as it was discussed for the case of oxygen doped Bi2212 and La124, the systems show a quasi first order phase transition as a function of doping [2,23,24].

The quantum critical point QCP is well defined at constant finite doping as a function of the micro-strain as it is shown in Fig. 3. Direct experimental evidence for quantum critical local lattice fluctuations has been obtained by measuring the dynamical fluctuations of the Cu-O bond at a high temperature $T_H > T_{co}$ in all families of cuprates ($T_H \sim 200K$).

In conclusion we have deduced a phase diagram for the superconducting phases where T_c depends from both doping and micro-strain. The anomalous normal phase of cuprate superconductors is determined by an inhomogeneous phases with co-existing polaronic stripes and itinerant carriers that appears for an electron lattice interaction larger than a critical value. Fluctuations of lattice-charge stripes appear in this critical regime. The micro-strain drives the electron lattice interaction to a QCP of a quantum phase transition [26-30]. Near this QCP the stripes get self organized in a superlattice of quantum wires of charges trapped into JT-LLD that co-exist with free carriers. This superlattice forms an array of superstripes where the chemical potential is tuned to a shape resonance. The plot $T_c(\varepsilon)$ reaches the highest temperature at the critical point ε_c.

ACKNOWLEDGEMENTS

This research has been supported by the Ministero dell'Università e della Ricerca Scientifica (MURST) under the Programmi di Ricerca Scientifica di Rilevante Interesse Nazionale coordinated by R. Ferro, by Istituto Nazionale di Fisica della Materia (INFM) and by Progetto 5% Superconduttività del Consiglio Nazionale delle Ricerche (CNR).

REFERENCES

1. Dolgov, O.V., and Maksimov, E.G., *Sov. Phys. JEPT* **56**, 1095 (1982); Dolgov, O.V., Kirzhmits, D.A., and Maksimov, E.G., *Rev. Mod. Phys.* **53**, 81 (1981).
2. Bianconi, A., *Sol. State Commun.* **91**, 1 (1994).
3. Bednorz, J.G., and Müller, K.A., *Rev. Mod. Phys.* **60** 565 (1988).
4. Keller, H. *et al. Physica* (Amsterdam) **185-189C**, 1089 (1991); Schneider, T., and Keller, H., *Phys. Rev. Lett.* **69**, 3374 (1992).
5. Sharma, R.P., Ogale, S.B., Zhang, Z.H., Liu, J.R., Chu, W.K., Veal, B., Pauliskas, A., Zheng, H., and Venkatesan, T., *Nature* **404**, 736-740 (2000).
6. Aeppli, G., Manson, T.E., Hayden, S.M., Mook, H.A., and Kulda, J., *Science* **278**, 1432 (1997).
7. Valla, T. *et al. Science* **285**, 2110 (1999).
8. Sachdev, S., *Quantum Phase Transitions*, Cambridge Univ. Press, New York, (1999).
9. Millis, A.J., Monien, H., and Pines, D., *Phys. Rev. B* **42**, 167 (1990).
10. Millis, A.J., Sachdev, S., and Varma, C.M., *Phys. Rev. B* **37**, 4975 (1988).
11. Bikers, N.E., and White, S.R., *Phys. Rev. B* **43**, 8044 (1991).
12. Micnas, R., and Robaszkiewicz, S., *Phys. Rev. B* **45**, 9900 (1992); ibid. *Phys. Rev. B* **52**, 6863 (1995); Micnas, R., and Robaszkiewicz, S., in *High T_c Superconductivity: Ten years after the Discovery* (Nato ASI, *Vol.* 343) edited by Kaldis, E., Liarokapis, E., and Müller, K.A., Kluwer, Dordrecht, (1996) pp. 31.
13. Perali, A., Castellani, C., Di Castro, C., and Grilli , M., *Phys. Rev. B* **54**, 16216 (1996).
14. Seino, Y., Kotani, A., and Bianconi, A., *J. Phys. Soc. Japan* **59**, 815 (1990).
15. Bianconi, A., Della Longa, S., Missori, M., Pettiti, I., and Pompa, M., in *Lattice Effects in High-T_C Superconductors*, edited by Bar-Yam, Y., Egami, T., de Leon, J.M., and Bishop, A.R., World Scientific Publ., Singapore, (1992) pp. 6.
16. Bianconi, A., in *Phase Separation in Cuprate Superconductors*, edited by Müller, K.A., and Benedek, G., World Scientific, Singapore, (1993) pp. 352; ibidem pp. 125.
17. Bianconi, A., Della Longa, S., Missori, M., Pettiti, I., and Pompa, M., and Soldatov, A., *Jpn. J. Appl. Phys.* **32** suppl. 2, 578 (1993).
18. Saini, N.L., Avila, J., Bianconi, A., Lanzara, A., Asensio, M.C., Tajima, S., Gu, G.D., and Koshizuka, N., *Phys. Rev. Lett.* **79**, 3464 (1997).
19. Capone, M., Stephan, W. and Grilli, M., *Phys. Rev. B* **56**, 4484 (1997)
20. Cataudella, V., De Filippis, G., and Iadonisi, G., *Phys. Rev. B* **60**, 15163 (1999) ; Cataudella, V., De Filippis, G., Iadonisi, G., Bianconi. A., and Saini, N.L., to be published
21. Rao, C.N.R., and Ganguli A.K., *Chem. Soc. Rev.* **24**, 1 (1995).
22. Goodenough, J.B., *Supercond. Science and Technology* **3**, 26 (1990); Goodenough, J.B., and Marthiram, A., *J. Solid State Chemistry* **88**, 115 (1990).
23. Bianconi, A., Missori, M., *J. Phys. I (France)* **4**, 361 (1994).
24. Bianconi, A., *Physica C* **235-240**, 269 (1994).
25. Bersuker, G.I., and Goodenough, J.B., *Physica C* **274**, 267 (1997)
26. Bianconi, A., Agrestini, S., Bianconi, G., Di Castro, D., and Saini, N.L., *Stripes and Related Phenomena*, edited by Bianconi, A., and Saini, N.L., Kluwer Academic/Plenum Publisher, New Your, 2000, pp. 9.
27. Saini, N.L., Bianconi, A., and Oyanagi, Y., *J. of Phys. Soc. Jpn.*, in press (2000).

28. Saini, N.L., Bianconi, A., Lanzara, A., Agrestini, S., Di Castro, D., and Oyanagi, Y., *Int. J. Modern Physics B*, in press (2000).

29. Bianconi, A., Di Castro, D., Saini, N.L., and Bianconi, G., *Phase transitions and self-organization in electronic and molecular networks,* edited by Thorpe, M.F., and Phillips, J.C., Kluwer Academic/Plenum Publisher, Fundamental Materials Research Series, Proc. of the meeting at Cambridge 11-14 July, 2000.

30. Bianconi, A., Agrestini, S., Bianconi, G., Di Castro, D., and Saini, N.L., *J. of Alloys and Compounds*, in press (2000).

Conducting Electron Strings in Oxides

F. V. Kusmartsev

School of MAP, Loughborough University, LE11 3TU, and
Isaac Newton Institute forMathematical Sciences
20 Clarkson Road, Cambridge, CB3 0EH, UK
Landau Institute, Moscow, Russia

Abstract. We found that insulating and conducting electron strings may arise in oxide materials with not very wide bands, like HTSC, due to electron-phonon and electron-electron interactions although antiferromagnetic correlations are also very important. We estimate the string length and the number of particles self-trapped into a single string taking into account the typical parameters of cuprates, like La_2CuO_4. The conducting strings typically have a small doping ($n <$ 0.3) and are created mostly due to the long-range electron-phonon interaction associated with ionic bonding of Oxides. Probably such strings play an important role in the creation of the stripe phase observed in HTSC.

INTRODUCTION

There is a growing body of experimental evidence [1-4] and theoretical arguments [5, 6] indicating an existence of complex inhomogeneous mesoscopic structures created from holes and spins in High Temperature Superconductors (HTSC) and other oxides. Such inhomogeneous structures ascribed to *a stripe phase*, were anticipated in theoretical papers [8-12] and have been originally observed in HTSC by Bianconi, Thurston, Tranquada, *et al.* [13-18]. In more recent papers the one dimensional character of charge and spin fluctuations [17] and the importance of lattice distortions in the creation of these mesoscopic stripe structures [3,2] have been clearly demonstrated. This stripe structure has also been observed in other oxides, manganites [4] and there the lattice effects have been seen. These experiments show that the stripe phenomenon is generic and there the lattice deformations arise.

Recently we have suggested that the lattice plays an important role in a self-creation of such mesoscopic structures of the stripe phase [5, 6]. We have proposed that in materials with narrow bands there may arise a many-particle self-trapping in the form of a long cigar shaped object named as *strings*. The string phenomenon is a many-particle generalization of a conventional self-trapping effect which usually leads to the formation of fluctuons, deformons, polarons and other single and two particle objects studied very intensively in the past (see, for example, the review [7] and references in). The electron strings may be created both by a long- and by a short-range electron-phonon interaction. In the limit of very narrow bands studied in Refs. [5, 6] the electron hopping vanishes and strings are insulating defects.

CP554, *Physics in Local Lattice Distortions,* edited by H. Oyanagi and A. Bianconi
© 2001 American Institute of Physics 1-56396-984-X/01/$18.00

In the present work we have extended our studies, taking into account the finite bandwidth and treat the kinetic energy of electrons on equal footing with their potential energy. We find that with the increase of the bandwidth (when the contribution of electron kinetic energy increases) there may arise a new type of strings – *conducting strings*. Since the creation of strings is related to atomic displacements they may be detected by all sorts of experiments, but it must primarily include Scanning Tunneling Microscopy (STM) experiments. The photoemission, X-ray and neutron scattering experiments may also be very useful although the mesoscopic size of the strings may create an obstacle. The strong lattice fluctuations, which may be associated with a highly conducting string proposed in the present paper, has been detected by *MeV* helium ion channeling, an ultrafast real-space probe of atomic displacements [2].

In general such strings may correspond to either a ground state or a metastable state. For a short range electron phonon interaction the strings are mostly insulating and typically correspond to a metastable state. However in a doped antiferromagnet the criterion for the formation (of both conducting and insulating) strings is strongly improved. Either type of strings may arise in a ground state. The stripe phase observed in HTSC [13-18, 3, 1] may correspond to a liquid crystal consisting of highly conducting electron strings. The distance between these strings the liquid crystal may be in two states: insulating and conducting. The conducting state arises due to a percolation through these strings. In other words the conducting phase is formed when at critical doping a conducting percolative path of an infinite length arises.

In the present paper we study a single highly conducting string. We employ the Hartree-Fock many body-wave function as a variational wave function to describe fermions self-trapped by a string potential well. Outside the self-trapped potential well the wave function is vanishing reflecting the fact that these electrons are self-trapped. We have calculated the number of particles and the length of the string having a lowest energy per particle for narrow band materials with short- and long-range electron phonon interactions also taking into account finite values of the bandwidth.

HAMILTONIAN

We consider the general Hamiltonian of spinless fermions interacting via a strong long-range electron-electron repulsion and with different types of phonons on a *d*-dimensional hypercubic lattice:

$$H = [-t \sum_{<i,j>} a_i^\dagger a_j + \sum_{q,i} \omega(q) n_i u_i(q) b_q + \sum_q \omega(q) b_q^\dagger b_q + hc] + \sum_{i<j} V(i-j) n_i n_j \quad (1)$$

where t is the electron hopping-integral, the operator $a_i^\dagger (a_i)$ creates (destroys) a fermion at a lattice site i, n_i is the occupation number operator $a_i^\dagger a_i$ and the operator $b_q^\dagger (b_q)$ is an operator of the creation (destruction) of a phonon. The summations in

Eq. (1) extend over the lattice sites i and -as indicated by $<i, j>$-over the associated nearest sites j. The matrix element of the electron-phonon interaction is equal to

$$u_n(q) = \frac{\gamma(q)\exp(iqn)}{\sqrt{2N}} \tag{2}$$

The function $\gamma(q)$ and the phonon dispersion relation $\varpi(q)$ are different for different types of electron-phonon interactions. For example, for optical longitudinal phonons (Pekar-Frölich interaction [19]) in the continuum limit the product $\gamma^2(q)\varpi(q) = 4\pi e^2/(\varepsilon^* q^2)$ with $1/\varepsilon^* = 1/\varepsilon_\infty - 1/\varepsilon_o$. In general for the short-range electron phonon interaction the $\gamma^2(q)\varpi(q) = const = c$. The long-range part of the Coulomb interaction has a conventional type [5]:

$$V(i-j) = \frac{e^2}{\bar{\varepsilon}|i-j|} \tag{3}$$

where it is plausible to assume that for spinless fermions the effective dielectric constant $\bar{\varepsilon}$ may be taken as $\bar{\varepsilon} = \varepsilon_\infty$. Such a Hamiltonian, we believe, may correctly describe the physics and that both the strong electron-electron repulsion and an electron-phonon interaction may lead to a formation of highly conducting electron strings.

HARTREE-FOCK APPROACH

The Hartree-Fock many-body wave function of the M self-trapped particles $\Psi(1, 2, ..., M)$ which we employ to calculate an expectation value of the Hamiltonian, Eq. (1) has a form of a Slater determinant (see, also in Refs. [5, 6])

$$\Psi(1, 2, ..., M) = \frac{1}{\sqrt{M!}}\det\|\psi_i(k_j)\| \tag{4}$$

consisting of single particle wave functions:

$$\psi_{m_x}(k_j) = \begin{cases} \frac{1}{\sqrt{N}}\exp(ik_j m_x) & \text{if} \quad 1 \leq m_x \leq N \\ \equiv 0, & \text{otherwise} \end{cases} \tag{5}$$

Each of these wave functions describes the electron(hole) trapped by N neighboring sites (string potential well) with equal probability, $1/N$. If the string is oriented in the **x** direction and is located on the sites $m_x = 1, ..., N$ the particle quasi-momentum k_j is determined by boundary conditions at the ends of the string. For simplicity we use periodical boundary conditions. As a first example we consider short-range electron-phonon interaction with acoustical phonons, *i.e.* with the lattice deformations. As in

135

Refs. [5, 6] we employ an adiabatic approximation considering a very slow motion of the lattice.

THE STRING ENERGY FOR JAHN-TELLER PHONONS

With the use of this many body wave function we have estimated an expectation value of the Hamiltonian H, Eq. (1). These calculations have been done in two steps. Using this many-body wave function, first, we calculated the one body and the pair correlation functions. Then with the use of the adiabatic approximation we have excluded slow (classical) phonon variables to get an expression for adiabatic potential E_S including the Coulomb and exchange energies (see, for details, Refs. [5, 6]). The calculated expression of the total energy E_S per particle has the form:

$$\frac{E_S}{M} = 2dt - 2t\frac{\sin \pi n}{\pi n} + 2t\frac{\sin \pi n}{\pi M} - \frac{cn}{2} + \varepsilon_c n \log M \qquad (6)$$

where d is a dimension of the hypercubic lattice, the value n is the electron(hole) doping inside the string: $n = M/N$ and the value $\varepsilon_c = e^2/(a\varepsilon_\infty)$ with a as an interatomic distance; in the Hamiltonian (1) the coupling constant of interaction with acoustical phonons $\gamma^2(q)\omega(q) = c = D^2/K$ where D is a deformational potential and K is an elastic modulus. The first three terms in the r.h.s. of this Eq. (6) are associated with electron kinetic energy while the last two terms in the r.h.s. of the Eq. (6) are associated with the energies of electron-phonon and electron-electron interactions, respectively (see, for comparison, in Ref. [5]). This expression, Eq. (6), represents a variational estimation of the total energy of M particles self-trapped into a string of length N valid for a wide range of values of c/t since it was obtained on the basis of an exact solution [5] found in the limit of very strong coupling $c/t \gg 1$. Therefore, in the framework of this variational approach we may get a reliable estimation of the number of particles, the length and the energy of an electron string valid for a wide range of the parameters of the Hamiltonian such as a coupling constant c, the bandwidth t and the characteristic Coulomb energy ε_c. Here the values M and n are variational parameters. The optimal number of particles trapped into the string of fixed length N is determined by a minimization of E_S/M with respect to M and is given by:

$$M = \frac{2t \sin \pi n}{\varepsilon_c \pi n} \qquad (7)$$

After a substitution of this expression into Eq. (6) for E_S we get the dependence $E_s = E_s(n)$ on the doping of the string $n = M/N$. Depending on the relation between the values of t, c and ε_c there may exist one or two types of solutions which correspond to two different types of strings: when $n = 1$ we define an insulating string [5], and when $n < 1$ we define a conducting string. When $c \sim t > \varepsilon_c$ the conducting string may be in a ground state. Then the number of particles trapped into the string is described by Eq. (7) and the value of the string doping must be determined numerically by next minimization. When the coupling constant c is very large ($c \gg t$ and $c > \varepsilon_c$) the Eq. (7) is not applicable, since the associated solution describing a conducting string disappears while the other solution associated with the marginal extremum $n = 1$ and describing insulating strings still exists. Thus, in extremly strong coupling limit the string is always insulating with $N = M$ and the number of particles in such a string is

$$M = \exp(\frac{c}{2\varepsilon_c} - 1) \qquad (8)$$

The minimum energy E_S associated with an insulating string is calculated using $N = M$ Eq. (8) and Eq. (6). The result is (see, for comparison, in the Ref. [5]):

$$E_{S-min} = (2dt - \varepsilon_c)M \qquad (9)$$

The comparison of Eq. (9) with the energy of M single polarons equal to $M(2dt - c/2)$ indicates that an insulating string with M trapped charged particles is in a metastable state. This metastable minimum may become an absolute minimum in a doped antiferromagnet. For a single hole in an antiferromagnet there is an increase in the exchange energy equal to $2dJ$, where J is an exchange constant. For M separated holes this energy increase is equal to $2dMJ$. On the other hand for M holes trapped in a string such an increase in exchange energy is equal to $J(2dM - M + 1)$. Therefore, the absolute minimum of the total energy of the deformational string in a doped antiferromagnet is described by the Eq.:

$$E_{S-min} = [2dt - \varepsilon_c + (2d - 1)J]M + J \qquad (10)$$

where the value M is defined by Eq. (8). The comparison of this expression with the total energy of M separated self-trapped particles indicates that the strings may have a lower energy if the following inequality holds:

$$2\varepsilon_c \le c \le 2\varepsilon_c + 2J \qquad (11)$$

Thus, the exchange interaction between anti-ferromagnetic spins significantly improves the physical conditions required for string formation in doped antiferromagnets. Therefore, at low temperatures, if this condition holds then M separated particles will condense into a string configuration.

STRINGS IN IONIC SOLIDS

All these string solutions found for deformational type of strings arise also in the case when the electron(hole) is interacting with Jahn-Teller phonons [20] or in general for, any type of short-range electron-phonon interaction as, for example, with Holstein optical phonons. The number of particles in the string is defined by Eqs. (7, 8) while the string length depends on the type of the string and for conducting strings must be estimated by a minimization of $E_S(n)$ with respect to n, numerically. For each type of phonons which have a short-range interaction with electrons(holes) the coupling constant in Eqs. (7, 8) must be defined, respectively, while the main Eqs. (7, 8, 6) remain the same (for more details, see Ref. [21]).

The case when a single electron or hole is interacting with polar phonons, *i.e.* with longitudinal optical phonons with frequency ω_0 (and with the constant of the electron-phonon interaction $\gamma^2(q) = 4\pi e^2 /(q^2 \varepsilon^* \hbar \omega_0)$) [19] is relevant and important to most oxides having a considerable amount of ionic bonding. Here the value of total energy including the Coulomb and exchange contributions from the long-range Coulomb forces between fermions may be calculated analogously to the case of short-range electron-phonon interaction presented above (see, also for example, in the Refs. [5, 6]). That is, first, with the aid of the Hartree-Fock many-body wave function of the M self-trapped particles $\Psi(1, 2, ..., M)$ (see, Eq. (4)) we have calculated the pair and off-diagonal correlation functions, and then with the use of these functions the dependence of the total energy on n and M having the form:

$$\frac{E_s}{tM} = \frac{2(n-M)\sin(\pi n)}{M^2 \sin(\pi n / M)} - E_p n(1.4 - \log n + \log M) + E_c n(0.6 + \log M) \quad (12)$$

where we have introduced the notations $E_c = \varepsilon_c / t, E_p = E_c(1 - \varepsilon_\infty / \varepsilon_0)$. The first two terms in the r.h.s. of this Eq. (12) are associated with the electron kinetic energy while the other terms in the r.h.s. of the Eq. (12) are associated with the energies of electron-phonon ($\sim E_p$) and electron-electron ($\sim E_c$) interactions, respectively.

A minimization of this expression with respect to M and n gives an estimation for the length of the string N and the number of particles M trapped into the string. In the limit of a low density $n \ll 1$ the values of M and N or n may be presented by the analytic formulae:

$$M \approx \frac{2\varepsilon_0 at}{e^2} \quad \text{and} \quad n = \frac{M}{N} \approx (\frac{2\varepsilon_0 at}{e^2})^\beta \exp(1.6\beta - .4) \quad (13)$$

where $\beta = \varepsilon_\infty /(\varepsilon_\infty - \varepsilon_0)$. The total energy of the string per electron equals $j_{string} = 2d - 2 + 2/N - nE_c$. To be in a ground state this string energy must be smaller than the energy of an individual polaron j_p equal to $2d - E_p$. The comparison of these two energies gives the precise criterion for the string formation. The conducting string corresponds to the ground state iff $E_p < 2 - 2/N + nE_c$, which roughly means that the polaron shift must be smaller than the string bandwidth $2t$.

Thus, we arrive at the conclusion that in oxide compounds with ionic bonding the formation of highly conducting electron strings created by a polarization potential is possible. The string length is typically much larger than the number of self-trapped holes, which is determined by the dielectric constants of the solid.

STRINGS AND STRIPES IN HTSC

Our findings are probably relevant to stripe formation observed in HTSC [13-18, 1-3]. To check if this criterion is satisfied for the HTSC and to estimate the length of the string we choose typical parameters of HTSC (for example, for La_2CuO_4, $\varepsilon_0 = 30$, ε_∞ = 5, the interatomic distance $a = 3.8$ Å and the hopping integral $t = .5\ eV$). Then by a minimization of Eq. (12) we obtain that 7 holes will be trapped into a string with a length equal to $N = 40$ interatomic distances or ~ 150 Å. The value of M may be also estimated with the use of Eq. (13) which approximately gives the same number. Note that in antiferromagnet this cigar-shape 7-particle string has a lower energy than 7 separated polarons and this string is highly conducting which is in contrast with an insulating string created by strong short-range electron-phonon interactions [5]. It is probable that in other HTSC compounds the size of the strings is not that much different.

Thus, we arrive at the conclusions that in polar oxide materials, like HTSC there may arise electronic strings which are linear multi-particle "electronic molecules". At low temperatures the electron strings may be ordered in CuO planes creating a nematic liquid crystal [22]. The striped phase in HTSC observed in numerous experiments [13, 15, 17, 18], may correspond to such a liquid crystal of conducting strings. With the doping of an antiferromagnet La_2CuO_4 there arises only the change in the distance between the strings while the structure of the strings (like, the string doping n or the length N) is not changed. The metallic stripe phase arises due to a correlated percolation over these strings when a density of such strings will be larger than the percolation threshold. For square lattices the percolation threshold is well known and is equal to $x_c \sim .5$ [23]. Then, using this value and our estimation for the string doping in La_2CuO_4 as $n = 7/40$ we may readily get the hole doping $\delta = nx_c \approx .09$ of the antiferromagnet La_2CuO_4 at which the metallic stripe phase may arise. The spin-spin and hole-spin correlations will of course slightly modify this result. It seems that our conclusion about the important contribution of the phonons into the origin of the stripe phase is confirmed in recent experiments which discover a huge influence of isotope effect on the critical temperature of the stripe ordering [3] and strong lattice fluctuations in YBCO [2] which may be associated with the dynamics of the strings. With the isotope changes [3] the structure of individual strings is changed (for example, the strings become shorter) and, therefore, the critical temperature of the stripe ordering must change.

In summary, we find that in oxides HTSC there may arise *highly-conducting* electron strings which are linear electronic molecules. Note that to find such molecules we have to treat the kinetic and potential energies of electrons on equal footing. A single electronic molecule has a cigar shape with the length of the order of 1 - 20 nanometers and consisting of 2 - 10 holes. For other oxides the string parameters will not be changed as much. It is also very natural that such "polymeric" electron molecules may form a liquid crystal which may be associated with the stripe phase of HTSC [22].

ACKNOWLEDGMENTS

I am very grateful to A. Bianconi, A. Bishop, S. Kivelson, D. Edwards, G. Gehring, V. Emery, E. I. Rashba, Danya Khomskii, H. S. Dhillon and other participants of the workshop on strongly correlated electrons in Isaac Newton Institute (Cambridge) for illuminating discussions. The work has been supported by Isaac Newton Institute, University of Cambridge.

REFERENCES

1. Mook, H.A, Dai, P., Dogan, F., Hunt, R.D., *Nature* **404**, 729 (2000).
2. Sharma, R.P. *et al.*, *Nature* **404**, 736 (2000).
3. Lanzara, A., Zhao, G., Saini, N.L., Bianconi, A., Conder, K., Keller, H., and Müller, K.A., *J. Phys. Cond. Matt.* **11**, L541 (1999).
4. Uehara, M., Mori, S., Chen, C.H., and Cheong, S.W., *Nature* **399**, 560 (1999); Mori, S., Chen, S. *et al.*, *Nature* **392**, 473 (1999).
5. Kusmartsev, F.V., *J. de Physique IV*, **9**, Pr10-321, (1999).
6. Kusmartsev, F.V., *Phys. Rev. Lett.* **84**, 530, 5026 (2000).
7. Rashba, E.I., in *Excitons*, edited by Rashba, E.I., and Sturge, M.D., North-Holland, Amsterdam, 1982, pp.543.
8. Nagaev, E.L., *Sov. Jour.- JETP Lett.*, **16**, 558 (1972); Kaschin, V.A., and Nagaev, E.L., *Zh. Eks. Teor. Fiz.*, **66**, 2105 (1974).
9. Zaanen, J.R., and Gunnarson, O., *Phys. Rev. B* **40**, 7391 (1989).
10. Löw, U., Emery, V.J., Fabricius, K., and Kivelson, S.A., *Phys. Rev. Lett.* **72**, 1918 (1994).
11. Emery, V.J., and. Kivelson, S.A., *Nature* (London) **374**, 434 (1995); Emery, V.J., Kivelson, S.A., and Zachar, O., *Phys. Rev. B* **56**, 6120 (1997).
12. Emery, V.J., Kivelson, S.A., and Lin, H.Q., *Physica B* **163**, 306 (1990); *Phys. Rev. Lett.* **64**, 475 (1990).
13. Bianconi, A., *Phys. Rev. B* **54**, 12018 (1996); v. Zimmermann. M. *et al.*, *Eur.Phys. Lett.* **41**, 629 (1998).
14. Thursten, T.R. *et al.*, *Phys. Rev. B* **40**, 4585 (1989).
15. Tranquada, J.M., *Nature* (London) **375**, 561 (1995).
16. Bianconi, A. *et al.*, *Phys. Rev. Lett.* **76**, 3412 (1996); and see references therein.
17. Mook, H.A., Dai, P.C., Hayden, S.M., Aeppli, G., Perring, T.G., and Dogan, F., *Nature* **395**, 580 (1998).
18. Saini, N.L., Avila, J., Bianconi, A., Lanzara, A., Asensio, M.C., Tajima, S., Gu, G.D., and Koshizuka, N., *Phys. Rev. Lett.* **79**, 3467 (1997).
19. Pekar, S.I., *Untersuchungen über die Elektronentheorie Kristalle*, Akademie Verlag, Berlin, 1954.
20. Gorkov, L.P., and Sokol, A.B., *Pisma Zh. Eksp. Teor. Fiz.* **46**, 333 (1987).
21. Kusmartsev, F.V., in preparation.
22. Emery, V.J., Kivelson, S.A., and Tranquada, J.M., *Proc. Natl. Acad. Sci. USA* **96**, 8814 (1999).
23. Shklovskii, B.I., and Efros, A.L., *Electronic Properties of Doped Semiconductors*, Springer-Verlag, New York, 1984.

Stripes, Lattice Distortions and High T_c Superconductivity

N. L. Saini[*], H. Oyanagi[§] and A. Bianconi[*]

[*]Unitá INFM and Dipartimento di Fisica, Università di Roma "La Sapienza"
P.le Aldo Moro 2, 00185 Roma, Italy
[§]Electrotechnical Laboratory, 1-1-4 Umezono, Tsukuba 305-8568 Japan

Abstract. We have used x-ray absorption spectroscopy as a tool to study local lattice distortions in the perovskite superconductors. Temperature dependent Cu K-edge extended x-ray absorption fine structure (EXAFS) has been used to measure the amplitude of local lattice distortions. The stripe formation temperature T_{so} is determined by the upturn of the Debye-Waller factor due to the onset of local lattice distortions. The x-ray absorption near edge structure (XANES) spectra show an abrupt change in the spectral transfer, revealed by the multiple scattering peak intensities, due to a particular change in the local geometry at the stripe ordering temperature. A study of local lattice distortions in different materials with increasing chemical pressure on the CuO_2 layers reveals that there are critical local lattice fluctuations in the high Tc superconductors. We find a decrease of the amplitude of lattice fluctuations at the superconducting transition. The present results provide clear evidence for a vital role of the electron-lattice interactions in the charge-stripe ordering phenomena and the high Tc superconductivity.

INTRODUCTION

A growing number of experiments have indicated active role of local lattice structure in high T_c superconductivity, however, it has been generally ignored in the theoretical models because solution of local lattice distortions in the cuprates is not easy (due to smaller disorder, giving ambiguity in the magnitude of atomic displacements, that depends also on the time scale of the experimental techniques used). Evidence for presence electronic and lattice instabilities are, however, beyond any doubt now. Recent experiments on the isotope effect [1-4] have further demanded the need to include the electron lattice interaction as one of the parameters to describe the complex metallic phase of these superconductors. Indeed, the experiments support the emerging idea that the electronic ground state of the cuprates is not a single component but it is inhomogeneaous involving lattice, charge and spin degrees of freedom.

There have been a number of arguments on coexisting two components [5-8], where one of them comprises of charges associated with local atomic displacements: polarons or charge density waves [5-11]. The segregation of the two component on a mesoscopic scale gives a complex phase with charge carriers ordered in striped domains [12-15]. Indeed, the approach based on inhomogeneous charge distribution

CP554, *Physics in Local Lattice Distortions,* edited by H. Oyanagi and A. Bianconi
© 2001 American Institute of Physics 1-56396-984-X/01/$18.00

in stripes is making its ground among conventional theories to understand the physics of high T_c superconductors and several experiments are found to be consistent with the presence of striped phases in these materials [1-17].

The contribution of x-ray absorption spectroscopy (XAS) [18] has been vital to explore local lattice distortions as well as local density of states in the high T_c cuprates due to recent developments of high brilliance and polarized x-ray synchrotron radiation sources allowing to get directional information around a selective site. The x-ray absorption coefficient $\mu(E)$ is given by the product of the matrix element and the joint density of states for the electronic transitions from the initial to final states. The dipole matrix element from the initial state, the core level of well defined symmetry, selects the partial density of final states for the allowed electronic transitions. The x-ray absorption near edge structure (XANES) probes the final states in an energy range of about 50 eV above the chemical potential. While at the Fermi level the mean free path is large, it decreases rapidly with increasing the energy of the final state since it is strongly scattered by the many body electronic excitations and its mean free path becomes of the order of 5 Å. The XANES spectra can be solved in the real space multiple scattering describing the final state as an outgoing spherical wave which interferes with the waves backscattered from the neighbouring atoms within a cluster of atoms in the intermediate range of the order of 5 Å. The extended x-ray absorption fine structure (EXAFS) region (> ~ 50 eV), wherein the photoelectron mean free is small at higher energy, provides direct information on the pair distribution function (PDF) with respect to the absorber atom. The EXAFS amplitude depends on several factors as can be seen from the following equation for polarized K-edge EXAFS [18]:

$$\chi(k) = \frac{m\pi}{h^2} \sum_i 3N_i \cos^2(\theta_i) \frac{S_o^2}{kR_i^2} f_i(k, R_i) e^{-2R_i/\lambda} e^{-2k^2\sigma_i^2} \sin\left[2kR_i + \delta_i(k)\right]$$

Here N_i is the equivalent number of neighboring atoms, at a distance R_i, located at an angle (θ_i) with respect to the electric field vector of the polarized synchrotron light. S_o^2 is an amplitude correction factor due to photoelectron correlation (also called passive electrons reduction factor), $f_i(k, R_i)$ is the backscattering amplitude, λ is the photoelectron mean free path, and σ_i^2 is the correlated Debye-Waller factor of the photoabsorber-backscatterer pairs. Apart from these, the photoelectron energy origin E_0 and the phase shifts δ_i should be known. Here it should be recalled that the Debye-Waller factor in the EXAFS measurements is not the same as the one observed by diffraction experiments. While the diffraction Debye-Waller factor represents the mean-square deviation of a given atom with respect to its average site (e.g., the mean square deviation of the copper (σ_{Cu}^2) and oxygen (σ_O^2) atoms about their average crystallographic positions), the EXAFS Debye-Waller factor measures the broadening of the distance between the two atomic sites (i.e., the σ_{CuO}^2 measures the correlated distribution between the absorber (Cu) and the backscatter (O)) [19].

Here we have exploited capabilities of the Cu K-edge XANES and EXAFS spectroscopy to study local lattice distortions in the inhomogeneous CuO_2 plane of the high T_c superconductors. Cu K-edge XAS (XANES and EXAFS) spectroscopy, a local tool with a time scale of $\sim 10^{-15}$ sec, provides information on the instantaneous atomic displacements around the Cu-site within a cluster of atoms and is well suited to investigate the local lattice displacements in the complex systems such as cuprate superconductors. Temperature dependent XANES and EXAFS spectra are measured on different systems showing a particular change in the local lattice displacements, revealed by an abrupt change in the spectral weight and a clear increase of the correlated Debye-Waller factor of the Cu-O bonds, across the charge stripe ordering temperature. On the basis of several model examples we argue that onset of this particular local lattice distortion is a response function to the charge-stripe order in the cuprate superconductors. We have explored the local lattice fluctuations in the CuO_2 plane as a function of the chemical pressure. The chemical pressure induces an anisotropic strain in the CuO_2 plane. The relative compression of the average in-plane $<R_{Cu-O}>$ bond-lengths, $\varepsilon = 2((d_0 - <R_{Cu-O}>)/d_0))$, has been found to be directly related to the chemical pressure. Here the d_0 is Cu-O equilibrium distance, i.e., Cu-O bond length for an unstrained CuO_2 plane. The d_0 is measured to be ~ 1.985 (±0.005) Å on an undoped model system $Sr_2CuO_2Cl_2$, which is consistent with others [20]. The d_0 is taken to be 1.97 Å throughout this paper considering the correction due to effect of hole doping on the Cu-O bonds [21] (~ 0.16 doped holes per Cu site). We have determined ε in different families of cuprates at a fixed charge density by measuring $<R_{Cu-O}>$. The temperature dependent distribution of the local lattice distortions (dynamic and static) is measured by the correlated Debye-Waller factor of the Cu-O bond lengths. The results provide a direct experimental evidence for the presence of a critical strain ε_c in the CuO_2 plane where a large enhancement of the local lattice fluctuations occur and the superconducting transition temperature reaches the maximum.

EXPERIMENTAL

Well-characterized superconducting single crystals of different superconducting systems were used for the x-ray absorption measurements. For some of the near edge absorption measurements, we have used powder samples to have access to both in-plane and out-of-plane local geometrical information in the same sample. The Cu K-edge x-ray absorption measurements were performed at the beamlines BM29 and BM32 at the European Synchrotron Radiation Facility (ESRF), Grenoble and BL13B of Photon Factory, Tsukuba. At the BM29 the synchrotron radiation emitted by a Bending magnet source at the 6 GeV ESRF storage ring was monochromatized by a double crystal Si(311) monochromator. For temperature dependent measurements at the BM29 the samples were mounted in a closed cycle two stage He cryostat. The Cu K_α fluorescence yield (FY) off the samples were collected using 13 Ge element solid state detector system to measure the absorption signal. A Si(111) was used as monochromator and a 30-element Ge x-ray detector array was used to measure the absorption spectra at the BM32. At the BL13B the synchrotron radiation emitted by a 27-pole wiggler source at the 2.5 GeV Photon Factory storage ring was monochromatized by a double crystal Si(111) and sagittally focused on the sample. The samples were mounted in a closed cycle refrigerator placed on a Huber 420 goniometer. The spectra were recorded by detecting the fluorescence photons using a

19-element Ge x-ray detector array [22]. The sample temperatures were controlled and monitored to within an accuracy of ±1 K. As our standard experimental approach, several absorption scans were collected to limit the noise level to the order of 10^{-4}. Standard procedure was used to extract the EXAFS signal from the absorption spectrum and corrected for the x-ray fluorescence self-absorption before the analysis. Further details on the experiments and data analysis could be found in our earlier publications [12, 19, 23-24].

LOCAL LATTICE DISTORTIONS AS RESPONSE FUNCTION OF THE STRIPE ORDERING

We have studied several cuprates that are found to show charge stripe ordering in the CuO_2 plane. Widely discussed example is the $La_{1.48}Sr_{0.12}Nd_{0.4}CuO_4$ (LNS125) [26] and hence we start our discussion with this model compound. The Fourier transforms (FT) of the EXAFS spectra (multiplied by k^2) recorded in the in-plane geometry (E//Cu-O-Cu) of the LNS125 at two temperatures, a temperature below the charge stripe ordering ($T<T_{so}$) and above it ($T>T_{so}$) are shown in Fig. 1. The FT represents the global atomic distribution of nearest neighbors around the absorbing Cu atom in the LNS125 system. The Fig. 1 represents raw experimental data and show the standard peaks with main peaks denoted by Cu-O, La(Nd, Sr) and Cu-O-Cu representing respectively scattering of the ejected photoelectron at the Cu site with the nearest in-plane oxygen atoms, La(Nd,Sr) atoms (sitting at ~ 3.2Å and 45° from the direction of the photoelectron) and a direct multiple scattering with the next Cu atom (at ~ 3.8Å), across the in-plane oxygen. The high signal to noise in the data could also be judged from the FT as one hardly expects any true signal above ~ 9 Å and above this range the FT amplitude merely gives an idea of the noise in the data. The FT across the charge stripe ordering temperature shows some changes. The amplitude of

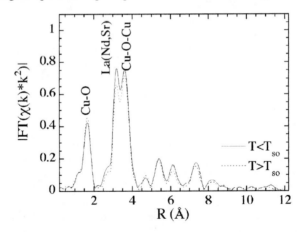

FIGURE 1. Fourier transforms of the Cu K-edge EXAFS spectra measured on the LNS125 system across the stripe ordering temperature.

the La(Nd, Sr) and Cu-O-Cu peaks increases below the charge stripe ordering temperature, however, this is partly due to decreased temperature. On the other hand, an anomalous change is observed in the amplitude of the Cu-O peak that gets decreased instead of a possible increase because of smaller thermal damping at lower temperature.

Here we focus our attention on the temperature dependent behavior of the EXAFS Debye-Waller factors of the Cu-O pairs σ_{CuO}^2 that takes into account both static and dynamic distortions of the CuO_2 plane. The temperature dependence of the Debye-Waller factor of the Cu-O pairs, σ_{CuO}^2, determined by the analysis of the Cu-O EXAFS, is shown in Fig. 2. The Debye Waller factor of the Cu-O bonds shows an anomalous upturn below 70 K. It is known that at the appearance of any charge density wave like instability the Debye-Waller factor shows an anomalous change as found in several density wave systems [27]. Therefore the anomalous increase in σ_{CuO}^2 is due to a charge density wave like instability in the CuO_2 plane of the LNS125. Considering evidences of stripe ordering in the LNS125, provided by several experimental techniques, we have assigned the anomalous upturn in the σ_{CuO}^2 to a charge instability giving stripe ordering in this system. We have earlier discussed that the anomalous upturn in the σ_{CuO}^2 is a good indicator of charge stripe ordering in the cuprates and particular local lattice displacements in the CuO_2 is a response function of the stripe order parameter [19, 23].

Let us now turn to discuss the XANES results on the LNS125. Fig. 3 shows representative Cu K-edge XANES spectra measured in the in-plane (**E//ab**) and out-of-plane (**E//c**) polarizations. The main absorption features are denoted by B_1 and B_2 in the **E//ab** spectrum and A_1 and A_2 in the **E//c** spectrum. The absorption features are due to multiple scattering of the photoelectron emitted at the Cu site in the direction of the electric field of the x-ray beam, and their physical origin has been revealed by XANES calculations for La_2CuO_4 [28]. The features A_1 and A_2 are determined by multiple scattering of the ejected photoelectron off apical oxygen and La, Nd and Sr

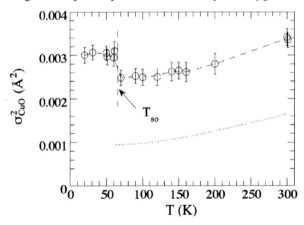

FIGURE 2. Temperature dependence of the Cu-O(planar) Debye-Waller factor σ_{CuO}^2, measured by EXAFS on the LNS125 system. The σ_{CuO}^2 shows anomalous increase below the stripe ordering temperature T_{so}.

atoms. On the other hand, the peak B_1 corresponds to multiple scattering of the photoelectron off the in-plane oxygen and Cu in the CuO_2 plane. The peak B_2 includes multiple scattering contribution, similar to the main peak B_1 and also contributed by many body "shake up" satellite of it. The unpolarized spectrum has been shown in the lower panel. The unpolarized spectrum contains information on the local and instantaneous displacements in the overall cluster around the central atom. In fact in complex systems, such as cuprate superconductors, the displacements are anisotropic with doping and temperature and hence the unpolarized spectrum is a useful tool to obtain important information on the instantaneaous geometrical displacements around the selected site (*i.e.* the Cu). Alongwith the unpolarized spectrum, we have shown an absorption difference between two temperatures across the stripe ordering temperature. The absorption difference shows temperature dependent high-energy spectral weight transfer across the stripe ordering temperature resulting in positive and negative differences with a maximum difference of ~ 2% of the normalized absorption. On cooling the sample, the spectral weight around the peak A_1 (A_2) is decreased with an increase around the peak B_1 (B_2).

We have made an attempt to understand this difference by simulating Cu K-edge XANES spectra for different distortions using multiple scattering calculations. The results of the calculations suggest that the spectral weight transfer is due to a particular octahedral distortion with one of the elongated in-plane Cu-O bonds and a decreased

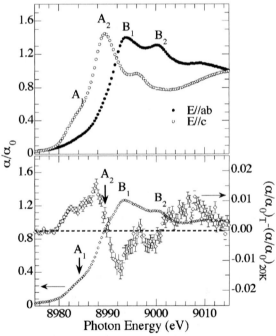

FIGURE 3. Polarized Cu K-edge XANES of the LNS125 (upper). An unpolarized spectrum measured on the LNS125 system has been shown (lower) alongwith an absorption difference at two temperatures across the stripe ordering temperature in the system.

axial Cu-O distance. This LTT-like rhombic distortion that has been previously determined by polarized Cu K-edge measurements, found to be common to the CuO_2 plane of the cuprate superconductors [12, 19, 25, 29]. Indeed this is the Q_2- type mode of the Jahn-Teller distortion [14].

The temperature evolution of the spectral weight transfer is shown in Fig. 4 where we have plotted the intensity ratio $R=(b_1-a_1)/(b_1+a_1)$. Here the b_1 and a_1 represents intensities of the peaks B_1 and A_1 probing high energy spectral weight transfer (from the core excitations $Cu1s \rightarrow \varepsilon p_z$ (peak A_1) to the $Cu1s \rightarrow \varepsilon p_{xy}$ (peak B_1). The ratio R shows a step-like increase below the charge-stripe ordering temperature. We have studied another model system, $La_{1.875}Ba_{0.125}CuO_4$ (LBC125) and measured temperature dependent ratio $R=(b_1-a_1)/(b_1+a_1)$ as a function of temperature. The result is shown in Fig. 5 revealing again a similar change across the stripe ordering temperature in the LBC125 indicating geometrical changes in the CuO_2 plane due to particular local lattice displacements giving rise to the Q_2-type distortions. These results further suggest that the local lattice distortions of the CuO_2 plane (Q_2-type of rhombic distortions), revealed by the ratio $R=(b_1-a_1)/(b_1+a_1)$ and an anomalous Debye-Waller factor of the Cu-O bonds, could be considered as a suitable response function of the charge stripe order parameter.

We have studied doping dependence of the stripe ordering temperature (defined by the abrupt local geometrical change as reflected by the anomalous increase of the XANES peak intensity ratio R) in the La-based cuprate superconductors [1, 30]. The stripe ordering temperatures, found by XANES measurements in different systems, are found to agree quite well with the stripe ordering temperature obtained by wipeout fraction of Cu NQR [31]. Recently, the response function of stripe order, determined by the XANES spectroscopy, has been used to study effect of oxygen isotope effect on the stripe ordering in the $La_{1.94}Sr_{0.06}CuO_4$ system [1]. The stripe ordering temperature was found to change from 110 ($\pm$10) K to 170 ($\pm$10) K while substituting ^{18}O in place of ^{16}O in this model compound. The large isotope shift of ~ 60 K not only suggests a

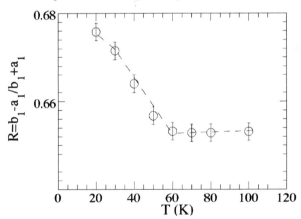

FIGURE 4. Temperature evolution of the peak intensity ratio $R=(b_1-a_1)/(a_1+b_1)$ for the $La_{1.48}Sr_{0.12}Nd_{0.4}CuO_4$. The ratio shows a clear change across the stripe ordering temperature. A temperature dependent contribution of normal thermal contraction of the sample has been subtracted.

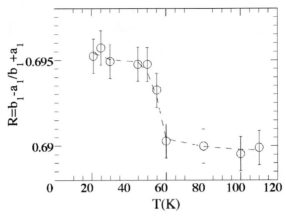

FIGURE 5. Temperature evolution of the peak intensity ratio $R=(b_1-a_1)/(b_1+a_1)$ for the $La_{1.875}Ba_{0.125}CuO_4$. The ratio shows an abrupt change across the stripe ordering temperature. A temperature dependent contribution of normal thermal contraction of the sample has been subtracted.

large change in the local structure of the CuO_2 plane due to isotope substitution but also ascertains important role of local lattice displacements in the stripe ordering phenomena in the cuprates.

To summarize this part, we have used Cu K-edge absorption (EXAFS and XANES) spectroscopy to explore response function of the stripe ordering in the perovskite superconductors. We find an order parameter like change in the instantaneous local lattice fluctuations across the stripe ordering temperature, revealed by an increase in the correlated Debye Waller factor of the Cu-O pairs (EXAFS) and by an abrupt change in the anisotropic and instantaneous local displacements around the Cu (XANES). The order parameter like change is found to be related to the Q_2 - type of Jahn-Teller distortion of the CuO_2 plane. On the basis of the studies on several model systems, we have argued that this local lattice distortion is a direct response function of the stripe order in the cuprates. The results constructs a clear evidence for importance of electron-lattice interactions in the stripe ordering. Moreover, a large isotope effect on the charge stripe ordering [1] and pseudogap temperature [2] suggest that electron-lattice interactions are important ingredients for the high T_c in the doped cuprates.

LOCAL LATTICE DISTORTIONS AS A FUNCTION OF CHEMICAL PRESSURE

Structurally the high T_c cuprate perovskites are heterogeneous materials made of alternated layers of metallic body centered cubic (bcc) CuO_2 layers and insulating rock-salt face centered cubic (fcc) M-O (M = Ba, Sr, La) layers [21, 32, 33]. The mismatch between the two sub-lattices is generally estimated by $1-t = [r(A-O)]/\sqrt{2}$ $[r(Cu-O)]$ where r(A-O) (i.e., r(La-O), r(Sr-O) and r(Ba-O)) and r(Cu-O) are the respective bond lengths and t is the Goldschmidt tolerance factor [32, 33]. Due to the lattice mismatch the CuO_2 sheets are under compression and (M-O) layers under tension. Here we have investigated local lattice distortions as a function of the local strain in the CuO_2 plane due to the lattice mismatch. The $La_2CuO_{4.1}$ (LCO) with $T_c \sim$ 40 K, $Bi_2Sr_2CaCu_2O_{8+\delta}$ (Bi2212) with $T_c \sim 87$ K and $HgBa_2CuO_{4+\delta}$ (Hg1201) with $T_c \sim 94$ K (and $HgBa_2CaCu_2O_{6+\delta}$ (Hg1212) with $T_c \sim 116$ K) systems are used as representatives for the La-based, Bi-based and Hg-based superconducting families containing respectively the La-O, Sr-O and Ba-O as rock-salt layers that sustain different chemical pressure on the CuO_2 planes, where the dopants are interstitial oxygen ions in the block layers. We have used in-plane polarized Cu K-edge EXAFS to obtain the information about the local lattice distortions in the CuO_2 plane.

The Cu-O Debye Waller factor (σ^2_{CuO}), is the parameter used to make a systematic comparison. Since the EXAFS Debye-Waller factors of the Cu-O pairs σ^2_{CuO} takes into account both static and dynamic distortions, it is an important structural parameter to study temperature evolution and compare the role of lattice fluctuations in the CuO_2 plane. The temperature dependence of the σ^2_{CuO} is shown in Fig. 6 for the LCO ($T_c \sim$ 40 K), Bi2212 ($T_c \sim 87K$) and Hg1201 ($T_c \sim 94K$) systems representing the three different families of the cuprates.

The σ^2_{CuO} shows abnormal temperature dependence revealing several well-defined structures. We can easily define at least two anomalous temperatures where the σ^2_{CuO} show anomalies. There is an anomalous increase below the stripe ordering temperature T_{so} followed by a decrease around the superconducting transition temperature T_c. The increase at T_{so} appears in the LCO and Bi2212 systems, however, the Hg1201 system does not show any evident up turn. On the other hand, the drop in σ^2_{CuO} at the superconducting transition temperature T_c appears common to all the systems (however, less evident in the LCO system). The Hg1201 system manifests a large decrease in the σ^2_{CuO} around the superconducting transition temperature. Here

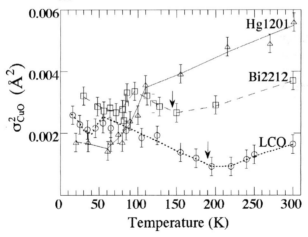

FIGUERE 6. Temperature dependence of the Cu-O(planar) Debye-Waller factor σ^2_{CuO} for the LCO (circles), Bi2212 (squares) and Hg1201(triangles). The line are guide to the eyes. The error bars represent the average estimated noise level. The σ^2_{CuO} shows abnormal temperature dependence with an increase below the stripe ordering temperature (down-headed arrows).

we should mention that, apart from the static and dynamic distortions of the CuO_2 lattice, σ^2_{CuO} contains contribution from the thermal vibrations. However, in the present case the thermal contribution to σ^2_{CuO} should be similar for all the systems and hardly affects the discussion in this paper.

As discussed above, the anomalous increase in σ^2_{CuO} is due to stripe ordering in the CuO_2 plane of the LCO ($T_{co} \sim 190$ K) and Bi2212 ($T_{co} \sim 140$ K) systems. X-ray diffraction measurements on the LCO system has clearly demonstrated onset of charge ordering to be ~ 190 K [34] which is consistent with the observation of the upturn in the σ^2_{CuO} in the system. Recently Sharma et al [35] have further confirmed our earlier results and found a clear upturn in the temperature dependence of the excess displacements (a parameter similar to the Debye-Waller factor measuring dynamic and static distortions) measured by ion-channeling on the YBCO system at the stripe ordering temperature. In fact, below this temperature the pair distribution function becomes larger than that due to thermal fluctuations and the formation of striped phase should give an asymmetric PDF due to splitting of the Cu-O bonds as demonstrated earlier [19, 23].

The lower temperature anomaly in σ^2_{CuO} appears around the superconducting transition temperature T_c. The correlated Debye-Waller factor σ^2_{CuO} shows an anomalous decrease around the superconduting transition temperature. This is a clear indication that the appearance of the superconducting state is accompanied by a decrease of the instantaneous local lattice distortions pointing towards a key role of electron lattice interactions in the superconducting pairing. Interestingly, the drop at the T_c is different for different systems and found to be maximum for the Hg-based compound where the block-layers are Ba-O with smaller local strain in the CuO2

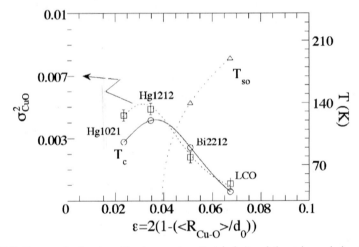

FIGUERE 7. Superconducting transition temperature T_c (circles) and the stripe ordering temperature T_{so} (triangles) as a function of the local strain ε. The Debye –Waller factors for the Cu-O bonds (squares) in the normal state are also shown. All the lines are guide to the eyes. There seems to be a critical value for the ε ($\sim 0.04 \pm 0.005$) where the T_c and the local lattice distortions appear to diverge indicating an intimate relationship between the two parameters.

plane than the case of Bi2212 (Sr-O) and LCO (La-O). Below the T_c the σ^2_{CuO} shows a small increase, but this is within the limits of experimental uncertainties. The anomalous decrease at the superconducting transition temperature and similar values around this characteristic temperature provides a direct evidence for an important role of critical lattice fluctuations in the superconducting state. At the superconducting transition the drop of the σ^2_{CuO} revealing decreased distortions at T_c could be due to decrease of the incoherent local lattice distortions to transfer electron lattice interaction energy in the pairing mechanism entering into a coherent state.

We have estimated the local strain in the CuO_2 plane ε, for different systems from the measured <Cu-O> bond-lengths by EXAFS. Fig. 7 shows ε dependence of the Cu-O distance broadening (σ^2_{CuO}) for different superconducting systems is their normal state (200 K). The ε dependence of the T_{so} and the superconducting transition temperature T_c is also shown. The σ^2_{CuO} and the superconducting transition temperature T_c seems to diverge around the $\varepsilon_c \sim 0.04 \pm 0.005$. At this critical value the superconducting transition temperature reaches the maximum value while the quantum lattice fluctuations diverge. This merely indicates an intimate relationship between the lattice fluctuations and the superconductivity.

In summary, we have determined local lattice distortions in high T_c superconductors at with variable chemical pressure on the electronically active CuO_2 plane at a fixed charge density ($\delta = 0.16$). Polarized Cu K-edge EXAFS has been used to measure the Cu-O bond distances, giving estimation of the local strain, and the distance broadening (given by the Debye-Waller factors) providing direct information on the static and dynamic lattice distortions. We find that the Debye-Waller factor shows a drop at the superconducting transition temperature revealing importance of electron-lattice interaction in the pairing mechanism. Furthermore, the Debye-Waller factor shows an upturn at the stripe ordering temperature. The results reveal that the superconducting T_c reaches at the maximum near a critical local strain in the CuO_2 plane.

In conclusion, the present work provides a direct evidence for an intimate relationship between the stripe ordering, local lattice fluctuations and high T_c superconductivity in the perovskite cuprates. Here, the local strain in the electronically active CuO_2 plane ε, that controls the local lattice fluctuations and hence the superconductivity and the normal state phenomena of the perovskite superconductors, is a key parameter to unlock the mystery of the phase diagram. The present findings suggest that, apart from the temperature and doping, the $\lambda(\varepsilon)$ should be included as a new axis to understand the the high T_c phase diagram [36-38].

ACKNOWLEDGMENTS

This research has been supported by the Ministero dell'Università e della Ricerca Scientifica (MURST) under the Programmi di Ricerca Scientifica di Rilevante Interesse Nazionale coordinated by R. Ferro, by Istituto Nazionale di Fisica della Materia (INFM) and by Progetto 5% Superconduttività del Consiglio Nazionale delle Ricerche (CNR).

REFERENCES

1. Lanzara, A., Zhao, G.-m., Saini, N.L., Bianconi, A., Conder, K., Keller, H., and Müller, K.A., *J. Phys., Condens. Matter* **11**, L541 (1999).

2. Rubio Temprano, D., Mesot, J., Janssen, S., Conder, K., Furrer, A., Mutka, H., and Müller, K.A., *Phys. Rev. Lett.* **84**, 1990 (2000).

3. Hofer, J., Conder, K., Sasagawa, T., Zhao, G.-M, Willemin, M., Keller, H., and Kishio, K., *Phys. Rev. Lett.* **84**, 4192 (2000) and references therein.

4. Zhao, G.-m., Hunt, M.B., Keller, H., Müller, K.A., *Nature* (London) **385**, 236-239 (1997).

5. Stevens, C.J., Smith, D., Chen, C., Ryan, J.F., Podobnik, B., Mihailovic, D., Wagner, G.A., and Evetts, J. E., *Phys. Rev. Lett.* **78**, 2212 (1997).

6. Kamimura, H., and Sano, A., *J. Superconductivity* **10**, 279 (1997).

7. Mihailovich, D., and Müller, K.A., *High Tc Superconductivity: Ten years after the Discovery*, Nato ASI, *Vol.* 343, edited by E. Kaldis, E. Liarokapis, K. A. Müller, Dordrecht, Kluwer, 1996, pp243.

8. Bozin, E.S., Billinge, S.J.L., Kwei, G.H., and Takagi, H., *Phys. Rev. B* **59**, 4445 (1999) and references therein.

9. Bi, X.-X., and Eklund, P.C., *Phys. Rev. Lett.* **70**, 2625 (1993).

10. Alexandrov, A.S., Bratkovsky, A.M., Mott, N.F., Salje, E.H., *Physica C* **215**, 359 (1993).

11. Perali, A., Castellani, C., Di Castro, C., and Grilli, M., *Phys. Rev. B* **54**, 16216 (1996).

12. Bianconi, A., Saini, N.L., Lanzara, A., Missori, M., Rossetti, T., Oyanagi, H., Yamaguchi, H., Oka, K., and Ito, T., *Phys. Rev. Lett.* **76**, 3412 (1996) and references therein.

13. Goodenough, J.B., and Zhou, J.S., *Nature* **386**, 229 (1997); Goodenough, J.B., and Zhou, J.S., *J. Superconductivity* **10**, 309 (1997); Zhou, J. S., Goodenough, J. B., *Phys. Rev. B* **56**, 6288 (1997) and references therein

14. See e. g., Müller, K.A., *Stripes & Related Phenomena*, ed. A. Bianconi and N.L. Saini, Kluwer / Plenum, New York, 2000, pp. 1.

15. Müller, K.A., Zhao, G.-M., Conder, K., Keller, H., *J. Phys. Condens. Matter* **10**, L291 (1998).

16. Bianconi, A., and Saini, N.L., "Stripes and High T$_c$ Superconductivity", special issue of *J. Superconductivity* **vol. 10**, 1997, No.4.

17. *Stripes & Related Phenomena*, eds. A. Bianconi, and N.L. Saini, Kluwer / Plenum, New York, 2000.

18. *X Ray Absorption: Principle, Applications Techniques of EXAFS, SEXAFS and XANES* edited by R. Prinz and D. Koningsberger, J. Wiley and Sons, New York, 1988.

19. Saini, N.L., Lanzara, A., Bianconi, A., Oyanagi, H., Yamaguchi, H., Oka, K., and Ito, T., *Physica C* **268**, 121 (1996).

20. Miller, L.L., Wang, X.L., Wang, S.X., Stassis, C., Johnston, D.C., Faber, J.,Jr and C.-K. Loong, *Phys. Rev. B* **41**, 1921 (1990).

21. Edwards, P.P., Peakok, G.B., Hodges, J.P., Asab, A., and Gameson, I., *High T$_c$ Superconductivity: Ten years after the Discovery*, Nato ASI, *Vol.* 343, ed E. Kaldis, E. Liarokapis, and K.A. Müller, Dordrecht, Kluwer, 1996, pp135.

22. Oyanagi, H., Martini, M., Saito, M., *Nuc. Int. Meth. Phys. Res. A* **403**, 58 (1998).

23. Saini, N.L., Lanzara, A., Oyanagi, H., Yamaguchi, H., Oka, K., Ito, T., and Bianconi, A., *Phys. Rev. B* **55**, 12759 (1997).

24. Saini, N.L., Lanzara, A., Bianconi, A., Oyanagi, H., *Phys. Rev.B* **58**, 11768 (1998).

25. Lanzara, A., Saini, N.L., Bianconi, A., Hazemann, J.L., Soldo, Y., Chou, F.C., and Johnston, D.C., *Phys. Rev. B* **55**, 9120 (1997).

26. Tranquada, J.M., Ichikawa, N., and Uchida, S., *Phys. Rev. B* **59**, 14712 (1999).

27. Grüner, G., "Density Waves in Solids", in *Frontiers in Physics Vol.* **89**, Addison-Wesley, USA, 1994.

28. Li, C., Pompa, M., Congiu Castellano, A., Della Longa, S., and Bianconi, A. *Physica C* **175**, 369 (1991).

29. Bianconi, A., Saini, N.L., Rossetti, T., Lanzara, A., Perali, A., Missori, M., Oyanagi, H., Yamaguchi, H., Nishihara, Y., and Ha, D.H., *Phys. Rev. B* **54**, 12018 (1996).

30. Saini, N.L., Oyanagi, H., and Bianconi, A., *J. Synch. Rad.*, to be published, 2000.

31. Hunt, A.W., Singer, P.M., Thruber, K.R., Imai, T., *Phys. Rev. Lett.* **82**, 4300 (1999); Singer, P.M., Hunt, A.W., Cederström, A.F., and Imai, T., *Phys. Rev. B* **60**, 15345 (1999).

32. Rao, C.N.R., and Ganguli, A.K., *Chem. Soc. Rev.* **24**, 1 (1995).

33. Goodenough, J.B., *Supercond. Science and Technology* **3**, 26 (1990); Goodenough, J.B., and Marthiram, A., *J. Solid State Chemistry* **88**, 115 (1990).

34. Bianconi, A., Di Castro, D., Bianconi, G., Pifferi, A., Saini, N.L., Chou, F.C., Johnston, D.C. and Colapietro, M., *Physica C*, 341-348, 1719-1722, (2000).

35. Sharma, R. P., Ogale, S. B., Zhang, Z. H., Liu, J. R., Wu, W. K., Veal, B., Paulikas, A., Zhang, H., and Venkatesan, T., *Nature* (London) **404**, 736 (2000).

36. Bianconi, A., Agrestini, S., Bianconi, G., Di Castro, D., and Saini, N.L., "Lattice charge stripes in high T$_c$ superconductors" in *Stripes and Related Phenomena*, edited by A. Bianconi and N.L. Saini, Kluwer Academic-Plenum Publisher, New Your, 2000, pp. 9.

37. Saini, N.L., Bianconi, A., Lanzara, A., Agrestini, S., Di Castro, D., and Oyanagi, H., *Int. J. Modern Physics B*, in press (2000).

38. Bianconi, A., Di Castro, D., Bianconi, G., and Saini, N.L., in this volume

Suppression of Superconductivity in La$_{1.85}$Sr$_{0.15}$Cu$_{1-y}$Ni$_y$O$_4$: The Relevance of Local Lattice Distortions

D. Haskel[*], E. A. Stern[†], V. Polinger[†] and F. Dogan[‡]

[*]Experimental Facilities Division, Advanced Photon Source, Argonne National Laboratory, Argonne IL 60439, USA
[†]Department of Physics, Box 351560, University of Washington, Seattle WA 98195, USA
[‡]Department of Materials Science and Engineering, Box 352120, University of Washington, Seattle WA 98195, USA

Abstract. The effect of Ni substitution upon the local structure of La$_{1.85}$Sr$_{0.15}$Cu$_{1-y}$Ni$_y$O$_4$ is commonly neglected when addressing the Ni-induced destruction of the superconducting state at $y \approx 0.03$ and a metal-insulator transition at $y \approx 0.05$. It is also sometimes assumed that direct substitution of a dopant into the CuO$_2$ planes has a detrimental effect on superconductivity due to in-plane lattice distortions around the dopants. We present here results from angular-dependent x-ray absorption fine structure (XAFS) measurements at the Ni, La and Sr K-edges of oriented powders of La$_{1.85}$Sr$_{0.15}$Cu$_{1-y}$Ni$_y$O$_4$ with $y = 0.01, 0.03, 0.06$. A special magnetic alignment geometry allowed us to measure pure $\hat{c}$ and $\hat{ab}$ oriented XAFS at the Ni K-edge in identical fluorescence geometries. Both the near-edge absorption spectra (XANES) and the XAFS unequivocally show that the NiO$_6$ octahedra are largely contracted along the c-axis, by $\approx$ 0.16 Å. Surprisingly, the Ni-O planar bonds and the Ni-O-Cu/Ni planar buckling angle are nearly identical to their Cu counterparts. The NiO$_6$ octahedral contraction drives the macroscopic $\hat{c}$-axis contraction observed with Ni-doping. The local $\hat{c}$-axis strongly fluctuates, due to the different NiO$_6$ and CuO$_6$ octahedral configurations and the much stronger bonding of a La^{+3} ion than a Sr^{+2} ion to the O(2) apical oxygens. We discuss the relevance of these findings to the mechanisms of T_c suppresion and hole-localization by Ni dopants.

INTRODUCTION

At first it might seem a rather negative approach: to try to learn about superconductivity by destroying it with Ni dopants. However, the work of Pan *et al.* demonstrated just the opposite [1]. In their scanning tunneling microscopy (STM) experiments on T_c-suppressed Zn doped Bi$_2$Sr$_2$CaCu$_2$O$_8$ at temperatures below T_c, no quasiparticle tunneling current was obtained at zero bias for most of the STM tip locations on the cleaved BiO surface, as expected due to the presence of a superconducting gap. However, a zero-bias tunneling current was measured when the STM tip was located precisely above a Zn dopant (two layers below the BiO surface) indicating a "metallic" or un-gapped state at the Zn dopant sites. By mapping the spatial extent of the Zn impurity-state away from its lattice site in different directions (by measuring the tunneling strength as function of tip displacement away from the

CP554, *Physics in Local Lattice Distortions*, edited by H. Oyanagi and A. Bianconi
© 2001 American Institute of Physics 1-56396-984-X/01/$18.00

dopant), a *d*-wave like dependence was obtained, the impurity-state extending the furthest along the *nodes* of the SC gap. This spatial dependence agrees with theoretical predictions [2,3]. However, an additional, weaker, tunneling signal with spatial extent that corresponds to an impurity state extending *along* the SC gap was also detected. It was proposed that this weaker signal is due to enhanced tunneling between layers along the Cu-O bond directions [3].

It is clear that the local structure around dopants will influence the nature of the impurity states. Electronic states at the Fermi level are derived from O $2p$ and Cu $3d$ atomic orbitals, which are largely affected by deviations in local symmetry through crystal-field splittings. Local distortions around dopants (and related changes in local electronic structure) will affect the tunneling matrix elements at a dopant site, compared to that at a neighboring Cu site. It is therefore of importance to obtain detailed information on the local environment of the dopants when interpreting information from other spectroscopies at the atomic level. It is interesting to note that doping Zn or Ni into the CuO_2 planes of SC cuprates have very similar effects: both contract the crystallographic $\hat{c}$-axis and suppress T_c, although Zn does the latter at a faster rate. In $La_{1.85}Sr_{0.15}Cu_{1-y}Ni_yO_4$, SC is destroyed at $y \approx 0.03$ and a M-I transition takes place at $y \approx 0.05$ [4].

EXPERIMENTAL

Powders of $La_{1.85}Sr_{0.15}Cu_{1-y}Ni_yO_4$ with $y = 0.01, 0.03, 0.06$ were obtained from nitrates by precipitation from solution [5]. Sintering of powder compacts (≈ 3 grams) took place at 1140°C for 24 hours. Lattice parameters were refined at room temperature using 14 Bragg reflections of the *I4/mmm* space group. Superconducting T_c's were obtained from zero field cooled magnetization curves measured by SQUID

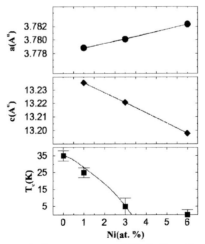

FIGURE 1. Room temperature lattice parameters and T_c's for $La_{1.85}Sr_{0.15}Cu_{1-y}Ni_yO_4$. The value of T_c at $y = 0$ is from Radaelli *et al.* [6].

magnetometry in a H = 1 Gauss applied field. Fig. 1 shows the results from these measurements.

Angular resolved XAFS measurements were taken on magnetically aligned powders. A general description of the alignment procedure is described in ref 5. This procedure results in powders align along their $\hat{c}$-axis but randomly oriented in the $\hat{a}b$-plane. The main advantage of this method over the use of single crystals is control over sample thickness, allowing the use of transmission geometry at the absorption edges of concentrated elements. It also allows controlling the orientation of the $\hat{c}$-axis relative to the sample's surface. The latter allowed us to measure pure $\hat{c}$-axis and $\hat{a}b$-plane orientations at the Ni K-edge in identical fluorescence geometries (Fig. 2). Measurements were done in fluorescence at the Ni K-edge and transmission at the Sr, La K-edges at beamline X11-A of the National Synchrotron Light Source using Si(111) (Ni, Sr K-edges) and Si(311) (La K-edge) double crystal monochromators. Ni K-edge XAFS is limited to about ≈ 13 Å^{-1} due to the appearance of Cu K-edge in the absorption spectra.

Absorption spectra at the Ni K-edge of La$_{1.85}$Sr$_{0.15}$Cu$_{1-y}$Ni$_y$O$_4$ with $y = 0.06$ are shown in Fig. 3 for both orientations of the electric field. Fig. 4 emphasizes the near-edge region of the same spectra.

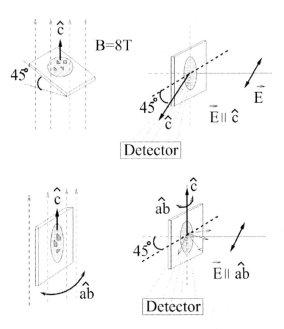

FIGURE 2. Magnetic alignment geometry. Having the crystallographic $\hat{c}$-axis forming 45° with the sample surface allowed measuring pure c-axis Ni K-edge fluorescence ($\hat{E} \parallel \hat{c}$) with the sample surface forming 45° with the x-rays' electric field (top panel). For the $\hat{E} \parallel \hat{a}b$ fluorescence measurements at the Ni K-edge (and transmission measurements in both orientations at the Sr an La K-edges) samples were aligned with the $\hat{c}$-axis in the plane of the sample (bottom panel).

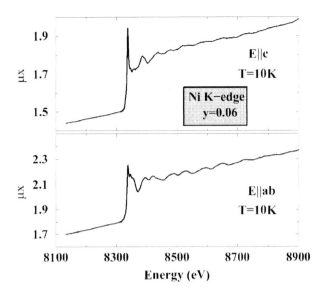

FIGURE 3. Ni K-edge absorption data measured in fluorescence geometry with $\hat{E} \parallel \hat{c}$ (top) and $\hat{E} \parallel \hat{a}\hat{b}$ (bottom) orientations.

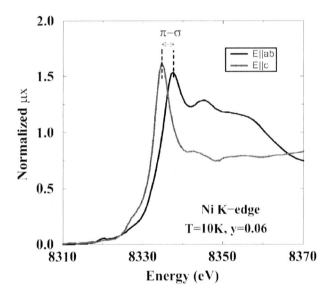

FIGURE 4. Orientation-dependent near-edge structure (XANES) at the Ni K-edge. The $4p_\pi$-$4p_\sigma$ energy splitting is determined by the distortion of the NiO_6 octahedron.

ANALYSIS AND RESULTS

Data analysis was carried out with the UWXAFS analysis package [7] together with theoretical standards from FEFF6 [8]. The orientation dependence of the XAFS signal was included in the FEFF6 calculation by performing the appropiate angular averaging for powder aligned along the $\hat{c}$-axis but randomly oriented in the $\hat{a}b$-plane. Data from both electric field orientations are analyzed concomitantly by constraining the structural parameters of scattering paths that contribute in both orientations, significantly reducing the number of fitting parameters relative to the number of independent points in the data. Coordination numbers were set to the values of the average structure determined by crystallography for undoped $La_{1.85}Sr_{0.15}CuO_4$ [6]. Fig. 5 shows fit results at the Ni K-edge for both electric field orientations. Fits are performed in real-space and multiple scattering (MS) paths that contribute to the XAFS in the fitted region are included in the fits.

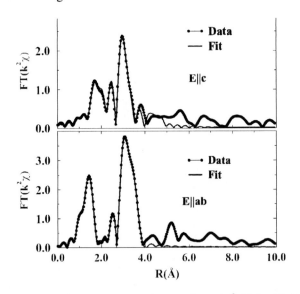

FIGURE 5. Magnitude of the complex Fourier transform of $k^2\chi(k)$ for Ni K-edge in $La_{1.85}Sr_{0.15}Cu_{1-y}Ni_yO_4$ with $y = 0.06$ at T=10K. The k-range used in the transform is $[2.5-12.0]Å^{-1}$. Number of fitting parameters in the constrained fit of both orientations is 16, compared to a total of 45 independent points in the fitted region. Figures of merit for these fits are $\chi_v^2 = 15.8$ and $R = 0.006$. For a definition of these figures of merit see ref. 7.

TABLE 1. Selected fit results at the Ni K-edge. Data corresponds to $y = 0.06$ at T=10K; $S_0^2 = 0.902 \pm 0.067$. Diffraction results are from Radelli *et al.* [6].

	XAFS	Diffraction
Ni-O(1)	1.882(8) Å	1.88773 Å
Ni-Cu	3.789(6) Å	3.7736 Å
Ni-O(2)	2.250(12) Å	2.4145 Å
Ni-La$_c$	4.701(16) Å	4.7604 Å

Structural parameters derived from the fits in Fig. 5 are summarized in Table 1. Although we show here fitting results for a Ni-doping level of $y = 0.06$, no significant differences were found in the fitted values for the lower Ni dopings. The Ni-O(2) apical distance is 0.16(1) Å shorter than the corresponding Cu-O(2) apical distance. The Ni-O(1) planar distance is nearly unchanged relative to the Cu-O(1) distance, although a relative expansion of the former by $<\sim 1$ Å cannot be ruled out based on the slightly larger Ni-Cu planar distance and the measured uncertainties. Ni-O(1)-Cu MS paths contribute to the XAFS at nearly the same distance as Ni-Cu single scattering (SS) paths and are included in the fits. As the O(2)-apical moves towards Ni it drags along its neighbor La atom, as seen by the ≈ 0.06 Å contraction in Ni-La$_c$ distance. Ni-O(2)-La$_c$ MS paths contribute at nearly the same distance as Ni-La$_c$ SS paths and are included in the fits.

The buckling angle α of the Ni-O(1)-Cu planar, nearly collinear, configuration was determined by fitting a parameterized form of the effective scattering amplitudes of double and triple scattering paths, $F_k(\alpha)$, to the data. The parameterization was done by simulating the buckling angle dependence of such amplitudes in FEFF6 calculations (for a detailed description of this procedure see ref. 5). The fitted value of $\alpha = 2.5 \pm 3°$ is in agreement with the average buckling angle determined by diffraction, $\langle \alpha \rangle = 3.61°$, indicating again that the structural disorder introduced by Ni in the CuO$_2$ planes is small. The large uncertainty in buckling angle α is due to the small variation of $F_k(\alpha)$ for small buckling angles near collinearity ($0 \leq \alpha \leq 5°$).

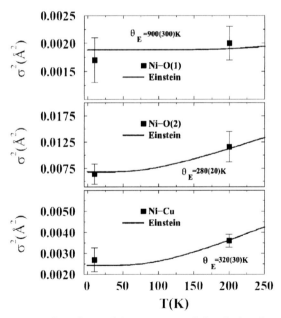

FIGURE 6. Temperature dependence of the mean-squared disorder in selected interatomic distances around Ni atoms together with their fits to Einstein models. The mean-squared disorder in these bonds is solely of thermal origin without any evidence for a static contribution.

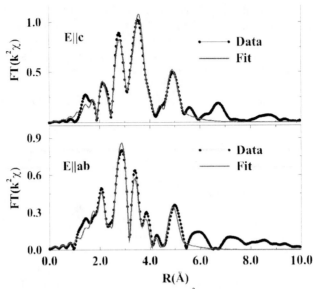

FIGURE 7. Magnitude of complex Fourier transform of $k^2\chi(k)$ for La K-edge at T=10K and $y = 0.06$. Data range used in the transform is $[3,14]$Å^{-1}. The constrained fit of both orientations uses 22 fitting parameters compared to 52 independent points in the fitted region. $S_0^2 = 0.93(6)$; figures of merit are $\chi_v^2 = 14.89$, $R = 0.008$.

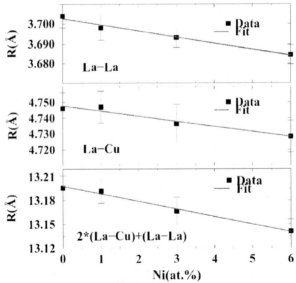

FIGURE 8. Ni-induced contraction of the local $\hat{c}$-axis, averaged over all La sites, at T=10K. The rate of contraction is in good agreement with the macroscopic $\hat{c}$-axis contraction determined from powder diffraction measurements at 300K. $y = 0$ results are from ref. 12.

Whether there is a spatial correlation between the positions of the two types of dopants (Ni,Sr) can be determined from Ni K-edge XAFS by fitting the Ni-(La/Sr) XAFS signal with varying amounts of (Sr/La) neighbors to Ni. Sr and La have very different backscattering amplitudes making plausible separating their isolated contributions to the XAFS. A fit to the Ni-(La/Sr) signal determines a relative Sr concentration of $x = 0.048 \pm 0.06$. For a random solution one expects $x = 0.075$, while if Sr avoids Ni one expects $x = 0$. Unfortunately the fitted value is consistent with either one of these scenarios. However, we can definitely rule out a strong tendency of Sr to occupy sites near Ni atoms. We note that determining the Sr-Ni correlation from Sr K-edge XAFS is not possible, due to the nearly identical backscattering amplitudes of Ni and Cu atoms.

Fitted values for the mean-squared disorder in selected interatomic distances at T=10, 200K together with their parameterization to an Einstein model are shown in Fig. 6. We found no evidence for a static contribution to the mean-squared disorder in these distances. The much weaker effective force constant of the Ni-O(2) apical bond compared to that of Ni-O(1) planar bond is evident from the much larger temperature dependence of the former. This weaker bonding allows the O(2) apical to relax towards the Ni^{+2} ion while the strongly covalent Ni-O(1) bonding opposes the relaxation of O(1) atoms, thus preventing a true *anti-Jahn-Teller* distortion at the Ni site to materialize, as discussed below.

The XANES spectra shown in Fig. 4 fully support the XAFS results regarding the distortion of the NiO$_6$ octahedra. The ratio of the out-of-plane, Ni-O(2) apical distance, to the in-plane, Ni-O(1) distance, $R = r_\parallel / r_\perp$, determines the $4p_\pi$-$4p_\sigma$ energy splitting shown in Fig. 4. This is because the energy of a $4p_\pi 3d^9 \underline{L}$ ($4p_\sigma 3d^9 \underline{L}$) final state scales as $1/r_\parallel^2 (1/r_\perp^2)$. Here $\underline{L}$ denotes a ligand shell hole which is created by transfering a ligand shell electron to the $3d$-shell in order to screen the core-hole in the excited final state. The assignment of the XANES features to the above mentioned final state configurations is based on extensive studies of La-cuprates and La-nickelates by Oyanagi *et al.* [9], Kosugi *et al.* [10] and Sahiner *et al.* [11]. Sahiner *et al.* summarized the measured $4p_\pi$-$4p_\sigma$ energy splitting for a series of cuprates and nickelates with known octahedral distortions. This splitting varies linearly with the distortion parameter $1/R^2$ [11]. By interpolating our measured splitting of 3.0 eV into their curve we obtain a distortion parameter $1/R^2 = 0.71$ or $R = r_\parallel / r_\perp = 1.19$. This is in excellent agreement with the XAFS result of $R = 1.19(1)$.

Fits to La K-edge XAFS data are shown in Fig. 7. The local $\hat{c}$-axis, as obtained by averaging over all La sites, is given by the sum $2\times$(La-Cu$_c$)+ (La-La$_c$), where La-Cu$_c$ and (La-La$_c$) bonds are nearly coincident with the c-axis. Fig. 8 shows the rate of contraction of the local $\hat{c}$-axis with Ni-doping, at T=10K. This rate is found to be $(dc/dy)_{10K} = -0.0095(25)$ [Å/at. %Ni]. This is comparable to the measured contraction rate at T=300K found from our powder diffraction measurements (fig. 1); i.e., $(dy/dc)_{300K} = -0.0080(5)$ [Å/at. %Ni].

Fits to Sr K-edge data are shown in Fig. 9. While the measured La-O(2) apical distance determined from La XAFS, $r = 2.36(1)$ Å, agrees with the value of crystallography [6] for the undoped structure (i.e., $y = 0$), $r = 2.354$ Å, the local Sr-O(2) apical distance is determined from the Sr K-edge analysis to be $r = 2.45(2)$ Å. In addition, Sr-distances along the $\hat{c}$-axis show a different response to Ni-doping than that found at the La sites. This is shown in Figs. 10-11 where the Ni-doping dependence of (La/Sr)-La$_c$ and (La/Sr)-(Cu/Ni)$_c$ distances and their mean-squared disorder, as obtained from La and Sr XAFS analysis, are shown. It is clear that the Sr distances along the $\hat{c}$-axis are less affected by the Ni substitution.

DISCUSSION

The local contraction of NiO$_6$ octahedra derived from the Ni XAFS and XANES measurements is driven by the non-Jahn-Teller nature of a Ni^{+2}-ion in its $3d^8$, high-spin, electronic configuration. This is schematically illustrated in Fig. 12, where the octahedral distortions measured by crystallography around Ni in undoped-La$_2$NiO$_4$ and around Cu in undoped-La$_2$NiO$_4$ are shown together with the electronic population/splitting of the $3d_{x^2-y^2}$ and $3d_{3z^2-r^2}$ energy levels in the e_g manifold. Since both cuprates and nickelates have a tetragonal crystal structure, this crystal field lifts the degeneracy of the e_g manifold. The *pseudo-Jahn-Teller* effect in La$_2$CuO$_4$ (Cu^{+2}, $3d^9$ configuration) arises due to the difference in population between $3d_{x^2-y^2}$ and $3d_{3z^2-r^2}$ energy levels; i.e., it is convenient to elongate the octahedron along the $\hat{c}$-axis (lower the energy of the doubly populated $3d_{3z^2-r^2}$ states) while compressing it in the plane (raising the energy of the singly populated $3d_{x^2-y^2}$ state by the same amount). The energy advantage in having a spontaneous elongation of the octahedra disappears in a Ni^{+2} ($3d^8$) ion in its high-spin state since the occupation of $3d_{x^2-y^2}$ and $3d_{3z^2-r^2}$ e_g levels is the same. The disappearance of the Jahn-Teller distortion is

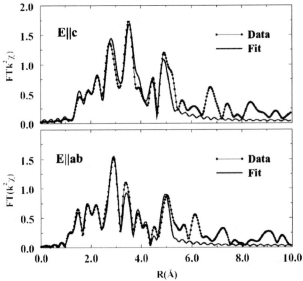

FIGURE 9. Magnitude of complex Fourier transform of $k^2\chi(k)$ for Sr K-edge at T=10K and $y = 0.06$. Data range used in the transform is [3,14]Å^{-1}. The constrained fit of both orientations uses 22 fitting parameters compared to 52 independent points in the fitted region. $S_0^2 = 1.08(10)$; figures of merit are $\chi_\nu^2 = 15.6$, $R = 0.022$.

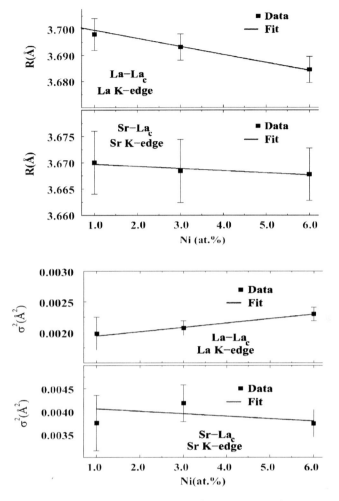

FIGURE 10. Ni-doping dependence of the (La/Sr)-La$_c$ distances (top) and mean-squared disorder (bottom) determined from La and Sr K-edge XAFS analysis. The La-distances respond more strongly to Ni-doping than the Sr-distances.

a direct proof of a divalent, high-spin Ni ion since a Ni^{+3} state ($3d^7$) or a Ni^{+2} low-spin state, would have resulted in measurable Jahn-Teller distortions. These findings severely limit models of T_c suppression which invoke trapping of doped holes by Ni, as that would imply a Ni^{+3} state, inconsistent with our measurements.

It is interesting to compare the NiO_6 octahedral distortion with that in pure La_2NiO_4. In the latter, a fully relaxed non-Jahn-Teller Ni^{+2} ion results in a Ni-O(1) planar distance ≈ 0.04 Å longer than the Cu-O(1) distance in JT-distorted La_2CuO_4 (despite Ni^{+2} being a smaller ion than Cu^{+2}). Similarly, the Ni-O(2) apical distance is $\approx$ 0.18 Å shorter than the Cu-O(2) one (Fig. 12).

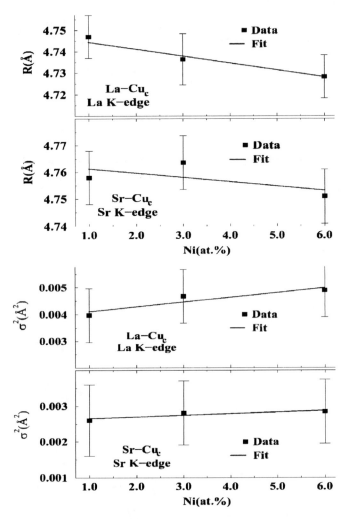

FIGURE 11. Ni-doping dependence of the (La/Sr)-Cu$_c$ distances (top) and mean-squared disorder (bottom) determined from La and Sr K-edge XAFS analysis. The La-distances respond more strongly to Ni-doping than the Sr-distances.

Our measurements indicate that although the $\hat{c}$-axis contraction of Ni-O(2) in La$_{1.85}$Sr$_{0.15}$Cu$_{1-y}$Ni$_y$O$_4$ is consistent with an almost fully-relaxed non-Jahn-Teller state (≈ 0.16 Å), the in-plane Ni-O(1) distance does not elongate by more than 0.01 Å. The most likely explanation for this is that the NiO$_6$ octahedra are embedded in the very rigid, highly covalent, CuO$_2$ planes, which prevents the in-plane Ni-O(1) distance from significantly expanding and deforming the surrounding CuO$_2$ network. The larger temperature dependence in mean-squared disorder of Ni-O(2) distance compared to Ni-O(1) is supportive of this notion. The softer Ni-O(2) bond can therefore contract without resulting in a significant elastic energy cost. The above

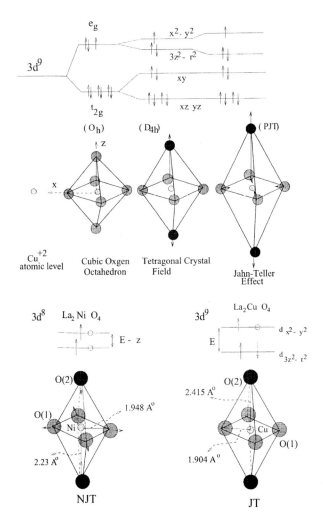

FIGURE 12. Top: schematic illustration of the electronic origin of the Jahn-Teller effect for a Cu^{+2} ion in an octahedral environment. The e_g and t_{2g} manifolds are split by the tetragonal crystal field and the pseudo-Jahn-Teller effect enhances the tetragonal distortion. Bottom: measured octahderal distortions in pure La_2NiO_4 and La_2CuO_4. The NiO_6 octahedral distortion in $La_{1.85}Sr_{0.15}Cu_{1-y}Ni_yO_4$ is very similar to that in pure La_2NiO_4, except that the rigid CuO_2 plane prevents a significant Ni-O(1) in-plane distortion to materialize. Sr doping contracts the Cu-O(1) distance in $La_{2-x}Sr_xCuO_4$ [6].

arguments imply that a fully relaxed non-Jahn-Teller state is not materialized for Ni in .$La_{1.85}Sr_{0.15}Cu_{1-y}Ni_yO_4$.

The larger response of La-La$_c$ and La-(Cu/Ni)$_c$ distances than Sr-La$_c$ and Sr-(Cu/Ni)$_c$ to Ni-doping can be understood in more than one way. One possibility is that Sr enters the lattice away from Ni sites and therefore is less sensitive to the enhanced local lattice deformation around Ni sites. As mentioned earlier, the fit results of the Ni-(Sr/La) XAFS signal are consistent with such hypothesis, due to the large uncertainty in the relative weight of La/Sr neighbors to Ni. Why would Sr avoid Ni? In a previous XAFS study of $La_{2-x}Sr_xCuO_4$ for a series of x-values below and above the insulator-metal (I-M) transition ($x \approx 0.06$ at T=10K), we detected a response of the Sr-O(2) apical distance to the delocalization of doped hole-carriers while no such response was found for the La-O(2) apical distance [13]. This testifies to the fact that doped-holes introduced by the Sr-dopants are sharply peaked in the vicinity of the Sr in the insulating phase while becoming more extended in the metallic phase. It is reasonable to argue that, if the I-M transition is caused by increased overlap of impurity states, the extended impurity states will remain peaked in the vicinity of the Sr even in the metallic state. A similar conclusion was reached by Hammel et al. in their nuclear quadrupole resonance study of $La_{2-x}Sr_xCuO_4$ [14].

It follows that if Sr likes to keep the doped-holes in its neighborhood due, e.g., to their attraction to the Sr^{+2} impurity potential, and, if Ni prefers to avoid them, a correlation in Ni-Sr positions could arise. It is clear that doped holes will be scattered strongly from the potential introduced by the local distortion around Ni. In fact doped-holes become localized with Ni-doping at $y \approx 0.05$ [4]. Our observation of a Ni^{+2}, $3d^8$ high-spin state for Ni even at $y = 0.06$ indicates that the doped holes are not localized at Ni sites. The localization of carriers at $y \approx 0.05$ must occur away from Ni sites. It could be that Ni "repels" holes to preserve its high-spin state and associated Hund's exchange energy.

An alternative explanation for the observed differences between Sr and La sites is the much weaker bonding between the O(2)-apical and a Sr^{+2} ion compared to a La^{+3} ion. The measured Sr-O(2) apical distance (longer by ≈ 0.1 Å) is a result of the much weaker Madelung potential felt by the negatively charged O(2) apical oxygen near a Sr ion, resulting in an equilibrium position further away from the Sr and closer to the CuO_2 planes. Such distortion was already reported in a previous study of $La_{2-x}Sr_xCuO_4$ [13]. When NiO_6 octahedra contract and the O(2) ion is displaced towards Ni, the La^{+3} ions follow the O(2) displacement, due to a significant degree of covalency in the La-O(2) bonding (the La-O(2) apical distance, 2.35 Å, is shorter than expected for a purely ionic bond). That La follows the O(2) apical is clearly seen by the measured contraction in Ni-La$_c$ distance (table 1) and in the Ni-induced contraction of the La-(Cu/Ni)$_c$ distance (fig. 11). However, since Sr ions are weakly bound to O(2) apicals, they might not follow the O(2) apicals in their journey towards the Ni ions in the contracted NiO_6 octahedra, resulting in a much smaller change in the Sr-(Cu/Ni)$_c$ distance relative to La-(Cu/Ni)$_c$ distance. A similar effect can also explain the behavior of La/Sr-La$_c$ distances. This explanation does not require a deviation from randomness in the solid solution.

Despite having a magnetic moment, Ni suppresses T_c at a similar rate (albeit smaller) than non-magnetic Zn. Zn doping in $La_{2-x}Sr_xCuO_4$ is expected to manifest a similar ZnO_6 octahedral distortion, as a filled-shell, Zn^{+2} $3d^{10}$ ion is also a non-Jahn-Teller ion. The very similar macroscopic $\hat{c}$-axis contraction observed for both dopants supports this notion. It is not clear how the local distortion around Ni would contribute to pair breaking, but it is reasonable to assume that doped holes will scatter strongly from such distorted centers. This scattering will modify the spatial distribution of hole carriers. If these hole carriers reside predominantly in the CuO_2 planes, their wavefunctions will be peaked in between Ni ions. The spatial extent of a superconducting state cannot be confined to a region smaller than its coherence length, as the resulting quasi-particle localization energy would be larger than the superconducting energy gap. It is interesting to note that T_c is completely suppressed at 3-4 at. %Ni, where the average distance between Ni dopants in the two-dimensional CuO_2 planes is $3.78/\sqrt{y}$ =19-22 Å. This is about the size of the in-plane superconducting coherence length, ζ_{ab} = 22.7 Å for $La_{1.85}Sr_{0.15}CuO_4$ [15]. It is then possible that superconductivity is destroyed when the doped holes are constrained to occupy regions of the CuO_2 plane in between Ni centers, which, for $y \geq 0.04$, are smaller than ζ_{ab}.

CONCLUSIONS

New tools are becoming available that allow imaging and spectroscopy of high T_c superconductors at the atomic level. Scanning tunneling microscopy, e.g., has already proven to be a powerful technique for unraveling some important details about the spatial symmetry and extent of impurity states around Zn dopants in a high T_c superconductor. Such experiments are bringing us closer to determining the symmetry of the superconducting order parameter. These new techniques can only be fully exploited if a quantitative understanding of the local structure at the atomic level is available. This is particularly important for dopants, where the local structure typically differs from the macroscopic structure.

We have shown here that a strong lattice distortion exists around the Ni atoms in $La_{1.85}Sr_{0.15}Cu_{1-y}Ni_yO_4$. This distortion manifests a large $\hat{c}$-axis contraction of the NiO_6 octahedra, which propagates to higher shells of neighboring atoms along the $\hat{c}$-axis. This has implications for the interpretation of spectroscopies that are sensitive to spatially-inhomogeneous local $\hat{c}$-axis distortions. For example, the local tunneling cross sections, which depend on the local inter-layer spacing, will vary from a (Ni/Zn) dopant site to a neighboring Cu site due to the local $\hat{c}$-axis fluctuations.

Enhanced scattering of the doped holes by the NiO_6 octahedral distortions will likely result in an inhomogeneous charge distribution of the doped holes, with hole-poor regions around Ni dopants. This is somewhat different than the "swiss cheese" model of Nachumi *et al.*, in that it implies an inhomogeneous hole distribution in the *normal* state, leading to a spatially inhomogeneous superconducting order parameter below T_c. The "swiss cheese" model predicts a uniform normal state charge distribution and the effect of dopants is to remove superconducting pairs from the superfluid in a $\pi\zeta_{ab}^2$ region around the dopants. Superconductivity is fully suppressed when the average distance between Ni dopants is on the order of the in-plane superconducting coherence length. This is expected if charge carriers are confined to the regions in between Ni dopants, as the spatial extent of a superconducting state cannot be confined to a region smaller than its coherence length.

The Sr and La environments respond differently to Ni doping. Although this could be explained as due to Ni entering the lattice away from Sr sites, a more likely explanation involves a weak Sr-O(2) apical bond compared to a stronger La-O(2) apical bond. The latter results in La ions "following" the O(2) apical distortion toward Ni ions, while the Sr ions are not much affected due to their weak link to the O(2) apicals.

ACKNOWLEDGMENTS

It is a pleasure to acknowledge Frank Perez and Mas Suenaga for their help with the magnetic alignment of powders and Arnie Moodenbaugh for the T_c measurements. Research done under auspices of DOE Grant No. DE-FG03-98ER45681 D. H. was also partially supported by DOE contract No. W-31-109-Eng-38.

REFERENCES

1. Pan, S. H., Hudson, E. W., Lang, K. M., Eisaki, H., Uchida, S., and Davis, J. C., *Nature* **403**, 746 (2000).
2. Byers, J. M., Flatte, M. E., and Scalapino, D. J., *Phys. Rev. Lett.* **71**, 3363 (1993).
3. Balatsky, A. V., *Nature* **403**, 717 (2000).
4. Cieplak, M. Z., Guha, S., Kojima, H., Lindenfeld, P., Xiao, G., Xiao, J. Q., and Chien, C. L., *Phys. Rev. B* **46**, 5536 (1992).
5. Haskel, D., Stern, E. A., Dogan, F., and Moodenbaugh, A. R., *Phys. Rev. B* **61**, 7055 (2000).
6. Radaelli, P. G. *et al.*, *Phys. Rev. B* **49**, 4163 (1994).
7. Stern, E. A., Newville, M., Ravel, B., Yacoby, Y., and Haskel, D., *Physica B* **208 & 209**, 117 (1995).
8. Zabinsky, S. I., Rehr, J. J., Ankudinov, A., Albers, R. C., and Eller, M. J., *Phys. Rev. B* **52**, 2995 (1995).
9. Oyanagi, H., Yokoyama, Y., Yamaguchi, H., Kuwuahara, Y., Katayama, and Nishihara, Y., *Phys. Rev. B* **42**, 10136 (1990).
10. Kosugi, N., Tokura, Y., Tajima, H., and Uchida, S., *Phys. Rev. B* **41**, 131 (1990).
11. Sahiner, A. *et al.*, *Phys. Rev. B* **51**, 5879 (1995).
12. Haskel, D., Ph. D Thesis, University of Washington, 1998.
13. Haskel, D., Stern, E. A., and Dogan, F., "Phase transitions and self organization in electronic and molecular networks", in *Fundamental Materials Research* series, edited by M. F. Thorpe and J. C. Phillips.
14. Hammel, P. C., Statt, B. W., Martin, R. L., Chou, F. C., Johnston, D. C., and Cheong, S. W., *Phys. Rev. B* **57**, R712 (1998).
15. Nachumi, M. E. *et al.*, *Phys. Rev. Lett.* **77**, 5421 (1996).

What Do We Need to Know about the Electronic Structure When Analyzing the Local Structure of Strongly Correlated Systems?

A. Yu. Ignatov

Department of Physics, Southern Illinois University, Carbondale, IL 62901, USA
Present address: Department of Physics, New Jersey Institute of Technology, Newark NJ 07102, USA

Abstract. One of the basic assumptions of modern XAFS analysis is that local structure is independent of electronic structure. This, however, is not the case for the strongly correlated compounds, where the electronic and lattice degrees of freedom are coupled through the electron-electron and electron-phonon interactions. The K-edge x-ray absorption spectrum can be obtained as a convolution product of the single-electron transition from the $1s$ core-level to the unoccupied electronics states, $I(\omega)$ and the spectrum of many-body excitations in the electronic states in the presence of the $1s$ core-hole, $S(\omega)$:, $\sigma(\varepsilon) = \int S(\omega) I(\omega - \varepsilon) d\omega$, $S(\omega) = \sum_f |\langle \Psi_f | \Psi_0 \rangle|^2 \delta(\omega - E_f + E)$ where Ψ_0 is the ground state with energy E_0 and Ψ_f is the i-th excited state with energy E_i referring to Hamiltonians in the initial (H_0) and final states (H_f) respectively.

A Generalized X-ray Analysis of Local Structure (GXALS) that accurately incorporates local electron-lattice and electron-electron correlations has been developed. Application to $BaBi_xPb_{1-x}O_3$ and $Ba_{1-x}K_xBiO_3$, $Nd_{2-x}Ce_xCuO_{4-\delta}$ and $La_{1-x}Ca_xMnO_{3+\delta}$ as well as the limitations of the approach are discussed.

I. INTRODUCTION

Strongly correlated compounds including high-T_c superconductors (HTSCs), colossal magneto-resistors (CMRs), several types of perovskites, and transition metal-oxides (TMOs) are in the focus of condensed matter physics today because of their unique physical properties, a variety of fundamental issues are raised in attempting to understand them, and promising technological applications. There is a growing realization that a intimate multiscale correlation between electronic spin and lattice degrees of freedom is fundamental to the properties of the strongly correlated compounds [1]. It's probably most evident in CMR compounds where the double-exchange model [2] was found insufficient to explain resistivity data. Millis, *et al.* [3] has argued that substantial lattice effect must be present to reduce the kinetic energy of the carriers at the metal to insulator transition. The correlations are not so obvious and, therefore, they are not commonly admitted in the HTSCs and TMOs. Whatever the lattice involvement in HTSCs is, the accurate determination of the electron-lattice coupling would essentially contribute to understanding the HTSC phenomena.

The key issue on the experimental front is *to establish the presence of local distortions and characterize them in detail.* Being direct and fast probe techniques,

CP554, *Physics in Local Lattice Distortions*, edited by H. Oyanagi and A. Bianconi
© 2001 American Institute of Physics 1-56396-984-X/01/$18.00

and providing structural information on atomic length scales, extended x-ray absorption fine structure (EXAFS) and x-ray absorption near edge structure (XANES) measurements have been widely used for characterization of local lattice distortions on the extremely short atomic scales, typically shorter than 5 Å. Often, even if the presence of local distortions leave no doubts, as for example, in $La_{1-x}Ca_xMnO_3$, the detail characterization is not straightforward. It is worth mentioning that one needs to be very careful not only with the quality of raw experimental data, but also with processing of these data because most structural analyses are *model-sensitive*. This means that raw experimental data are fitted to specific model inherited certain limitations.

As for the basic assumption taken in conventional EXAFS, it is the so-called phase transferability [4] implying the local structure to be independent of the electronic structure that has been shown to be incorrect for the strongly correlated compounds.

The purpose of this paper is to include spectroscopic information into the structural x-ray absorption study by accurate incorporating of the local electron-lattice and electron-electron correlations. In Section II I will critically review several drawbacks of conventional EXAFS analysis making some contacts to the complementary synchrotron (neutron) diffraction pair distribution function (PDF) analysis. *Ab-initio* formulation of a Generalized X-ray Analysis of Local Structure (GXALS) is given in Sect. III. I will briefly consider fragments of the GXALS applied to $BaBi_xPb_{1-x}O_3$ and $Ba_{1-x}K_xBiO_3$ perovskites, $Nd_{2-x}Ce_xCuO_{4-\delta}$ superconductor, and $La_{1-x}Ca_xMnO_{3+\delta}$ CMR oxides in Sections IV. Section V is devoted to summary.

II. SEVERAL DRAWBACKS OF CONVENTIONAL LOCAL STRUCTURAL ANALYSIS

Despite a significant progress in the local structure analysis of strongly correlated compounds [5-7] one can point out several drawbacks of conventional EXAFS and synchrotron (neutron) diffraction PDF analysis. Those can be summarized in following four problems:

(1) Interplay between the valence fluctuation and local structure.

A back Fourier transformed EXAFS spectrum is usually fitted to the model: $\chi(k) \propto \Sigma_i A(k, r_i) sin[2kr_i + \phi(k, r_i)]$, where the sum is taken over the interatomic distances, r_i. $A(k, r_i)$ and $\phi(k, r_i)$ are amplitude and phase shifts, respectively. The beat in the XAFS may be due not only to different r_i but also to different $\phi(k, r_i)$ which depend on the valence states of the constituent ions such as Cu $3d^9$ and Cu $3d^{10}\underline{L}$ for the Cu-based HTSCs or Mn $3d^4$ and Mn $3d^5\underline{L}$ for the CMRs. If the amplitudes and phase shifts are not corrected for valence fluctuations the interatomic distances derived from the fit differ from the real ones. Egami and Billinge have initially formulated this problem as follows [6]: "...the atomic distance derived from the EXAFS may be influence by the local valence through the EXAFS phase shift, ϕ, as well as by the positions of the neighboring atoms. The EXAFS phase shift has a contribution from the central atom and the backscattering atom, and may depend upon the valence states of these ions when the electron-electron repulsion is strong, as in the transition metals and, to a lesser extent, oxygen. If ϕ is different for Cu^{3+} and Cu^{2+} for example, or for O^{2-} and O^-, then the presence or absence of a hole on copper or oxygen could result in beat in the EXAFS, even when no atomic displacements are present. .... it is possible that the relationship between the electronic structure and the EXAFS may not be as simple as generally believed."

Apparently, resonance synchrotron PDF analyses experience the similar problem of interplay between the valence fluctuations and local structure. Due to strong electron-electron and electron-lattice coupling that are endemic properties of strongly correlated compounds they may be several *inelastic* scattering channels. Kao *et al.* [8] reported large increase in the inelastic scattering in NiO when the incident x-ray energy was tuned at Ni K-edge. They described additional inelastic excitations in terms of charge-transfer editions from a $3d^8$ ground state to a $3d^9\underline{L}$ exited state, where $\underline{L}$ denotes a hole on the ligand site. Recently, Hill [9] explored resonant inelastic x-ray scattering in electron-doped $Nd_{2-x}Ce_xCuO_{4-\delta}$. Hard x-ray regime or neutron-diffraction PDF data are less subjected to the valence fluctuations. However, whether the inelastic channels are suppressed depends on a few electronic levels near the ground state.

(2) Systematic classification of the observed distortions.

Staying in the framework of pure structural studies (conventional XAFS or neutron-diffraction PDF analysis) it is not possible to distinguish between different type of polaronic distortions such as polarons, bi-polarons, spin-polarons, etc. Currently "classification" depends on what the investigators believe in or advocate for at the moment of the assignment.

(3) Search for global minimum.

Near the transition temperature the system may be frustrated between several displacement configurations (domains), so that a sharp minimum in the local structure is very unlikely. It has been mentioned in the literature [10,11] that several displacement configurations could provide almost the same low agreement factors. Consequently, the question of unique, meaningful fits to data is very complex if it's raised in purely structural sense.

(4) Sensitivity to the correlated displacements of the neighboring atoms.

In XAFS analysis the sensitivity to the correlated displacements of the neighboring atoms is achievable from careful analyses of the fourth order terms of the MS expansion. However, vibration frequencies in strongly correlated compounds do not obey a simple Debye law, resulting in a total number of varying parameters to be far above those available in the fit. The neutron-diffraction PDF technique has better capabilities in tracing the correlated displacements, except for the breathing displacements.

III. POSSIBLE SOLUTION

To provide insights into unusual properties of strongly correlated compounds it is important to carefully study both electron-electron and electron-phonon interactions and their signatures in various response functions. Notable examples of such response functions shown in Fig. 1 includes soft x-ray techniques (UPS, IPS, XPS, XAS, and RPES), which provide reliable information on energy parameters of the extended Hubbrad Hamiltonian generally used for description of electron-electron correlation. Usually, Hamiltonian parameters are considered to be independent of local structure. Those who deal with local structural probes including EXAFS or neutron diffraction PDF analysis would probably disagree on this obvious simplification. Indeed, it has been recognized as results of many experimental measurements that the structure of most HTSCs [6] and CMRs [7] are locally distorted from perfect crystallographic structure as determined by long range order techniques. The onset of these distortions reflects the tight relationship between local and electronic degrees of freedom. To what extent the local structural distortions affect the energy balance is still to be answered. It turns out that some correlated local distortions, which are compatible to those observed experimentally, could alter the ground state energy by 0.2 - 0.5 eV. Notice, that it is a much larger energy than most of critical temperatures in HTSCs and CMRs and, therefore, there is a serious reason to include local structural configurations (domains) into the total energy balance of strongly correlated compounds.

In this endeavor, I would propose three amendments to current soft x-ray spectroscopy techniques, which for the reasons mentioned above will be beneficial and as we shall see below will offers natural solutions of all four problems.

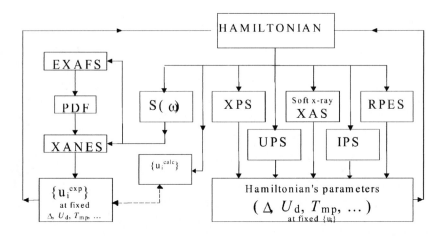

FIGURE 1. Scheme of the GXALS that accurately incorporates local electron-lattice and electron-electron correlations.

Firstly, we need to include *implicitly* the local structural distortions into the Hamiltonian. The simplest way to do so is to consider the Peierls-Hubbard type of Hamiltonian. We might start from the generalization of the 2D Peierls-Emery and Anderson- Hasegawa Hamiltonians in the hole representation. A quantum approximation for the electronic part and classical one for the lattice part will be assumed. The Hamiltonian for GXALS can read as follows:

$$H^{total} = H^{elect} + H^{lat} + H^{e-ph}$$

$$H^{elect} = \sum_{i,\sigma} \varepsilon_i n_{i,\sigma} + U_m \sum_i n_{i,\uparrow} n_{i,\downarrow} + \sum_{k,\sigma} \varepsilon_k n_{k,\sigma} + U_p \sum_k n_{k,\uparrow} n_{k,\downarrow}$$

$$+ V_{mp}(u_k) \sum_{<i,k>\sigma,\sigma'} n_{i,\sigma} n_{k,\sigma'} + V_{pp}(|\mathbf{r}_k - \mathbf{r}_{k'}|) \sum_{<k\,k'>\sigma,\sigma'} n_{k,\sigma} n_{k',\sigma'}$$

$$- T_{mp}(u_k) \sum_{<i\neq k>\sigma,\sigma'} (c_{i,\sigma}^+ p_{k,\sigma'} + h.c) - T_{pp}(|\mathbf{r}_k - \mathbf{r}_{k'}|) \sum_{<k\neq k'>\sigma,\sigma'} (p_{k,\sigma}^+ p_{k',\sigma'} + h.c) \quad (1)$$

$$+ J \sum_{<i,i'>} S_i \cdot S_{i'} - 2J_H \sum_i S_i \cdot s_i - J_{pd}(u_k) \sum_{k,i} (p_{k,\uparrow}^+ p_{k,\downarrow} - p_{k,\uparrow}^+ p_{k,\downarrow}) S_i^z$$

$$H^{lat} = \sum_k \frac{1}{2M_k} p_k^2 + \sum_{k,k'} \frac{1}{2!} K_{k,k'} u_k u_{k'} + \sum_{k,k',m,m'} \frac{1}{4!} W_{k,k',m,m'} u_k u_{k'} u_m u_{m'}$$

$$H^{e-ph} = -\sum_i \lambda_i^{e-ph} \sum_{k,\sigma} n_{k,\sigma} u_k$$

The model is restricted to the motion of the conduction holes in the M = (metal, for example, Cu 3d, Mn 3d, Bi 6s) and O 2p bands in the lattice of localized spins, S. $\hat{c}_{i,\sigma}^+ (\hat{p}_{k,\sigma}^+)$ creates a hole with spin σ at site $i(k)$ in the M 3d/6s (O 2p) orbitals respectively; $n_{i,\sigma} = \hat{c}_{i,\sigma}^+ \hat{c}_{i,\sigma}$ and $n_{k,\sigma} = \hat{p}_{k,\sigma}^+ \hat{p}_{k,\sigma}$ are hole number operators; ε_i and ε_k are M 3d/6s and O 2p// energies; U_m and U_p are M and O on-site Coulomb repulsions; V_{mp} and V_{pp} are off-site M-O and O-O Coulomb repulsions. Electron-lattice coupling is introduced through the M-O and O-O nearest-neighbor hops and Coulomb repulsions. These parameters are assumed *to be modified by the local displacements*, u_k, of the oxygen atoms depicted in Fig. 2 so that both the breathing and tilting components of the oxygen displacements are included.

Among a obvious limitation of this model is omission of long range interactions ("inta-atomic" interactions [1,12]) that appear to give rise to the stripes. Being extremely local sensitive probe (typical cut off is ~ 5 Å) x-ray absorption spectroscopy could not distinguish between randomly oriented domains on one hand and ordered into the stripes domains on the other hand. DAFS [13] is expected to a promising generalization of EXAFS to account for the long range order interactions. Small cut off distances in x-ray absorption spectroscopy allows the use a cluster of 12 atoms (4 metal + 8 oxygen sites) with periodic boundary conditions. The Lanczos technique can be employed for numerical diagonalization and spectral calculations for such a small cluster.

Secondly, I would proposed consideration of local structural configurations, $\{u_k\}_i$, that minimizes the ground state energy of the system as one of *the response functions* (see Fig. 1). Being directly available from the Hamiltonian it relates to the pair distribution function reconstructed from EXAFS or neutron-diffraction PDF measurements offering a variety of advantages from adjusting the energy balance to test various theoretical models (regimes) of a given Hamiltonian.

Thirdly, to ensures that the experiments do not suffers from the so-called "final state" effects when the appearance of a core-hole results in strong perturbations in the many-body electronic structure one needs to introduce an additional spectral function, $S(\omega)$, that is assumed to be structural dependent as well as other spectral functions in our approach. Two static limits of the so-called final state rules should be considered. In the unrelaxed Z-approximation the *N-hole final state* is affected by a $1s$ core-hole in the case of K edge absorption. The excitation spectra $S(\omega)$ is given by the sum over all f-th final N-hole states projected onto the ground state N-hole wavefunction without the core-hole:

$$S(\omega) = \sum_f \left| \left\langle \Psi_f(N) \middle| \Psi_0(N) \right\rangle \right|^2 \delta(\omega - E_f(N) + E_0(N)) \tag{2}$$

In the fully relaxed $Z+1$ approximation, the Z atom is replaced by the $Z+1$ that adds one electron to the absorption site creating the *N-1 hole final state* with the $1s$ core-hole:

$$S(\omega) = \sum_f \left| \left\langle \Psi_f(N-1) \middle| \hat{c}_{i,\sigma} \middle| \Psi_0(N) \right\rangle \right|^2 \delta(\omega - E_f(N-1) + E_0(N)) \tag{3}$$

In Eqs. (2, 3) the Ψ_0 is the ground state with energy E_0 and Ψ_f is the f-th excited state with energy E_f referred to Hamiltonians in the initial, H_0 (Eq. 1) and final states, H_f, respectively. In the final states $H_f = H_0 + H_c$, where H_c is subjected to the $1s$-$3d$ inter-hole Coulomb repulsion, Q.

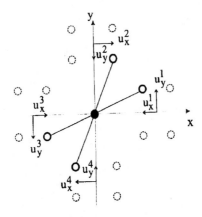

FIGURE 2. Sketch of the oxygen atom displacements allowed in the 2D model.

The final state rule [14] has been developed from the model calculations of the dynamic response of the conduction electrons to the suddenly switched on core-hole potential. These model calculations may be applicable directly to simple metals, but in more complex systems the final state rules probably need to be systematically tested. Such tests deserve separate consideration. They are beyond the scope of the present paper. In all examples given below I will employ the Z-approximation.

Now let me specifically elaborate on how four challenges for x-ray absorption may be resolved in the framework of the proposed approach.

3.1 Interplay between the valence fluctuation and local structure

The K-edge absorption spectrum can be obtained as a convolution product of the single-electron transition from the $1s$ core-level to the unoccupied electronics states, $I(\omega)$ and the $S(\omega)$ [15]:

$$\sigma(\varepsilon) = \int S(\omega)I(\omega - \varepsilon)d\omega \qquad (4)$$

that is valid as long as one can neglect the interaction between N-hole states and np photoelectron.

Recalling that EXAFS is just the second term in the MS expansion of the absorption cross-section, the amplitudes and phase shifts for each displaced configuration calculated, for instance, by FEFF code [16], should be corrected for the excitation spectrum as prescribed by Eq. (4). This correction introduces the *multiple-channel absorption* providing a proper reply to the question raised by Egami and Billinge. As revealed from Eqs. (3, 4) the correction reduces to the appearance of several sine waves with smaller magnitudes shifted by $\sim$ 2 - 8 eV with respect to the major sine wave. In fact, each amplitude and phase shift has its own scaling factors S^2_{0i} and energy shifts ΔE_i, respectively.

It worth mentioning that typical energy shift of two major final states reached in Cu K-edge absorption of HTSCs is about 7 eV that is 2 - 3 times larger that the splitting in the states. Since $k \propto \sqrt{E}$, the correction is not important at $k > 4$ - 5 1/Å. Simple estimate over the range 4 - 18 1/Å shows that critical EXAFS finding reported on different Cu-O distances in the CuO_2 planes of $La_{1.85}Sr_{0.15}CuO_4$ [5] and $Bi_2Sr_2Ca_2O_8$ [17] are not altered. The spitting is not suppressed, but it possible reduced to 0.01 - 0.02 Å that is in the error bars of EXAFS analysis. However, as we shall see in the next Section the multiple channel correction (Eq. 4) is very important at low k, *i.e.* in the XANES region.

3.2 Systematic classification of the observed distortions

GXALS would allow systematic classification of the observed distortions. As is well known from neutron-diffraction PDF analysis [6] most of the domains are characterized by *correlated displacements*, that, in turn, allow a systematic classification for the observed distortions in terms of polarons, bi-polarons, spin-polarons, or JT polarons by counting the total number of the carriers confined within the deformed area. The proposed microscopic approach enables one to relate calculated displacements $\{u_k\}^j$ with those extracted from the EXAFS analysis resulting in the optimization of the Hamiltonian parameters and finally, to a better understanding of the energy balances and the effective interactions between carriers.

3.3 Search for global minimum

The new approach is expected to simplify a search for the local distortions corresponding to energy minimum. Fast and accurate minimization of the ground state energy enables one to choose between displaced configurations characterizing by similar PDF (and, therefore, to be hardly distinguishable by purely structural means) but different ground state energies. Notice, what a difference of 0.1 - 0.9 eV usually coming up from the ground energy consideration is a huge value on thermodynamic scale. This provides some grounding in searching for "global" minimum. The problem of the unique refinement is not completely worked out. As usual, for the strongly-non-linear minimization, every specific case must be addressed separately.

3.4 Sensitivity to the correlated displacements of the neighboring atoms

Introduction of the excitation spectra, allowing for multi-channel absorption improves the XANES dramatically. For 2D systems including most of HTSCs the components of O atom displacements from their undistorted positions can be derived from *a direct nonlinear least square fit* of K-edge calculated spectrum to the experimental one. Only three components of the oxygen atom displacements ($\Delta O_{||}$, $\Delta O_{\perp}$, ΔO_z) that are compatible to a pair distribution function obtained from the conventional EXAFS are the subject to fit. Unfortunately, the direct fit seems hardly to be possible for 3D systems where there are nine independent components of the oxygen atom displacements in strongly distorted domains. However, XANES calculations along with accurate experimental spectra could be use to test several possible displaced configurations (domains).

To summarized, the most involved description is to proceed with self-consistent calculations those algorithm is shown in Fig. 1. The self-consistent GXALS would provide a full coverage for EXAFS and possibly XANES refinement as well as systematic classification of observed local distortions. If completed, it's equivalent to establishing a relationship between the electronic and local structure on the atomic length scales. However, to the best of my knowledge, the whole scheme has never been implemented so far. In next Section, I will briefly consider fragments of the scheme.

IV. SEVERAL EXAMPLES

4.1 BaBi$_x$Pb$_{1-x}$O$_3$ and Ba$_{1-x}$K$_x$BiO$_3$ perovskites

4.1.1 Valence fluctuations

I have found that to add two holes on half of the Bi-O$_6$ complexes (Bi^{+3}+ Bi^{+3}L^2) costs less energy than to put one hole on each of the Bi-O$_6$ complexes (Bi^{+3}L+ Bi^{+3}L). This provides some grounding in the model of charge disproportionation: 2Bi^{+4}→ Bi^{+3} + Bi^{+3}L^2 proposed earlier [18]. Here L^2 denotes two holes spread around four (six) oxygen atoms, surrounding the Bi ion in 2D (3D) compounds.

Around a Bi(1) site (with the larger charge density), the four surrounding O atoms are displaced toward it symmetrically. They gain energy by increasing bond-charge density and thus enhancing covalency on the four Bi-O bonds around the central Bi site. The electron-electron interaction forms a Bi^{+3}L^2-O$_4$ complex that consists of four stronger Bi-O bonds around a Bi(1) site. The on-site repulsion at Bi sites U_m is too weak to have magnetic moments there.

4.1.2 Classification of distortions

The ground state in BaBiO$_3$ may be regarded as a regular lattice of covalent molecules, where each molecule consists of two holes spread around four (for the 2D model) or six, (for real compounds) oxygen atoms surrounding the Bi ion. This covalent molecule can be viewed, as *bipolaron* so far as the two holes are self-trapped within the Bi^{+3}L^2-O$_{4(6)}$ complex.

4.1.3 Sensitivity to correlated displacements

Bi L_3 edge is very broad. A direct XANES fit is impossible. I found a set of oxygen atom displacements, $\{u_k\}$ which minimizes the ground state energy of the cluster. The energy minimum was achieved for the breathing displacements of $u_x^1 = u_y^2 = -u_x^3 = -u_y^4 = 0.106$Å and tilting displacements $u_y^1 = u_x^2 = -u_y^3 = -u_x^4 = 0.167$Å. The calculated local structure of BaBiO$_3$ agrees reasonably well with those supplied by EXAFS and XRD measurements [19].

4.1.4 Broader impact

Microscopic model enables one [20]:
- calculate IPES, XPS, and XAS taking into account the local structure distortions;
- evaluate local structure as a function of electron(hole) doping [18];
- give a qualitative description of the low-temperature anomalies in the local structure and the two-gap problem;
- explain the results of Seebeck, upper critical field, EPR measurements, and some other "anomalies" properties.

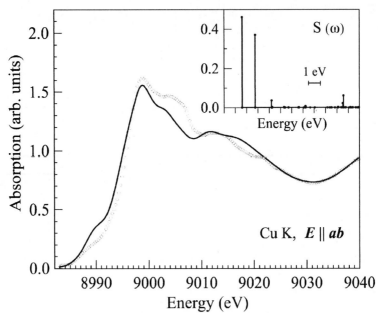

FIGURE 3. Comparison of experimental (dots) and calculated (solid line) XANES spectra. The solid line represent a best model consisting a 60 to 40% mixture of *2x2* undistorted antiferromagnetic and the MCDW domains. The agreement factor over the range of the figure is only 6.3 %. The excitation spectrum of the MCDW domains is shown in the insert. Notice that second peak is ~ 1 eV above the main peak. This picture sheds some light on typical energy shift and the spectral weight transfer that could be expected in the response functions (Fig. 1) when realistic local structural distortions are taken into account.

4.2 $Nd_{2-x}Ce_xCuO_{4-\delta}$ superconductor

Three components of O1 atom displacements in $Nd_{1.85}Ce_{0.15}CuO_{4-\delta}$ superconductor have been derived from the direct fit of the polarized $E\|ab$ Cu K-edge XANES spectrum shown in Fig. 3 [11,21]. Only O1 displacements that are compatible with the Cu-O1 pair distribution function reconstructed from Cu K-edge EXAFS were allowed into the refinement. The best agreement is achieved for a 60 to 40% mixture of *2x2* undistorted antiferromagnetic and modified charge density wave (MCDW) domains. The local structure of the MSDW is characterized by correlated displacements of $\Delta O1_\| = 0.12 \pm 0.04$, $\Delta O1_\perp = 0.25 \pm 0.08$, and $\Delta O1_z = 0.12 \pm 0.1$ Å. An agreement factor over the range of 8980-9040 eV is 2-3 times smaller than one may obtain from single-electron multiple-scattering calculations. This improvement comes from accounting for many-body excitations and the relationship between electronic and local structure in the CuO_2 planes. This picture of the local distortions supports a two-component model for superconductivity.

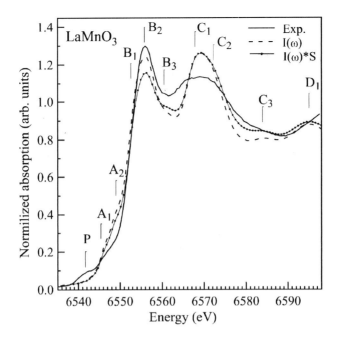

FIGURE 4. Comparison of the Mn K-edges of LaMnO$_3$. The solid line corresponds to the experimental spectrum at 300 K; dashed line shows the result of the single-electron MS calculations for the 87-atom cluster; solid with dots represents the convolution of the single-electron spectrum with the excitation spectrum shown in the insert. Applying Eq. (4) gives rise to the shake-up peak B_3, lowers the intensities of the A_1 and A_2 features and improves the agreement in the extended area, E > 6575 eV.

4.3 La$_{1-x}$Ca$_x$MnO$_{3+\delta}$ CMR

A small feature, B_3, standing approximately 6 eV above the main absorption peak in Fig. 4 is beyond single-electron calculations including the spin-polarized ones. Experimental observation of the shake-up peak and Hamiltonian parameters used in the calculations both imply that LaMnO$_3$ should be viewed as a charge-transfer-type insulator with a *substantial O 2p component in the ground state*. These findings contradict the "intermediate" Mn valence [22] and conventional DE mechanism [2] both implying $3d$ character of doped states. We argue [23] that the disproportionation may be understood as a mixture of the charge-transfer many-body electronic configurations: $\alpha_1|3d^5> + (\alpha_2|3d^4> + \alpha_3|3d^5L> + \alpha_4|3d^6L^2>) + (\alpha_5|3d^3> + \alpha_6|3d^4L> + \alpha_7|3d^5L^2>) + \alpha_8|3d^4L^2>+ \ldots$ coupled with spin and lattice degrees of freedom.

V. CONCLUSIONS

Traditional local structural analysis solves the structure in the adiabatic approximation that is relevant for most of EXAFS and neutron-diffraction PDF applications. However, for the strongly correlated systems the adiabatic approximation is oversimplified as soon as *quantitative* characterization of electronic and local structure is implied.

Answering the question raised in the headline, *we need to know a structural-dependent excitation spectrum* when analyzing the local structure of the strongly correlated compound. As has been shown in Sect. 4 the inclusion of $S(\omega)$ into the XANES calculations do improve the agreement between experimental and calculated spectra even if the energy balance is not completely settled. This suggests that the relatively simple non-self consistent approach could provide a reasonable approximation for $S(\omega)$: calculated $S(\omega)$ is going to be a much better resemblance of realistic many-body excitations than the single channel absorption.

A Generalized X-ray Analysis of Local Structure that accurately incorporates local electron-lattice and electron-electron correlations has been developed. If performed in the self-consistent manner it might finally evolve into the *unified approach* to such different x-ray spectroscopes utilizing synchrotron radiation as the XPS, UPS, IPES, and XAS. In the present work I have demonstrated how the fragments of this scheme work based on our recent findings on $BaBi_xPb_{1-x}O_3$ and $Ba_{1-x}K_xBiO_3$ perovskite [18,20], $Nd_{2-x}Ce_xCuO_{4-\delta}$ superconductor [11,21], and $La_{1-x}Ca_xMnO_{3+\delta}$ CMR [23]. The important remaining questions in the field concern validation of either the Z or $Z+1$ approximation and improvement in single-electron calculations of XANES spectra.

ACKNOWLEDGMENTS

Author is indebted to F. G. Bridges and T. A. Tyson for enlightening discussions. In addition it is great pleasure to acknowledge the contributions of K. Attenkofer, V. A. Chernov, J. Feldhaus, A. A. Ivanov, S. Khalid, K. V. Klementev, A. P. Menushenkov, A. P. Rusakov, and M. Tischer who were involved at the different stages of this work. This work was supported in part by the Consortium for Advanced Radiation Source, University of Chicago and the NSF grant no. DMR-9733862. Data acquisition was done at VEPP-3m and HASYLAB (for the $BaBi_xPb_{1-x}O_3$ and $Ba_{1-x}K_xBiO_3$), HASYLAB and VEPP-3m ($Nd_{2-x}Ce_xCuO_{4-\delta}$), and NSLS ($La_{1-x}Ca_xMnO_3$). NSLS is supported by the US DOE, Division of Materials Sciences and Division of Chemical Sciences under contract no. DE-AC02-98CH10886.

REFERENCES

1. See, for example, Bishop, A.R. in the Proc. of this Conference and references therein.
2. Ziner, C., *Phys. Rev.* **82**, 403 (1951).
3. Millis, A.J., *Phys. Rev. Lett.* **74**, 5144 (1995).

4. Hayes, T.M., and Boyce, J.B., in *Solid State Physics*, edited by H. Ehrenreich, F. Seitz, and D. Turnbull, Academic, New York, 1982, Vol. 37, p. 173.

5. Bianconi, A., Saini, N.L., Lanzara, A., Missori, M., Rossetti T., Oyanagi, H., Yamaguchi, H., Oka K., and Ito, T., *Phys. Rev. Lett* **76**, 3412 (1996).

6. Egami, T., and Billinge, S.J.L., in " *Physical Properties of High- Temperature Superconductors*", edited by D.M. Ginsberg, World Scientific, Singapore, 1996, Vol. 5, in the *review "Lattice Effect in High- Tc Superconductors*"

7. Booth, C.H., Bridges, F., Kwei, G.H., Lawrence, J.M., Cornelius, A.L., and Neumeier, J.J., *Phys. Rev. B* **57**, 10440 (1998).

8. Kao, C.-C., Caliebe, W.A.L., Hastings, J.B., and Gillet, J.-M., *Phys. Rev. B* **54**, 16361 (1996).

9. Hill, J.P., in *X-Ray and Inner Shell Processes*, edited by R.W. Dunford *et al.*, AIP 2000, p. 312.

10. Billinge, S., and Egami, T., *Phys. Rev. B* **47**, 14386 (1993).

11. Ignatov, A.Yu., Feldhaus J., Chernov, V.A., and Ivanov, A.A., *Nucl. Instrum. & Methods A* **448** (2000).

12. Yonemitsu, K., Bishop, A.R., and Lorenzana, J, *Phys. Rev. B* **47**, 8065 (1993).

13. Sorensen, L.B. *et al*, in *Resonant Anomalous X-Ray Scattering: Theory and Applications*, edited by G. Materlik, C.J. Sparks, and K. Fischer, North-Holland, 1994, p. 389-420.

14. von Barth, U., and Grossman, G., *Phys. Rev. B* **25**, 5150 (1982).

15. Mahan, G.D., in *Solid State Physics*, edited by H. Ehrenreich, F. Seitz, and D. Turnbull Academic, New York, 1974, Vol. 29, p. 75.

16. Ankudinov, A.L., Ravel, B., Rehr, J.J., and Conradson, S.D., *Phys. Rev. B* **58**, 7565 (1998).

17. Bianconi, A., Saini, N. L., Rossetti, T., Lanzara, A., Perali, A., Missori, M., Oyanagi, H., Yamaguchi, H., Nishihara, Y., and Ha, D.H., *Phys. Rev. B* **54**, 12018 (1996).

18. Ignatov, A.Yu., Ph. D. thesis, (1995); Ignatov, A.Yu., *Nucl. Instrum. Methods A* **405**, 359 (1998).

19. See, for example, Boyce, J.B., Bridges, F.G., Claeson, T., Geballe, T.H., and Remeika, J.M., *Phys. Rev. B* **41**, 6306 (1990) and references therein.

20. Ignatov, A.Yu., *Nucl. Instrum. Methods A* **448**, 332 (2000).

21. Ignatov, A.Yu., *J. of Synchrotron Rad.* (2001) to be published.

22. Subías, G., García, J., Proietti, M.G., and Blasco, J., *Phys. Rev. B* **56**, 8183 (1997).

23. Ignatov, A.Yu., and Khalid, S., *Phys. Rev. B* (2001) to be published.

Competing Quantum Orderings in Cuprate Superconductors: A Minimal Model

I. Martin, G. Ortiz, A. V. Balatsky and A. R. Bishop

Theoretical Division, Los Alamos National Laboratory, Los Alamos, NM 87545

Abstract. We present a minimal model for cuprate superconductors. At the unrestricted mean-field level, the model produces homogeneous superconductivity at large doping, striped superconductivity in the underdoped regime and various antiferromagnetic phases at low doping and for high temperatures. On the underdoped side, the superconductor is intrinsically inhomogeneous and global phase coherence is achieved through Josephson-like coupling of the superconducting stripes. The model is applied to calculate experimentally measurable ARPES spectra.

INTRODUCTION

We are witnessing an increase of experimental evidence indicating that (charge and magnetic) incommensuration characterize the low-energy physics of underdoped cuprate superconductors, both above and below the critical superconducting temperature T_c [1]. This poses a challenging problem to theorists since these compounds appear to be at the verge of a multitude of different quantum ordered states that can be tuned by varying physical parameters of the system. Theorists like to use the word "quantum criticality" to refer to this phenomenon. The truth is, however, that so far there is no rigorous theoretical framework that can explain unambiguously this variety of complex phenomena (superconductivity, magnetism, incommensuration, etc.), characterized by intrinsic nonlinearities producing large-scale sensitivities to small perturbations.

In this paper we present a minimal model of high-T_c superconductors that clearly displays a variety of commensurate and incommensurate competing thermodynamic phases. The advantage of our approach is that it is simple, not subject to vague argumentation, and it allows one to rigorously and exhaustively explore a variety of physical observables.

Crucial to the experimental findings have been neutron scattering techniques which probe the spin dynamics of the high-T_c compounds and suggest that different families of cuprate superconductors share inhomogeneously spin and charge textured phases as their quantum states [1]. It is well-known that the stoichiometric (half-filled) compounds are antiferromagnetic (AF) Mott insulators as a result of strong electron interactions and it is upon charge doping that they display incommensuration. Indeed, in a recent paper [2] we have presented a unified theory for the commensurate resonance peak and low-energy incommensurate response observed in neutron

CP554, *Physics in Local Lattice Distortions*, edited by H. Oyanagi and A. Bianconi
2001 American Institute of Physics 1-56396-984-X

scattering experiments. We ascribe both features to be purely magnetic in origin: They represent universal features signaling the existence of an incommensurate spin state both below and above the superconducting transition temperature. Our interpretation indicates that superconductivity *is not* the reason for the resonance peak, and that the incommensurate quantum state provides a reference state for the underdoped cuprates.

In previous work [3,4] we introduced two classes of microscopically *inhomogeneous* models which captured the magnetic and pairing properties of underdoped cuprates. Starting from a generalized *t-J* model Hamiltonian in which appropriate terms mimic stripes, we found that inhomogeneous interactions that locally break magnetic $SU(2)$ symmetry can induce substantial pair binding of holes in the thermodynamic limit. We showed that these models qualitatively reproduce the ARPES and neutron scattering data seen experimentally. Moreover, based on the phenomenology of our microscopic model we developed a mean-field ("Josephson spaghetti") model which provides a scenario for the macroscopic superconducting state. From our model Hamiltonian of random stripe separation r and associated inter- and intra-stripe random Josephson coupling $J(r) \sim 1/r$ we obtained the experimentally observed relation $T_c(x) \simeq \langle J(r) \rangle \propto [\langle r \rangle]^{-1} = \delta(x)$, where x represents charge doping and δ is the inverse of a characteristic length scale associated with the incommensuration.

Our previous numerical simulations have helped to elucidate a certain fraction of the underdoped cuprate puzzle. Here we assume a different strategy complementary to the previous approach: We propose a minimal *homogeneous* model based on the one-band repulsive Hubbard Hamiltonian on a square lattice. The attractive particle-particle singlet channel is included through the nearest neighbor attraction V [5], which produces predominantly *d*-wave pairing close to half-filling. We solve this model at the mean-field level allowing all physical quantities to vary from one lattice site to another. In this way, quantum morphologies characterized by a certain correlation length ξ will appear when the minimum length of our supercell is larger than ξ.

The basic question we address in this paper is whether antiferromagnetic striped ordering [6] and *d*-wave superconductivity can coexist in a certain parameter range of our model. In our opinion, this question is crucial for the understanding of the superconducting state in the underdoped materials.

In the next Section we introduce the model and briefly describe the way we solve it. We then summarize the resulting competing quantum states in a phase diagram and discuss the spectral density. At the end of the paper, we review our main findings.

Some of the results presented in this paper have been introduced in Ref. [7]; the analysis of the spectral density is described here for the first time.

MINIMAL MODEL

We consider here a minimal model that brings together stripes and superconductivity. The model is the two-dimensional one-band Hubbard Hamiltonian with an on-site repulsion U [8]. Pairing correlations are introduced by including the nearest neighbor attraction V [9]. The effective minimal Hamiltonian is thus

$$H_{t-t'-U} = -\sum_{i,j,\sigma} t_{ij} c_{i\sigma}^{\dagger} c_{j\sigma} + U \sum_i n_{i\uparrow} n_{i\downarrow} \tag{1}$$

$$H = H_{t-t'-U} + V \sum_{\langle ij \rangle} n_i n_j , \tag{2}$$

where the operator $c_{i\sigma}^{\dagger}$ ($c_{j\sigma}$) creates (annihilates) an electron with spin σ on the lattice site i, and $n_i = c_{i\uparrow}^{\dagger} c_{i\uparrow} + c_{i\downarrow}^{\dagger} c_{i\downarrow}$ represents the electron density on site i. The hopping t_{ij} equals t for nearest neighbors and t' for for the second-nearest neighbor sites i and j. For our computations, we use the unrestricted mean-field approximation to this Hamiltonian,

$$H_{MF} = -\sum_{\langle ij \rangle \sigma} t_{ij} c_{i\sigma}^{\dagger} c_{j\sigma} + U \sum_i n_{i\uparrow} \langle n_{i\downarrow} \rangle + \langle n_{i\uparrow} \rangle n_{i\downarrow} + \sum_{\langle ij \rangle} c_{i\downarrow} c_{j\uparrow} \Delta_{ij}^* + \text{H.c.}, \tag{3}$$

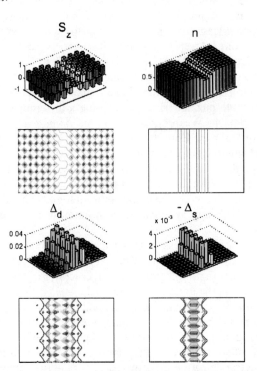

FIGURE 1. Typical example of density and superconducting order parameter profiles in a stripe state (here, period 17). The top two bar charts represent the site-dependent spin and charge densities, respectively. The contour plots indicate the sites with low (blue) and high (red) values of the corresponding densities. The bottom four plots show the values of the superconducting order parameters, defined as $\Delta_i^{d(s^*)} = (\Delta_{i,right} + \Delta_{i,left} \mp \Delta_{i,up} \mp \Delta_{i,down})/4$ for d-wave (extended s-wave) order parameter on site i ($U = 4t$, $V = -0.9t$, $t' = 0$). Different choices of parameters lead to qualitatively similar patterns, with stronger U leading to a stronger AF order and more attractive V causing the superconducting stripes to become wider and larger in amplitude. The doping level is 5.9%.

where $\Delta_{ij} = V\langle c_{i\downarrow}c_{j\uparrow}\rangle$ is the MF superconducting order parameter. The effect of V in our model is limited to the generation of superconducting correlations. We do not explicitly address the important issue of the microscopic origin of the attraction V [5].

A typical zero-temperature MF inhomogeneous solution is shown in Fig. [1]. In the lowest energy configuration, the spin density develops a soliton-like AF anti-phase domain boundary — a stripe — at which the AF order parameter changes sign. At the domain boundary, the electronic charge density is depleted. The width of the domain wall, ξ_{DW}, decreases with increasing on-site repulsion U. However, for values of U that are not much larger than the hopping t, the charge per unit length of the optimal (the lowest energy) stripe remains the same and is close to unity near half-filling for t' = 0. The bond-centered stripes are favored relative to the site-centered ones, although the energy difference in our case is small due to the smooth charge distribution.

Stripe formation is the result of the competition between antiferromagnetism (which can lead to charge confinement) and delocalization (driven by kinetic energy). Non-linear feedback is responsible for these complex patterns. A half-filled (one electron per site) antiferromagnet is the state with the lowest energy per electron. Upon small doping, the energy can be optimized by segregating the excess electrons or holes into domain walls or charged clusters, and keeping the bulk antiferromagnetism unperturbed. The anti-phase domain walls are favored over the simple charged stripes as they allow the charge carriers to optimize their transverse fluctuations in the direction across stripes, thereby lowering their kinetic energy. The linear filling of the emergent stripes vary depending upon the specific nature of the band structure.

For different band structures the exact relation between the doping x and inter-stripe distance, $L(x)$, may change; however, any model whose ground state is AF at zero doping, can be expected to have AF stripes for a finite doping, with incommensuration proportional to the doping, $1/L(x) \propto x$, near half-filling. For example, negative next-nearest neighbor hopping t' (relevant in the hole-doped cuprates [10]), modifies the stripe filling without compromising the stripe phase stability relative to commensurate

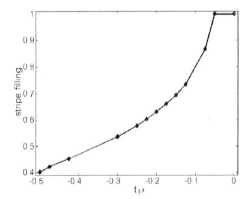

FIGURE 2. Linear filling of an isolated vertical stripe as a function of next-nearest neighbor hopping, t'. Here $U = 4t$ and $V = 0$.

AF at the MF level [11]. The stripe filling is a monotonically decreasing function of the magnitude of t', with the filling 1/2 occurring when $t' = -0.35t$ (Fig. 2). While filling-one stripes correspond to a correlated insulator, the fractional filling stripes in the t-t'-U model (Eq. (1)) are metallic, which can be understood in terms of the partial occupancy of the mid-gap band formed due to the stripes [11]. In the case of insulating stripes ($t' = 0$) a threshold value of attraction V should be exceeded to generate superconductivity and hence to overcome the insulating gap [12]; however, in the case of metallic stripes one would expect that any attraction would yield superconductivity through the Cooper instability [13]. Indeed, this is what we find [7]. On the contrary, the diagonal stripes which can also be the ground states of the Hubbard model, particularly at low dopings, always have a filling of one electron per Cu site, and hence are insulators. This makes diagonal stripes antagonistic to superconductivity [11], and also agrees with the experimental observations [14].

From Fig. 1 it is clear that the superconducting order parameter $\Delta_{ij}^{d(s^*)}$ is maximized on the stripes and is not smooth (even within the stripe) due to the presence of the AF background. In addition to the dominant d-wave component, there is a small extended s-wave (s^*) component generated on the stripe, which can be interpreted as a distortion of the d-wave at the level of about 10%. This happens because the symmetry of the lattice has been spontaneously broken by the stripes. For dopings less than about 10% (corresponding to $L(x) > 10$ lattice sites) the stripes have negligible overlap. In this regime, the amplitude of the superconducting order parameter on the stripes no longer depends upon the stripe-stripe separation. For higher doping levels, an overlap between the superconducting order parameters on adjacent stripes is established, and for even higher doping the stripes "melt" and superconductivity becomes homogeneous, of a classical BCS type [13].

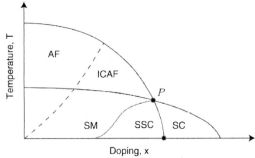

FIGURE 3. Schematic phase diagram obtained by superimposing the antiferromagnetic (AF) / striped (ICAF) and the d-wave superconducting (SC) phase diagrams. In the intersection region we distinguish the subregions of Josephson-coupled striped superconductor (SSC), and non-superconducting "strange metal" (SM), which is neither a superconductor, nor a simple insulator. The upper boundary of the AF/ICAF corresponds to the weak pseudogap crossover, and the line between the pure AF/ICAF and the SM marks the strong pseudogap crossover. A detailed finite-temperature study is required to precisely locate the left boundary of the SSC region, and hence to determine the order of the critical point P.

Phase Diagram

From our zero-temperature analysis of the coexistence of AF stripes (ICAF) and superconductivity [7], a simple qualitative thermodynamic phase diagram emerges. In the conjectured phase diagram, we utilize the finite-temperature AF/ICAF phase diagram of the Hubbard model together with the superconducting (SC) phase diagram of the t-V model. The SC phase diagram is obtained in the homogeneous MF [9], while the AF/ICAF phase boundary is constructed under the assumption of the second order phase transition between the homogeneous and inhomogeneous states [8]. For a suitable choice of parameters, for instance $U = 2t$ and $V = -t$, the SC and the AF/ICAF regions in the phase diagram intersect, as shown in Fig. 3. The energy scale associated with the AF/ICAF region of the phase diagram is much larger than that of the SC part. Thus, one expects that only the SC phase boundary is modified when it passes through the AF/ICAF region. The central result of our work is that the superconductivity *does not* disappear in the region of the AF stripes, but rather becomes striped, with anisotropic superfluid stiffness.

Based on familiar Josephson coupling physics, in the region of coexistence of superconductivity and stripes, we can expect a part that is a globally coherent striped superconductor (SSC). The rest of the intersection region is covered by an exotic phase which, if it were perfectly orientationally ordered, would be a superconductor in one direction and a strongly-correlated insulator in the other. In reality, due to the meandering of the stripes and their break-up into finite segments [15], the state is likely to be highly inhomogeneous and neither an insulator, nor a superconductor, but also not a simple metal. In agreement with the experimental attribution, we refer to this region as a "strange metal" (SM). The line separating the SM from the rest of the AF/ICAF region, in the context of the experiments, can be associated with the crossover to the strong pseudogap regime, and corresponds to the opening of the local superconducting gap. The high-temperature boundary separating AF/ICAF phases from the homogeneous state, marks the onset of the weak pseudogap. For small dopings, there is also a possibility of a transition from the vertical to diagonal stripes [11].

Spectral Density

In this Section we demonstrate how our model can be applied to compute the energy spectrum of the system, which can then be compared with the experimental data [16]. At any temperature, the MF solution yields a self-consistent spectrum $\{E_n\}$ of Bogoliubov quasiparticles, which diagonalize the MF Hamiltonian in Eq. (3). Knowing how the electron operators are related to the Bogoliubov quasiparticles, one can compute the electronic spectral densities for positive (particle) and negative (hole) biases. Experimentally, angle resolved photoemission (ARPES) measures the electronic spectral density integrated in the window of $\pm\Delta\varepsilon$ around the Fermi energy.

In Fig. 4 we show the computed ARPES spectra for various energy integration windows. The symmetry of the spectrum is spontaneously broken due to the presence of stripes (the stripes run along the horizontal k direction). For small integration window near the Fermi surface, the spectral weight is concentrated around the $(0, \pi)$ point. The reason is that the stripes gap out the flat parts ("diagonals") of the Fermi surface, while keeping the quasiparticles around $(0, \pi)$ gapless if $t' < 0$ or weakly gapped if $t' = 0$. Since these are the quasiparticles that are primarily responsible for the formation of d_{x2-y2} superconductivity, it is this particular structure of the "stripe gap" that allows for the peaceful coexistence of stripes and superconductivity in our model. Indeed, in the absence of superconductivity ($V = 0$), the spectral patterns remain essentially unchanged, except for the enhanced weight around $(0, \pi)$.

For larger energy integration windows, the Fermi surface gradually "reconstructs", with the energy states around the diagonal reappearing when $\Delta\varepsilon$ exceeds the "stripe gap".

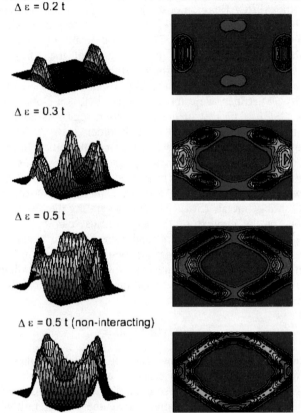

$\Delta\varepsilon = 0.2\, t$

$\Delta\varepsilon = 0.3\, t$

$\Delta\varepsilon = 0.5\, t$

$\Delta\varepsilon = 0.5\, t$ (non-interacting)

FIGURE 4. Computed ARPES spectra for various energy integration windows, $\Delta\varepsilon$. The Brillouin zone is defined as $(0, 2\pi) \times (0, 2\pi)$, with the horizontal direction being along the stripes and the vertical direction perpendicular to the stripes. Notice that for small $\Delta\varepsilon$ the spectral weight is concentrated around $(0, \pi)$, with the Fermi surface being gradually reconstructed with the increasing window of integration. The parameters are $U = 4t$, $V = -t$, $t' = -0.2t$, and doping 8.3%. The system size is 16×16, with two collinear stripes. For comparison we also show the case of free electrons with the same non-interacting band structure and doping.

Notice that due to twinning and the expected presence of stripe domains in the real experimental systems the computed spectra have to be symmetrized. Similar results for the striped spectra (but without superconductivity) have been obtained previously [17,11,4]. Similarly, one can calculate temperature-dependent specific heat, entropy, spin susceptibility, among other experimentally measurable quantities, as we will report elsewhere.

CONCLUSIONS

In summary, we have presented a minimal model supporting the coexistence of incommensurate antiferromagnetism ("stripe" order) and global anisotropic superconductivity. Contrary to the common belief, these two order parameters can coexist and our calculation is a faithful realization of such a physical situation. At the same time, the stripe order provides a natural competing order parameter limiting the increase of the superconducting transition temperature on the underdoped side of cuprates.

Our model displays a variety of other competing homogeneous and inhomogeneous thermodynamic phases. Based on the model, we constructed a phase diagram that captures many features of the superconducting cuprates. Finally, we computed the photoemission spectra which give a clear interpretation of experimental data.

This work was supported by the U.S. DOE.

REFERENCES

1. For a review see Mason, T.E., in *Handbook on the Physics and Chemistry of Rare Earths*, eds. Gschneidner, K.A., Jr., Eyring, L., and Maple, M.B..

2. Batista, C.D., Ortiz, G., and Balatsky, A.V., *cond-mat*/0008374.

3. Eroles, J., Ortiz, G., Balatsky, A.V., and Bishop, A.R., *Europhys. Lett.* **50**, 540 (2000); *cond-mat*/0001430; *cond-mat*/0003322.

4. Eroles, J. *et al.*, *cond-mat*/0008341.

5. For instance, an effective attraction in the singlet channel is generated as a result of reduction of the 3-band Hubbard model of CuO planes to a one-band model with correlated hopping. See M. E. Simón, M.E., and Aligia, A.A., *Phys. Rev. B* **52**, 7701 (1995). A possible origin of the attractive interaction generated *locally* can be found in Ref. [3].

6. Henceforth, we will assume that the incommensurate ordering is "striped". This ordering is characterized by clustering of charge carriers (i.e., holes or electrons whose motion occurs primarily in two-dimensional CuO_2 layers) into fluctuating quasi-one-dimensional channels and channel segments which act as anti-phase domain walls for the otherwise unfrustrated AF background.

7. Martin, I., Ortiz, G., Balatsky, A.V., and Bishop, A.R., *cond-mat*/0003316.

8. Schulz, H.J., *Phys. Rev. Lett.* **64**, 1445 (1990).

9. Micnas, R., *et al.*, *Rev. Mod. Phys.* **62**, 113 (1990).

10. White, S.R, and Scalapino, D.J., *Phys. Rev. B* **60**, R753 (1999).

11. Machida, K., and Ichioka, M., *J. Phys. Soc. Jpn.* **68**, 2168 (1999).

12. Zaanen, J., preprint *cond-mat*/9811078.

13. Bardeen, J., Cooper, L.N., and Schrieffer, J.R., *Phys. Rev.* **106**, 162 (1957).
14. Wakimoto, S. *et al.*, *Phys. Rev. B* **60** R769 (1999).
15. Kivelson, S.A. *et al.*, *Nature* **393**, 550 (1998);
16. Zhou, X.J. *et al.*, *Science* **286**, 268 (1999).
17. Salkola, M.I. *et al.*, *Phys. Rev. Lett.* **77**, 155 (1996).

Local Lattice Distortion and Charge-Spin Stripe Structure of Superconducting YBa$_2$Cu$_3$O$_{6+x}$

M. Arai^1, Y. Endoh^2, S. Tajima^3 and S. M. Bennington^4

^{1} Institute of Materials Structure Science, KEK, Oho, Tsukuba, 305-0801, Japan

^{2} Institute for Material Research, Tohoku Univ., Katahira, Aoba-ku, Sendai, 980-8577, Japan

^{3} Superconductivity Research Laboratory, ISTEC, Shinonome, Koutou-ku, Tokyo, 135-0062, Japan

^{4} Rutherford Appleton Laboratory, Chilton, Didcot, Oxon, OX11 0QX, UK

Abstract. We have clarified the incommensurate spin dynamics of YBCO. The incommensurability is 1/8 and 1/6 for the sample in the 60K-phase and one in the 90K-phase respectively, and is proportional to the transition temperature, resemble to that of LSCO. Superlattice peaks at (1/4 0 0) and (1/3 0 0) were also found, which are expected from a scenario of stripe domain structure. Those have temperature evolution relevant to occurrence of superconductivity. Paper also discusses the behaviour of the lattice parameters. In this paper we speculate a possible scenario of the mechanism of the superconductivity based on the stripe model.

INTRODUCTION

The incommensurate spin dynamics seem to have an important role for the pairing mechanism in oxide high-T$_c$ superconductivity, showing linear relation between incommensurability and transition temperature T$_c$ [1]. On the other hand the stripe domain structure model was proposed by Tranquada *et al.* on La$_{2-x-y}$Nd$_y$Sr$_x$CuO$_4$ [2]. One dimensionally segregated charges deform lattice locally and forms charge-spin stripe domain structure. Therefore, it has become more crucial to give an information about local spin correlation and lattice deformation to elucidate the concrete role of the stripe domain structure for superconductivity.

Quite recently we have performed inelastic neutron scattering experiments on underdoped YBa$_2$Cu$_3$O$_{6.7}$ (YBCO$_6$, T$_c$ ~ 67K, nh ~ 0.12) [3] and optimally doped YBa$_2$Cu$_3$O$_7$ (YBCO$_7$, T$_c$ ~ 90K, nh ~ 0.25) [4]. Both of the sample showed incommensurate spin correlation with the same symmetry as that of LSCO, and the incommensurability is almost proportional to T$_c$ as shown in LSCO [1]. Hence the results prompted possibility of existence of stripe domain structure as a common feature in high-T$_c$ materials. In this report we show results from diffraction experiments on superlattice peak of YBCO$_6$ and YBCO$_7$, which are expected from the incommensurability of the spin dynamics and supportive for the stripe domain scenario. The superlattice peaks exist at (1+1/4 0 0) for YBCO$_6$ and (1+1/3 0 0) for YBCO$_7$. The results suggest that 3 leg spin ladder for underdoped YBCO$_6$ in the 60K phase and sudden step-up to the 90K phase of the optimally doped YBCO$_7$ associated

CP554, *Physics in Local Lattice Distortions*, edited by H. Oyanagi and A. Bianconi
© 2001 American Institute of Physics 1-56396-984-X/01/$18.00

with a structural phase transformation to two leg spin ladder. The results suggest us a speculative scenario of the superconducting paring mechanism analogous to a spin singlet formation in the two-leg ladder system.

INCOMMENSURATE SPIN DYNAMICS

In Fig. 1 we illustrate observed spin dynamical structure factor $S(Q, E)$ of $YBCO_6$ along the (h, h, 0) (specified Q2D) direction at 20K. At the low energy, it has two legs recognized as incommensurate spin correlation. The profile was fit by two Gaussian functions, and the results are shown in Fig. 2. Two peaks at low energy merge to form an intense single peak, refereed as the resonance peak, at the resonance energy ER and split again with broad and dispersive feature above ER. Optimally doped $YBCO_7$ has similar features in the spin dynamics. There are a more prominent intensity spot at ER = 41meV and very weak two legs below ER with a wider separation in the incommensurability than that of $YBCO_6$ [4].

In Fig. 3 we plot the incommensurate peak separation along (1 1 0) as a function of oxygen content, along with the results of Regnault *et al.*, which are peak widths, not peak separations [5]. We conjecture here that the peak widths previously reported can be recognized as a smeared peak of two separated incommensurate peaks due to coarse instrumental resolution or poor mosaicity of sample. We also plotted the superconducting transition temperature (T_c) [6] as a function of oxygen content. We immediately see that the superconducting transition temperature T_c can be scaled by the incommensurability, in the same manner as Yamada's plot on LSCO [1]. It is also remarkable that the so-called 60K-plateau in the phase diagram corresponds to the incommensurability $d \sim 1/8$ ($d = \Delta Q_{2D}$) around hole concentration $n_h \sim 1/8$, and then a rapid increase of T_c to the 90K-plateau correspond to the rise to $d \sim 1/6$ around $n_h \sim 1/6$. Here n_h was estimated from an empirical formula $n_h = 0.187 - 0.21(1-x)$ obtained by the thermoelectric power measurement [7].

We also plot T_c as a function of the incommensurability in Fig. 4, using the relation in Fig. 3 and the results for LSCO [1] by scaling T_c at $d = 1/8$ (scaling factor is about 3/2). T_c increases almost linearly up to $d = 1/8$ for both systems, but the behaviour differs drastically above $d = 1/8$. The increase in T_c above $d = 1/8$ for YBCO is associated with a continuous increase in d. On the other hand, T_c is suddenly suppressed above $d = 1/8$ for LSCO.

DIFFRACTION MEASUREMENT OF (1/4 0 0) AND (1/3 0 0) SUPERLATTICE PEAK

From the results of incommensurate spin dynamics of 1/8 and 1/6 for $YBCO_6$ and $YBCO_7$, it is a natural extension of measurements to find possible superlattice peaks at (1/4 0 0) and (1/3 0 0) according to stripe scenario.

Fig. 5(a) shows the diffraction peak at nominally (1+1/4 0 0) at 10K and 250K of $YBCO_6$. There is a large evolution in the intensity with temperature. Fig. 5(b) is the temperature dependence of the integrated intensity of the peak obtained independently

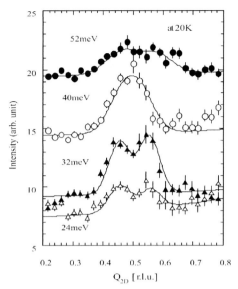

FIGURE 1. Profiles at 24±2meV, 32±2meV, 40±2meV and 52±2meV are shown with shifted base lines. There are double peaks below and above ER=40meV. At around ER the two peaks abruptly merge into a single broad peak.

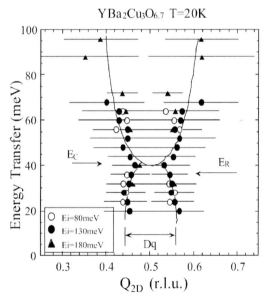

FIGURE 2. The energy dependence of the peak position of the incommensurate peaks projected along Q_{2D}. The separation (Δq) is about 0.12 r.l.u. in the low-energy region, it shrinks at E_C and then gradually increases above E_R. The solid line is just a guide to the eye. The horizontal bars stand for FWHM of peak.

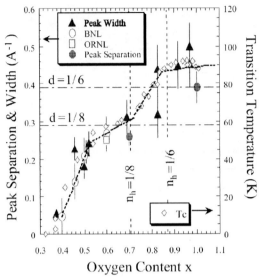

FIGURE 3. Peak separation along Q_{2D}, of incommensurate peaks are plotted as a function of oxygen content. T_C is also displayed, scaled appropriately. The incommensurate peak separation is strongly correlated with T_C. The plot includes data from the present work plus peak width from Regnault et al.(specified as Peak Width and BNL)[Regn] and peak separation from Mook et al (ORNL)[Mook] .

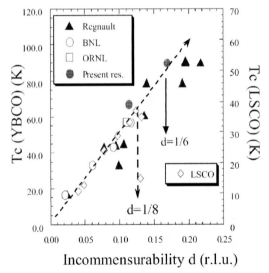

FIGURE 4. The superconducting transition temperature T_C is plotted as a function of the estimated incommensurability. T_C of LSCO is also shown by scaling T_C at $d = 1/8$. The plot includes data from the present work plus estimated incommensurability from the peak width from Regnault et al.(specified as Peak Width and BNL)[Regn] and peak separation from Mook et al (ORNL)[Mook] .

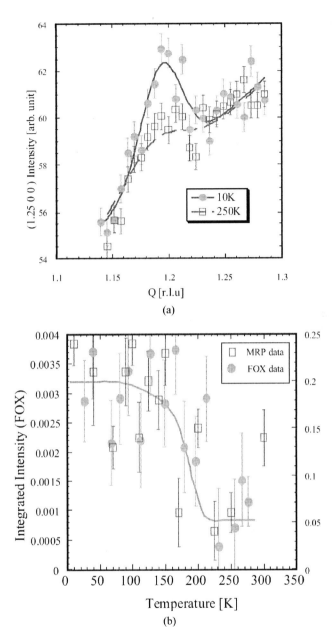

FIGURE 5. (a) Peak profile of (5/4 0 0) along a* at 10K and 250K. (b) Integrated intensity of (5/4 0 0) after subtracting background as a function of temperature. Intensity increase start below 200K, much above Tc=65K.

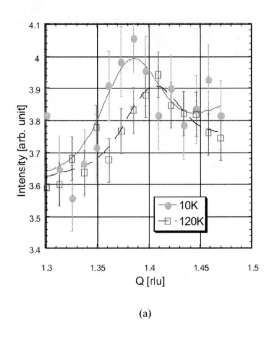

(a)

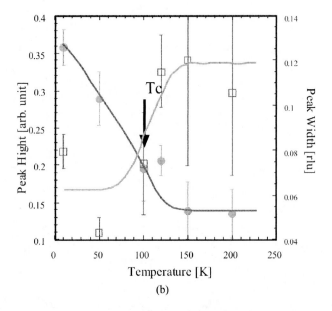

(b)

FIGURE 6. (a) Peak profile of (4/3 0 0) along a* at 10K and 120K. (b) Peak intensity and FWHM of the peak are plotted as a function of temperature. The evolution start around Tc=90K.

by two different diffractometers MRP and FOX in the KENS facility of High Energy Accelerator Research Organization in Japan. The intensity suddenly starts to increase at about 200K, much above $T_c = 65K$. This result reminds us a pseudo-spin gap opening around that temperature for underdoped YBCO. Fig. 6(a) shows peak around nominally (1+1/3 0 0) of YBCO$_7$ at 10K and 120K. In Fig. 6(b) the peak height and the width, fit by Gaussian function, are also plotted. There is peak structure even at high temperature with a shift of the peak position. The peak height has a steep increase below $T_c = 90K$. On the other hand the peak width seems to decrease below T_c. These results mean that there is a modulated structure even at high temperature, and the correlation develops longer to be $16Å = 1/FWHM$ below T_c. We have also observed peaks at (2/3 0 0) and (1+2/3 0 0). The intensity of the three peaks nearly obeys $I \sim Q^2$. Hence, the origin of the peaks is related to a lattice distortion of about 0.01 r.l.u. We should, however, remind that there are oxygen ordering in the chain site, which gives similar superlattice peaks. Although, it cannot neglect such a possibility, however, the intensity of the superlattice peaks from oxygen ordering has one to two order of magnitudes stronger intensity, and the observed superlattice peak has strong temperature dependence related with the superconductivity. Hence, we believe that the origin of the observed peaks is attributed to a lattice distortion by charge stripe in the two-dimensional Cu-O$_2$ plane. We could not confirm the direction of the stripe structure. However, considering an influence from the charge reservoir at the oxygen chain site to the stripe structure and preliminary results on detwined sample [8], we expect the direction of the stripe structure is along the b-axis, parallel to the oxygen-ordering axis.

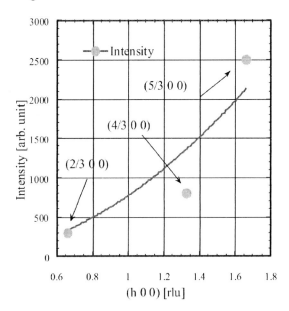

FIGURE 7. The momentum dependence of the integrated intensity at (2/3 0 0), (4/3 0 0) and (5/3 0 0).

ANOMALY OF LATTICE CONSTANT

Fig. 7 shows the temperature evolution of the b-lattice constant of $YBCO_6$. There are chattering in the data, but these are not statistics error but are intrinsic fluctuation. The solid line in Fig. 7 is simple calculation according to Einstein model. In Fig. 8 the ration between the observed data and the Einstein model is depicted. There is a humps between 200K and $T_c = 65K$. This can be attributed to superconducting fluctuation because of strong electron-lattice coupling [9]. The shape of the double structured hump is resemble to the electron specific heat [10]. More interesting thing is the up-turn below T_c. This behaviour is only observed on the b-lattice constants including optimally doped $YBCO_7$ [11]. A similar anomaly was observed on the c-axis in case of LSCO [12]. This reminds us the lattice evolution (spontaneous strain) associated with spin-singlet formation of a spin-Peierls transition observed on $CuGeO_3$ [13]. Actually if spin singlet was formed in the stripe structure, elongation of the b-axis helps this formation.

DISCUSSION

T_c and Resonance Energy

From the observed results shown in the previous sections, we illustrate expected stripe structures, in Fig. 8(a) and (b), for $YBCO_6$ and $YBCO_7$, which has *4a* or *3a* lattice spacing respectively. We call these "4a-Stripe" and "3a-Stripe", hereafter, which also correspond to 3-leg ladder and 2-leg ladder spin system. The correlation length ξ_{INC} ($= 1/\Delta q_{INC}$), estimated from the incommensurate peak width is also consistent with the extent of spin-stripe. On the other hand, the peak width Δq_{RES} of the resonance peak is unchanged with hole concentration as is reported [14], and the correlation length ξ_{RES} ($= 1/\Delta q_{RES}$) is about the distance over a neighbouring spin pair ($\sim 3.8\text{Å}$). This is very consistent with an idea that the resonance peak is created from a singlet-to-triplet excitation of spin-singlet pair in spin-stripe structure owning spin gap below the resonance energy E_R as was proposed [15]. Singlet formation gives very strong constraint to the behaviour between spin states and charge states. Hence, it is easily expected that there is a very strong correlation between singlet pair formation and charge coherence. This scenario speculates that the Cooper pair is formed in the adjacent charge stripe, both sides of a singlet pair, which is directly governed by spin singlet formation, hence very high-Tc is expected. The coherence lengths, ξ_{COH}, estimated from Hc2 is consistent to this scenario, which becomes shorter for the optimally doped $YBCO_7$[16]. Tc should coincide with the energy decrease associated with spin singlet formation. Hence, Tc can be high as about 100K based on BCS theory ($2\Delta \sim 3/4 E_R (= 41\text{meV})) = 3.5 kBTC: T_c = 100K$).

There are several theories to explain the Resonance peak at 41meV for optimally doped $YBCO_7$. However, by the stripe model it can be estimate the resonance energy, 41meV, by applying theory for two-leg ladder system [17], which can be expressed as a spin gap energy $\Delta = J_R(rung) - J_C(chain)$ at the AF zone center. We can estimate the

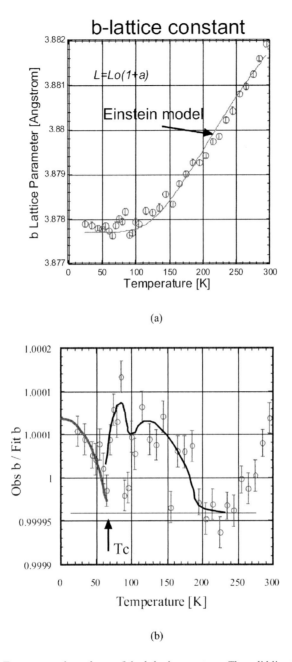

(a)

(b)

FIGURE 8. (a) Temperature dependence of the b-lattice constant. The solid line is a calculated result from Einstein model. (b) The ratio between the observed b-lattice constant and the calculated one. Solid lines are just guides for eye.

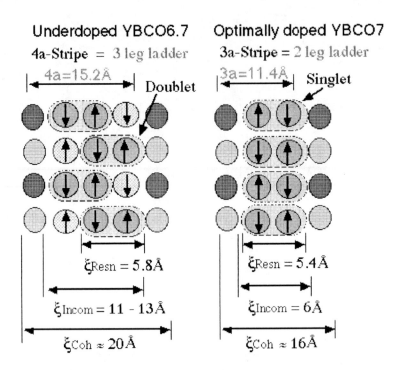

FIGURE 9. Proposed stripe structure for underdoped YBCO6 and optimally doped YBCO7. They are 4a-Stripe (3-leg ladder) and 3a-Stripe (2-leg ladder), respectively. The correlation lengths, ξ_{INC} $(=1/\Delta q_{INC})$ estimated from the incommensurate peak width and ξ_{RES} $(=1/\Delta q_{RES})$ estimated from resonance peak, and coherence lengths, ξ_{COH}, estimated from Hc2 are displayed.

exchange energy of the chain direction by using $J_C = J_R(a/b)^{12}$ [18], where $a = 3.83\text{Å}$ and $b = 3.88\text{Å}$ and the exchange energy of rung $J_R = 240\text{meV}$[19], respectively. Hence, we get $\Delta = 41\text{meV}$. This result suggests that orthorhombicity is essentially important character for the superconducting mechanism to form spin gap. This may also suggest why H_{c2} is so high as 300T. This is a natural consequence of field strength to destruct a singlet pair, *i.e.* $H_{c2} = (41\text{meV} * 8.6\text{T}) \sim 350\text{T}$.

CONCLUSION

We observed incommensurate spin dynamics and superlattice peaks of underdoped YBCO$_6$ and optimally doped YBCO$_7$, and proposed 4a- and 3a-stripe structure for those materials. The spin singlet formation in the spin stripe has a role of attracting force and produce a coherent charge states in the charge stripe structure. From this structure we can explain the resonance energy, T_c, H_{c2} and phase diagram by taking a model for two-leg ladder system. For this scenario, two-dimensional S = 1/2 system with orthorhombic structure is essential to form stripe structure and superconductivity.

ACKNOWLEDGEMENT

The authors acknowledge Drs K. Tomimoto, Y. Shiohara, J. W. Jang, A. I. Rykov, M. Kusao, and S. Koyama for preparing the sample. This work was supported by Grant-in-Aid on Scientific Research on Priority Areas "Anomalous Metallic State near the Mott Transition" (07237102) of Ministry of Education, Science, Sports and Culture, Japan, and done under the collaboration with NEDO.

REFERENCES

1. Yamada, K. *et al.*, *Phys. Rev. B* **57**, 6165 (1998).
2. Tranquada, J.M. *et al.*, *Naure* (London) **375**, 561 (1995).
3. Arai, M. *et al.*, *Phys. Rev. Lett.* **83**, 608 (1999).
4. Arai, M. *et al.*, *Cond-mat*/9912233
5. Regnault, L.P. *et al.*, *Physica B* **213&214**, 48 (1995).
6. Graf, T. *et al.*, *T. Less-Common Met.* **159**, 349 (1990).
7. Tallon, J.L. *et al.*, *Phys. Rev. B* **51**, 12911 (1995).
8. Mook, H.A. *et al.*, *Naure* (London) **404**, 729 (2000).
9. Millis, A.J., and Rabe, K.M., *Phys. Rev. B* **38**, 8908 (1988).
10. Loram, J.W. *et al.*, *Phys. Rev. Lett.* **71**, 1740 (1993).
11. You, H. *et al.*, *Phys. Rev. B* **43**, 3660 (1991); Meingast, C. *et al.*, *Phys. Rev. B* **67**, 1634 (1991)
12. Arakawa, T. *et al.*, *Czech. J. Phys.* **46**, Suppl. S3, 1239 (1996).
13. Fujita, M. *et al.*, *Physica B* **213&214**, 288 (1995).
14. Bourges, P. *et al.*, *Phys. Rev. B* **56**, R11439 (1997).
15. Morr, D.K., and Pines, D., *Phys. Rev. Lett.* 81, 1086 (1998).
16. Vandervoort, K.G. *et al.*, *Phys. Rev. B* **43**, 13042 (1991).
17. Dagotto, E., and Rice, T.M., *Science* **271**, 618 (1996).
18. de Jongh, L.J., and Block, R., *Physica* **79B**, 568 (1975).
19. Hayden, S.M. *et al.*, *cond-mat*/9710181

Impurity Effects on the Stripes in the La-214, Bi-2212 and Y-123 Phases

Y. Koike[*], M. Akoshima[*], M. Aoyama[*], K. Nishimaki[*], T. Kawamata[*],
T. Adachi[*], T. Noji[*], I. Watanabe[†], S. Ohira[†],
W. Higemoto[¶], and K. Nagamine[†¶]

[*]Department of Applied Physics, Tohoku University, Aoba-yama 08, Aoba-ku, Sendai 980-8579, Japan
[†]Muon Science Laboratory, The Institute of Physical and Chemical Research (RIKEN),
2-1 Hirosawa, Wako 351-0198, Japan
[¶]Meson Science Laboratory, Institute of Materials Structure Science, High Energy Accelerator
Research Organization (KEK-MSL), 1-1 Oho, Tsukuba 305-0801, Japan

Abstract. Our recent experimental studies on the 1/8 anomaly in the Bi-2212 and Y-123 phases and on the new anomaly at $x = 0.21$ in $La_{2-x}Sr_xCu_{1-y}Zn_yO_4$ are reviewed. In the Zn-substituted Bi-2212 and Y-123 phases, we have found anomalous suppression of superconductivity at p (the hole concentration per Cu in the CuO_2 plane) $\sim 1/8$, and it has been found from the muon spin relaxation (μ SR) measurements that the magnetic correlation between Cu spins is enhanced singularly at p $\sim 1/8$. Furthermore, we have found marked suppression of superconductivity singularly at $x = 0.21$ in the 1% Zn-substituted single-crystals of $La_{2-x}Sr_xCu_{1-y}Zn_yO_4$, and it has been found from the μ SR measurements that the magnetic correlation is also enhanced at $x = 0.21$. These results suggest that the dynamical stripe correlations of holes and spins exist in the Bi-2212 and Y-123 phases as well as in the La-214 phase and that they tend to be pinned by Zn, leading to the suppression of superconductivity. If this tendency is the case in a wide range of p, it is possible to understand the universal results in the high-T_c cuprates that Zn is a strong scatterer and that the Zn substitution markedly suppresses superconductivity.

INTRODUCTION

Recently, it has been pointed out from the inelastic neutron-scattering experiments that stripe correlations of holes and spins exist in a wide range of hole concentration in the La-214 phase [1] and also in the Bi-2212 [2] and Y-123 [3, 4] phases. The stripe correlations have attracted great interest in relation to the mechanism of the high-T_c superconductivity (SC). However, when the dynamical stripe correlations are pinned to become a static stripe-order at p (the hole concentration per Cu in the CuO_2 plane) $\sim$ 1/8 in the La-214 phase, the SC is markedly suppressed [5]. This is called the 1/8 anomaly. Accordingly, the 1/8 anomaly is expected to appear in the other high-T_c cuprates, when adequate pinning centers are introduced into a sample. In fact, we have found the 1/8 anomaly in the Bi-2212 [6-9] and Y-123 [9-11] phases by introducing a small amount of Zn into a sample. Moreover, we have found a new anomaly at $x = 0.21$ in the overdoped region of $La_{2-x}Sr_xCu_{1-y}Zn_yO_4$, which suggests the existence of a static stripe-order as well as at $x = 0.115$ [12-15].

CP554, *Physics in Local Lattice Distortions*, edited by H. Oyanagi and A. Bianconi
© 2001 American Institute of Physics 1-56396-984-X/01/$18.00

In this paper, we review our experimental studies from the transport and μ SR measurements on the 1/8 anomaly of the Bi-2212 and Y-123 phases and on the new anomaly at x = 0.21 in $La_{2-x}Sr_xCu_{1-y}Zn_yO_4$.

EXPERIMENTAL

Sintered samples of $Bi_2Sr_2Ca_{1-x}Y_x(Cu_{1-y}M_y)_2O_{8+\delta}$ (M = Zn, Ni) of the Bi-2212 phase and those of $Y_{1-x}Ca_xBa_2Cu_{3-2y}Zn_{2y}O_{7-\delta}$ of the Y-123 phase were prepared by the solid-state reaction method. Single crystals of $La_{2-x}Sr_xCu_{1-y}Zn_yO_4$ were grown by the TSFZ method. T_c was determined from the resistivity and magnetic susceptibility measurements. μ SR measurements were carried out at the RIKEN-RAL Muon Facility in the UK and at the KEK-MSL in Japan, using a spin-polarized pulsed surface muon beam with a momentum of 27 MeV/c.

RESULTS AND DISCUSSION

1/8 Anomaly in the Bi-2212 Phase

In order to pin the possible dynamical stripe correlations, Zn or Ni was partially substituted for Cu. As shown in Fig. 1, anomalous suppression of SC has been found at x = 0.30 - 0.35, where p ~ 1/8, in the Zn-substituted $Bi_2Sr_2Ca_{1-x}Y_x(Cu_{1-y}Zn_y)_2O_{8+\delta}$ with y = 0.02 - 0.03 [6]. In these samples, transport properties are also anomalous [6, 9]. Recently, our results have been confirmed by Ilonca and co-workers [16].

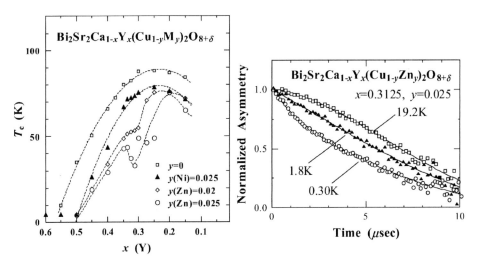

FIGURE 1. x dependence of T_c in $Bi_2Sr_2Ca_{1-x}Y_x$ $(Cu_{1-y}M_y)_2O_{8+\delta}$ (M = Zn, Ni).

FIGURE 2. μ SR time spectra of $Bi_2Sr_2Ca_{1-x}Y_x$ $(Cu_{1-y}Zn_y)_2O_{8+\delta}$ with x = 0.3125 and y = 0.025.

Fig. 2 shows μSR time spectra of the Zn-substituted sample with x = 0.3125 and y = 0.025 [7, 8]. The depolarization behavior is of a Gaussian-type at high temperatures, indicating that the muon spins depolarize by only the nuclear dipole field. It changes to an exponential-type at low temperatures, while it is still of a Gaussian-type at low temperatures for the non-Zn-substituted sample with x = 0.3125 and y = 0. This suggests that the magnetic correlation between Cu spins is enhanced at low temperatures for the Zn-substituted sample so that the Cu-spin fluctuations slow down. On the other hand, it is well known that antiferromagnetic (AF) order appears in the underdoped region of the high-T_c cuprates. In fact, an increase of the muon-spin depolarization rate λ affected by the AF order has been observed in both the Zn-substituted and non-substituted samples with x > 0.5 in the underdoped region. As shown in Fig. 3, however, the increase of λ at x = 0.3125 in y = 0.025 is singular.

In order to investigate the magnetic state of Cu spins, longitudinal field (LF) was applied in the direction of the initial muon-spin polarization [17]. The time spectrum of the Zn-substituted sample with x = 0.3125 and y = 0.025 at 0.30 K has been found to exhibit a long-time depolarization behavior even in LF of 3950 G. This means that the Cu spins are still dynamically fluctuating even at 0.30 K, though the Cu-spin fluctuations slow down to some extent. The value of λ has been found to decrease with increasing LF with a square-root field dependence, suggesting the existence of one-dimensional (1D) diffusion of a magnetically excited state of the Cu spins [18]. Therefore, this LF dependence supports the existence of the 1D stripe-correlations in the Bi-2212 phase.

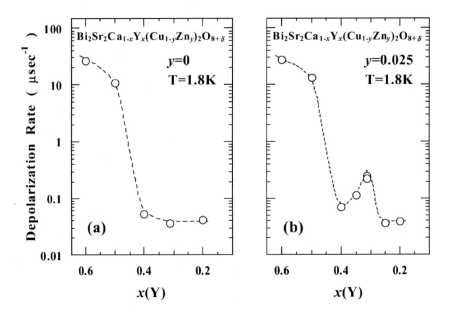

FIGURE 3. x dependence of the muon-spin depolarization rate at 1.8 K for (a) y = 0 and (b) y = 0.025 in $Bi_2Sr_2Ca_{1-x}Y_x(Cu_{1-y}Zn_y)_2O_{8+\delta}$.

As shown in Fig. 1, the Ni substitution is not effective for the suppression of SC at p ~ 1/8 in the Bi-2212 phase as in the case of the La-214 phase [19]. This would be reasonable, supposing that Zn^{2+} with the spin S = 0 pins the hole-rich domain in the stripe order more effectively than Ni^{2+} with S = 1. If this is the case, Zn will effectively scatter holes in the 1D path of the hole-rich domain and markedly decreases T_c. Furthermore, if this tendency is the case in a wide range of p, it is possible to understand the universal results in the high-T_c cuprates that Zn is a strong scatterer in the unitarity limit and that the Zn substitution decreases T_c more markedly than the Ni substitution. Accordingly, the above supposition does not seem irrelevant.

In the long run, it is concluded that there is a possibility that the dynamical stripe correlations tend to be pinned by a small amount of Zn at p ~ 1/8 in the Bi-2212 phase also, leading to the enhancement of the magnetic correlation and the suppression of SC.

1/8 Anomaly in the Y-123 Phase

In the Y-123 phase, the so-called 60-K plateau of T_c was known as being due to the ordering of oxygen atoms in the $CuO_{1-\delta}$ chain. We have investigated dependences of T_c in $Y_{1-x}Ca_xBa_2Cu_3O_{7-\delta}$ (x = 0, 0.2) on both the oxygen content 7−δ and p [10]. Consequently, it has been found that the 60-K plateau is not correlated with 7−δ but is correlated with p. Therefore, it has been concluded that the appearance of the 60-K plateau is interpreted as being due to the suppression of SC at p ~ 1/8 rather than the ordering of the oxygen atoms.

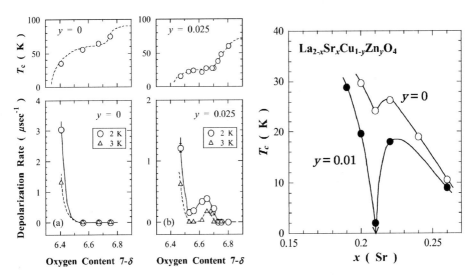

FIGURE 4. Oxygen-content dependences of T_c and the muon-spin depolarization rate at 2 and 3 K for (a) y = 0 and (b) y = 0.025 in $YBa_2Cu_{3-2y}Zn_{2y}O_{7-\delta}$.

FIGURE 5. x dependence of T_c in the single-crystal $La_{2-x}Sr_xCu_{1-y}Zn_yO_4$ with y = 0 and 0.01.

The μSR time spectrum of the Zn-substituted $YBa_2Cu_{3-2y}Zn_{2y}O_{7-\delta}$ with $7-\delta = 6.65$ and $y = 0.025$, where $p \sim 1/8$, has been found to change from a Gaussian-type to an exponential-type at low temperatures, while it is still of a Gaussain-type for the non-substituted sample with $p \sim 1/8$ [9]. This suggests that the magnetic correlation between Cu spins is enhanced at low temperatures for the Zn-substituted sample. As in the case of the Bi-2212 phase, λ is enhanced in the underdoped region, owing to the influence of the AF order. As shown in Fig. 4, however, the increase in λ in the Zn-substituted sample with $7-\delta = 6.65$ is singular [11]. In the former μSR measurements of the heavily Zn-substituted samples of the Y-123 phase, no slowing-down behavior was reported at $p \sim 1/8$ [20]. This suggests that a large amount of Zn operates to destroy the magnetic correlation. Therefore, it is concluded that there is a possibility that the dynamical stripe correlations tend to be pinned by a small amount of Zn at $p \sim 1/8$ in the Y-123 phase as well as in the Bi-2212 phase, leading to the enhancement of the magnetic correlation and the suppression of SC.

New Anomaly at x = 0.21 in the La-214 Phase

Although we have already found a new anomaly in the overdoped region of $La_{2-x}Sr_xCu_{1-y}Zn_yO_4$ [12, 13], we have recently confirmed it using single crystals of good quality [14]. Fig. 5 displays the x dependence of T_c in the 1% Zn-substituted and non-Zn-substituted single-crystals. It is found that the SC is markedly suppressed at x = 0.21 in the 1% Zn-substituted single-crystals and that it is a little suppressed at x = 0.21 in the non-Zn-substituted ones. In the 1% Zn-substituted crystal with x = 0.21, the electrical resistivity exhibits upturn at low temperatures below ~ 80 K and the temperature dependence of the thermoelectric power is also anomalous. From the μSR measurements, it has been found that the magnetic correlation is enhanced at low temperatures below 3.5 K in the 1% Zn-substituted crystal with x = 0.21, while it is enhanced below 0.8 K in the non-Zn-substituted crystal with x = 0.21 [15]. Moreover, incommensurate magnetic peaks around (π, π) have been observed by Kimura et al. [21] from the elastic neutron-scattering experiment, which is similar to those observed from the inelastic neutron-scattering experiment in the overdoped region. These results strongly suggest that the dynamical stripe correlations of holes and spins tend to become static at x = 0.21 and are pinned by Zn, leading to a static stripe-order and the marked suppression of SC, as in the case of x = 0.115 [19].

CONCLUSIONS

The 1/8 anomaly has been found not only in the La-214 phase but also in the Bi-2212 and Y-123 phases. Furthermore, a new anomaly has been found at x = 0.21 in the La-214 phase. In conclusion, it is very likely that the dynamical stripe correlations of holes and spins exist in all high-T_c cuprates and that they tend to be pinned by a small amount of Zn at $p \sim 1/8$ and also at x = 0.21 in the La-214 phase, leading to the enhancement of the magnetic correlation and the suppression of SC. If this tendency is the case in a wide range of p, it is possible to understand the universal results in the high-T_c cuprates that Zn is a strong scatterer in the unitarity limit and that the Zn substitution markedly suppresses SC. Finally, it may be pointed out that the dynamical stripe correlations of hole and spins may play an important role in the appearance of high-T_c SC, while the static stripe-order are not favorable for SC.

ACKNOWLEDGMENTS

This work was supported by a Grant-in-Aid for Scientific Research of the Ministry of Education, Science, Sports and Culture, Japan, and also by CREST of Japan Science and Technology Corporation.

REFERENCES

1. Yamada, K., Lee, C.H., Kurahashi, K., Wada, J., Wakimoto, S., Ueki, S., Kimura, H., Endoh, Y., Hosoya, S., Shirane, G., Birgeneau, R.J., Greven, M., Kastner, M.A., and Kim, Y.J., *Phys. Rev. B* **57**, 6165-6172 (1998).
2. Mook, H.A., Dogan, F., and Chakoumakos, B.C., *cond-mat*/9811100.
3. Mook, H.A., Dai, P., Hayden, S.M., Aeppli, G., Perring, T.G., and Dogan, F., *Nature (London)* **395**, 580-582 (1998).
4. Arai, M., Nishijima, T., Endoh, Y., Egami, T., Tajima, S., Tomimoto, K., Shiohara, Y., Takahashi, M., Garrett, A., and Bennington, S.M., *Phys. Rev. Lett.* **83**, 608-611 (1999).
5. Tranquada, J.M., Sternlieb, B.J., Axe, J.D., Nakamura, Y., and Uchida, S., *Nature (London)* **375**, 561-563 (1995).
6. Akoshima, M., Noji, T., Ono, Y., and Koike, Y., *Phys. Rev. B* **57**, 7491-7494 (1998).
7. Watanabe, I., Akoshima, M., Koike, Y., Ohira, S., and Nagamine, K., *J. Low Temp. Phys.* **117**, 503-507 (1999).
8. Watanabe, I., Akoshima, M., Koike, Y., and Nagamine, K., *Phys. Rev. B* **60**, R9955-R9958 (1999).
9. Akoshima, M., Noji, T., Koike, Y., Nishizaki, T., Kobayashi, N., Watanabe, I., and Nagamine, K., *J. Low Temp. Phys.* **117**, 1163-1167 (1999).
10. Akoshima, M., and Koike, Y., *J. Phys. Soc. Jpn.* **67**, 3653-3654 (1998).
11. Akoshima, M., Koike, Y., Watanabe, I., and Nagamine, K., *Phys. Rev. B* **62**, 6761-6765 (2000).
12. Kakinuma, N., Ono, Y., and Koike, Y., *Phys. Rev. B* **59**, 1491-1496 (1999).
13. Koike, Y., Kakinuma, N., Aoyama, M., Adachi, T., Sato, H., and Noji, T., *J. Low Temp. Phys.* **117**, 1157-1161 (1999).
14. Kawamata, T., Adachi, T., Noji, T., and Koike, Y., *Phys. Rev. B* **62**, No. 17 (2000).
15. Watanabe, I., Aoyama, M., Akoshima, M., Kawamata, T., Adachi, T., Koike, Y., Ohira, S., Higemoto, W., and Nagamine, K., *Phys. Rev. B* **62**, No. 17, 2000.
16. Ilonca, G., Pop, A.V., Jurcut, T., Tarta, G., and Deltour, R., *Physica B* **284-288**, 1099-1100 (2000).
17. Watanabe, I., Akoshima, M., Koike, Y., Ohira, S., and Nagamine, K., *Phys. Rev. B* **62**, No. 21 (2000).
18. Pratt, F.L., Blundell, S.J., Pattenden, P.A., Hayes, W., Chow, K.H., Monkman, A.P., Ishiguro, T., Ishida, K., and Nagamine, K., *Hyperfine Interactions* **106**, 33-38 (1997).
19. Koike, Y., Kobayashi, A., Kawaguchi, T., Kato, M., Noji, T., Ono, Y., Hikita, T., and Saito, Y., *Solid State Commun.* **82**, 889-893 (1992).

20. Mendels, P., Alloul, H., Brewer, J.H., Morris, G.D., Duty, T.L., Johnston, S., Ansaldo, E.J., Collin G., Marucco, J.F., Niedermayer, C., Noakes, D.R., and Stronach, C.E., *Phys. Rev. B* **49**, 10035-10038 (1994).
21. Kimura, H., Hirota, K., Aoyama, M., Adachi, T., Kawamata, T., Koike, Y., Yamada, K., and Endoh, Y., to be published in *J. Phys. Soc. Jpn.*

Correlated versus Uncorrelated Stripe Pinning: The Roles of Nd and Zn Co-Doping

C. Morais Smith[a,b], N. Hasselmann[b], and A. H. Castro Neto[c]

[a] Institut de Physique Théorique, Université de Fribourg, Pérolles, CH-1700 Fribourg, Switzerland
[b] I Institut für Theoretische Physik, Universität Hamburg, D-20355 Hamburg, Germany
[c] Dept. of Physics, University of California, Riverside, CA, 92521, USA

Abstract. We investigate the stripe pinning produced by Nd and Zn co-dopants in cuprates via a renormalization group approach. The two dopants play fundamentally different roles in the pinning process. While Nd induces a correlated pinning potential that traps the stripes in a flat phase and suppresses fluctuations, Zn pins the stripes in a disordered manner and promotes line meandering. We obtain the zero temperature phase diagram and compare our results with neutron scattering data. A good agreement is found between theory and experiment.

The existence of stripes in doped Mott insulators has engendered a great debate recently. While the presence of stripes in the manganites and nickelates has been firmly established, uncertainties remain concerning whether they are present in the cuprates. In manganites and nickelates stripes are static and can be easily observed [1, 2]. In cuprates, on the other hand, they form a collective fluctuating state and their detection is more subtle. Co-doping of cuprates has been extremely important for unveiling the modulated charge states. However, the inclusion of co-dopant usually reduces the critical superconducting temperature, T_c, raising doubts about the coexistence of superconductivity and the striped phase [3, 4]. The first experimental detection of stripes in the cuprates was achieved in a Nd co-doped compound $La_{2-x-y}Nd_ySr_xCuO_4$. For $y = 0.04$ and $x = 0.12$ Tranquada et al. [3] found that the commensurate magnetic peak at $Q = (\pi/a, \pi/a)$ splits by a quantity δ, giving rise to four incommensurate peaks. In addition, the Bragg peaks split by 2δ, indicating that the charges form domain walls and that the staggered magnetization undergoes a π-phase shift when crossing them. The study of co-doped cuprates has also involved other elements, such as Zn, Ni, Fe, Co, etc. [4-7].

In this paper we study the problem of co-doping within the stripe scenario by performing a renormalization group calculation on a model of quantum elastic strings under the influence of lattice and disorder potentials. We determine the zero temperature pinning phase diagram and compare it with experimental data. The different roles played by rare earth (Nd, Eu) and planar (Zn, Ni) impurities led us to establish a parallel between the stripe- and vortex-pinning problem in high-T_c superconductors.

CP554, *Physics in Local Lattice Distortions*, edited by H. Oyanagi and A. Bianconi
© 2001 American Institute of Physics 1-56396-984-X/01/$18.00

By doping the antiferromagnetic insulator La_2CuO_4 with Sr, *i.e.*, by replacing La^{3+} with Sr^{2+}, charge carriers are introduced into the CuO_2 planes. In the stripe scenario the carriers, instead of forming a homogeneous quantum fluid, arrange themselves into a highly anisotropic charge-modulated state with one-dimensional (1D) characteristics. The ionized Sr dopants are a source of disorder since they are located randomly in the neighborhood of the CuO_2 planes. Hence, the number of holes is intrinsically connected with the number of pinning centers, and the stripes can be collectively pinned by these point-like impurities. Despite the correspondence between the number of impurity pinning centers (Sr) and the charge carriers (holes) within the CuO_2 planes, it is experimentally possible to control these two parameters independently. By co-doping the superconducting material with Nd or Zn, for instance, one can alter the disorder without modifying the number of charge carriers [3, 4, 6, 7]. On the other hand, by growing the superconducting film over a ferroelectric substrate and using an electrostatic field as a control parameter, the number of charge carriers in the plane can be increased for a fixed doping concentration [8]. Hence, the treatment of these two parameters independently is an important theoretical problem.

Calculations of the pinning energy within a model in which stripes are regarded as elastic strings have shown that the problem can be described by the Collective Pinning Theory [9], with a critical Larkin pinning length $L_c \sim 100$ Å for doping of order $x \approx 10^{-2}$ [10]. Another possible source of pinning for the stripes is lattice distortion such as the tilt of the oxygen octahedra. We have recently studied the role played by lattice and dopants and have generated a phase diagram in terms of the incommensurability, δ, and the ratio between kinetic and elastic stripe energy, μ (this parameter measures the strength of quantum fluctuations) [11]. Three different phases were identified: at large values of μ and δ, the stripes form a collective fluctuating state or quantum membrane phase; as μ is reduced the stripes become pinned by the underlying lattice and decoupled from each other leading to the so-called flat phase; finally, at small values of δ and μ disorder becomes relevant and the system can be described in terms of a disordered phase. In this paper we will generalize our earlier approach in order to incorporate the differences between different co-dopants.

In general, impurities will pin the stripes, leading to the formation of a static charge order, which is usually accompanied by a reduction of T_c. This statement holds for co-doping with several types of impurities, such as Zn, Ni, Nd, Eu, independently of the intrinsic characteristics of each dopant [3, 4]. Moreover, a special reduction of T_c takes place when the effective number of charges in the CuO_2 plane is $n \sim 1/8$ [5-7]. At this doping value, the striped structure becomes commensurate with the underlying lattice and the effective pinning potential for collective motion of the stripes is at a maximum [12]. A second important feature of doping within the stripe model is that the average separation L between neighboring stripes is not expected to change upon co-doping if the substitution element has the same valence as the replaced one. Therefore, co-doping will simply pin the stripes without changing their overall number or separation. By replacing La^{3+} with Nd^{3+}, for instance, one does *not* change the number of holes introduced into the plane. The same argument holds if one replaces Cu^{2+} by Zn^{2+} or Ni^{2+}. Hence, the average stripe separation L and consequently the incommensurability $\delta = a/2L$ are not altered by the introduction of the co-dopant, as is experimentally observed (and trivially inferred) [3, 13, 14]. We classify the pinning generated by co-dopants as *uncorrelated* or *correlated*. In the former case the statistical mechanics of the stripes is characterized by line wandering, whereas in the latter case the characteristic feature is localization. The situation here is analogous to the case of a vortex line pinned by weak point-like impurities (uncorrelated disorder) or by extended defects, like 1D screw dislocations or artificially produced columnar defects (correlated disorder). For extended defects the pinning energy grows linearly with the distance along the vortex for the case in which the vortex system is properly aligned with the defect structure. This strong anisotropic pinning is in contrast with the weak isotropic pinning produced by point-like defects that compete with one another, leading to a square-root growth of the pinning energy along the vortex line [9] or the stripe [10].

We consider the transverse motion of stripes embedded in an antiferromagnetic background with lattice constant a. This is possible because the longitudinal and transverse motions decouple due to magnetic confinement [15]. We restrict our studies to the underdoped regime, where the stripe-stripe interaction is weaker than the interaction of each stripe with the lattice and disorder pinning potentials. Hence, we assume that the stripe-stripe interaction is merely restricting the motion of one stripe to a "box" of size $2L$ limited by the next neighboring stripes. This assumption simplifies the analysis to the case of a single stripe which interacts with the lattice and impurity potentials. The phenomenological Hamiltonian describing the system is

$$H = \sum_n \left[\frac{J}{2a^2} (\hat{u}_{n+1} - \hat{u}_n)^2 - 2t \cos\left(\frac{\hat{p}_n a}{\hbar} \right) + V_n(\hat{u}_n) \right],$$

where $\hat{u}_n$ denotes the displacement of the n-th hole from its equilibrium position, $|\hat{u}_n|$ < L, $\hat{p}_n$ is the canonically conjugate momentum, t is the hopping parameter, J is the stripe stiffness, and V_n is a random pinning potential with Gaussian average over the disorder ensemble, $\langle V_n(u) V_{n'}(u') \rangle_D = D\delta(u - u')\delta_{n,n'}$. The parameter D measures the strength of disorder.

The calculations can be simplified by going to the continuum limit and introducing replicas. The replicated zero temperature action reads [11] ($\phi^i = \sqrt{\pi} u^i / a$)

$$S^r = \sum_i S_0[\phi^i] + \frac{G}{a} \sum_i \int_0^\infty d\tau \int dy \, \cos(2\sqrt{\pi}\phi^i)$$
$$+ \frac{D}{2aL\hbar} \sum_{i,i'} \int dy \int_0^\infty d\tau d\tau' \, \cos[2\sqrt{\pi}\delta(\phi^i(y,\tau) - \phi^{i'}(y,\tau'))], \tag{1}$$

with the Gaussian action S_0 given by

$$S_0[\phi(y,\tau)] = \frac{\hbar}{2\pi u} \int_0^\infty d\tau \int dy \left[\frac{1}{c}(\partial_\tau\phi)^2 + c(\partial_y\phi)^2 \right].$$

The stripes are oriented along the y-direction and τ is imaginary time. The free stripe velocity is $c = a\sqrt{2tJ}/\hbar$ and the dimensionless parameter $\mu = \sqrt{2t/J}$ measures the competition between kinetic and confining energies. The parameter G accounts for the lattice effects and i counts the replicas. The one-loop RG equations for G, D, and μ were obtained in Ref. 11. The phase diagram can be divided into a flat phase (correlated pinning), a disordered phase (uncorrelated pinning) and a membrane phase. Here we focus on the pinned phases only. In these two phases the single stripe approach is expected to be a reasonably good approximation and we can obtain quantitative results on the effects of the different co-dopants.

Correlated pinning: Let us first analyze the limit of vanishing disorder $D = 0$. The RG equations then read:

$$\frac{d}{d\ell}\Gamma^2 = 2(2-\pi\mu)\Gamma^2 \tag{2}$$

$$\frac{d}{d\ell}\mu^{-1} = \frac{1}{2}\Gamma^2. \tag{3}$$

where $\Gamma = \pi^{3/2}Ga/(\hbar c)$. Near the critical region, we can define the small parameter $\varepsilon_g = 2 - \pi\mu$ which measures the distance from the critical line $\mu_{c1} = 2/\pi$, where the roughening transition (Kosterlitz-Thouless) takes place in the absence of co-doping. Hence, $\mu^{-1} \approx (\pi/2)+(\pi/4)\varepsilon_g$ and $d\varepsilon_g/d\ell = (4/\pi)d\mu^{-1}/d\ell$. Using Eq. (3) we obtain $d\varepsilon_g/d\ell = (2/\pi)\Gamma^2$, which can then be combined with Eq. (2) yielding

$$\frac{d}{d\ell}\left(\varepsilon_g^2 - \frac{2}{\pi}\Gamma^2\right) = 0 \tag{4}$$

Eq. (4) implies that there is a transition at the critical value $\varepsilon_g^c = \pm\sqrt{2/\pi}\,\Gamma$. The critical value of $\mu_c(\Gamma)$ is thus

$$\mu_{c1}(\Gamma) = \frac{2}{\pi} + \sqrt{\frac{2}{\pi^3}}\,\Gamma \tag{5}$$

For $\mu < \mu_{c1}(\Gamma)$ the stripes are pinned by the underlying lattice in the so called "flat phase". The excitation spectrum is gaped and quantum fluctuations are strongly suppressed. On the other hand, for $\mu > \mu_{c1}(\Gamma)$ the stripes are fluctuating freely. Consider a fixed doping concentration δ for which the stripe system is in the free phase for $\Gamma = 0$. Upon increasing the lattice parameter Γ, the system moves along the thick line in Fig. 1 and eventually enters the pinned "flat phase" after crossing the surface $\mu_{c1}(\Gamma)$.

This result can describe the effects of Nd co-doping of the lanthanum cuprate. The introduction of Nd (or any other rare earth element) induces a structural transition in the material from a low temperature orthorhombic (LTO) to a low temperature tetragonal (LTT) phase, corresponding to a buckling of the oxygen octahedra. Hence, although the Nd randomly replaces the La atoms which are located out of the plane, they indirectly produce a *correlated* pinning potential along the copper lattice that will act to pin the stripes in the so called flat phase, strongly suppressing thermal or

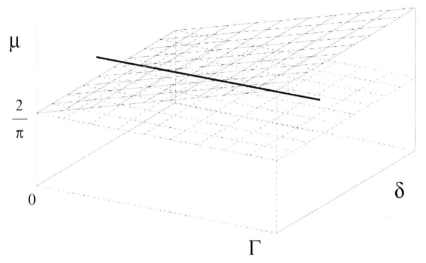

FIGURE 1. Pinning phase-diagram of the striped phase in the presence of correlated pinning.

quantum fluctuations. This is analogous to the case of pinning of vortices by artificially introduced columnar defects. Notice, however, that in the vortex-problem the columnar defects are randomly distributed, whereas here the "correlated"' pinning potential is actually a periodic lattice potential, which is enhanced by the tilting of the oxygen octahedra within the LTT phase. Despite of this difference, the analogy is helpful, because it emphasizes the linear character of the pinning potential in both problems.

Since the Nd pins the stripe in an ordered configuration, we expect the width of the incommensurate (IC) peaks measured by neutron scattering to be *reduced* and the 1D behavior to be reinforced by co-doping. These conclusions are supported by experimental data: ARPES [16] and Hall transport [17] measurements of $La_{2-x-y}Nd_ySr_xCuO_4$ show strong evidence for a 1D striped structure in the underdoped regime. Besides, neutron scattering data taken for both compounds (with and without Nd) indicate a reduction of the half width at half maximum (HWHM) IC peak in the presence of Nd for all the investigated Sr compositions. [3, 13]. This is an indication that lattice pinning and hence commensuration effects are enhanced through Nd doping.

Uncorrelated pinning: A completely different scenario is presented for the Zn doping case: Zn^{2+} replaces Cu^{2+} directly on the CuO_2 planes. They are located *randomly* and act as uncorrelated point-like pinning centers in a way very similar to the pinning of vortices by oxygen vacancies. Again, they do not alter the position of the IC peaks observed in neutron scattering, since they do not change the hole density. Moreover, the randomly distributed Zn atoms induce stripe meandering and pin the stripe in a fuzzy phase, similar to Sr doping. The RG equations in the limit of negligible lattice pinning but relevant disorder are [11]

$$\frac{d}{d\ell}\Delta = (3 - \gamma\mu)\Delta \tag{6}$$

$$\frac{d}{d\ell}\mu = -\frac{1}{2}\mu^2\Delta \tag{7}$$

where $\gamma = 2\pi\delta^2$ and $\Delta = 4\pi^2 D\delta^2 a^2 /(\hbar^2 c^2 L)$. Close to the critical region we define $\varepsilon_d = 3 - \gamma\mu$, and following a similar procedure as done for the lattice pinning case, we obtain

$$\frac{d}{d\ell}\left(\Delta - \frac{\gamma}{9}\varepsilon_d^2\right) = 0 \tag{8}$$

indicating that $\Delta - (\gamma/9)\varepsilon_d^2$ is preserved under the RG flow. The transition then happens at the critical value

$$\mu_{c2} = \frac{3}{2\pi\delta^2} + \frac{3\sqrt{\Delta}}{(2\pi\delta^2)^{3/2}} \ . \tag{9}$$

The corresponding phase diagram is shown in Fig. 2. The thick line indicates how the system undergoes a transition for a constant δ from a "free phase" ($\mu > \mu_{c2}$) at $\Delta = 0$ to a "fuzzy phase" ($\mu < \mu_{c2}$) at finite Δ. An inspection of the phase diagram indicates that uncorrelated pinning is more relevant at low doping values. Hence, we expect the effects of Zn co-doping to decrease with doping. Moreover, in contrast to Nd doping, Zn pinning destroys the 1D behavior and *increases* the width of incommensurate neutron scattering peaks implying that the stripes are pinned within a broader region. This is indeed observed experimentally [13, 14, 18]. Neutron scattering measurements in $La_{2-x}Sr_xCu_{1-y}Zn_yO_4$ for $y = 0.012$ and $x = 0.14$ show that Zn produces no relevant effect and that the width κ_s of the IC peaks remains practically unaltered. The IC peak width for $La_{1.85}Sr_{0.15}CuO_4$ is $\kappa_s = 0.020 \pm 0.006$ Å ($E = 8$ meV, $T = 8$ K) (see Ref. 13) and the Zn-doped compound with a similar Sr-concentration ($x = 0.14$ and $y = 0.012$) displays the same features within the experimental error bars: $\kappa_s = 0.014 \pm 0.002$ Å for $E = 5$ meV, $T = 10$ K (Ref. 18). The scenario changes quite a bit in the underdoped regime. For $x = 0.12$ and $y = 0.03$ (a composition for which superconductivity is completely suppressed) the elastic IC peaks were observed at the same position as for the Zn-free material, but κ_s was increased due to the doping: $\kappa_s < 0.005$ Å^{-1} for the Zn-free material, whereas $\kappa_s = 0.013(1)$ Å^{-1} for $y = 0.03$ (Ref. 14), reflecting the random character of the pinning centers. Although the commensurability at $x = 0.12$ makes this point special, we expect this trend (increase of κ_s upon Zn-doping) to continue, especially at lower values of x. We emphasize that the pinning energy grows sub-linearly with the length of the stripe in the case of uncorrelated disorder [10]. Hence, the Zn-pinned phase is analogous to the vortex-glass phase discussed in the context of vortex creep [9].

In conclusion, we have shown that the main experimental features of co-doping in cuprates can be understood within models of lattice-pinned or disorder-pinned stripes. We divide the co-dopants into two classes: those which produce correlated and those which produce uncorrelated pinning. Correlated pinning is produced through rare earth co-doping. The problem is analogous to the pinning of vortices by columnar defects or screw dislocations. In this case the stripes are pinned in a flat phase and the fluctuations are strongly suppressed. The effective stripe width is reduced and consequently the IC neutron scattering peaks become sharper after the introduction of the co-dopant. On the other hand, in-plane Zn- or Ni-doping provides randomly distributed point-like pinning centers, similar to the oxygen vacancies in the vortex-creep problem. Within our model, in which the stripe is regarded as a quantum elastic string, the effect of randomness is to "disorder" the string, increasing the effective stripe width and broadening the IC peaks. We expect this kind of pinning to be relevant only at low doping, as indicated in the phase diagram shown in Fig. 2, in agreement with the experimental results.

We are indebted with G. Blatter, D. Baeriswyl, A. O. Caldeira, R. Noack, S. Uchida, and K. Yamada for fruitful discussions. N. H. is financially supported by the Graduierten Kolleg "Physik nanostrukturierter Festkörper". A. H. C. N. acknowledges support from a LANL CULAR grant.

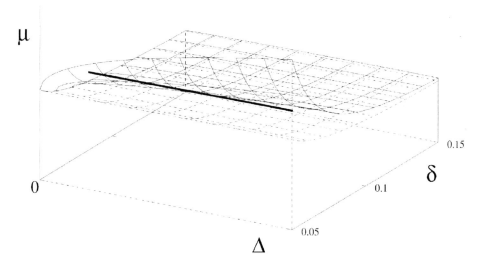

FIGURE 2. Pinning phase-diagram of the striped phase in the presence of uncorrelated pinning. At low doping values the stripes are pinned in a fuzzy phase, whereas above the critical surface they are fluctuating freely.

REFERENCES

1. Mori, S., Chen, C. H., and Cheong, S. -W., *Nature* **392**, 473 (1998); *Phys. Rev. Lett.* **81**, 3972 (1998).

2. Tranquada, J. M. *et al.*, *Phys. Rev. Lett.* **73**, 1003 (1994); Lee, S. -H. and Cheong, S. -W., *Phys. Rev. Lett.* **79**, 2514 (1997); Yoshizawa , H. *et al.*, cond-mat/9904357.

3. Tranquada, J. M. *et al.*, *Nature* **375**, 561 (1995); *Phys. Rev. Lett.* **78**, 338 (1997); *Phys. Rev. B* **54**, 7489 (1996).

4. Koike, Y. *et al.*, *Sol. State. Commun.* **82**, 889 (1992); *J. Low. Temp. Phys.* **105**, 317 (1996); Adachi, T. *et al.*, *idem* **117**, 1151 (1999).

5. Xiao, G., Cieplak, M. Z., Xiao, J. Q., and Chien, C. L., *Phys. Rev. B* **42**, 8752 (1990).

6. Koike, Y. *et al.*, *J. Phys. Soc. Jpn.* **67**, 3653 (1998); *J. Low Temp. Phys.* **117**, 1163 (1999).

7. Akoshima, M. *et al.*, *Phys. Rev. B* **57**, 7491 (1998).

8. Ahn, C. H. *et al.*, *Science* **284**, 1152 (1999).

9. Larkin, A. I. and Ovchinnikov, Y. N., *J. Low Temp. Phys.* **34**, 409 (1979); Blatter, G. *et al.*, *Rev. Mod. Phys.* **66**, 1125 (1994).

10. Morais Smith, C., Dimashko, Yu., Hasselmann, N., and Caldeira, A. O., *Phys. Rev. B* **58**, 453 (1998).

11. Hasselmann, N., Castro Neto, A. H., Morais Smith, C., and Dimashko, Yu., *Phys. Rev. Lett.* **82**, 2135 (1999).

12. Notice that the commensurability or incommensurability of the stripe array with the underlying copper lattice is a different issue than the incommensurability δ measured with neutron scattering. When $\delta \propto x = 1/8$ the stripe array is commensurate with the lattice.

13. Yamada, K. *et al.*, *Phys. Rev. B* **57**, 6165 (1998).

14. Kimura, H. *et al.*, *Phys. Rev. B* **59**, 6517 (1999).

15. Chernyshev, A. L., Castro Neto, A. H., and Bishop, A., *Phys. Rev. Lett.* **84**, 4922 (2000).

16. Zhou, X. J. *et al.*, *Science* **286**, 268 (1999).

17. Noda, T., Eisaki, H., and Uchida, S., *Science* **286**, 265 (1999).

18. Hirota, K., Yamada, K., Tanaka, I., and Kojima, H., *Physica B* **241-243**, 817 (1998)

Polarization Dependent EXAFS Study of Bi$_2$Sr$_2$Ca$_{0.4}$Pr$_{0.6}$Cu$_2$O$_{8-\delta}$ Insulating Single Crystal

S. Dalela[1], J. F. Lee[2], S. Venkatesh[2], J. -H. Choy[3], B. Dalela[1], D. C. Jain, R. K. Singhal[1] and K. B. Garg[1]

[1] Physics Department, Rajasthan University, Jaipur – 302 004, India.
[2] Synchrotron Radiation Research Centre, Hsinchu 300, Taiwan.
[3] Chemistry Department, National Seoul University, Seoul 151-742, Korea.

Abstract. Pr doping studies occupy a special place for determining the role played by different cations in evolution of superconductivity in the cuprate perovskites. It is known to quench superconductivity in different systems through different mechanisms that still remain to be firmly established. Reporting of a pure Pr123 superconducting crystal has served to further heighten this debate. Polarised EXAFS measurements have been made on the Pr L$_{III}$ edge in a Bi$_2$Sr$_2$Ca$_{0.4}$Pr$_{0.6}$Cu$_2$O$_{8-\gamma}$ insulating single crystal to examine the site geometry of the Pr cations and the mechanism of quenching superconductivity in this system. Measurements have been confined to within the ab-plane in E//a, E//a^45 and E//b polarisations. Results show that Pr has a valence of 3+ only, occupies only the Ca sites and hence quenches the itinerant holes rendering the CuO$_2$ planes insulating. No evidence is found of presence of any distortion in the Pr/Ca plane.

INTRODUCTION

Doping of cations has been a very popular way of trying to learn about the mechanism of superconductivity in the cuprate perovskites. Of these, Pr occupies a special place as, in contrast to heavier rare-earths, it is known to quench superconductivity [1,2]. What is more interesting is the fact that it does so in different ways in different systems even though unlike the other light rare-earths, Ce and Tb [3] it yields orthorhombic structure. As Y in YBCO(123) is replaced by Pr or Ca in BSCCO(2212) it depresses T$_c$ finally turning the system insulating. Abrikosov-Gor'kov pair breaking mechanism was first thought to be responsible for the quenching of the superconductivity [4]. However, this model could not explain the localization of charge carriers resulting into insulating behavior of the system. Later Fehrenbacher and Rice developed a theory based on hybridization of localized states [5] which accounts for the semiconducting behavior of Pr123 but this model failed to explain that the suppression of T$_c$ with Pr doping concentration. Later on, it was found that in the YBCO system it localizes the holes through Pr 4f-O 2p hybridisation [6]. However, in the BSCCO(2212) system it appears to quench holes as Pr^{3+} replaces the Ca^{2+} [7] cations. The final consensus on its role in these is however, yet to emerge. Recently it was reported that there are two Pr-O distances in the YBCO system and increasing Pr concentration does not alter these but instead increases the statistical

CP554, *Physics in Local Lattice Distortions*, edited by H. Oyanagi and A. Bianconi
© 2001 American Institute of Physics 1-56396-984-X/01/$18.00

weight of the shorter distance [8]. The recent reporting of synthesis of pure PrBCO(123) single crystals [9] has given a new dimension to this discussion. More recently, V. N. Nazhosky [10] has reported from his magnetic susceptibility studies that the superconductivity in the pure PrBCO(123) system perhaps derives from the fact that only about one half of the nominal concentration of Pr occupies the Y-site and the other half Y vacancies are most probably filled by non-magnetic Ba cation.

In view of these developments we, therefore, decided to verify if in the case of BSCCO(2212) Pr^{3+} substitution quenches holes resulting in local destruction of superconductivity of the CuO_2 planes. We have carried out polarized EXAFS study at the Pr L_{III} edge in different orientations but within the ab plane as most of the action is here. In this paper, we report and discuss our results from the EXAFS data analysis of the Pr L_{III} edge E~//a, E~//b and E~//a^45 orientations.

EXPERIMENTAL

The pristine BSCCO(2212) and the $Bi_2Sr_2Ca_{0.4}Pr_{0.6}Cu_2O_{8-\delta}$ were grown by the usual flux method through the process of solid state fusion. These were then characterized by XRD and the c-axis lattice parameters obtained by least square fitting analysis of the (001) XRD reflection peaks. The c-axis length turned out to be 30.584 Å for the pristine crystal and 30.357 Å for the Pr-doped system. The values are identical to those reported earlier [11]. The Pr-doping content analyzed by electron probe micro analysis (EPMA) turned out to be consistent with the nominal composition $Bi_2Sr_2Ca_{0.4}Pr_{0.6}Cu_2O_{8-\delta}$ within the experimental error. The polarized Pr L_{III} absorption edge EXAFS measurements were made at SRRC, Taiwan at the BL 17 C (wiggler-C) beamline employing fluorescence detection.

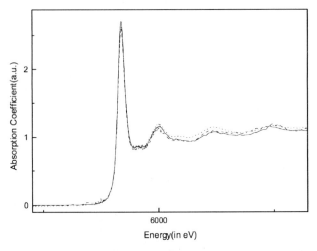

FIGURE 1. Pr-L_{III} EXAFS in the Pr-doped BSCCO(2212) insulating single crystal with different orientations E//a (-----), E//a^45 (solid line) and E//b (........).

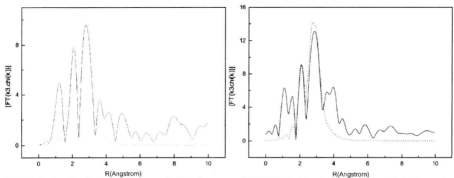

FIGURE 2. Fourier Transform of Pr L_{III} edge in the E//a^45 E//a^45 orientation (solid line) and the fitting in R-space (dotted line).

FIGURE 3. Fourier Transform of Pr L_{III} edge in the E//a orientation (solid line) and the fitting in R-space (dotted line).

RESULTS AND DISCUSSION

Fig. 1 shows the Pr L_{III} edge EXAFS spectra in the polarization E//a, E//a^45 and E//b. The spectra are more or less identical in appearance indicating isotropic environment around the Pr cation. These spectra have been analyzed by first subtracting the pre-edge and post-edge atomic-absorption background using the AUTOBK program which provides for subtraction of an energy dependent atomic-absorption. We have then used ATOMS 3.00 program [12] to generate the coordinates etc. of the various atoms in the cluster with Pr as the central absorbing atom. Thereafter, we replaced 60 percent of the Ca sites by Pr in a random fashion to generate a cluster that may correspond to the assumed picture that all Pr goes to the Ca site. Then with FEFF800 [13] using a cluster of 87 atoms we generated the potentials, phase-shifts, amplitude etc. and the various scattering paths that contribute to the EXAFS.

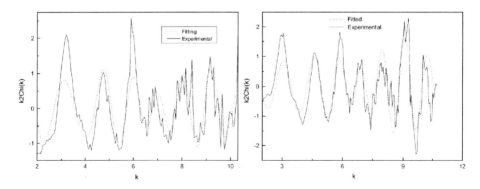

FIGURE 4. $k^2 \chi(k)$, experimental and fitted, for the polarization directions E//a (left) and E//a^45 (right).

Final fitting in the R-space was done using the FEFFIT 2.54 [14]. Prior to it the FEFFIT was used in NO Fit mode to select the paths that contribute to the first two peaks corresponding to the first two shells.

Fig. 2 and 3 show the resulting Fourier transforms for the E//a and E//a^45 Pr L_{III} polarised spectra. From these figures we can see that the Fourier Transforms for the two orientations are virtually identical. Fig. 4 shows the fitting in the k-space for both the polarization. The program was made to iterate on e_0, S_0^2, σ^2 and Δr and finally on R_{nn} while the coordination number N was taken from the crystallographic data. The results of our EXAFS analysis are shown in the Table 1 for the fitting in case of the first two peaks in the E//a and E//a^45 spectra.

It gives the values for the bond lengths R_{nn}, the mean square variations σ^2 for the Pr-oxygen and Pr-copper distances. From this table it is further confirmed that the first nearest neighbour for Pr are eight O atoms at a distance equal to ~ 2.459 Å in both the polarisations. Next nearest neighbours are four + four Cu atoms at two distances 3.1153 and 3.1184 Å which are again the same for the two polarisations. For Pr, the EXAFS clearly shows a single well defined environment essentially identical to that in BSCCO(2212) system. It is thus amply clear that the near-neighbour environment of Pr along the both polarization direction E//a and E//a^45 are identical to each other as well as to that around the Ca atom in the pristine system which shows that Pr is indeed occupying only the Ca sites in this system.

Earlier, we had made high resolution measurements on the Pr L_{III} edges in these orientations for the same system [15] and reported that the huge white line has only one component and its energy clearly corresponds to the 2p → 5d dipole transition. Also, the edges had the same appearance and structure in the two cases. From this we had concluded that Pr is present in 3^+ valence state only ruling out the presence of Pr^{4+} in any detectable proportion. Read with the present EXAFS results we can confidently assert that Pr occupies only the Ca site and is in 3+ state only, meaning thereby that the superconductivity in the BSCCO(2212) system is indeed a result of the hole quenching by the Pr cations.

TABLE 1. Data on fitting of the first two peaks in the Fourier Transform for E//a and E//a^45 polarization for the Pr L_{III} edge in BSCCO(2212)

Polarization Direction	F.T. Peak	Type of n-n	Coordination number N	R_{n-n} (Å)	σ^2 Å^2
E//a	First	O	8	2.459$_6$	0.004$_5$
	Second	Cu	4	3.115$_2$	0.002$_0$
			4	3.118$_4$	
E//a^45	First	O	8	2.459$_2$	0.0008$_6$
	Second	Cu	4	3.115$_1$	0.004$_2$
			4	3.118$_4$	

CONCLUSION

Results from our polarised EXAFS measurements on the L$_{III}$ edge clearly indicate our earlier results from XANES study that Pr exists solely in 3+ valence state in BSCCO(2212) system and solely occupies the Ca site quenching the holes that destroys superconductivity in the Cu-O$_2$ planes, eventually resulting into loss of bulk superconductivity as well as conductivity.

ACKNOWLEDGMENTS

This study was supported by the U.G.C. COSIST and the C.S.I.R. Emeritus Scientist programmes. We acknowledge CSIR and UGC, New-Delhi for granting the scholarship to S. Dalela and B. Dalera respectively.

REFERENCES

1. Venkatesh, S., Khaled, M., Saini, N.L., Studer, F., and Garg, K.B., *Solid State Commun.* **100**, 773, (1996).
2. Khaled, M., Saini, N.L., Gurman, S.J., Amiss, J.C., and Garg, K.B., *J. de Physique IV*, (France), **7**, C2, 1121, (1996).
3. Hor, P.H., Meng, R.L., Wang, Y.Q., Gao, L., Huang, Z.J., Becktold, J., Forster, K., and Chu, C.W., *Phys. Rev. Lett.* **58**, 1891, (1987); Solderholm, L., Zhang, K., Hinks, D.G., Beno M.A., Jorgensen, J.D., Segre, C.U., and Schuller, I.K., *Nature* **328**, 604, (1987).
4. Quitmann, C. *et al.*, *Phys. Rev. B* **46**, 11813, (1992).
5. Fehrenbacher, R., and Rice, T. M., *Phys. Rev. Lett.* **70**, 3471, (1993).
6. Nishihara, Y., Zou, Z., Ye, J., Oka, K., Minawa, T., Kawanaka, H., and Bando, H., *Bull. Mater Sci.* **22**, 257, (1999).
7. Singh, R., Gupta, A., Agarwal, S.K., Singh, D.P., and Narlikar, A.V., *Supercond. Sci. Technol.* **11**, 311, (1998).
8. Gurman, S., Amiss, J., Venkatesh, S., Singhal, R.K., and Garg, K.B., *J. Phys. Cond. Matter* **11**, 1847, (1999).
9. Merz, M., Nucker, N., Pelligrin, E., Schweiss, P., and Schuppler, S., *Phys. Rev. B* **55**, 9160, (1997); Zou, Z., Ye, J., Oka, K., and Nishihara, Y., *Phys. Rev. Lett.* **80**, 1074, (1998).
10. Naroznhyi, V.N., Eckert, D., Fuchs, G., Nenkov, K., Uvarova, T.G., and Müller, K.-H., *cond.mat/9909110*, (1999).
11. Sun, X.F., Zhao, X., Li, X.-G., and Ku, H.C., *Phys. Rev. B* **59**, 8978-8983,(1999).
12. Ravel, B., Newville, M., Cross, J.O., and Bouldin, C.E., *Physica B* **208 & 209**, 145, (1995).
13. Rehr, J.J., Zabinsky, S.I., and Albers, R.C., *Phys. Rev. Lett.* **69**, 3397, (1992).
14. Stern, E.A., *Phys. Rev. B* **48**(13), 9825, (1993); Newville, M., Livins, P., Yacoby, Y., Rehr, J.J., and Stern, E.A., *Phys. Rev. B* **47**(21), 14126, (1993).
15. Garg, K.B., Dalela, S., Dalela, B., Venkatesh, V., Lee, J.F., Choy, J.H., Chaturvedi, D., Singhal, R.K., and Garcia-Ruiz, J.J., *Syn. Rad.* (to be published).

Competition between Superconductivity and SDW in the 2D Hubbard Model and Possibility of Superconductivity along a Stripe

K. Yamaji, T. Yanagisawa, S. Koike, and M. Miyazaki

Electrotechnical Laboratory, 1-1-4 Umezono, Tsukuba 305-8568, Japan

Abstract. The variational Monte Carlo calculation gives a finite bulk-limit value of superconducting (SC) condensation energy in the titled model with next-nearest-site transfer energy t' in an appropriate parameter region. The obtained SC condensation energy around the optimal doping is of the same order as the experimental value of about 0.26 meV/site for YBCO. This assures the appropriateness of the 2D Hubbard model for cuprate high-T_c superconductors. However, the SDW condensation energy computed by the same method was found to overwhelm the former in the most parameter region where the superconductivity is expected to appear. The SC parameter region seems to be restricted to the highly hole-doped region around the SDW region in a narrow parameter window if it exists. This makes a conspicuous contrast to the experimentally observed wide SC region extending down into the low hole-doping region. We look for the possibility of an inhomogeneous superconducting state coexisting with striped SDW in this low hole-doping SC region. In such a striped SDW state we have obtained an indication that superconductivity is brought about along the stripe in a mid-gap band presumably due to the two-band effect.

INTRODUCTION

Our two-dimensional (2D) Hubbard model is defined by

$$H = -t \sum_{<jl>,\sigma} \left(c_{j\sigma}^{\dagger} c_{l\sigma} + \text{H.c.} \right) - t' \sum_{<<jl>>,\sigma} \left(c_{j\sigma}^{\dagger} c_{l\sigma} + \text{H.c.} \right) + U \sum_{j} c_{j\uparrow}^{\dagger} c_{j\uparrow} c_{j\downarrow}^{\dagger} c_{j\downarrow}, \quad (1)$$

where $-t$ and $-t'$ are the transfer energies between the nearest neighbor (n.n.) sites and the next nearest sites, respectively, on the rectangular lattice; U is the on-site Coulomb energy; $c_{i\sigma}$ is the annihilation operator for the site i and spin σ; other notations are standard.

This model is one of the most basic ones which can possibly drive superconductivity (SC) in them only due to electronic interactions and plausible candidates for explaining SC in cuprates and organics. There has been a long controversy if SC occurs in this model or not [1-7]. We have been investigating this problem using the variational Monte Carlo method, which allows to treat realistically large U and to employ lattices of sufficiently large sizes enabling finite size scaling [8-11]. We take account of t', which is known to enhance the tendency for SC to occur [5, 12].

CP554, *Physics in Local Lattice Distortions*, edited by H. Oyanagi and A. Bianconi
© 2001 American Institute of Physics 1-56396-984-X/01/$18.00

In this paper first we report on a finite bulk limit of SC condensation energy in an appropriate parameter region, the parameter values in which are close to experimental values. Next we examine the competition of the SC with commensurate SDW and find that the SDW overwhelms the SC in the most parameter space. We argue that the SC can survive only in a narrow parameter space in the high doping region in the case of hole doping. Finally in the low-hole-doping region doped holes are known to be segregated into stripes in the SDW phase. Along a stripe we show an indication of the occurrence of SC in a two-band situation using the variational Monte Carlo method.

BULK-LIMIT SC CONDENSATION ENERGY

We have been studying the SC condensation energy E_{cond} using the variational Monte Carlo method [13, 14]. First we observed the Gutzwiller-projected BCS-type wave function gives a minimum of the total energy only in the case of d-wave SC, $i.e.$, when we take the d-wave gap function $\Delta_k = \Delta(\cos k_x - \cos k_y)$. When we decrease Δ close to zero, the total energy approaches that of the variational normal state.

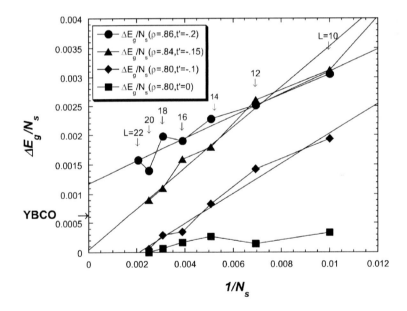

FIGURE 1. Energy gain per site $\Delta E_g/N_s$, or condensation energy E_{cond}, in the SC state in reference to the normal state is plotted as a function of the inverse $1/N_s$ of the number of sites for four sets of parameters. ρ is the electron number per site. L is the length of the edge of the square lattice. YBCO attached to the vertical axis indicates the experimental value of the SC E_{cond} for $YBa_2Cu_3O_7$.

Therefore, this difference normalized per site gives the SC condensation energy. Technically, we obtained this value by using a parabola fit as a function of Δ to the minimum energy and the energy for an appropriately small value of Δ; both energies were determined by applying the correlated measurements method to the variational Monte Carlo calculations. A difficult problem with this kind of numerical computation is to eliminate a doubt that the result may be spurious due to the small size of the treated lattice and may not give the true property of the bulk system. Therefore, we tried to extrapolate the finite-size results to the bulk limit by finite-size scaling.

Lattice sizes we treated are $8\times8 \sim 22\times22$. System parameters range in $0.80 \leq \rho \leq 0.86$ and $-0.30 \leq t' \leq 0$ with $U = 8$ (our energy unit is t); ρ is electron density per site. Bulk limit of SC condensation energy E_{cond} was obtained by plotting as a function of the inverse of the number of sites, $1/L^2$; here L is the length of the square lattice. Success of linear fitting allows to get the bulk limit. The bulk-limit E_{cond} was found finite, when $-0.25 \leq t' \leq -0.10$ and $\rho \geq 0.84$ (Fig. 1). It was found to vanish, when $t' = 0$ and $\rho \leq 0.86$ [11].

When $\rho = 0.86$, $t' = -0.20$ and $U = 8$, we obtained $E_{cond} = 0.00117$/site $\cong 0.60$ meV/site, where $t = 0.51$ eV was used [15]. This value is remarkably close to experimental values $0.17 \sim 0.26$ meV/site estimated for optimally doped Y123 from specific heat data [16] and 0.26 meV/site from the critical magnetic field H_c [17]. This good agreement assures the soundness of the 2D Hubbard model for applying to the cuprate high-T_c superconductors.

Incidentally, the t-J model without taking account of t' gives $E_{cond} = 0.026t \cong 13$ meV/site [18], which is 50 times larger than the experimental values and indicates a serious quantitative problem with this model.

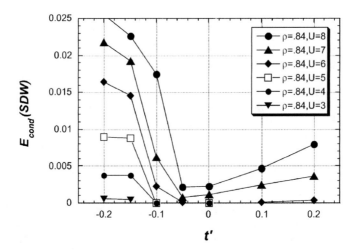

FIGURE 2. t'-dependence of the SDW E_{cond} in the case of $\rho =0.84$ and the 10×10 lattice for six values of U

COMPETITION WITH SDW

In order to determine the boundary between the SC and SDW phases, we computed the condensation energy E_{cond} for the SDW state with commensurate wave vector $Q = (\pi, \pi)$ using the same VMC method and compared it with the SC E_{cond} in the same parameter region (our length unit is the lattice constant). In an earlier work on the 10×10 lattice [10] both quantities were found to be nearly equal to each other at $\rho = 0.84$ when $t' = 0$ and further the SC E_{cond} increased while the SDW E_{cond} was observed to decrease with increase of $|t'|$, when t' is small and negative [19], so that the former wins against the latter. The tendency of the SDW E_{cond} for small t' is in accord with the t'-dependence of SDW instability obtained from the divergence of magnetic susceptibility [12].

When we slightly improved the trial SDW wave function so that the occupation in the k-points reflects the t'-dependence of the bare band ε_k, we observed a sharp increase of SDW E_{cond} when $|t'|$ exceeds a certain critical value (Fig. 2). Until this value SDW E_{cond} decreases with decrease of t' as before. This is ascribed to the fact that the k-points around the van Hove singularity start to be occupied and to help the formation of commensurate SDW. As a result unfortunately in the case of $\rho \geq 0.84$ with $U = 8$ and $-0.25 \leq t' \leq -0.10$, we found that SDW E_{cond} is larger than SC E_{cond}.

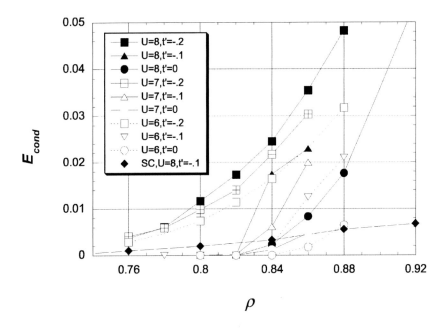

FIGURE 3. SDW and SC condensation energies E_{cond} as functions of electron density ρ calculated by the variational Monte Carlo method (unmodified) for the 10×10 lattice. The SDW E_{cond} is given for three values of t' and three values of U indicated by the labels shown in the inset. (Note that the filled triangle for $\rho = 0.82$ lies close to the absissa and is behind the empty circle.)

As seen in Fig. 3, for fixed ρ and U the SDW E_{cond} increases very quickly with increase of $|t'|$ when $t' \leq -0.10$. We are interested in the region around $t' = -0.10$ since the cuprates are considered to take such values [15].

After this finding, we restarted the search for the SC parameter region where the SC E_{cond} is larger than the SDW E_{cond}. We examined the region of higher hole doping where SDW vanishes and allows us to expect SC to appear, not defeated by SDW. As seen in Fig. 3, for $\rho = 0.82$ the SDW E_{cond} almost vanishes when $-0.10 \leq t' \leq 0$ in the case of 10×10 lattice. Therefore, we searched for a t' region with $\rho = 0.82$. With $\rho = 0.82$, since SDW E_{cond} increases much faster than SC E_{cond} with increase of $|t'|$, we looked for a t' region where the bulk-limit SC E_{cond} is finite in such a t' region where SDW $E_{cond} = 0$.

We chose $U = 7$, since the SDW E_{cond} was found largest as a function of U in the preceding work and is close to the theoretical value 6. 7 [15].

Even with $U = 6$ and $t' = 0$, the SC E_{cond} seems to be unable to win against the SDW E_{cond} when $\rho \geq 0.88$, which clearly shows the impossibility of the pure SC state in the cuprates with low hole doping.

In the following investigation carrying out the variational Monte Carlo calculations in this section, we brought in another slight improvement of the trial wave functions for both SC and SDW, optimizing n.n. correlation of electrons. The improved wave function for SC state is as follows:

$$\Psi_s = P_{Ne} \bullet \prod_{<ij>} h^{n_i n_j} \bullet \prod_l \left(1-(1-g)n_{l\uparrow}n_{l\downarrow}\right) \bullet \prod_k \left(u_k + v_k c_{k\uparrow}^\dagger c_{k\downarrow}^\dagger\right)|0\rangle, \tag{2}$$

where variational parameter h is introduced, which is similar to Gutzwiller parameter g and controls the n.n. correlation. Here multiplication over k goes over all k-points. For the SDW state

$$\Psi_{SDW} = \prod_{<ij>} h^{n_i n_j} \bullet \prod_l \left(1-(1-g)n_{l\uparrow}n_{l\downarrow}\right)$$
$$\bullet \prod_k \left(u_k c_{k\uparrow}^\dagger + v_k c_{k+Q\uparrow}^\dagger\right) \bullet \prod_{k'} \left(u_{k'} c_{k'\downarrow}^\dagger - v_{k'} c_{k'+Q\downarrow}^\dagger\right)|0\rangle. \tag{3}$$

Here multiplication over k and k' is performed only over the occupied k-points [9]. By these modifications the total energy was substantially improved for both orderings. However, the SC E_{cond} slightly increased, while the SDW E_{cond} slightly decreased, shifting the balance to the SC side slightly. Since the variational Monte Carlo method seems to lead to overestimation of the SDW E_{cond} [20], this modification is considered to be an improvement.

With $\rho = 0.82$ and $U = 7$, we first decided a critical value t_c' at which SDW vanishes. For this purpose we computed the SDW E_{cond} for $L = 10 \sim 18$ and determined the best fit parameters for a scaling function $t_c' = a - b/L^d$, with fitting parameters a, b and d, obtaining the bulk-limit $t_c' = -0.60$ (Fig. 4). Then, with $t' = -0.055$ the SC E_{cond} was computed on the lattices with $L = 10 \sim 20$ as shown in Fig. 5. This time the size scaling does not give a finite bulk limit of SC E_{cond} due to the fall

226

around $L = 20$. Since this failure to get a finite bulk limit may be due to another type of size effect (abrupt decrease of density distribution of ε_k was observed in the neighborhood of the Fermi energy below a dense distribution due to the van Hove singularity, which appears discretely in the ε_k-k plane for finite lattices), we reserve the conclusion. However, this situation indicates the difficulty for obtaining a SC region in the region $\rho \leq 0.82$. The difficulty is due to facts that in most cases in this region the SDW E_{cond} increases faster than the SC E_{cond} with increase of $|t'|$, and that with small $|t'|$ the bulk-limit of SC E_{cond} tends to vanish.

We carried out a similar calculation for $\rho = 0.84$ as well. $t' = -0.05$ was chosen since there the SC E_{cond} was expected to win against the SDW E_{cond}. Unfortunately,

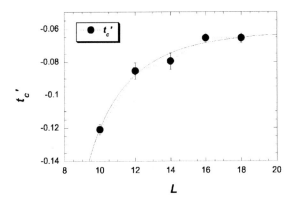

FIGURE 4. t_c' for the $\rho = 0.82$ and $U = 7$ is plotted as a function of edge length L of the lattice. Fitting curve $t_c' = a - b/L^d$, with $a = -0.0603$, $b = -1097$ and $d = 4.26$, gives the bulk-limit of $t_c' = -0.0603$.

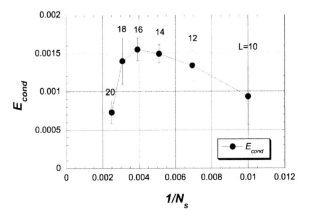

FIGURE 5. SC E_{cond} for the $\rho = 0.82$, $U = 7$ and $t' = -0.055$ is plotted as a function of the inverse of the number N_s of sites. L is the edge length of the square lattice.

the SC E_{cond} showed a sharp drop when $L \geq 20$ in a similar way as for $\rho = 0.82$. Since we observe such an abrupt change of the undressed one-particle energy distribution around the van Hove singularity that depends strongly on the system size, we refrain from drawing the conclusion here again. However, both from results with $\rho = 0.82$ and 0.84 we suppose that a pure uniform SC region is restricted to a narrow parameter space in a high-doping region with $\rho \leq 0.84$ if it exists.

Quantum Monte Carlo results showing no dominant SC correlations [2, 3, 7] are considered to be due to this narrowness of the SC parameter region and the dominance of SDW over SC in most parameter regions.

STRIPE STATE AND SC IN LOW-HOLE-DOPING REGION

Preceding Theories of Stripes and Metallic Mid-Gap Bands

The results in the preceding section suggest that the pure superconductivity exists only relatively high-doping region concerning the hole doped case. This is considered to explain the SC region in the high-hole-doping region with $\rho \leq 0.84$ and may be in accord with the spin-fluctuation mediation theory of SC in the high-T_c cuprates [21, 22]. In the low-hole-doping region it is impossible for a pure uniform SC phase to win against the commensurate SDW phase, as seen above. Since the 2D d-p model gives a similar feature [23] as well, this feature must be true, independent of the model. This

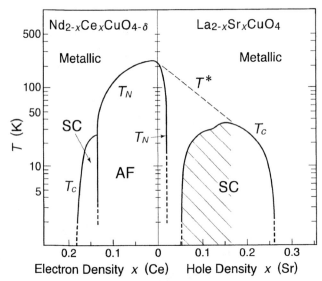

FIGURE 6. Experimental phase diagram of cuprate high-T_c superconductors in the plane of temperature versus doped hole density. Hatched part of the SC phase is argued here as inhomogeious SC state strongly influenced by stripes. Dashed line labeled T^* indicates the boundary of the pseudo-gap region. From [24].

makes a clear discrepancy with the well-known phase diagram (Fig. 6) in the temperature-doping plane, where the SC region extends down to a low-hole-doping region. In our viewpoint there must be a new type of SC state, not pure d-wave one as considered above but inhomogeneous one, in this parameter region. Such a state may be closely related with such amonalies as pseudogap and stripes observed in this region.

In the low-doping region incommensurate SDW structures with stripes are known to form states with a lower total energy than uniform commensurate or incommensurate SDW states. This fact was clarified by HF theories by Machida [25] by Zaanen et al. [26], by H. J. Schulz [27] and by Poilblanc and Rice [28] and variational Monte Carlo studies by Giamarchi et al. on the Hubbard model [29] and by Yanagisawa *et al.* on the d-p model [23]. The stripes are formed as a boundary between commensurate undoped SDW domains. Both neighbor sites straddling the boundary have spins of opposite directions. The stripe accommodates the doped holes. Many features, including lattice distortions due to electrostatic effects, are understandable in terms of the stripe [30].

Machida and Ichioka [31] showed that the Hartree-Fock one-particle solution gives two mid-gap bands lying in the SDW gap in the presence of stripes. They originate from the non-bonding state around the boundary and the non-bonding band is folded into two mid-gap bands because of the periodicity in unit of two lattice units. The mid-gap bands are flat in the case of the diagonal stripes, leading to non-metallic state. In the case of the vertical stripes an appreciable k-dependence apprears in the lower mid-band gap, leading to metallic conduction due to the doped holes in it. This result allows to interpret the neutron diffraction observation [32] that $La_{2-x}Sr_xCuO_4$ has diagonal stripes and is non-metallic when $x \leq 0.05$, while it has vertical stripes and is metallic when $x=0.06$. The appearance of SC in the latter case tempts us to try to find SC occurring along the conductive mid-gap band along the stripes.

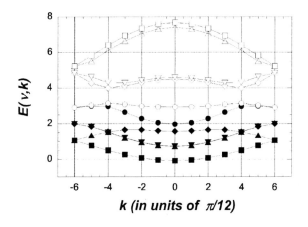

FIGURE 7. Hartree-Fock bands $E(v, k)$ on the 5×24 lattice. v is the band index and k is the wave number along the stripe. Central two bands in the energy range from 2 to 3.2 are the mid-gap bands.

Possibility of Superconductivity Along The Stripe

The k-dependence of the lower mid-gap band becomes stronger with increase of $|t'|$. It can have a semimetallic band overlap with the bands below the SDW gap when t' and hole doping are appropriate. We presume that in such a situation SC occurs due to the two-band mechanism due to the interband pair-wise Coulomb scattering of electrons if this coupling constant is sufficiently large. Here the two bands are the lower mid-gap band and the fundamental bands just below it. These bands can play the role of the bonding and antibonding bands in the two-leg ladder system [33].

We tried the 2D Hubbard model on the (even number)×(odd number) lattice with appropriate hole doping. The periodic boundary condition was imposed along both directions. When the parameters are such that the SDW dominates scheme the system has a stripe along the even-number edge. As is seen in Fig. 7, we have midgap bands. We have examined if SC occurs or not presumably due to the two-band mechanism, using the following trial variational MC wave function:

$$\Psi_s = P_{Ne} \cdot \prod_l \left(1 - (1-g) n_{l\uparrow} n_{l\downarrow}\right) \cdot \prod_{v,k} \left(u_{vk} + v_{vk} c^{\dagger}_{vk\uparrow} c^{\dagger}_{vk\downarrow}\right) |0\rangle, \tag{4}$$

where the second multiplication goes over all the HF one-particle states specified by band index v and wave number k in the direction of the stripe; $c_{vk\sigma}^{\dagger}$ is the annihilation operator for the electron in the vth band with wave number k. u_{vk} and v_{vk} are defined as in the BCS wave function with a SC gap parameter Δ_{vk} which takes a constant absolute value but changes its sign between the filled bands and other bands. For some lattices we got clear positive results with a finite value of $|\Delta_{vk}|$ at least for small lattices. Therefore, the SC along stripes are considered to be possible and to form the SC in the low-hole-doping region basically.

SUMMARY

(1) Carrying out variational Monte Carlo calculations we demonstrated that the 2D Hubbard model has a finite bulk-limit of SC condensation energy E_{cond} in an appropriate parameter region ($U = 8$, $\rho \geq 0.84$, $-0.25 \leq t' \leq -0.10$); the values are close to experimental values.

(2) A pure uniform SC region of this model was shown to be restricted to the high-doping region with parameter sets in a narrow window if it exists.

(3) In the low-hole-doping region, the SDW E_{cond} is overwhelmingly larger than the SC E_{cond}. The stripes were considered to be formed along the boundary between undoped SDW domains, which allows to interpret lattice anomalies etc. The mid-gap bands accompanying the stripe accommodate doped holes and are conductive. SC was indicated to occur here presumably due to the two-band mechanism working between the mid-gap and the lower fundamental SDW bands.

ACKNOWLEDGMENTS

Computations were performed on the massively parallel computer, Center for Computational Physics, University of Tsukuba, a parallel computer provided by the ACT-JST program of Japan Science and Technology Corporation, and computers of Tsukuba Advanced Computing Center, AIST, Japan.

REFERENCES

1. Hirsch, J.E., *Phys. Rev. Lett.* **54**, 1317-1320 (1985).
2. Furukawa, N., and Imada, M., *J. Phys. Soc. Jpn.* **61**, 3331-3354 (1992).
3. White, S.R., Scalapino, D.J., Sugar, R.L., Loh, E.Y., Gubernatis J.E., and Scalettar, R.T., *Phys. Rev. B* **40**, 506-509, (1989).
4. Dagotto, E., *Rev. Mod. Phys.* **66**, 763-840 (1994).
5. Husslein, T., Morgenstein, I., Newns, D.M., Pattnaik, P.C., Singer, J.M., and Matuttis, H.G., *Phys. Rev. B* **54**, 16179-16182 (1996).
6. Kuroki, K., Aoki, H., Hotta, T., and Takada, Y., *Phys. Rev. B* **56**, R14287-14290 (1997).
7. Zhang, S., Carlson, J., and Gubernatis, J.E., *Phys. Rev. Lett.* **78**, 4486-4489 (1997).
8. Nakanishi, T., Yamaji, K., and Yanagisawa, T., *J. Phys. Soc. Jpn.* **66**, 294-297 (1997).
9. Yamaji, K., Yanagisawa, T., Nakanishi, T., and Koike, S., *Physica C* **304**, 225-238 (1994).
10. Yamaji, K., Yanagisawa, T., Nakanishi, T., and Koike, S., in *Advances in Superconductivity XI*, edited by Koshizuka, N. *et al.*, Springer-Verlag, Tokyo, 1999, pp. 343-346.
11. Yamaji, K., Yanagisawa, T., and Koike, S., *Physica B* **284-288**, 415-416 (2000).
12. Raimondi, R., Jefferson J.H., and Feiner, L.F., *Phys. Rev. B* **53**, 8774-8788 (1996).
13. Yokoyama, H., and Shiba, H., *J. Phys. Soc. Jpn.* **57**, 2482-2493 (1988).
14. Gros, C., *Ann. Physics* **189**, 53-88 (1989).
15. Feiner, L.F., Jefferson J.H., and Raimondi, R., *Phys. Rev. B* **53**, 8751-8772 (1996).
16. Loram, J.W., Mirza, K.A., Cooper, J.R., and Liang, W.Y., *Phys. Rev. Lett.* **71**, 1740-1743 (1993).
17. Hao, Z., Clem, J.R. *et al.*, *Phys. Rev. B* **43**, 2844-2852 (1991).
18. Yokoyama, H., and Ogata, M., *J. Phys. Soc. Jpn.* **65**, 3615-3629 (1996).
19. Kondo, H., and Moriya, T., *J. Phys. Soc. Jpn.* **68**, 3170-3173 (1999).
20. Yanagisawa, T., Koike, S., and Yamaji, K., *J. Phys. Soc. Jpn.* **67**, 3867-3874 (1997).
21. Moriya, T., and Ueda, K., *Adv. Physics* **48**, 555-606 (2000).
22. Pines, D., *Physica C* **282-287**, 273-278 (1977)
23. Yanagisawa, T., Koike, S., and Yamaji, K., these proceedings.
24. Tokura, Y., *Kotaibutsuri* (in Japanese) **25**, 618-636 (1990); Takagi, H., ibid. pp. 736-745.
25. Machida, K., *Physica C* **158**, 192-196 (1989).
26. Schulz, H. J., *J. Phys. France B* **50**, 2933-2849 (1989).
27. Poilblanc, D., and Rice, T.M., *Phys. Rev. B* **39**, 9749-9752 (1990).
28. Zaanen, J., and Gunnarson, O., *Phys. Rev. B* **40**, 7391-7394 (1990).
29. Giamarchi, T., and Lhuillier, C., *Phys. Rev. B* **42**, 10641-10647 (1990).
30. Papers in these proceedings.
31. Machida, K., and Ichioka, *J. Phys. Soc. Jpn.* **68**, 2168-2171 (1988).
32. Wakimoto, S. *et al.*, *Phys. Rev. B* **60**, R769-772 (1990).
33. Yamaji, K. *et al.*, *Physica C* **222**, 349-360 (1994).

Stripes and Phase Diagram of *d-p* Model

T. Yanagisawa, S. Koike and K. Yamaji

Superconductivity theory group, Electrotechnical Laboratory, 1-1-4 Umezono, Tsukuba, Ibaraki 305-8568, Japan

Abstract. We study the ground state of the two-dimensional *d-p* model in the oxide superconductors by using the variational Monte Carlo method. We employ the Gutzwiller-projected BCS and SDW wave functions in search for possible ground states with respect to dependence on parameters included in the *d-p* model. Near half-filling the strong antiferromagnetic correlations exist and the SDW phase extends up to 20 percent doping. It is shown that the *d*-wave state is possible away from half-filling for both the hole and electron doping cases. Overall structure of the phase diagram obtained by our calculations coincides with experimental indications. The superconducting condensation energy agrees well with the experimental value obtained from specific heat and critical magnetic field measurements. A possibility of stripe phase is also examined by using the SDW wave functions with incommensurate spin structures. The distance between stripes depends on the hole concentration in the underdoped region. We show that a stability of the stripe state is dependent on the transfer parameter t_{pp} between oxygen sites.

INTRODUCTION

The mechanism of superconductivity of high-T_c cuprates is not still clarified after the intensive efforts over a decade [1]. An origin of the anomalous metal with pseudo-gap is also attracting many physicists as a challenging problem. It is important to examine the ground state of the two-dimensional CuO_2 planes which are contained commonly in the crystal structures of high-T_c superconductors. A basic model for the CuO_2 plane is the 2D *d-p* model with *d* and *p* orbitals. We can expect that the 2D *d-p* model contains the essential features of high-T_c cuprates [2,3]. It is not an easy task to clarify the ground state properties of the 2D *d-p* model because of strong correlations among *d* and *p* electrons. We must treat the strong correlations properly to examine the phase diagram of the high-T_c cuprates. The variational Monte Carlo method is an established tool to investigate the overall structure of phase diagram form weakly to strongly correlated regions. In this paper we investigate a possibility of superconductivity (SC) for the *d-p* model and a symmetry of Cooper pairs, and then discuss the phase structure of the ground state, i.e. a competition between antiferromagnetism (AF) and SC. A *d*-wave pairing state is stabilized due to the correlation effects contained in the Gutzwiller projection. It is shown that the SC condensation energy obtained from our evaluations is in reasonable agreement with available experimental data. A possibility of incommensurate stripe states is also investigated with respect to dependence on the hole density δ and transfer t_{pp} between oxygen orbits.

CP554, *Physics in Local Lattice Distortions*, edited by H. Oyanagi and A. Bianconi
© 2001 American Institute of Physics 1-56396-984-X/01/$18.00

2D *d* - *p* MODEL AND WAVE FUNCTIONS

The 2D *d-p* model is a three-band Hubbard model given as [4-6]

$$H = \varepsilon_d \sum_{i\sigma} d_{i\sigma}^\dagger d_{i\sigma} + U_d \sum_i d_{i\uparrow}^\dagger d_{i\uparrow} d_{i\downarrow}^\dagger d_{i\downarrow} + \varepsilon_p \sum_{i\sigma}(p_{i+\hat{x}/2,\sigma}^\dagger p_{i+\hat{x}/2,\sigma} + p_{i+y/2,\sigma}^\dagger p_{i+\hat{y}/2,\sigma})$$

$$+ t_{dp} \sum_{i\sigma}[d_{i\sigma}^\dagger(p_{i+\hat{x}/2,\sigma} + p_{i+\hat{y}/2,\sigma} - p_{i-\hat{x}/2,\sigma} - p_{i-\hat{y}/2,\sigma}) + h.c.]$$

$$+ t_{pp}\sum_{i\sigma}[-p_{i+\hat{y}/2,\sigma}^\dagger p_{i+\hat{x}/2,\sigma} + p_{i+\hat{y}/2,\sigma}^\dagger p_{i-\hat{x}/2,\sigma} + p_{i-\hat{y}/2,\sigma}^\dagger p_{i+\hat{x}/2,\sigma} - p_{i-\hat{y}/2,\sigma}^\dagger p_{i-\hat{x}/2,\sigma} + h.c.].$$

$$(1)$$

x and y with a hat represent a unit vector along x and y directions, respectively.

$p_{i+\hat{x}/2,\sigma}^\dagger$ and $p_{i+\hat{x}/2,\sigma}$ denote the operators for the *p* electrons at site $R_{+\hat{x}/2}$. Similarly $p_{i\pm\hat{y}/2,s}$ and $p_{i+\hat{y}/2,\sigma}^\dagger$ are defined. Other notations are standard and energies are measured in units of t_{dp}. For simplicity we neglect the Coulomb interaction among *p* electrons. We consider the standard wave functions for strongly correlated electron systems, which are given by the normal state, BCS and SDW wave functions with the Gutzwiller projection. They have been investigated considerably for the Hubbard model [7-14]. In Refs. [13,14] it has been discussed that they can be improved systematically by operating correlation factors. For the *d-p* model they are written as

$$\psi_n = P_G \prod_{|k| \le k_F, \sigma} \alpha_{k\sigma}^\dagger |0\rangle, \tag{2}$$

$$\psi_{BCS} = P_{N_c} P_G \prod_k (u_k + v_k \alpha_{k\uparrow}^\dagger \alpha_{-k\downarrow}^\dagger)|0\rangle, \tag{3}$$

$$\psi_{SDW} = P_G \prod_{|k| \le k_F, \sigma} \beta_{k\sigma}^\dagger |0\rangle, \tag{4}$$

where $\alpha_{k\sigma}$ is the linear combination of $d_{k\sigma}$, $p_{xk\sigma}$ and $p_{yk\sigma}$ constructed to express an operator for the lowest band of a non-interacting Hamiltonian in the hole picture. For $t_{pp} = 0$, $\alpha_{k\sigma}$ is expressed in terms of a variational parameter Δ_v:

$$\alpha_{k\sigma}^\dagger = \frac{1}{\sqrt{2}}\left(1 + \frac{\Delta_v}{2E_k}\right)^{1/2} d_{k\sigma}^\dagger + \frac{i}{\sqrt{2}}\left(1 - \frac{\Delta_v}{2E_k}\right)^{1/2}\left(\frac{w_{xk}}{w_k} p_{xk\sigma}^\dagger + \frac{w_{yk}}{w_k} p_{yk\sigma}^\dagger\right). \tag{5}$$

where $w_{xk} = 2t_{dp}\sin(k_x/2)$, $w_{yk} = 2t_{dp}\sin(k_y/2)$, $w_k = (w_{xk}^2 + w_{yk}^2)^{1/2}$ and $E_k = [\Delta_v^2/4 + w_k^2]^{1/2}$. Coefficients u_k and v_k, which are determined to minimize the energy, are expressed as the BCS form given by

$$\frac{v_k}{u_k} = \frac{\Delta_k}{\xi_k + \sqrt{\xi_k^2 + \Delta_k^2}}, \qquad (6)$$

for $\xi_k = \varepsilon_k - \mu$ where ε_k is the energy dispersion for the lowest band. For the commensurate SDW $\beta_{k\sigma}$ is given by a linear combination of $d_{k\sigma}$, $p_{xk\sigma}$, p_{yk}, $d_{k+Q\sigma}$, $p_{xk+Q\sigma}$ and $p_{yk+Q\sigma}$ for $Q = (\pi, \pi)$. We can easily generalize it to the incommensurate case by diagonalizing the Hartree-Fock Hamiltonian. P_G is the Gutzwiller projection operator for the Cu d site:

$$P_G = \prod_i (1 - (1 - g) n_{di\uparrow} n_{di\downarrow}) \qquad (7)$$

P_{Ne} is a projection operator, which extracts only the states with a fixed total electron number. The parameters in our calculations are the following: $\varepsilon_d = -2$, $\varepsilon_p = 0$, $U_d = 8$ and $0 \le t_{pp} \le 0.4$ in units of t_{dp}. We investigate 6×6 and 8×8 square lattices and 16×4 rectangular lattice, which are systems of larger size compared to previous works. We examine overall structure of the phase diagram with respect to dependence on the hole density.

CONDENSATION ENERGY AND PHASE DIAGRAM

First let us consider the projected-BCS wave function, where the Gutzwiller parameter g, effective d level Δ_v, chemical potential and superconducting order parameter Δ_s are considered as variational parameters. We employ the correlated measurements method in the process of searching optimal parameter values minimizing energy [15,16]. In Fig.1 we show the energy as a function of Δ_s for (a) $\delta = 0.111$ and $t_{pp} = 0$, (b) $\delta = 0.333$ and $t_{pp} = 0$, and (c) $\delta = 0.333$ and $t_{pp} = 0.2$. The d-wave state is stabilized because of the correlation effect induced by the Gutzwiller projection. Thus an origin of attractive interaction lies in the Gutzwiller projection.

The d-wave superconductivity is most stable among various possible SC states such as extended-s wave and isotropic-s wave states as for the case of Hubbard model. The superconducting condensation energy is defined as the energy difference between the minimum and the intercept of the E/N-Δ_s curve with the vertical axis. Its dependence on t_{pp} is weak as seen from Fig.1. The SC condensation energy by our calculations is estimated as $E_{cond} \sim 0.0005 \, t_{dp} = 0.75$meV per site for $\delta = 0.333$ and $t_{pp} = 0.2$ where we have used $t_{dp} = 1.5$eV as estimated by cluster calculations [17-19]. Now we can evaluate E_{cond} from several experiments such as specific heat and critical field measurements. They are given as $0.17 \sim 0.26$meV by specific heat data [20] and 0.26meV by critical magnetic field value $H_c^2/8\pi$ [21]. Our value is in reasonable agreement with the experimental data as was already shown for the Hubbard model where the SC energy gain in the bulk limit is given by $0.00117t/\text{site} = 0.59$meV/site [12]. Although the estimate of SC condensation energy is regarded as a first approximation to the exact value since the wave functions are of the Gutzwiller type, we expect that this value is near the correct value according to the evaluations for

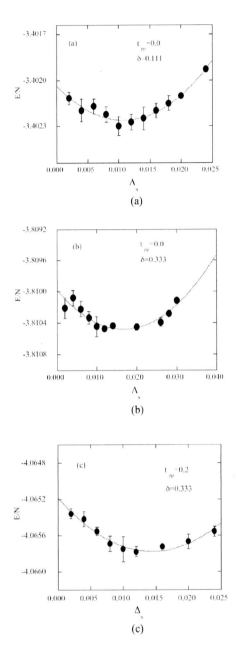

FIGURE 1. Ground state energy per site as a function of Δ_s on 6×6 lattice for (a) $\delta = 0.111$ and $t_{pp} = 0.0$, (b) $\delta = 0.333$ and $t_{pp} = 0.0$, and (c) $\delta = 0.333$ and $t_{pp} = 0.2$. Parameters are given by $U_d = 8$ and $\varepsilon_p - \varepsilon_d = 2$ in units of t_{dp}.

improved wave functions with multiplicative correlation factors [14]. This agreement between the theoretical and experimental condensation energy is highly remarkable and should be compared with the orresponding t-J model evaluations where the SC condensation energy is larger than the experimental value by almost two orders of magnitude [22].

Second, we can similarly evaluate the energy for SDW state for optimum parameters g, Δ_v and AF order parameter Δ_{AF}. The condensation energies are shown in Fig. 2 for $t_{pp} = 0.2$, where the negative delta indicates the electron-doping case. Solid symbols indicate the results for 8×8 and open symbols for 6×6. The Fig. 2 shows that the dependence of F_{cond} on the system size is weak. The energy is lowered considerably by the antiferromagnetic long-range order up to 20 percent doping for both the hole and electron doping cases. The d-wave state exists for $0.25 < \delta < 0.44$ for the hole-doping case. The structure for electron doping case is quite similar to the phase diagram indicated by experiments, which means that a large SDW phase exists near half-filling and there is a small superconducting phase. Our calculations predict d-wave symmetry away from half-filling, which is consistent with recent experiments on $Nd_{1.85}Ce_{0.15}CuO_{4-y}$ [23]. Our calculations are greatly supported by this agreement between experiments and theoretical predictions. In Fig.2 the condensation energy for an ansatz of extended-s wave is also shown. The extended s-wave state is more stable than the d-wave state for the electron-doping case near half-filling.

The phase structure obtained by our calculations agrees well with the available phase diagram indicated by experiments for the electron-doping case, while the agreement with experiments for the hole-doping case is not so good. The d-wave

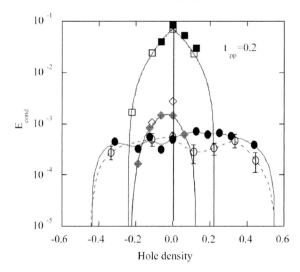

FIGURE 2. Condensation energy per site as a function of hole density δ for $t_{pp} = 0.2$. Circles, squares and diamonds denote the energy gain per site in reference to the normal state energy for d-wave, SDW and extended-s wave states, respectively. Solid symbols are for 8×8 and open symbols are for 6×6. Curves are guide for eyes.

phase extends up to $\delta \sim 0.44$ which is larger than the experimental limit, which may be due to size effect. The phase diagram indicated from experiments is more complicated for the hole doping, *i.e.* the pseudo-gap phase exists and AF region is rather small near half-filling. The phase diagram from variational calculations assuming uniform states indicates that the hole and electron doping states are basically symmetric with respect to electron-hole transformation. We expect that asymmetry between the hole and electron-doping cases may result from inhomogeneity in the ground state. Thus we investigate the wave function of SDW state with incommensurate spin modulation as candidate for the ground state.

STRIPES

In this section the stripe wave function is examined in order to consider unusual properties of the high-T_c superconductors. Recent neutron-scattering measurements have revealed incommensurate structures suggesting stripes [24-28]. The SDW state can be possibly stabilized further in lightly doped systems if we take into account a spin modulation in space which has been studied for the one-band Hubbard model [29-32]. We can introduce a stripe in the uniform spin density state so that doped holes occupy new levels close to the original Fermi energy keeping the energy loss of antiferromagnetic background minimum. The wave function with a stripe can be

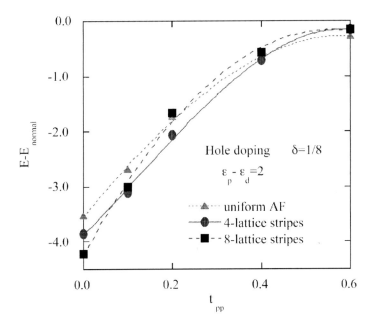

FIGURE 3. Energies per site as a function of t_{pp} for 16×4 rectangular lattice for the doping ratio $\delta = 1/8$. Circles, squares and triangles denote the energy in reference to the normal state energy for 4-lattice stripes, 8-lattice stripes, and commensurate SDW, respectively.

taken of the Gutzwiller type: $\Psi_{stripe} = P_G|\Psi_{stripe}\rangle$. $|\Psi_{stripe}\rangle$ is the Slater determinant made from solutions of the Hartree-Fock Hamiltonian given as

$$H_{stripe} = H^0_{dp} + \frac{U}{2}\sum_{i\sigma}[n_{di} - \sigma(-1)^{x_i+y_i}m_i]d^\dagger_{i\sigma}d_{i\sigma} \qquad (8)$$

where H^0_{dp} is the non-interacting part of the Hamiltonian H. n_{di} and m_i are expressed by modulation vectors Q_s and Q_c for spin and charge part, respectively. In this paper n_{di} and m_i are assumed to have the form [33]

$$n_{di} = 1 - \sum_j \frac{\alpha}{\cosh((x_i - x^{str}_j)/\xi_c)} \qquad (9)$$

$$m_i = m\prod_j \tanh((x_i - x^{str}_j)/\xi_c) \qquad (10)$$

with parameters α, m, ξ_c and ξ_s, where x^{str}_j denotes the position of a stripe.

Recently, an incommensurate ground state has been studied intensively for the two-dimensional t-J model [34-36]. It suggests that the incommensurate states obtained for these models do not necessarily coincide with the Tranquada's proposal [24] which indicates that modulation vectors are given by $Q_s = (\pi-2\pi\delta, \pi)$ and $Q_c = (4\pi\delta, 0)$ at the doping ratio $\delta = 1/8$. It will be shown that the d-p model actually gives the results, which are consistent with neutron-scattering measurements, implying the importance of oxygen orbitals, which are neglected in the process of mapping to the t-J model.

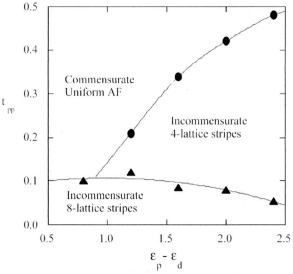

FIGURE 4. Phase diagram in the plane of t_{pp} and the level difference $\varepsilon_p-\varepsilon_d$ on the 16×4 lattice for $\delta = 1/8$.

In Fig.3 we show the energies for commensurate and incommensurate SDW states on the 16×4 lattice at $\delta = 1/8$, where the incommensurability is given by $\pi/4$ ($= 2\pi\delta$) for 4-lattice stripes and $\pi/8$ for 8-lattice stripes, respectively. The 4-lattice stripe is stable for the parameters corresponding to real cuprates, which are estimated as $\varepsilon_p - \varepsilon_d \approx 2$ and $t_{pp} \approx 0.4$ [17-19] as is shown in Fig.4 where the stable phases are presented in the plane of t_{pp} and the level difference $\varepsilon_p - \varepsilon_d$. The energies for $\delta = 1/16$ are also shown in Fig.5 where the 8-lattice stripe state has the lowest energy indicating that the incommensurability is proportional to δ when δ is small. For the electron doping case the commensurate SDW region extends and stripe state cannot be stable for $\delta = 1/8$, implying asymmetry between the hole and electron doping.

V SUMMARY AND DISCUSSION

We have presented our evaluations for the 2D d-p model by using the variational Monte Carlo method. We have evaluated the SC condensation energy for both the underdoped and overdoped regions. The strength of U_d is also important to determine the phase boundary between the SDW and superconducting phases because when U_d is extremely large, the SDW region extends to $\delta \approx 0.5$ for which the d-wave region hardly exists [6]. Our results indicate that the superconducting phase exists for the intermediate value of U_d. The d-wave symmetry pairing results from the strong correlation between d electrons and the SC condensation energy agrees reasonably

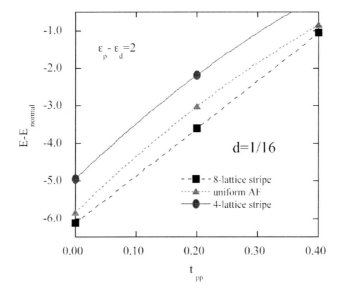

FIGURE 5. Energies per site as a function of t_{pp} for 16×4 rectangular lattice for the doping ratio $\delta = 1/16$. Circles, squares and triangles denote the energy in reference to the normal state energy for 4-lattice stripes, 8-lattice stripes, and commensurate SDW, respectively.

with the experimental data. The phase diagram for the electron-doping case is consistent with the available experimental indications. Thus the properties of electron-doped materials may be understood within our approach. Instead the hole-doping case has left many issues for future studies. Asymmetry between the hole and electron doping may be related to the anomalous properties of high-T_c cuprates. The inhomogeneous states are expected to play an important role in the underdoped region for the hole-doped materials. In fact, the stripe phase is stabilized for low-doping case in our calculations. The incommensurability of stripe states is dependent on t_{pp}, ε_p-ε_d, and δ being highly non-trivial. The structure of phase diagram depends on a competition among the uniform SDW state, SDW state with stripes, coexistence state of stripes and SC, and pure d-wave SC. A picture for the hole-doping case following from our evaluations is that a stripe state is stable at low doping and changes into the d-wave SC away from half-filling.

REFERENCES

1. See, for example, *Proceedings of 22nd International Conference on Low Temperature Physics* (LT22, Helsinki, Finland, 1999) Physica B **284-288** (2000).
2. Emery, V.J., *Phys. Rev. Lett.* **58**, 2744 (1987).
3. Tjeng, L.H., Eskes, H., and Sawatzky, G.A., *Strong Correlation and Superconductivity* edited by Fukuyama, H., Maekawa, S., and Malozemoff, A.P., Springer, Berlin Heidelberg, 1989, pp.33.
4. Guerrero, M., Gubernatis, J.E., and Zhang, S., *Phys. Rev. B* **57**, 11980 (1998).
5. Yanagisawa, T., Koike, S., and Yamaji, K., *Physica B* **284**, 467 (2000).
6. Asahata, T., Oguri, A., and Maekawa, S., *J. Phys. Soc. Jpn.* **65**, 365 (1996).
7. Yokoyama, H., and Shiba, H., *J. Phys. Soc. Jpn.* **56**, 1490 (1987)
8. Gros, C., *Ann. Phys.* **189**, 53 (1989).
9. Giamarchi, T., and Lhuillier, C., *Phys. Rev. B* **43**, 12943 (1991).
10. Nakanishi, T., Yamaji, K., and Yanagisawa, T., *J. Phys. Soc. Jpn.* **66**, 294 (1997).
11. Yamaji, K., Yanagisawa, T., Nakanishi, T., and Koike, S., *Physica C* **304**, 225 (1998).
12. Yamaji, K., Yanagisawa, T., and Koike, S., *Physica B* **284**, 415 (2000).
13. Yanagisawa, T., Koike, S., and Yamaji, K., *J. Phys. Soc. Jpn.* **67**, 3867 (1998).
14. Yanagisawa, T., Koike, S., and Yamaji, K., *J. Phys. Soc. Jpn.* **68**, 3867 (1999).
15. Umrigar, C.J., Wilson, K.G., Wilkins, J.W., *Phys. Rev. Lett.* **60**, 1719 (1988).
16. Kobayashi, K., and Iguchi, K., *Phys. Rev. B* **47**, 1775 (1993).
17. Eskes, H., Sawatzky, G.A., Feiner, L.F., *Physica C* **160**, 424 (1989).
18. Hybertson, M.S., Stechel, E.B., Schlüter, M., Jennison, D.R., *Phys. Rev. B* **41**, 11068 (1990).
19. McMahan, A.K., Annett, J.F., and Matrin, R.M., *Phys. Rev. B* **42**, 6268 (1990).
20. Loram, J.W., Mirza, K.A., Cooper, J.R., and Liang, W.Y., *Phys. Rev. Lett.* **71**, 1470 (1993).
21. Hao, Z. *et al.*, *Phys. Rev. B* **43**, 2844 (1991).
22. Yokoyama, H., and Ogata, M., *J. Phys. Soc. Jpn.* **65**, 3615 (1996).
23. Tsuei, C.C., and Kirtley, J.R., *Phys. Rev. Lett.* **85**, 182 (2000).
24. Tranquada, J. *et al.*, *Phys. Rev. B* **54**, 4596 (1996).
25. Suzuki, T. *et al.*, *Phys. Rev. B* **57**, 3229 (1998).
26. Mook, H.A. *et al.*, *Nature* **395**, 580 (1998).
27. Arai, M. *et al.*, *Phys. Rev. Lett.* **83**, 608 (1999).
28. Wakimoto, S. *et al.*, *Phys. Rev. B* **61**, 3699 (2000).

29. Poilblanc, D., and Rice, T.M., *Phys. Rev. B* **39**, 9749 (1989).

30. Kato, M., Machida, K., Nakanishi, H., and Fujita, M., *J. Phys. Soc. Jpn.* **59**, 1047 (1990).

31. Schulz, H., *Phys. Rev. Lett.* **64**, 1445 (1990).

32. Ichioka, M., and Machida, K., *J. Phys. Soc. Jpn.* **68**, 4020 (1999).

33. Giamarchi, T., and Lhuillier, C., *Phys. Rev. B* **42**, 10641 (1990).

34. White, S., and Scalapino, D.J., *Phys. Rev. Lett.* **80**, 1272 (1998).

35. White, S., and Scalapino, D.J., *Phys. Rev. Lett.* **81**, 3227 (1998).

36. Hellberg, C.S., and Manousakis, E., *Phys. Rev. Lett.* **83**, 132 (1999).

Bound States of Holes in Distorted Spin Ladder

A. Mishima

Faculty of Engineering, Kanazawa Institute of Technology, P.O. Box 921-8501, Nonoichi, Ishikawa, Japan

Abstract. The effects of the lattice distortions along the rung on the bound states of two holes doped away from half-filling in terms of electrons have been studied in the two-leg-spin ladders. As the electron-lattice interaction increases, the binding energy of two holes with negative value increases. Two holes along the rung form bound states with the spin singlet. It turns out that the electron-lattice interaction along the rung plays an important role in the bound states of two holes in the two-leg-spin ladders.

INTRODUCTION

Since the discovery of high-T_c cuprates, many physicists have actively studied the two-leg ladders [1]. The following reasons are considered. A ladder is intermediate between a chain and a plane. High-T_c superconductivity in lightly doped cuprates occurs in the CuO_2 plane with the antiferromagnetic (AF) order for nondoping, while the long-range order does not exist in a Heisenberg $S = 1/2$ chain. Thus the study on the two-leg ladders is very important for the elucidation of the mechanism of high-T_c superconductivity in cuprates and understanding the spin-1/2 Heisenberg antiferromagnets in one and two dimensions and will advance the explore of new superconductors.

It has been theoretically pointed out that two coupled spin-1/2 chains have a spin-gap and hole-pair formation upon doping within a t-J-t'-J' model and the model exhibits superconducting pairing correlations away from half-filling [2] and furthermore they show superconductivity, when they become metals by doping carriers into these systems [1]. It has been observed experimentally that the susceptibility shows the spin singlet state in the igorganic compounds of $Sr_{14-x}Ca_xCu_{24}O_{41+\delta}$ at low temperatures and the systems show superconductivity under the pressure at $x \approx 13.6$ [3]. These substances are insulators of two-leg-spin ladders with spin gaps and half-filled bands.

Previously, I studied the bound states of two holes in the one-dimensional (1D) two-band (2B) model [4], the 2D-2B model [5], the double-layer 2D-2B model [6], and the Peierls d-p model [7] for high-T_c superconductivity in the cuprates. I studied the bound states of two holes in the 1D spin-electron-phonon model for new compounds [8]. In order to research the possibility of superconductivity in the organic compounds of two-leg ladders with spin gaps, I studied the bound states of two holes in the two chains within an Su-Schrieffer-Heeger Hubbard model [9] and in the new two π-d chains composed of localized d electrons and itinerant π electrons [10].

CP554, *Physics in Local Lattice Distortions*, edited by H. Oyanagi and A. Bianconi
© 2001 American Institute of Physics 1-56396-984-X/01/$18.00

The study on the roles of the electron-lattice interaction in the two-leg-spin ladders is necessary for the sake of the elucidation of the mechanism of high temperature superconductivity. In this paper, I investigate the effects of the lattice distortions along the rung on the bound states of two holes doped away from half-filling in terms of electrons in the two-leg-spin ladders.

MODEL AND METHOD

The Hamiltonian for the two-leg-spin ladders in the electron picture is defined as

$$H = -(t \sum_{i,m,\sigma} c^\dagger_{i,m+1,\sigma} c_{i,m,\sigma} + \text{H.c.}) - \sum_{m,\sigma} \{[t_v + \tau(u_{1,m} - u_{2,m})]c^\dagger_{1,m,\sigma} c_{2,m,\sigma} + \text{H.c.}\}$$

$$+ J \sum_{i,m} s_{i,m} s_{i,m+1} + J_v \sum_m s_{1,m} \cdot s_{2,m} + \frac{k}{2} \sum_{j,m} (u_{1,m} + u_{2,m})^2 ,$$

(1)

where $c^\dagger_{i,m,\sigma}$ and $c_{i,m,\sigma}$ create and annihilate an electron on leg i ($= 1, 2$) and site m with spin σ ($= \alpha$ up, β down), respectively. The t and t_v are the transfer energies between nearest-neighbor sites along the leg and the rung, respectively. The $u_{i,m}$ is the displacement of the m-th lattice site in the i-th leg along the rung from the equilibrium position and τ is the transfer energy per difference of the displacements of the m-th lattice sites in the 1-th and 2-th legs. The k is the spring constant between nearest-neighbor lattice sites along the rung. The $s_{i,m}$ is the spin operator of the electron. J and J_v are the AF exchange interactions between the electron spins along the leg and the rung, respectively.

The size of system is 2×80 sites with periodic boundary conditions parallel to the legs. All energies are measured in units of t. The dimensionless electron-lattice interaction is defined by $\lambda = 2\tau^2/\pi kt$. For simplicity, an Ising-spin interaction rather than a Heisenberg-spin interaction, a mean-field approximation for electrons and an adiabatic approximation for lattices have been used, i.e.,

$<s_{i,m} \cdot s_{i,m+1}> \rightarrow s_{i,m,z}<s_{i,m+1,z}> + <s_{i,m,z}>s_{i,m+1,z} - <s_{i,m,z}><s_{i,m+1,z}>,$ where

$s_{i,m,z} = \dfrac{n_{i,m,\alpha} - n_{i,m,\beta}}{2}$ with $n_{i,m,\sigma} = c^\dagger_{i,m,\sigma} c_{i,m,\sigma}$ and $< \cdots >$ denotes the ground-state

average.

The dimerization parameter is defined by $y_m = \dfrac{\tau(u_{1,m} - u_{2,m})}{t}$. The dimerization

parameter has been calculated by using the Feynman-Hellmann theorem and minimizing the total energy of the system, self-consistently. The spin and charge densities at each lattice site have been also computed self-consistently.

RESULTS

I have computed the binding energy of two holes doped away from half-filling in terms of electrons, E_B. E_B is defined by $E_B = E_2 - E_0 - 2(E_1 - E_0)$, where E_0, E_1 and E_2 $(= N_\alpha - 1 + N_\beta - 1)$ are the total energies of the systems with 0, 1 and 2 holes, respectively. I have investigated the bound state of holes for large J / t and J_v / t as an example.

Figure 1 shows the binding energy normalized by the transfer energy between nearest-neighbor sites along the leg, E_B / t, as a function of electron-lattice interaction, λ, for $t_v/t = 0.1$, $J / t = 6.0$ and $J_v / t = 6.0$. As λ increases from 0 to 0.9, the binding energy of two holes with negative value increases. The reason is that the charge densities and the amplitude of spin densities increase at sites with holes and decrease at neighbor sites.

The holes on rungs form bound states as $E_B < 0$. Thus two holes in the two-leg-spin ladders bind in the spin-singlet state. It turns out that the electron-lattice interaction along the rung plays an important role in the bound states of two holes doped away from half-filling in terms of electrons in the two-leg-spin ladders. This result suggests that the electron-lattice interaction plays an important role in the bound states of two holes in the high-T_c cuprates.

I have used a mean-field approximation. In the future, I will study the bound states of holes using the density matrix renormalization group.

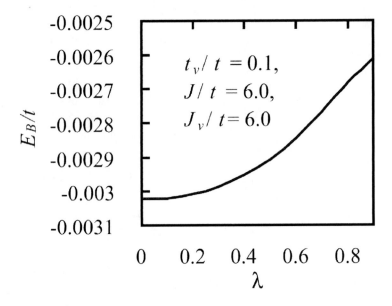

FIGURE 1. The normalized binding energy of two holes, E_B / t, vs. electron-lattice interaction, λ

REFERENCES

1. Dagotto E., and Rice, T. M., *Science* **271**, 618-623 (1996).
2. Dagotto, E., Riera, J., and Scalapino, D., *Phys. Rev.* **B 45**, 5744-5747 (1992).
3. Uehara, M., Nagata, T., Akimitsu, J., Takahashi, H., Môri, N. and Kinoshita, K., *J. Phys. Soc. Jpn.* **65**, 2764 –2767 (1996).
4. Mishima, A., "Electronic States of the One-Dimensional Two-Band Model" in *The Physics and Chemistry of Oxide Superconductors*, edited by Iye, Y., and Yasuoka, H., Proceedings in Physics, Vol. 60, Springer-Verlag, Berlin, Heidelberg, 1992, pp. 147-151.
5. Mishima, A., *Physica B* **194-196** 1379-1380 (1994).
6. Mishima, A., *Physica C* **235-240** 2223-2224 (1994).
7. Mishima, A., *Physica C* **282-287** 1769-1770 (1997).
8. Mishima, A., *Czechoslovak J. Phys.* **46** 941-942 (1996).
9. Mishima, A., *Synthetic Metals* **101** 316-317 (1999).
10. Mishima, A., *J. Low Temperature Phys.* **117** 1759-1763 (1999)

Parameters for Systems Exhibiting Local Lattice Distortions, Charge and Spin Ordering

S. Alam, T. Yanagisawa and H. Oyanagi

Physical Science Division, Electrotechnical Laboratory, Tsukuba, Ibaraki 305, Japan

Abstract. Keeping in mind the experimental results that indicate local lattice distortions, charge and spin orderings, we have developed a phenomenological approach which allows us to describe the electronic phase diagram of cuprates and related systems in terms of few parameters. In the present work we consider a third-order parameter theory which characterize charge, spin and superconductivity orderings. We are thus led to a theory of three scalar fields. By coupling these scalars to gauge fields we are naturally led to string-like solutions, which we interpret as stripes. This ties nicely with our quantum group conjecture that 1d systems play an important role in the physics of cuprates and related materials. We show that this simple approach can give rough values for two-order parameters which can be naively be interpreted as charge and spin orderings. We also report our attempt to understand how local lattice distortions are involved and what role they play in terms of these two order parameters.

INTRODUCTION

In a previous work one of us [1] has advanced the conjecture that one should attempt to model the phenomena of antiferromagnetism and superconductivity by using quantum symmetry group. Following this conjecture to model the phenomenona of antiferromagnetism and superconductivity by quantum symmetry groups, three toy models were proposed [2], namely, one based on $SO_q(3)$ the other two constructed with the $SO_q(4)$ and $SO_q(5)$ quantum groups. Possible motivations and rationale for these choices were outlined [2]. In [3] a model to describe quantum liquids in transition from 1d to 2d dimensional crossover using quantum groups was outlined. In [5] the classical group $SO(7)$ was proposed as a toy model to understand the connections between the competing phases and the phenomenon of psuedo-gap in High Temperature Superconducting Materials [HTSC]. Then we proposed in [6] an idea to construct a theory based on patching critical points so as to simulate the behavior of systems such as cuprates. To illustrate our idea we considered an example discussed by Frahm *et al.*, [4]. The model deals with antiferromagnetic spin-1 chain doped with spin-1/2 carriers. In [7] the connection between Quantum Groups and 1-dimensional [1-d] structures such as stripes was outlined. The main point of [7] is to emphasize that *1-d structures play an important role in determining the physical behaviour [such as the phases and types of phases these materials are capable of exhibiting] of cuprates* and related materials.

In this note we examine a phenomenological and classical model to understand the phenomenon of "ordering" in HTSC and related materials.

CP554, *Physics in Local Lattice Distortions,* edited by H. Oyanagi and A. Bianconi
© 2001 American Institute of Physics 1-56396-984-X/01/$18.00

SYMMETRY BREAKING

It is well-known that the Higgs-model in particle physics has its origins in condensed matter physics [CMP] and that Higgs-model is a model of spontaneous symmetry breaking [SSB]. A well-known example of SSB in CMP is ferromagnetism near the Curie temperature. When the temperature is greater than Curie temperature all dipoles are randomly oriented and the ground-state is *rotationally* invariant, whereas when the temperature is below the Curie temperature all the dipoles are aligned along an arbitrary direction giving rise to spontaneous magnetization and rotational symmetry is *hidden*. In other words we can say that we have SSB when the symmetry of Hamiltonian or Lagrangian is not explicitly shared by the ground-state or the vacuum. Thus the symmetry breaking condition is the non-invariance of the vacuum (ground-state)

$$U\,|\,0> \neq \,|\,0>, \tag{1}$$

where U is an element of the symmetry group.

Ginzburg-Landau theory [GL] is a phenomenological model for mainly second-order phase transitions. The main-point of GL can be easily understood by resorting to ferromagnetism as an example. The essential point is that for temperatures near the Curie temperature, the magnetization $\mathbf{M}$ is assumed to be small and one can Taylor expand the free energy density as

$$
\begin{aligned}
f(\mathbf{M}) &= (\partial_i \mathbf{M})^2 + V(\mathbf{M}), \\
V(\mathbf{M}) &= a_1(T)(\mathbf{M} \cdot \mathbf{M}) + a_2(T)(\mathbf{M} \cdot \mathbf{M})^2,
\end{aligned}
\tag{2}
$$

assuming slowly varying field. By construction in 2 the energy densities f and V are clearly rotationally invariant (symmetric). Higher powers of $\mathbf{M}$ are neglected, the term $a_2(T)(\mathbf{M \cdot M})^2$ is kept since at Curie temperature $a_1 = a(T - T_C)$, $a > 0$, vanishes. The kinetic energy term $(\partial_i \mathbf{M})^2$ is non-negative as usual, thus to obtain ground-state magnetization we must extremize $V(\mathbf{M})$. This yields for $T < T_C$ the magnitude of magnetization

$$|\,\mathbf{M}\,| = (-\frac{a_1}{2a_2})^{1/2}, \tag{3}$$

which is the *order - parameter*.

ORDER-PARAMETERS

Following the discussion of the previous section we can easily set-up a GL type of model for cuprates. The GL type of phenomenological approach is not uncommon and has been adopted by many authors, for example see [8][1]. The basic physical components of HTSC materials are

- Charge density.
- Magnetization.
- Superconductivity.

Thus generically we may write the free energy density of such a system in terms of three order parameters, namely, ρ [charge density], ϕ [complex superconducting order parameter] and σ [order parameter for magnetization]. The free energy density is the sum of the contributions from each of these order parameters [each with functional form as in 2] plus an interaction term,

$$f = f_\phi + f_\rho + f_\sigma + f_{int},$$
$$f_\phi = (\partial_i \phi)^2 + \alpha_1^\phi \phi^* \phi - \alpha_2^\phi (\phi^* \phi)^2,$$
$$f_\sigma = (\partial_i \sigma)^2 + \alpha_1^\sigma \sigma^* \sigma - \alpha_2^\sigma (\sigma^* \sigma)^2, \qquad (4)$$
$$f_\rho = (\partial_i \rho)^2 + \alpha_1^\rho \rho^* \rho - \alpha_2^\rho (\rho^* \rho)^2,$$
$$f_{int} = f_{\phi\sigma} + f_{\phi\rho} + f_{\rho\sigma}.$$

The simplest choice of the crossed interaction terms if higher powers than fourth are neglected are of the form $\alpha_2^{\phi\sigma}(\phi^*\phi)(\sigma^*\sigma)$ for the $\phi - \sigma$ term and similarly for the rest.

What we have is essentially three scalar fields, which represent the order parameters. One way to proceed is to consider the interaction of scalar fields with gauge fields and subject this to some symmetry group. As it is known from particle physics such considerations lead to some interesting non-perturbative solutions such as t'Hooft-monopole, Nielsen-Olsen vortices, instantons and several others. Clearly we have many choices for gauge group, but we consider the most basic ones which can cover the essential physics. We have to broad choices abelian and non-abelian groups. The abelian group of direct relevance is $U(1)$ or $O(2)$ and the one of the simplest non-abelian choice is $SU(2)$ or $SO(3)$. We may also consider higher groups such as $SU(N)$, $SO(N)$, CP^{n-1} and others.

The Lagrangian [9] of gauge-field A_μ interacting with a scalar field ϕ may be written as for abelian case

$$L = \frac{1}{2}(D_\mu \phi)^\dagger (D^\mu \phi) + \mu^2 \phi^\dagger \phi - \lambda (\phi^\dagger \phi)^2 - \frac{1}{4} F_{\mu\nu} F^{\mu\nu},$$
$$D_\mu \phi = \partial_\mu \phi - ig A_\mu \phi, \qquad (5)$$
$$F_{\mu\nu} = \partial_\mu A_\nu - \partial_\nu A_\mu,$$

and

---

[1] The main differences between reference [8] and our model is that unlike them we don't assume that the scalar field corresponding to charge is real and that we deal with couplings of scalar fields with gauge fields.

$$L = \frac{1}{2}(D_{\mu}^{ij}\phi_j)^{\dagger}(D_{ij}^{\mu}\phi^j) + \mu^2\phi_i^{\dagger}\phi^i - \lambda(\phi_i^{\dagger}\phi^i)^2 - \frac{1}{4}F_{\mu\nu}^i F_i^{\mu\nu},$$

$$D_{\mu}^{ij}\phi_j = \partial_{\mu}\phi^i - igF^{ijk}A_{\mu}^j\phi_k,$$

$$F_{\mu\nu}^i = \partial_{\mu}A_{\nu}^i - \partial_{\nu}A_{\mu}^i + gF^{ijk}A_{\mu}^j A_{\nu}^k, \qquad (6)$$

for the non-abelian case. The group indices are denoted by $i, j, k, ...$, whereas $\mu, \nu, ...$ are the space-time indices.

Let us consider SSB in the abelian case. If the scalar potential is extremized and the scalar field acquires a vacuum expectation value [VEV]

$$|<0|\phi|0> = v/\sqrt{2},$$

$$|\phi| = v/\sqrt{2}, \qquad (7)$$

$$v = (\mu^2/\lambda)^{1/2},$$

the symmetry is spontaneously broken. To see this we first write the ϕ field in terms of two real fields ϕ_1 and ϕ_2, and choose an arbitrary direction to achieve SSB, as discussed above, viz.

$$\phi = \frac{1}{\sqrt{2}}(\phi_1 + i\phi_2),$$

$$|<0|\phi_1|0> = v, \qquad (8)$$

$$|<0|\phi_2|0> = 0.$$

After shifting the fields ϕ_1 and ϕ_2 about their respective vacuum expectation values we will end up with the gauge-boson A_{μ} having acquired a mass $M = gv$, and a scalar field with mass $m = \sqrt{2}\mu$. The degrees of freedom [DOF] must correctly add up. We started with massless gauge boson [2 DOF] and complex ϕ field [2 DOF] and ended up with a massive gauge-boson [3 DOF] and one real scalar field [1 DOF] and so the DOF correctly add up. The same procedure easily carries over to the non-abelian case. For example if we consider the group $SU(2)$, we would start with three massless gauge fields [6 DOF]. Choosing ϕ to be a complex doublet [4 DOF] and assuming SSB we will arrive at 3 massive gauge bosons [9 DOF] and one real scalar field [i.e. Higgs field] [1 DOF] as one should.

The Lagrangian in 5 is called the Abelian-gauge model [AGM]. The connection between the static solution of AGM and GL model was elegantly pointed out by Nielsen and Olsen [11]. Using the *vortex* solution of AGM the connection with the dual model of strings [Nambu action] was also pointed out in this work. Thus by looking at particular solution of a field theory one arrives at a string model. It is rather unfortunate that although lot of people in CMT use vortices in context of GL, reference to [11] is almost never made, to our knowledge.

In the above scenario of SSB there is a physical Higgs, however we may consider other scenarios where there is *no physical Higgs* and symmetry breaking is non-linearly realized [10] [and references therein]. This approach is same as non-linear sigma model, which is discussed in [5]. For our purposes we can think of non-linear sigma model as a type of scalar field theory, whose structure is different from ϕ^4 theory discussed above and which gives an alternative description of SSB.

RESULTS

Our phenomenological model given in Eq. 4 consists of three scalar fields. Even without including gauge fields it is non-trivial to solve this model. We thus break it into simpler logical parts which we have discussed above. For example we can choose to introduce a single Abelian gauge field which interacts with charged complex field, clearly there are several interesting choices one can make. We now outline our results for some of our choices.

- Consider the choice that an Abelian gauge field interacts with a complex scalar. In the usual manner we can obtain the string-like vortex solution. We interpret the string-like solution as charged stripes. We note that so far the treatment is classical. The interaction of the other two scalar fields [which represent superconductivity and magnetism] with these charged strings can be studied. In a simple sense we can take the charged scalar field to represent the holons. The string-like solution [11] can be obtained by assuming that the scalar field approaches a constant value at large distance. It is straightforward to see that the minimum of potential is at

$$|\phi| = \phi_0 = (\alpha_1/(2\alpha_2))^{1/2}. \tag{9}$$

If we expand about this VEV, and denote fluctuations by ϕ', it can be shown that the fluctuations are defined by a characteristic length ξ,

$$\phi' = e^{-r/\xi}, $$
$$\xi = 1/\sqrt{(2\alpha_1)}. \tag{10}$$

ξ measures the distance before $|\phi|$ reaches its asymptotic value, it is the measure of the transverse extenstion of the string. Another relevant parameter is directly related to the ratio [r] of the strength of the self-coupling of scalar field to the interaction strength of scalar field (α_2) with the gauge field (e)

$$r = (2\alpha_2/(e^2\alpha_1))^{1/2} \tag{11}$$

r is a measure over which the electromagnetic field is appreciably different from zero. A well-defined string results if ξ and r have roughly the same magnitude. Thus the deviation from an exact string structure or deformed string may be got by choosing suitable values for these parameters. Such a parameterization is useful for modeling realistic stripes. Although the string-like solution of the AGM is well-known, our interpretation is new, namely to interpret the string-like solution as charge stripe. We note that Nielsen and Olsen [11] found the solution by realizing that AGM is the relativistic generalization of GL Lagrangian which has a vortex solution as for example found by Abrikosov. Yet another new feature is to use the deviation of the parameters ξ and r from each other for modelling realistic stripes.

- If we choose a non-abelian group such as $SU(2)$, $SU(3)$ or $SO(3)$ we can still obtain a string-like solution provided we have at least two scalar fields [which are "vectors" under the symmetry groups]. The same arguments go through as discussed for the abelian case. We may use these solutions to models spin stripes or regions of spins or spinons. By considering the interactions of charge and spin stripes we can hope to obtain a phenomenological model for cuprates and related materials.

- It is also known [see [11]] that in 2+1 dimensions the simplest non-trivial string-model [or vortex] arises from the sine-Gordon theory, with Lagrangian, equation of motion and static solution [with string lying along the x-axis],

$$L = \frac{1}{2}(\partial_\mu \phi)^2 + c\cos(d\phi),$$

$$\Box \phi + cd\sin(d\phi) = 0, \tag{12}$$

$$\phi = \frac{1}{4}\arctan e^{\sqrt{cd}\,y},$$

where $\Box \equiv \partial_\mu \partial^\mu \equiv (1/v^2)\frac{\partial^2}{\partial t^2} - \nabla^2$ is the familiar wave operator.

We may thus interpret this solution as giving rise to charge stripes. The above are classical field theory results. On the other hand it is known in CMP according to Luttinger liquid theory of 1d metals that at large wavelengths the relevant degrees of freedom are *collective charge and spin oscillations* which are governed by the quantum sine-Gordon field theories. Our results are supported by the work of Zaanen *et al.* [12], who have used sine-Gordon theory for separation of spin, charge and string fluctuation.

CONCLUSIONS

By considering a set of scalar fields without [global theory] or with [local theory] which we take to define order-parameters of cuprates we are led naturally led to a phenomenological model of strings. These strings and their interactions with the relevant degrees of freedom of systems constitute an interesting direction to pursue. This approach nicely ties with and supports our quantum group conjecture [2], namely that 1d structures are relevant to the understanding of the physics of the cuprates and related materials. It must be pointed out that almost all works have used the argument that since quantum groups were restricted to 1d, it was a problem to use them. On the contrary we have argued that it is precisely the opposite, namely 1d systems or collective excitations are the ones relevant [2] to the physics of the cuprates and related materials.

ACKNOWLEDGMENTS

The Sher Alam's work is supported by the Japan Society for Technology [JST].

REFERENCES

1. Alam, S., "*A Conjecture for possible theory for the description of high temperature superconductivity and antiferromagnetism*" in *Proceedings of Quantum Phenomena in Advanced Materials at High Magnetic Fields*, 4th International Symposium on Advanced Physical Fields [APF-4]. KEK-TH-607, KEK Preprint 98-xxx, *cond-mat/9812060*.
2. Alam, S., "*Quantum Group based Modelling for the description of high temperature superconductivity and antiferromagnetism*", *Phys. Lett. A* **272**, 107-112 (2000). KEK-TH-613, KEK Preprint 98-xxx, *cond-mat/9903038*.
3. Alam, S., "*Theoretical modeling for quantum liquids from 1d to 2d dimensional crossover using quantum groups*" KEK-TH-619. KEK Preprint 99-xxx, *cond-mat/990345*.
4. Frahm, H., *et al.*, *Phys. Rev. Lett.* **81**, 2116 (1998).
5. Alam, S. *et al.*, "*The Choice of symmetry group for cuprates*", *cond-mat/0004269*.
6. Alam, S. *et al.*, "*The patching of critical points using quantum groups*", *cond-mat/0004350*.
7. Alam, S., "*Quantum group conjecture and Stripes*", *cond-mat/0005168*.
8. Timm, C., and Bennemann, K.H., *J. of Low Temp. Phys.* **117**, 205-209 (1999).
9. Cheng, T.-P., and Li, L.-F., *Gauge Theory of Elementary Particle Physics*, Clarendon Press, Oxford. 1982.
10. Alam, S., *et al.*, *Phys. Rev.* **D57**, 1577-1590 (1998).
11. Nielsen, H.B., and Olsen, P., *Nucl. Phys.* **B61**, 45-61 (1973).
12. Zaanen, J. *et al.*, "*Metallic stripes:separation of spin, charge and string fluctuation*", *cond-mat/9804300*.

The Local Distortions and Spin Glass Mechanism in High-T_c Cuprates

I. Kanazawa

Department of Physics, Tokyo Gakugei University,
Koganei-shi, Tokyo 184-8501, Japan

Abstract. It is proposed that the spin glass phase in a high T_c cuprate might correspond to the quasi-two dimensional chiral spin glass phase, which is introduced by the hole doping.

INTRODUCTION

Holes introduced into La$_{2-x}$Sr$_x$CuO$_4$ destroy the three dimensional long range antiferromagnetic order with $x \sim 0.02$. A so-called spin-glass phase appears between $x \sim 0.02$ and ~ 0.05. In this phase, the zero-field-cooled magnetic susceptibility shows a cusp at around the spin freezing temperature. Recently the relationship between the spin-glass phase in La$_{2-x}$Sr$_x$CuO$_4$ and the stripe order has been discussed [1]. Machida and co-workers [2-4] have introduced the stripe order as the charged soliton's picture by using mean-field treatment for a two-dimensional Hubbard model. Thus it is important to study the spin-glass phase from the standpoint of the soliton's view.

In the quasi-two dimensional high-Tc cuprates, the doped carrier near the antiferromagnetic insulation phase is regarded as a kind of quasi-particle "carrieron", which is composed of the hole and the cloud of the massive gauge fields A_μ^1, A_μ^2, and the massless gauge field A_μ^3 around the hole [5-10]. In other words, we can think that a "carrieron" is a complex particle composed of the hedgehog-like (monopole-like) soliton and the hole trapped into the mobile soliton. In this study, we will discuss the spin-glass behavior in high-Tc cuprates from the model of the spin-disorderd hedgehog-like soliton, based on the gauge-invariant effective Lagrangian (the gauge-invariant carrieron model).

A MODEL SYSTEM

Taking into account that the symmetry in the undoped (2+1) dimensional quantum antiferromagnet is invariant under local SU(2) symmetry [11]. Then it is suspected that SU(2) gauge field A_μ^a are spontaneously broken through the Anderson-Higgs mechanism in a way similar to the breaking of the antiferromagnetic symmetry around the hole. We set the symmetry breaking $= \langle 0, 0, \mu(k_F) \rangle$ of the Bose field ϕ_a. The value, $\mu(k_F)$, of the symmetry breaking depends strongly on an angle of Fermi momentum, (k_F), on the Fermi surface. The value, $\mu(k_F)$, is higher around the hot

CP554, *Physics in Local Lattice Distortions*, edited by H. Oyanagi and A. Bianconi
© 2001 American Institute of Physics 1-56396-984-X/01/$18.00

spot. Thus, at small doping of holes, we can obtain the effective Lagrangian density [5-10],

$$L_{eff} = \frac{1}{2}(\partial_i N_c^j - g_1 \varepsilon_{abc} \varepsilon_{jik} A_i^b N_a^k)^2$$

$$+ \psi^+(i\partial_0 - g_2 T_a A_0^a)\psi$$

$$- \frac{1}{2m}\psi^+(i\nabla - g_2 T_a A_{(\mu \neq 0)}^a)^2 \psi$$

$$- \frac{1}{4}(\partial_v A_\mu^a - \partial_\mu A_v^a + g_3 \varepsilon_{abc} A_\mu^b A_v^c)^2$$

$$+ \frac{1}{2}(\partial_\mu \phi_a - g_4 \varepsilon_{abc} A_\mu^b \phi_c)^2 \tag{1}$$

$$+ \frac{1}{2}m_1^2[(A_\mu^1)^2 + (A_\mu^2)^2] + m_1[A_\mu^1 \partial_\mu \phi_2 - A_\mu^2 \partial_\mu \phi_1]$$

$$+ g_4 m_1 \{\phi_3[(A_\mu^1)^2 + (A_\mu^2)^2] - A_\mu^3[\phi_1 A_\mu^1 + \phi_2 A_\mu^2]\}$$

$$- \frac{m_2^2}{2}(\phi_3)^2 - \frac{m_2^2 g_4}{2m_1}\phi_3(\phi_a)^2 - \frac{m_2^2 g_4^2}{8m_1^2}(\phi_a \phi_a)^2,$$

where N_a^i is the spin parameter, ψ is Fermi field of the hole, $m_1 = \mu \cdot g_4$, $m_2 = 2\sqrt{2}\,\lambda \cdot \mu$. The effective Lagrangian describes two massive vector field A_μ^1 and A_μ^2, and one massless U(1) gauge field A_μ^3.

The generation function $Z[J]$ for Green functions is shown as follows,

$$Z[J] = \int DADBDNDCD\overline{C}D\psi^+ D\psi D\phi$$

$$\cdot \exp i \int d^4x (L_{eff} + L_{GF+FP} + J \cdot \Phi). \tag{2}$$

$$L_{GF+FP} = B^a \partial^\mu A_\mu^a + \frac{1}{2}\alpha B^a B^a + i\overline{C}^a \partial^\mu D_\mu C^a, \tag{3}$$

where B^a and C^a are Nakanishi-Lautrup (NL) fields and Faddeev-Popov fictitious fields, respectively.

$$J \cdot \Phi \equiv J^{a\mu} A_\mu^a + J_B^a B^a + J_N^a \cdot N_a + \overline{J}_C^a \cdot C^a + J_{\overline{C}}^a \overline{C}^a + \overline{\eta}\psi + \eta\psi^+ + J_\phi^a \phi_a \tag{4}$$

BRS-quartet [12,13] in the present theoretical system are $(\phi_1, B^1, C^1, \overline{C}^1)$, $(\phi_2, B^2, C^2, \overline{C}^2)$, and $(A_{L,\mu}^3, B^3, C^3, \overline{C}^3)$. Where $A_{L,\mu}^3$ is the longitudinal component of A_μ^3.

Because masses of A_μ^1 and A_μ^2 are formed through the Higgs mechanism by introducing the hole, the fields A_μ^1 and A_μ^2 exist around the hole within the length of ~ $1/m_1 \equiv R_c$.

The quantized gauge fields A_μ^a are expressed in

$$A_\mu^a = (2\pi)^{-3/2} \int [a^a(p) e_\mu^a(p) \exp(ipr) + a^{a+}(p) e_\mu^a(p) \exp(-ipr)] d^3 p \big/ \sqrt{2\omega_p^a},$$

where $\omega_p^a = \sqrt{p^2 + m_1^2}, (a = 1,2)$ and $\omega_p^a = \sqrt{p^2}, (a = 3), a^{a+}(p)$ and $a^a(p)$ are the creation and annihilation operators of the gauge particle A_μ^a with momentum p, respectively, and $e_\mu^a(p)$ are the polarization vectors. The masses, m_1, of the gauge fields A_μ^1 and A_μ^2 induced by the hole depend strongly on an angle of Fermi momentum of the hole on the Fermi surface. The value, m_1, is higher in the case of the hole around the hot spot. It is thought that the interaction between two holes for the Cooper pair formation is derived from the exchange of the fields A_μ^a [5-7]. From the first term in Eq. (1), the spin N_a^i is much distorted from the antiferromagnet state within the length of $\sim R_c$ around the hole. Furthermore, the spin order will be distorted in the long-range by the massless U(1) gauge field A_μ^3 [8]. When $N(i)$, $N(j)$ and $N(k)$ are spins on triangle sites i, j and k within $\sim \pi R_c^2(i)$ around the hole at the site $\tilde{i}$, the chiral spin liquid parameter $q_{\tilde{i}}$ is introduced as follows [14],

$$q_{\tilde{i}} \equiv \sum_{(ijk) \in \pi R_c^2(\tilde{i})} N(i) \cdot (N(j) \times N(k)), \quad (ijk)$$ are local triplet sites of spins. Because the hole state trapped into the hedgehog-like soliton is thought as that the instanton-like fluctuation [15] is stabilized by the hole [16], we assume that $q_{\tilde{i}}$ is approximately proportional to the topological number of the instanton $\sim c/4\pi \int dx dy (N \cdot \partial_x N \times \partial_y N) \sim c/2\pi \int dS_{\mu\nu}(\partial_\mu A_\nu^3 - \partial_\nu A_\mu^3)$ [15]. When the carrieron is located at the position $r_{\tilde{i}}$ and $|r - r_{\tilde{i}}| \gg 1/|m_1| \sim R_c$ is assumed, the gauge field $A_\mu^3(r, r_{\tilde{i}})$ at the position r is represented as $A_\mu^3(r, r_{\tilde{i}}) \propto q_{\tilde{i}}/|r - r_{\tilde{i}}|$. Thus, we can introduce the interaction between the chiral spin-disordered hedgehog-like solitons at positions $r_{\tilde{i}}$ and $r_{\tilde{j}}$ as $V_{\tilde{i}\tilde{j}} \propto (q_{\tilde{i}} \cdot q_{\tilde{j}})/|r_{\tilde{i}} - r_{\tilde{j}}|$.

For the mean-field approximate, it is assumed that $V_{\tilde{i}\tilde{j}}$ describes N hedgehog-like solitons's interaction, which mediated by the massless U(1) A_μ^3 fields in pairs $(\tilde{i}, \tilde{j})$ via infinite-range Gaussian-random interaction for simplifying discussion [17],

$$P(V_{\tilde{i}\tilde{j}}) = \frac{1}{(2\pi \langle V_{\tilde{i}\tilde{j}}^2 \rangle)^{1/2}} \exp\left(\frac{-V_{\tilde{i}\tilde{j}}^2}{2\langle V_{\tilde{i}\tilde{j}}^2 \rangle} \right) \tag{5}$$

Now we can get the spin-glass like behavior from the analogy of the Sherrington-Kirkpatrik (SK) formula [18] by using the replica method [19]. Thus we can estimate the free energy βf for one chiral spin-disordered hedgehog-like soliton in the method of steepest decent and replica symmetry condition as follows,

$$\beta f = -\frac{1}{4}(\beta \tilde{V})^2 \cdot (1 - G)^2 - \frac{1}{\sqrt{2\pi}} \int e^{-(1/2)z^2} \cdot \log z \cosh \beta \tilde{V} \sqrt{G} z dz, \tag{6}$$

where $G \equiv \langle q_{\tilde{i}}^\alpha q_{\tilde{i}}^\beta \rangle = Q^{\alpha\beta}$, $\beta = 1/T$ and $\tilde{V} = \sqrt{N \langle V_{\tilde{i}\tilde{j}}^2 \rangle}$, α and β are replica indices, $\langle q_{\tilde{i}}^\alpha q_{\tilde{i}}^\beta \rangle$ represents the canonical average with weight of $\exp(-\beta \tilde{V} \sum_{(\alpha,\beta)} q_{\tilde{i}}^\alpha q_{\tilde{i}}^\alpha Q^{\alpha\beta})$. Then we can introduce G self-consistently from $\partial f/\partial G = 0$ as follows,

$$G = \frac{1}{\sqrt{2\pi}} \int_{-\infty}^{\infty} e^{-(1/2)z^2} \tanh^2(\beta \widetilde{V} \sqrt{G} z) dz. \tag{7}$$

In the temperature region below $T_c = \widetilde{V}/k_B$, we can obtain the phase of $G \equiv \langle q_{\bar{i}}^{\alpha} q_{\bar{i}}^{\beta} \rangle = Q^{\alpha\beta} \neq 0$ and $\langle q_{\bar{i}}^{\alpha} \rangle = 0$. It is thought that this phase corresponds to the quasi-two dimensional chiral spin glass phase. More exactly we must define the order parameter $\overline{G} \equiv \int_0^1 dx Q(x)$, where $Q(x)$ is Parisi order parameter and is derived from $Q^{\alpha\beta}$, in Parisi's theoretical formula [20]. In the temperature region below $\widetilde{V}/k_B$, we got the phase of the order parameter $\overline{G} \neq 0$ and $\langle q_{\bar{i}}^{\alpha} \rangle = 0$, which corresponds to the chiral spin glass phase.

CONCLUSION

We have discussed the spin glass mechanism in a high Tc cuprate (LaSrCuO) by the gauge-invariant carrieron (GIC) model, which is based on the the gauge-invariant effective Lagrangian density in the strongly-correlated electron system. It has been proposed that the spin glass phase in a high Tc cuprate might correspond to the quasi-two dimensional chiral spin glass phase.

REFERENCES

1. Yamada, K., Lee, C. H., Kurahashi, K., Wada, J., Kimura, H., Ueki, S., Endou, Y., Hosoya, S., Shirane, G., Birgeneau, R. J., Greven, M., Kastner, M. A., and Kim, Y., *Phys. Rev.* **B57**, 6165 (1998).
2. Machida, K., *Physica C* **158**, 192 (1989).
3. Kato, M., Machida, K., Nakanishi, H., and Fujita, M., *J. Phys. Soc. Jpn.* **59**, 1047 (1990).
4. Machida, K., and Ichioka, M., *J. Phys. Soc. Jpn.* **68**, 2168 (1999).
5. Kanazawa, I., *The Physics and Chemistry of Oxide Superconductors,* Springer, Berlin, 1992, pp.481.
6. Kanazawa, I., *Physica C* **185-189**, 1703 (1991).
7. Kanazawa, I., *Synth. Met.* **71**, 1641 (1995).
8. Kanazawa, I., *Superlattice Microst.* **21**, 279 (1997).
9. Kanazawa, I., *Advances in Superconductivity XII*, Springer, Berlin, 2000, pp.275.
10. Kanazawa, I., *Physica B.* **284-288**, 409 (2000).
11. Affleck, I., Zou, Z., Hsu, T., and Anderson, P. W., *Phys. Rev. B* **38**, 745 (1988).
12. Becchi, C., Rouet A., and Stora, R., *Comm. Math. Phys.* **42**, 127 (1975).
13. Kugo, T., and Ojima, I., *Prog. Theor. Phys Suppl.* **66**, 1 (1979).
14. Wen, X. G., Wilczek, F., and Zee, A., *Phys. Rev.B* **39**, 11413 (1989).
15. Haldane, F. D. M., *Phys. Rev. Lett.* **61**, 1029 (1988).
16. Wiegmann, P., *Prog. Theor. Phys. Suppl.* **107**, 243 (1992).
17. Kanazawa, I., *Prog. Theor. Phys. Suppl* **126**, 393 (1997).
18. Sherrington, D., and Kirkpartic, S., *Phys. Rev. Lett.* **35**, 1792 (1975).
19. Edwards, S. F., and Anderson, P. W., *J. of Phys. F* **5**, 965 (1975).
20. Parisi, G. *Phys. Lett. A* **73**, 203 (1979).

Impurity Scattering in Metallic Carbon Nanotubes with Superconducting Pair Potentials

K. Harigaya

Electrotechnical Laboratory, Umezono 1-1-4, Tsukuba 305-8568, Japan

Abstract. Effects of the superconducting pair potential on the impurity scattering processes in metallic carbon nanotubes are studied theoretically. The backward scattering of electrons vanishes in the normal state. In the presence of the superconducting pair correlations, the backward scatterings of electron- and hole-like quasiparticles vanish, too. The impurity gives rise to backward scatterings of holes for incident electrons, and it also induces backward scatterings of electrons for incident holes. Negative and positive currents induced by such the scatterings between electrons and holes cancel each other. Therefore, the nonmagnetic impurity does not hinder the supercurrent in the regions where the superconducting proximity effects occur, and the carbon nanotube is a good conductor for Cooper pairs. Relations with experiments are discussed.

INTRODUCTION

Recent investigations [1,2] show that the superconducting proximity effect occurs when the carbon nanotubes contact with conventional superconducting metals and wires. The superconducting energy gap appears in the tunneling density of states below the critical temperature T_c. On the other hand, the recent theories discuss the nature of the exceptionally ballistic conduction [3] and the absence of backward scattering [4] in metallic carbon nanotubes with impurity potentials at the normal states.

In this communication, we study the effects of the superconducting pair potential on the impurity scattering processes in metallic carbon nanotubes, using the continuum $k \cdot p$ model for the electronic states. We find the absence of backward scatterings of electron- and hole-like quasiparticles in the presence of superconducting proximity effects, and the nonmagnetic impurity *does not hinder the supercurrent* in the regions where the superconducting proximity effects occur. Therefore, the carbon nanotube is a good conductor for Cooper pairs as well as in the normal state. This finding is interesting in view of the recent experimental progress of the superconducting proximity effects of carbon nanotubes [1,2]. We note that the details in this short communication have been reported elsewhere [5].

CP554, *Physics in Local Lattice Distortions*, edited by H. Oyanagi and A. Bianconi
© 2001 American Institute of Physics 1-56396-984-X/01/$18.00

IMPURITY SCATTERING IN NORMAL NANOTUBES

We will study the metallic carbon nanotubes with the superconducting pair potential. The model is as follows:

$$H = H_{tube} + H_{pair},$$ (1)

H_{tube} is the electronic states of the carbon nanotubes, and the model based on the $k \cdot p$ approximation [4] represents electronic systems on the continuum medium. The second term H_{pair} is the pair potential term owing to the proximity effect. The hamiltonian of a graphite plane by the $k \cdot p$ approximation [4] in the secondly quantized representation has the following form:

$$H_{tube} = \sum_{k,\sigma} \Psi_{k,\sigma}^{\dagger} E_k \Psi_{k,\sigma},$$ (2)

where E_k is an energy matrix:

$$E_k = \begin{pmatrix} 0 & \gamma(k_x - ik_y) & 0 & 0 \\ \gamma(k_x + ik_y) & 0 & 0 & 0 \\ 0 & 0 & 0 & \gamma(k_x + ik_y) \\ 0 & 0 & \gamma(k_x - ik_y) & 0 \end{pmatrix},$$ (3)

$k = (k_x, k_y)$, and $\Psi_{k,\sigma}$ is an annihilation operator with four components: $\Psi_{k,\sigma}^{\dagger} = (\psi_{k,\sigma}^{(1)\dagger}, \psi_{k,\sigma}^{(2)\dagger}, \psi_{k,\sigma}^{(3)\dagger}, \psi_{k,\sigma}^{(4)\dagger})$. Here, the fist and second elements indicate an electron at the A and B sublattice points around the Fermi point K of the graphite, respectively. The third and fourth elements are an electron at the A and B sublattices around the Fermi point K'. The quantity γ is defined as $\gamma \equiv (\sqrt{3}/2)a\gamma_0$, where a is the bond length of the graphite plane and $\gamma_0 (\simeq 2.7 \, eV)$ is the resonance integral between neighboring carbon atoms. When the above matrix is diagonalized, we obtain the dispersion relation, $E_{\pm} = \pm\gamma\sqrt{k_{v\phi}^2(n) + k_y^2}$ where k_y is parallel with the axis of the nanotube, $\kappa_{v\phi}(n) = (2\pi/L)(n + \phi - v/3)$, L is the circumference length of the nanotube, $n (= 0, \pm1, \pm2, ...)$ is the index of bands, ϕ is the magnetic flux in units of the flux quantum, and $v (= 0, 1, \text{ or } 2)$ specifies the boundary condition in the y-direction. The metallic and semiconducting nanotubes are characterized by $v = 0$ and $v = 1$ (or 2), respectively. Hereafter, we consider the case $\phi = 0$ and the metallic nanotubes $v = 0$.

The second term in Eq. (1) is the pair potential:

$$H_{pair} = \Delta \sum_{k} (\psi_{k,\uparrow}^{(1)\dagger}\psi_{-k,\downarrow}^{(1)\dagger} + \psi_{k,\uparrow}^{(2)\dagger}\psi_{-k,\downarrow}^{(2)\dagger} + \psi_{k,\uparrow}^{(3)\dagger}\psi_{-k,\downarrow}^{(3)\dagger} + \psi_{k,\uparrow}^{(4)\dagger}\psi_{-k,\downarrow}^{(4)\dagger} + h.c.)$$ (4)

where Δ is the strength of the superconducting pair correlation of an s-wave pairing. We assume that the spatial extent of the regions where the proximity effect occurs is as long as the superconducting coherence length.

Now, we consider the impurity scattering in the normal metallic nanotubes. We take into account of the single impurity potential located at the point r_0:

$$H_{imp} = I \sum_{k,p,\sigma} e^{i(k-p)\cdot r_0} \Psi_{k,\sigma}^{\dagger} \Psi_{p,\sigma} , \qquad (5)$$

where I is the impurity strength.

The scattering t-matrix at the K point is

$$t_K = I[1 - \frac{2}{N_s} \sum_k G_K(k,\omega)]^{-1} , \qquad (6)$$

where G_K is a propagator of a π-electron around the Fermi point K. The discussion about the t-matrix at the K' point is qualitatively the same, so we only look at the t-matrix at the K point. The sum for $k = (0, k)$, which takes account of the band index $n = 0$ only, is replaced with an integral:

$$\frac{2}{N_s} \sum_k G_K(k,\omega) = \rho \int d\varepsilon \frac{1}{\omega^2 - \varepsilon^2} \begin{pmatrix} \omega & -i\varepsilon \\ i\varepsilon & \omega \end{pmatrix} \simeq -\rho\pi i \operatorname{sgn}\omega \begin{pmatrix} 1 & 0 \\ 0 & 1 \end{pmatrix}, \qquad (7)$$

where $\rho = a/2\pi L\gamma_0$ is the density of states at the Fermi energy. Therefore, we obtain

$$t_K = \frac{1}{1 + I\rho\pi i \operatorname{sgn}\omega} \begin{pmatrix} 1 & 0 \\ 0 & 1 \end{pmatrix}. \qquad (8)$$

The transformation into the energy-diagonal representation where the branches with $E = \pm\gamma|k|$ are diagonal has the same form of t_K.

The scattering matrix t_K in the energy-diagonal representation is diagonal, and the off-diagonal matrix elements vanish. This means that only the scattering processes from k to k and from $-k$ to $-k$ are effective. The scatterings from k to $-k$ and from $-k$ to k are cancelled. Such the absence of the backward scattering has been discussed recently [4].

EFFECTS OF SUPERCONDUCTING PAIR POTENTIAL

We consider the single impurity scattering when the superconducting pair potential is present. In the Nambu representation, the scattering t-matrix at the K point is

$$\tilde{t}_K = [1 - I\frac{2}{N_s}\sum_k \tilde{G}_K(k,\omega)\tilde{I}]^{-1}, \tag{9}$$

where $\tilde{G}_K$ is the Nambu representation of G_K and

$$\tilde{I} = I\begin{pmatrix} 1 & 0 & 0 & 0 \\ 0 & 1 & 0 & 0 \\ 0 & 0 & -1 & 0 \\ 0 & 0 & 0 & -1 \end{pmatrix} \tag{10}$$

The sign of the scattering potential for holes is reversed from that for electrons, so the minus sign appears at the third and fourth diagonal matrix elements.

The sum over k is performed as in the previous section, and we obtain the scattering t-matrix (with the same form in the energy-diagonal representation):

$$\tilde{t}_K = \frac{I}{1+(I\rho\pi)^2}\begin{pmatrix} 1+\alpha\omega & 0 & -\alpha\Delta & 0 \\ 0 & 1+\alpha\omega & 0 & -\alpha\Delta \\ -\alpha\Delta & 0 & -1+\alpha\omega & 0 \\ 0 & -\alpha\Delta & 0 & -1+\alpha\omega \end{pmatrix} \tag{11}$$

where $\alpha = I\rho\pi i/\sqrt{\omega^2 - \Delta^2}$.

Hence, we find that the off-diagonal matrix elements become zero in the diagonal 2×2 submatrix. This implies that the backward scatterings of electron-line and hole-like quasiparticles vanish in the presence of the proximity effects, too. Off-diagonal 2×2 submatrix has the diagonal matrix elements whose magnitudes are proportional to Δ. The finite correlation gives rise to backward scatterings of the hole of the wavenumber $-k$ when the electron with k is incident. The back scatterings of the electrons with the wavenumber $-k$ occur for the incident holes with k, too. Negative and positive currents induced by such the two scattering processes cancel each other. Therefore, the nonmagnetic impurity *does not hinder the supercurrent* in the regions where the superconducting proximity effects occur. This effect is interesting in view of the recent experimental progress of the superconducting proximity effects [1,2].

SUMMARY

We have investigated the effects of the superconducting pair potential on the impurity scattering processes in metallic carbon nanotubes. The backward scattering of electrons vanishes in the normal state. In the presence of the superconducting pair correlations, the backward scatterings of electron- and hole-like quasiparticles vanish, too. The impurity gives rise to backward scatterings of holes for incident electrons, and it also induces backward scatterings of electrons for incident holes. Negative and positive currents induced by such the scatterings between electrons and holes cancel each other. Therefore, the carbon nanotube is a good conductor for the Cooper pairs coming from the proximity effects.

REFERENCES

1. Kasumov, A. Y. *et al*, *Science* **284**, 1508 (1999).
2. Morpurgo, A. F., Kong, J., Marcus, C. M., and Dai, H., *Science* **286**, 263 (1999).
3. White, C. T., and Todorov, T. N., *Nature* **393**, 240 (1998).
4. Ando, T., and Nakanishi, T., *J. Phys. Soc. Jpn.* **67**, 1704 (1998).
5. Harigaya, K., *J. Phys. Soc. Jpn.* **69**, Letters 1958 (2000).

Quantum Order and Disorder as A Signature of Competing Interactions: A Brief Report on Recent Theoretical Developments

D. Feinberg and F. V. Kusmartsev

The increasing importance of local lattice distortions in the physics of oxide and related materials calls for further theoretical developments. Several aspects were discussed in this Conference, and here we try to pinpoint a few basic problems which call for urgent developments. An essential one is that structures appear at the micro- or nanoscale which locally break the translational symmetry of the lattice, or lower it if they can form a periodic pattern. Evidence for such a patterning behavior stands from direct or indirect observation of stripes in cuprates and manganites, systems which bear in common the presence of many competing (Coulomb, magnetic, electron-lattice) interactions. Anomalous behaviors of local lattices in cuprates and manganites have been recognized by many researchers. But other degrees of freedom such as charge and spin are also involved in such modulated structures. In particular, charge segregation may arise as instability of a metallic phase due to competing interactions, but it calls for the consideration of long-range Coulomb interactions, often overlooked in theoretical models.

The role of stripes in high temperature superconductivity, in particular the distinction between "diagonal" and "horizontal-vertical" stripes was pointed out [1], although, the general tendency of static stripes to conspire against superconductivity has been recognized which seems to be true so far only for magnetic stripes. The situation for dynamic stripes is still more complex, which is a very busy area of research. Even the nature of the dynamics (charge, lattice, spin) is not clear yet. Egami et al. have reported an appearance of charged inhomogeneities which may be ascribed to strings or droplets associated with local charge segregation [1]. Their studies of LO phonons in cuprates by inelastic neutron scattering indicate an appearance of two different modes with doping providing a clear evidence of charge inhomogeneity.

Local lattice distortions are a signature of such inhomogeneities, and X-ray absorption spectroscopy (XAS) among other techniques provides a privileged tool for studying them. XAS studies of perovskite oxides where clearly demonstrated that a charge lattice stripe is stabilized by LTT-like tilting of the octahedra MO_6 causing Q_2-like local distortion below T^* [2]. Recently, even an anomalous local lattice fluctuations prior to stripe ordering above T^* has been observed [3], which indicates the importance of electron-lattice interaction.

CP554, *Physics in Local Lattice Distortions*, edited by H. Oyanagi and A. Bianconi
© 2001 American Institute of Physics 1-56396-984-X/01/$18.00

Compared to a number of research claiming stripe ordering, properties of stripe have been theoretically studied only recently. Kusmartsev emphasized the prominent role of long-range interactions and the large variety of electron droplet or chain structures which can be stabilized, as "pieces" of stripes for instance [4]. The lattice is very important in making some of theses objects stable or metastable. Beyond ground state considerations, understanding their excitations is a fully open area.

Further, it has been shown that the strings may also coexist with itinerant current carriers [5]. The matter is that with doping the energy associated with the strings increases and at some doping becomes equal to the bottom of the conduction band. With the next doping most of electrons (holes) are generated in a free itinerant state. These new generated free current carriers coexist with the localized electron clusters-electron strings. In other words this coexistence means that even in a metallic type phase with large Fermi surface there may arise local lattice distortions. Such type of inhomogeneous lattice distortions has been reported by many papers in this book. Most important of these reports was an observation by Bianconi and co-authors of the strings and stripes using a synchrotron radiation. Their X-ray diffraction studies indicate on the coexistence of stripes and itinerant current carries in complete consistence with theoretical predictions [2].

Recently, Bishop concluded on stripes and other microstructures by putting the questions on three stages: i) Where do stripes come from? How the interactions compete together, and what is the role of the lattice structure. ii) What are their local consequences, in particular on the lattice (role of buckling of octahedra for instance in perovskites)? [6] Answers can provide clues for their observation. In particular one might see them only if one really looks for them on a "local and dynamical" structural level, by going beyond the usual analysis based on translational invariance. iii) What are their global consequences regarding properties such as phase transitions and collective excitations?

In conclusion we also would like to stress the following important points which have been discussed on the conference:

1) It is very natural to ask the question: What is a stripe? And what does it consist of ? At the present time there are different views on this phenomenon. In general it is a form of electronic phase separation (EPS) although a microscopic structure remains a subject of an intensive investigation. One point which is clear is that the stripe shape of EPS arises due to a competition of a long-range electron-electron and electron-phonon and antiferromagnetic interactions. The electron-phonon and antiferro-magnetic interactions produce effectively a form of a short-range attraction between electrons (holes). Such attraction may be generated not only by electron-phonon interaction but also by spin-spin and electron-electron exchange correlations. Thus the stripe structure may be easily produced by many different ways and, in fact, indicate on the novel type of physics. On the other hand, the long-range Coulomb repulsion between electrons gives rise to splitting of an individual stripe into strings-segments of the stripe having a finite length. Each string consists of a finite number of particles. More precisely, there are many different states which corresponds to strings of different length but associated with very close energies.

2) The last fact is very important and characterizes the system with these string states as a complex and glassy system. The complexity and glassiness arise due to the fact that the EPS may have many different forms or shapes and sizes of localized regions. The creation energy of these regions per particle differ each other by very small amount, less than 1%. Each of such states like in a glass are separated from the other one by very large barrier. This fact constitutes the creation of the new type of a glassy state as a novel electron string glass. Such a glass consists of many different forms of electronic phase separation such as polaron, bipolaron, tri-polaron, four-polaron and other types of many particle-generalization of bi-polaron concept such as electron strings, electron blobs, droplets and electron (hole) stripes. The most important here is the possible coexistence of different forms of the EPS.

3) The glassy state created in such a manner may display many exotic properties. For example, inside of individual continuous stripes there may arise a Luttinger liquid, between the stripes there may arise an Anderson confinement with the creation of different type of condensed states such as superconductivity, charge density waves and so on. On the other hand, the individual continuous stripes will probably split into strings with not well defined length (this effect arises due to a long-range Coulomb interaction). The last fact indicates the glassy character and the dependence on any perturbations. Any new perturbation leads to a new structure.

4) The complex meso-scopic and nano-scopic structures are not unique and also are not well defined. In any new way there a new different type of structures may be created. Any small perturbation may give rise to a new structure consisting of stripes, strings, blobs and droplets. Of course with an increase of doping, at some stage, there will arise a collective effect leading to one or another type of a condensed state.

5) Depending on the doping these structures may be connected or disconnected. With doping there must always arise a percolation transition. On the other hand this percolation transition may be both a metal-insulator transition and may have nothing to do with conducting properties. The latter depends on the conducting character of the strings which may always serve as a basis for any microscopic EPS.

6) The formation of strings or any other forms of electronic phase separation is always related to a creation of lattice distortion inside or around of the region with localized electrons (as shown in Refs [4]). The anti-ferromagnetic correlations are playing an important role in the string and stripe formation. Without the anti-ferromagnetic correlations taken into account the string having an insulating character are in most cases correspond to a metastable state. However with the anti-ferromagnetic correlation taken into account such glassy strings may correspond to a glassy ground state.

7) A dynamics of strings is a very important aspect. The electrons self-trapped into the string do create a multi-particle polaron state – a multi-electron molecule. These electron molecules in a comparison with atoms have not very large masses and,

therefore, they are mobile. Moreover, the strings similar to normal molecules have normal vibronic modes.

8) On the other hand in a system consisting of parallel stripes there may arise "stripe" plasmons. Probably, such stripe fluctuations or the high frequency string vibrons discussed above may be important for the creation of the high-temperature superconducting state.

9) The most important aspects which lead to the theoretical prediction of the string formation was a consideration of discrete equations both for the electron and the lattice motion. The discrete nonlinear physics is much more reach than a continuum physics. Recently, there the similar phenomenon of the breathers formation has been reported both theoretically and experimentally. The breather is by a definition a localized, one-site or one-bond long-living excitation.

10) The breathers may form an array or a pattern. The different patterns are equivalent. They arise with nearly equal probability and are having nearly the same energies. A very small dynamical perturbation may easily change the formation of one pattern into another form.

In general, we face now a new type of phenomena mainly based on the discreteness of the considered systems, i.e., both breathers and EPS which have many novel features and can not be neglected and must be exploited in a full measure. Answering these questions requires to develop new theoretical methods and concepts, as well as heavy numerical work.

REFERENCES

1. Egami,T., this issue; Egami, T., Ishihara, S. and Tachiki, M., *Science* **55** 3163 (1993); McQueeney, R.J., Petrov, Y., Egami, T., Yethiraj, M., Shirane, G. and Endoh, Y., *Phys. Rev. Lett* **18** 628 (1999).
2. Bianconi, A., Saini, N.L., Lanzara, A., Missori, M., Oyanagi, H., Yamaguchi, H., Oka, K. and Ito, T., *Phys. Rev. Lett.* **76** 3412 (1996).
3. Oyanagi, H., this issue; Oyanagi, H. and Zegenhagen, J., *J. Superconductivity* **10** 415 (1997).
4. Kusmartsev, F.V., this issue also in *J. de Physique IV* **9**, Pr10-321,(1999); *Phys.Rev.Lett.* **84**, 530, 5026 (2000).
5. Kusmartsev, F.V., Di Castro, G., and Bianconi, A., *Physics Letters* **A275** 118 (2000).
6. Bishop, A.R., this issue; Eroles, J., Otiz, G., Balatsky, A.V. and Bishop, A.R., *Europhys. Lett.* **50**, 540 (2000).

2. PHASE TRANSITIONS

Local Lattice Distortions and the Nature of Superconductivity in Ba(K)BiO$_3$-BaPb(Bi)O$_3$

A. P. Menushenkov, K. V. Klementev, A. V. Kuznetsov
and M. Yu. Kagan[*]

Moscow State Engineering Physics Institute, 115409 Moscow, Russia
[*]*P.L. Kapitza Institute for Physical Problems, Kosygin Str. 117334 Moscow, Russia*

Abstract. Based on the temperature dependent XAFS- studies of bismuthates BaPb$_{1-x}$Bi$_x$O$_3$ (BPBO) and Ba$_{1-x}$K$_x$BiO$_3$ (BKBO) the model of the relationship of the local crystal and the local electronic structures have been proposed. We found that it is the local structure distortions that are mostly responsible for the main macroscopic properties of these compounds. The proposed model explains the long list of contradictions observed in transport measurements, inelastic neutron and electron scattering, UPS, XPS, EXAFS and Raman spectroscopy of the BPBO-BKBO systems. The likeness of the local peculiarities of BiO$_6$ and CuO$_n$ complexes allows us to hope that similar approach can be applied to the copper-based superconductors.

INTRODUCTION

The local structure of the superconducting alloys Ba$_{1-x}$K$_x$BiO$_3$ with $x > 0.37$ dramatically differs from the average one. The integral methods (x- ray and neutron scattering) showed a simple cubic structure for these compounds [1] while the Raman scattering pointed out the local deviations from a simple cubic symmetry [2], EXAFS studies showed the existence of the local oxygen octahedra tilting up to 4°–5° [3], XPS spectra found the strong splitting of the Bi 4f spectral lines which pointed on the existence of two different Bi valence states in the superconducting compositions [4].

All the above experimental facts lead us to the necessity to reconsider the theoretical and the first principle calculation conclusions based on the average structure analysis. Taking into account the similarities of the main local structure peculiarities for both the cuprates and bismuthates, one can suggest that a deeper physical understanding of bismuthates will be relevant to understanding other high-T$_c$ materials.

Our resent EXAFS - studies of the nearest oxygen environment of Bi in BKBO evidenced that oxygen ions vibrate in the double-well potential and their vibration are correlated with the charge carriers transfer due to the existence of the two different electronic structures of the octahedral complexes BiO$_6$ and BiL^{2}O$_6$ [5,6] where L^2 denotes the hole pair in the antibonding Bi$6s$O$2p_{\sigma^*}$ orbital of the octahedral complex. We showed that these two different types of complexes exist in the full range of the K-doping of BaBiO$_3$ including the superconducting compositions and these local lattice distortions are responsible for the main macroscopic properties of bismuthates.

CP554, *Physics in Local Lattice Distortions*, edited by H. Oyanagi and A. Bianconi
© 2001 American Institute of Physics 1-56396-984-X/01/$18.00

It is the purpose of this report to present the empirical model of the relationship between the local crystal and the local electronic structures which describes qualitatively an insulator-metal transition and the superconductive state in bismuthtes in the framework of the one common approach.

RELATIONSHIP BETWEEN THE LOCAL CRYSTAL AND LOCAL ELECTRONIC STRUCTURES

The model includes the following key positions:

1. The parent compound $BaBiO_3$ represents a system with the initially preformed local electron and hole pairs spatially and energetically localized inside the octahedral complexes BiO_6 and BiL^2O_6 [see Fig. 1(a)]. The BiL^2O_6 octahedra have stiff (quasi-molecular) Bi-O bonds and the smaller radius, and BiO_6 octahedra represent non-stable molecules with filled antibonding orbitals and the larger radius. Because the sum of the nearest octahedron radii exceeds the lattice parameter a, the octahedral system is tilted around [110] axis, producing a monoclinic distortion in $BaBiO_3$ as a combination of the static breathing and tilting distortions.

2. The conductivity of $BaBiO_3$ occurs only at the local pair transfer due to the dynamic exchange $BiL^2O_6 \leftrightarrow BiO_6$. The localization energy of a pair determines the transport activation gap E_a. The binding energy of a pair E_b becomes apparent as the optical gap.

3. The two types of carriers are present in the metallic phase of $Ba_{1-x}K_xBiO_3$ with $x > 0.37$: the itinerant electrons from the infinite BiL^2O_6 - $\cdots$ -BiL^2O_6 Fermi-cluster (fermions) and the delocalized local electron pairs from the BiO_6 complexes (bosons) [see the stripe structure picture in Fig. 1(b)]. The former produce the Fermi-liquid state and are responsible for the insulator-metal transition observed at $x \approx 0.37$. The latter provide the superconductivity at $T < T_c$ due to the free moving of the local electron pairs in the process of the dynamic exchange $BiL^2O_6 \leftrightarrow BiO_6$ (see Fig. 2). The both Fermi and Bose subsystems are spatially separated in these compounds since belong to the different clusters.

4. The pairing mechanism in the bismuthates is more probably of the electronic than of the phonon-mediated origin However the lattice plays an important role in superfluid providing the phase coherence of the local pair moving. It is the phase coherence destruction that limits the critical temperature in BKBO-BPBO systems.

The coexistence in bismuthates of the different types of the octahedra with the two Bi-O bond lengths and strengths reflects their different electronic occupation: the octahedron BiL^2O_6 contains 18 electrons and has one free level or a hole pair L^2 in the upper antibonding $Bi6sO2p_{\sigma^*}$ orbital, while in the octahedron BiO_6 with 20 electrons this antibonding orbital is filled [6].

The local electronic structure of $BaBiO_3$ combined with the realspace local crystal structure is presented in Fig. 3(a). In such a system there are no free Fermi carriers, and the conductivity occurs only due to the transfer of the carrier pairs with activation energy $2E_a$ [8] which is defined by the combined effect of the Coulomb intersite repulsing and the deformation potential between the neighboring octahedral

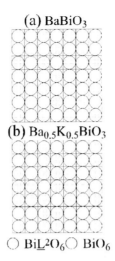

FIGURE 1. A sketch of the structure of BiO_2-plane a) charge-ordered state in $BaBiO_3$, b) stripe structure of the Bose-Fermi mixture in $Ba_{0.5}K_{0.5}BiO_3$.

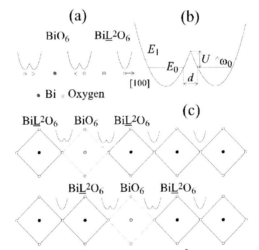

FIGURE 2. A sketch of the dynamic exchange $BiO_6 \leftrightarrow Bi\underline{L}^2O_6$ is shown in the BiO_2 plane of the octahedra. (a) A breathing mode of the vibrations along [100]-type direction of two neighboring octahedra with the different electronic structures. The BiO_6 octahedron transforms to the $Bi\underline{L}^2O_6$ one and vice versa due to the electron pair tunneling between the octahedra. An oxygen ion belonging to such octahedra oscillates in a double-well potential. An oxygen ion belonging to the equivalent neighboring $Bi\underline{L}^2O_6$ octahedra oscillates in a simple parabolic potential. (b) A double-well potential with the energy levels for the vibration of the oxygen ion. The following parameters describe the tunneling barrier between the wells in $Ba_{0.6}K_{0.4}BiO_3$ at low temperatures [6]: the tunneling frequency $\omega_0 \simeq 200K$, the height of the barrier $U \simeq 500K$, the width of the barrier $d \simeq 0.07$Å. (c) A motion of the local electron pair centered on the BiO_6 octahedron through the $Bi\underline{L}^2O_6 \cdots Bi\underline{L}^2O_6$ Fermi cluster.

complexes [6,12]. The dissociation of the pairs and the hopping of a single electron from one octahedron to another in similarity with Varma's [9] suggestion cost an energy: $E_b = 2E(BiL^1O_6) - [E(BiO_6) + E(BiL^2O_6)]$, and is observed experimentally as an optical conductivity peak at the photon energy $hv = 1.9$ eV [8] [see the excited band F' on Fig. 4(a)].

Experimentally, $BaBiO_3$ shows a semiconductor-like behavior with an energy gap $E_a = 0.24$ eV, which can be explained only as a two particle transport with the activation energy $2E_a$ due to the delocalization of the pairs, while the nature of the optical gap is just the pair binding energy.

The substitution of the each two K^+ for the two Ba^{2+} modifies the BiO_6 complex to the BiL^2O_6 one. As a result, the number of the small stiff BiL^2O_6 octahedra increases as $n_0(1+x)/2$, while the number of the large soft BiO_6 octahedra decreases as $n_0(1+x)/2$ (here $n_0 = 1/a^3$ is the number of unit cells) (see Fig. 1(b)). The spatial overlap of the BiL^2O_6 complexes appears at the finite doping levels, which, taking into account their

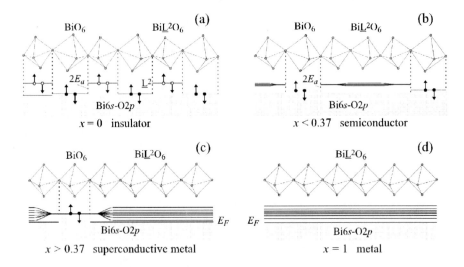

FIGURE 3. The scheme of the insulator-metal phase transition for the K-doping of $Ba_{1-x}K_xBiO_3$ in the framework of the relationship between the local crystal and the local electronic structures. The local crystal structure of the octahedral complexes (at the top) and the local electronic structure (at the bottom) are shown on the pictures (a)-(d). The occupied states of the $Bi6sO2p$ valence band are marked by gray. $2E_a$ is the activation gap. Black and white circles with arrows denote, correspondingly, the electrons and the holes with the opposite spin orientations. (a) A monoclinic phase of an insulator $BaBiO_3$. (b) An orthorhombic phase of a semiconducting BKBO at $0 < x < 0.37$. The splitting of free level L^2 at a spatial overlap of the BiL^2O_6 octahedra is sketched. (c) An undistorted cubic phase of a superconducting metal at $x > 0.37$. The formation of a Fermi-liquid state is shown arising due to the overlap of an unoccupied fermionic band F with an occupied $Bi6sO2p$ valence band when the percolation threshold is reached. (d) An undistorted cubic phase of a nonsuperconducting metal at $x = 1$. A Fermi liquid state with Fermi level E_F is shown.

small radii and the rigid bonds, contracts the lattice, despite on the practically the same ion radius of K^+ and Ba^{2+}.

The structural changes are accompanied by the essential changes in the local electronic structure [see Fig. 3(b), Fig. 4(b)]. In the doping range $0 < x < 0.37$ the BKBO compounds demonstrate a semiconducting-like conductivity changing from a simple activation type to the variable-range-hopping Mott's law [11]. Moreover the activation energy lowers with the doping down to $E_a \approx 0$ at $x \approx 0.37$.

At the doping level $x \approx 0.37$ (see Fig. 3(c) and Fig. 4(c)) the following cardinal changes take place: (i) Both the breathing and the rotational static lattice distortions of the parent $BaBiO_3$ transform to the dynamic ones. At the cluster borders, where all oxygen ions belongs to BiO_6 and BiL^2O_6 octahedra, the local breathing dynamic distortion is observed as a vibration in a double-well potential of $(1-x)$ part of the oxygen ions [5,6] but cannot be detected by the integral methods of the structural analysis such as an x-ray and a neutron diffraction. (ii) The infinite percolating Fermi

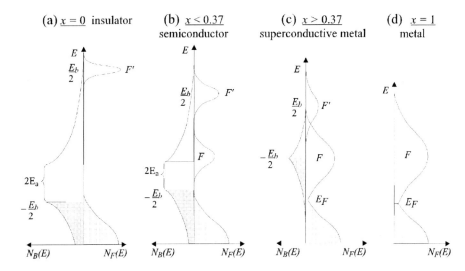

FIGURE 4. A sketch of the one-particle density of states for $Ba_{1-x}K_xBiO_3$. The contributions from the bosons $N_B(E)$ and the fermions $N_F(E)$ are depicted separately because the bosonic and fermionic states are spatially separated. The filled (dark gray) and the unoccupied (transparent) bosonic bands correspond, respectively, to the contributions from the electron and the hole pairs. The bands are separated by the activation gap $2E_a$ which is lowered with a doping level x. An empty fermionic band F', corresponding to the destruction of the pairs, is separated from an occupied bosonic band by a binding energy E_b. An empty fermionic band F is formed from an unoccupied bosonic band due to the splitting of the free level $\underline{L}^2$, which arises from a spatial overlap of the BiL^2O_6 octahedra. A filled fermionic band (gray) represents the $Bi6sO2p$ valence band. A band F' and bosonic bands grow downwards by the doping because of a decrease of the number of the electron pairs, while a band F grows due to the increase of the number of free levels. A Fermi liquid state is formed (c), (d) as a result of the overlap between the band F and the $Bi6sO2p$ valence band.

cluster (formed from the spatially overlapped BiL^2O_6 octahedra) appears, which leads to the overlap of an empty fermionic band with a filled one and produces a conduction band. Overcoming of the percolation threshold provides the insulator-metal phase transition and the formation of the Fermi-liquid state for $x > 0.37$. The valence electrons of the BiL^2O_6 complexes previously localized become itinerant. (iii) The pair localization energy disappears $E_a \approx 0$ so the local electron pairs (from the BiO_6 complexes) can freely move in the real space providing a bosonic contribution into the conductivity.

Thus, in the metallic phase the two types of carriers are present: the itinerant electrons from the BiL^2O_6 - $\cdots$ -BiL^2O_6 infinite Fermi-cluster (fermions) and the delocalized local electron pairs from the BiO_6 complexes (bosons).

Note, that the stripe structure in bismuthates [Fig. 1(b)] is dynamic due to the electron pair moving from one octahedron to a neighboring one. It observes as the double-well vibration potential of the oxygen ions at the border of the clusters [6] (see Fig. 2). The positions of the minima of the double-well potential in our model are the *average* positions. They are spaced widely at maximal octahedra tilting and spaced closely at minimal octahedra tilting. In the latter case the probability of interwell tunnelling is maximal. In this sense the tunnelling frequency ω_0 is bounded by a soft-rotation-mode frequency.

It should be stressed that the fermions and the bosons belong to the complexes with the different electronic structure, therefore *the Fermi and the Bose subsystems are spatially separated at any doping level* [12]. These subsystems are connected by the relations $2n_B + n_F = 2n_0$ and $2n_B / \hat{n}_F = (1-x)/(1+x)$. The high enough value of the binding energy, which in the superconductive compositions becomes apparent as a pseudo-gap $E_b \approx 0.5eV$, [13] is the guarantee against the pair destruction. The unpairing is possible only under the optical excitation to the band F' (see Fig. 4), and does not play any role in the charge transport.

At $x = 1$ all the $BiOd_6$ octahedra are transformed to the BiL^2O_6 ones. The Bose system disappears ($n_B = 0$) together with an excited fermionic band F'. As a result, $KBiO_3$ should behave as a simple Fermi-liquid metal without superconductive properties [see Fig. 3(d) and Fig. 4(d)].

Practically the same changes in the electronic structure arise upon doping $BaBiO_3$ with lead. The electronic structure of octahedral PbL^2O_6 complex is entirely equivalent to that of BiL^2O_6 complex. Unoccupied levels of overlapping octahedra in metallic $BaPbO_3$ form the conduction band, as when the overlapping of BiL^2O_6 in BKBO occurs. Upon doping $BaPbO_3$ with bismuth ($0 < x < 0.37$), the case is similar to that for BKBO ($x > 0.37$) and superconductive properties of BPBO in our model are connected with the dynamic exchange $Pb(Bi)L^2O_6 \leftrightarrow Bi(Pb)O_6$ [6].

SUPERCONDUCTIVITY IN BKBO-BPBO PEROVSKITES

An interplay between the Bose and the Fermi subsystems is illustrated in Fig. 2.

An oxygen ions belonging to the two neighboring octahedra BiO_6 and BiL^2O_6 vibrates in a double-well potential, and hence the tunneling of the electron pair between the neighboring octahedra occurs simultaneously with the ion tunneling through the potential barrier between the wells. So one can estimate the matrix element of the pair tunneling as $t_B \sim \omega_0 e^{-D}$ where ω_0 is the tunneling frequency, $D = (1/\hbar)\int_{x_0}^{x_1} |p|\,dx \simeq (d/\hbar)\sqrt{2MU}$ is a quasiclassical transparency of the barrier in the double-well potential, U and d are the barrier height and width, and M is the oxygen ion mass.

The tunneling of the pairs helps to establish a macroscopic long range order (a phase coherence) in the bosonic system. On the language of the spatially separated Fermi-Bose mixture, a local pair is transferred from one Bose cluster to a nearest one over the Fermi-cluster, which, depending on the doping level, consists of several octahedra. The pairs overcome the Fermi-cluster step by step and the tunneling frequency ω_0 is the same for each step. If one assumes that the steps are independent events, the matrix element of the pair tunneling through the Fermi-cluster can be estimated as $\hat{t}_B \sim \omega_0 e^{-\langle N \rangle / D}$ where an average number of steps $\langle N \rangle$ can be obtained from the ratio of the concentrations of $Bi\underline{L}^2 O_6$ and BiO_6 octahedra: $\langle N \rangle \simeq \{(1+x)/(1-x)\}^{1/3}$.

Of course, it is naturally to assume that the critical temperature of superconductivity is of the order of the temperature of the Bose-Einstein condensation $T_c \sim \hat{t}_B a^2 n_B^{2/3}$ in the bosonic system with a large effective mass $m_B \sim 1/\hat{t}_B a^2$. For $x = 0.4$ and the parameters of the double-well potential obtained in the Ref. [6] (see also Fig. 2) we estimated $T_c \sim 50$ K. This value is larger than the measured $T_c \simeq 30$ (Ref. [1]) in BKBO.

However this estimation does not take into account the problem of the phase coherence arising due to the interconnection between the vibrations of the oxygen ions and the transfer of the pairs. When the pair is transferred from one octahedron to another, the lattice has sufficient time to relax, forming each time a new configuration before the next tunneling event occurs. As a result the pair's "deformed" environment (the BiO_6 octahedron) may follow the tunneling processes without the retardation. If the local pair motion is slow compared to the frequencies of the optical phonons associated with the dynamic deformations of the octahedra, the so-called *anti-adiabatic limit* is fulfilled in our system [10].

The observed dispersion of the longitudinal phonon modes studied in $Ba_{1-x}K_xBiO_3$ by an inelastic neutron scattering [14] showed that the breathing-type vibrations with the wave vector $q_b = (\pi/2a,0,0)$ are energetically favorable since an energy of the longitudinal stretching phonons is the lowest at the Brillouin band edge. That is why a breathing mode of each octahedron along [100] axes should be coordinated with its neighbors to guarantee a resonant tunneling in the system. Hence a long-range correlation of the vibrations should occur at low temperatures when only the low-energy states are occupied. The bandwidth of the longitudinal stretching mode is of the order of 100 K, and thus a temperature $T \sim T_c$ is high enough to excite the non-breathing-type longitudinal stretching phonons with the wave vectors shorter than q_b. The thermal excitation of the phonons with such short wave-vectors leads to the destruction of the long-range correlation between the breathing-type vibrations, and hence play a destructive role for the long-range phase coherency of the local pair transfer. Thus from our viewpoint T_c is bounded above by a phase coherence destruction.

CONCLUSION

In conclusion we briefly summarize the main results.

Finally, we would like to emphasize that the presented model of the relationship between the local lattice and the local electronic structures allows us to describe qualitatively an insulator-metal phase transition and a superconductive state in BKBO in the framework of the one common approach. To some extent this model explains the contradictions observed experimentally by the UPS and XPS, [4] the EXAFS and XANES, [5,6,3] and the Raman [2,7] spectroscopy, as well as by an inelastic neutron scattering, [14] the transport and the optical measurements [8,11,13].

Summarizing the discussion above we point out that the two processes are important for the superconductivity in bismuthates. The vibrations of the oxygen ions in the double-well potential provides the mechanism for a transfer of the local pairs from the one Bose-cluster to the other. At the same time the pair motion is strongly affected by the stretching longitudinal vibrations of the oxygen ions in the octahedra which the pair is passing by. We suppose that the last process should be taken into account to estimate correctly the critical temperature.

It is worth to notice, that a similar dispersion of the longitudinal stretching phonons, which leads to the dominant role of the [100] breathing-type phonons at low temperatures, has been observed also for the high T_c cuprates [15]. Taking into account the recent experimental evidence by Müller et al. [16] of the coexistence of the small bosonic and fermionic charge carriers in $La_{2-x}Sr_xCuO_4$, we suppose also to apply our model to HTSC cuprates.

ACKNOWLEDGMENTS

The authors acknowledge LURE program committee for giving possibilities to provide EXAFS-experiments, R.Cortes for the help in experiments, Yu.Kagan for the fruitful discussion. This work was supported by Russian Foundation for Basic Research (Grant 99-02-17343) and Program "Superconductivity" (Grant 99010).

REFERENCES

1. Pei, S., Jorgensen, J.D., Dabrowski, B. et al., *Phys. Rev. B* **41**, 4126 (1990).
2. Anshukova, N.V., Golovashkin, A.I., Gorelik, V.S. et al., *J. Molecular Structure* **219**, 147 (1990).
3. Yacoby, Y., Heald, S.M., and Stern, E.A., *Solid State Commun.* **101**, 801 (1997).
4. Qvarford, M., Nazin, V.G., Zakharov, A.A. et.al., *Phys. Rev. B* **54**, 6700 (1996).
5. Menushenkov, A.P., Klementev, K.V., Konarev, P.V., and Meshkov, A.A., *Pis'ma Zh. Eksp. Theor. Fis.* **67**, 977 (1998); *Sov. Phys.-JETP Lett.* **67**, 1034 (1998).
6. Menushenkov, A.P., and Klementev, K.V., *J. Phys.: Condens. Matter* **12**, 3767 (2000).
7. Sugai, S., *Jpn. J. of Appl. Phys.* **26**, Suppl. 3, 1123 (1987).
8. Uchida ,S., Kitazawa, K., and Tanaka, S., *Phase Transition* **8**, 95 (1987).
9. Varma, C.M., *Phys. Rev. Lett.* **61**, 2713 (1988).
10. De Jongh, L.J., *Physica C* **152**, 171 (1998).

11. Hellman, E.S., Miller, B., Rosamilia, J.M. *et al.*, *Phys. Rev. B* **44**, 9719 (1991).

12. Kagan, M.Yu., Menushenkov, A.P., Klementev, K.V., and Kuznetsov, A.V., *cond-mat/0003504*.

13. Blanton, S.H., Collins, R.T., Kelleher, K.H. *et al.*, *Phys. Rev. B* **47**, 996 (1993).

14. Braden, M., Reichardt, W., Schmidbauer, W. *et al.*, *J. of Supercond.* **8**, 595 (1995).

15. Pintschovius, L. and Reichardt, W., in *Physical Properties of High Temperature Superconductors IV*, edited by Ginsberg, D.M., World Scientific, 1994, pp.344.

16. Müller, K.A., Zhao, G., Conder, K., and Keller, H., *J. of Phys.: Cond. Matter* **10**, L291 (1998).

The Role of Local Distortions in Phase Transitions of Pure and Mixed Perovskites

Y. Yacoby and Y. Girshberg

Racah Institute of Physics, Hebrew University, Jerusalem, Israel 91904

Abstract. We present a phenomenological model of ferroelectricity that quantitatively accounts for both displacive like and order-disorder like properties of pure perovskite ferroelectrics. The model is based on the interplay between ions at off-center positions hoping among equivalent positions and a soft vibrational mode. We further show that a similar model with tunneling instead of hoping also accounts for the dielectric and phase transition properties of incipient ferroelectrics doped with off-center impurities. Finally, we present a microscopic model that quantitatively shows that the off-center displacements in oxygen perovskite ferroelectrics is a manifestation of ground state polarons.

INTRODUCTION

Perovskite crystals undergo various structural phase transitions as a function of temperature and pressure. The crystals involved can be either pure or mixed and the transitions can be ferroelectric, antiferroelectric or antiferrodistortive. Strictly speaking the ferroelectric transitions are first order as a result of the interaction of the order parameter with acoustic phonons [1,2]. However, away from the transition temperature their dielectric and other properties are actually governed by a second order phase transition with an extrapolated transition temperature T_c that is smaller than the first order transition temperature T_p. Here, we shall confine ourselves to the properties associated with the second order transition only.

A great deal of experimental work during the past 30 years has established that structural phase transitions in perovskites and other crystal families have both displacive like and order-disorder like properties. The existence of a soft mode in the paraelectric phase that softens as the temperature decreases and a large Curie-Weiss constant of the order of 10^5K support a displacive like picture [1]. On the other hand, the fact that the soft mode frequency does not extrapolate to zero at T_c [3-5] suggests that the displacive model is inadequate. On the basis of diffuse x-ray scattering experiments [6-8], Comes *et al.* were the first to suggest that the ions in the cubic phase reside at off-center positions and that the transition has at least an element of order-disorder. Raman experiments on both pure and mixed crystals [9-11] showed that the Raman selection rules break down in the high temperature phase further supporting the model of off-center ions in the cubic phase. A number of years latter the central peak was discovered. Its existence in mixed crystals was explained by Halperin and Varma [12] but it has never been satisfactorily explained in pure crystals.

CP554, *Physics in Local Lattice Distortions*, edited by H. Oyanagi and A. Bianconi
© 2001 American Institute of Physics 1-56396-984-X/01/$18.00

The experiments mentioned above have qualitatively established the existence of off-center ionic displacements in a variety of perovskites crystals. However, these experiments did not provide quantitative information on the size of the displacements and their temperature dependence. These have been recently measured using X-ray Absorption Fine Structure (XAFS) experiments [13-17]. These experiments have shown that the magnitude of the off-center displacements is at least 50% of their value at very low temperatures far below T_c and they persist hundreds of degrees above T_c. Furthermore, they are observed not only in crystals undergoing ferroelectric transitions but also in crystals undergoing antiferrodistortive [18] and antiferroelectric [19] transitions. In mixed crystals consisting of an incipient ferroelectric, $KTaO_3$ and a ferroelectric component $KNbO_3$ the XAFS results have shown that the Nb ions are displaced to off-center positions while the Ta ions are within the experimental accuracy on center. XAFS provides also the position probability distribution function. Two examples are shown in Fig. 1. One for a pure crystal, $PbTio_3$, the second for a mixed crystal $KTa_{1-x}Nb_xO_3$. If the fluctuations are quasi-harmonic, one expects the position probability peak to be at the center of the unit cell. Instead as seen in Fig. 1, the peaks are clearly displaced to off-center positions even at temperatures far above T_c. This shows that the off-center displacements are not simply quasi-harmonic fluctuations about the center but a manifestation of a local multi-well potential.

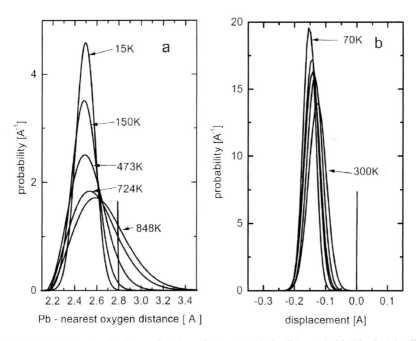

FIGURE 1. Probability distribution function of a) nearest $Pb-O_2$ distance in $PbTiO_3$, b) Nb off-center displacement in [111] type directions in $KTa_{0.94}Nb_{0.06}O_3$. Vertical lines represent peak position for a quasi-harmonic probability distribution function.

In this paper we present a model that accounts explicitly and quantitatively for the displacive like and order-disorder like properties of pure perovskite ferroelectrics. We further show that the same model with no additional parameters accounts for the phase transition and dielectric properties of incipient ferroelectrics with off-center ions. In particular we show that ionic tunneling among equivalent off-center positions plays a crucial role in determining the critical off-center ions concentration that leads to $T_c = 0$. Finally, we show that in a number of materials, the off-center displacements are manifestations of ground state polarons and show that the model leads to quantitative agreement with the experimental values of the off-center displacements, the transition temperature and the Curie-Weiss constant.

THE PHENOMENOLOGICAL MODEL [20]

The model assumptions for pure ferroelectric systems are as follows:
- In the cubic phase, the cations are locally displaced to off-center positions at a distance b from the central positions and are treated as spins. The ions hop among the symmetry equivalent position with a rate v. At high temperatures, the hopping is dominant and tunneling is neglected.

- The material has a soft mode with a frequency

$$\tilde{\omega}_{0t}^2 = A(T - T_0) \tag{1}$$

This soft mode is associated with the long wavelength vibrations of the displaced ions around their respective local minima.

- The soft mode and the spin-like off-center displaced ions interact with each other with an interaction constant f_q.
This interaction gives rise to an effective spin-spin interaction mediated via the soft mode.

- The direct spin-spin interaction constant is neglected in comparison to that mediated via the soft mode.

These assumptions lead to the following Hamiltonian:

$$H = \sum_q [\tilde{\omega}_{qt} \hat{b}_{qt}^+ \hat{b}_{qt} - \frac{\alpha_q}{\tilde{\omega}_{qt}^2} \sigma_q^z \sigma_{-q}^z] \tag{2}$$

$$\hat{b}_{qt} = b_{qt} + \sqrt{\alpha_0} \frac{\sigma_{-q}^z}{\tilde{\omega}_{qt}} \tag{3}$$

Where,

$$\alpha_q = 2 \mid f_g \mid^2; \tag{4}$$

and $\hat{b}$ is the renormalized phonon annihilation operator. The first term represents the soft mode branch phonons after renormalization and the second, the spin-spin interaction mediated by these phonons. Notice that this interaction constant is inversely proportional to the frequency $\tilde{\omega}_{qt}$ which at $q = 0$ is the soft mode frequency. Thus at $q = 0$, this interaction constant increases as the temperature decreases and becomes the dominant spin-spin interaction.

The physical situation described by this Hamiltonian is implicitly contained also in Hamiltonians with a local quartic anharmonic potential ('ϕ^4-model' see for example [21]). In those Hamiltonians the position of each atom is represented by a single spatial vector. In the present Hamiltonian the atomic positions are represented by a combination of a spin and a spatial vector measuring the position relative to the off-center potential energy minimum. The advantage of the present Hamiltonian in contrast to the ϕ^4-model is that it can be analytically solved in the mean field approximation and still maintain the displacive like and order-disorder like character. In contrast if the ϕ^4-Hamiltonian is solved in the mean field approximation it can have either displacive or order-disorder like character. Solutions beyond the mean field approximation can only be achieved numerically.

Using the density matrix formalism we solved the dynamical equations and obtained the dielectric function of the system as a function of temperature and frequency:

$$\varepsilon(T,\omega) = \frac{e^{*2}}{M^* V_c \varepsilon_0} \frac{(\iota\omega + v) + 2(\tilde{\omega}_{0t}^2 - \omega^2 + 2i\omega\Gamma_D)(vM^* b^2 / T) + 4vb \mid f_0 \mid \sqrt{M^*} / T}{(\iota\omega + v)(\tilde{\omega}_{0t}^2 - \omega^2 + 2i\omega\Gamma_D) - v\alpha_0 / T}. \tag{5}$$

Here e^* and M^* are the effective charge and mass of the critical vibrational mode respectively, V_c is the cell volume, $\tilde{\omega}_{0t}$ is the soft mode frequency at $q = 0$. The relaxation rate v is loosely related to the hopping rate among equivalent off-center positions and its temperature dependence is given by:

$$v = v_0 e^{-u/T}, \tag{6}$$

where u is the effective potential barrier and v_0 is the attempt rate. Finally, Γ_D is the soft mode damping constant resulting from third and fourth order anharmonicities.

Notice that if the interaction constant $\alpha_0 = 0$, the dielectricfunction decomposes into two terms, a relaxor-like and an oscillator-like and the system undergoes a ferroelectric transition at $T = T_0$ namely when the soft mode goes to zero. In contrast if $\alpha_0 \neq 0$ the denominator has zeros at three frequencies. Two are complex, $\tilde{\omega}_t$ and the third is pure imaginary $i\tilde{v}$. The extrapolated second order phase transition takes place when the zero frequency dielectric constant has a pole. This happens only when $\tilde{v} \rightarrow 0$.

The transition temperature is given by:

$$T_c = \frac{1}{2}(T_0 + \sqrt{T_0^2 + 4\alpha_0 / A}) \tag{7}$$

Strictly speaking since the dielectric constant has a pole due to the vanishing of an imaginary frequency, the transition is order-disorder. This is true for any finite value of α_0. However, the soft mode plays a dominant role in coupling the spins and therefore plays a major role in the material properties. This can be seen in the Curie-Weiss constant. Close to $T = T_c$ the dielectric constant obeys the Curie-Weiss law:

$$\varepsilon \cong \tilde{C} / (T - T_c); \tag{8}$$

$$\tilde{C} = C \frac{1 + 2\omega_0^2 M^* b^2 / T}{1 + |T^*| / T} \tag{9}$$

$$C = \frac{e^{*2}}{M^* V_c \varepsilon_0 A} \tag{10}$$

$$T^* = \frac{1}{2}(-T_0 + \sqrt{T_0^2 + 4\alpha_0 / A}) \tag{11}$$

Notice that the Curie-Weiss constant $\tilde{C}$ is proportional to C, the constant for purely displacive materials, multiplied by a factor of the order of unity, namely the Curie-Weiss constant is displacive-like.

At frequencies $\omega \ll \tilde{\omega}_{0t}$, the imaginary part of the dielectric constant can be expressed in the form:

$$\varepsilon_i = \frac{\omega S \tilde{v}}{\omega^2 + \tilde{v}^2} \tag{12}$$

$$\tilde{v} = v \frac{(T - T_c)(T + |T^*|)}{(T - T_0)T} \tag{13}$$

$$S = C \frac{T_c |T^*|}{(T + T^*)(T - T_c)(T - T_0)} \tag{14}$$

Notice that the imaginary part of the dielectric constant has in this case the frequency and temperature dependence of a central peak. The intensity S diverges and the width $\tilde{v} \to 0$ as $T \to T_c$.

We have compared the theoretical results discussed here with experimental measurements in $KNbO_3$ and $PbTiO_3$. These include comparisons with the soft mode frequency data obtained from IR [3] and neutron [5] measurements, temperature dependence of the low frequency dielectric constant, Raman measurements of the central peak [22] and hyper Raman measurements of the temperature and frequency dependence of the imaginary part of the dielectric function [23]. We found that for a proper choice of the parameters (all of which are physically reasonable) the theoretical and experimental results are in excellent agreement.

OFF-CENTER IONS IN INCIPIENT FERROELECTRICS

The phase transition properties of incipient ferroelectrics with impurities that make the system ferroelectric has been extensively investigated both experimentally and theoretically[17, 24-31]. Examples include $KTaO_3$:Nb and $SrTiO_3$:Ba where, $KTaO_3$ and $SrTiO_3$ are incipient ferroelectrics and Nb and Ba drive these systems into the ferroelectric phase. Experiments showed [26,27] that a minimum dopant concentration, x^*, is needed to make the system ferroelectric at $T_c = 0$ and that T_c rises approximately as $(x - x^*)^{1/2}$. Theoretically, the problem was treated by numerous authors. Two approaches have been used. The first [24] assumes that the effect of the impurities is to change the average properties of the host system. Specifically to change the linear and nonlinear polarizabilities of the oxygen atoms in, for example, $KTaO_3$:Nb. With proper parameters this approach is able to reproduce some experimental results but this approach is purely phenomenological. It seems to us that the parameters involved have no real physical meaning. The second approach [12,25] takes into account off-center impurities and their interaction with the host lattice soft mode. We believe that these models are basically correct but their treatment is limited to relatively high temperatures where zero point vibrations and tunneling can be ignored. Recently, Kleemann et al. [28] have suggested a number of models including one that takes into account off-center impurities, spin-phonon interactions, tunneling, direct spin-spin interaction and zero point vibrations and calculates expressions for the zero frequency dielectric function.

The impurity concentration dependence of T_c of a system undergoing a displacive phase transition close to zero temperature ("quantum ferroelectricity") have been studied in references [29-31]. The authors show that indeed $T_c \propto (x - x^*)^{1/2}$. However, the theory does not explain what determines the critical concentration x^*.

We have extended our model for pure ferroelectrics to the case of incipient ferroelectrics with off-center impurities. As in $KNbO_3$ we neglect the direct spin-spin interaction and assume that the spin-spin interaction mediated by the soft mode is dominant. The existence of a finite spin-spin interaction means that for any impurity concentration, the system will undergo a phase transition at a finite temperature unless the ions tunnel among symmetry equivalent positions. We therefore in agreement with Kleemann et al. [28] include tunneling in our Hamiltonian. The resulting Hamiltonian is as follows:

$$H = 1/2 \sum_q [p_q p_{-q} + \tilde{\omega}_{qt}^2 Q_q Q_{-q}] + \sum_{m,m',i} f(m - m') \eta_m^i Q_m \sigma_{-m'}^z - \sum_{m,i} \eta_m^i \Omega \sigma_m^x \quad (15)$$

Here η_m^i is one or zero at a site with an impurity or a host atom respectively. Using the density matrix formalism we have calculated the temperature and frequency dependence of the dielectric function.

$$\varepsilon = C / \{\tilde{\omega}_{0t}^2(T) - \omega^2 - \frac{|f_0|^2 \, \Omega x \tanh(\Omega/T)}{4\Omega^2 - \omega^2 - i\omega\gamma}\} \qquad (16)$$

Where $\tilde{\omega}_{0t}$ is the frequency of the pure $KTaO_3$ soft mode, x is the impurity concentration and γ is a damping factor that in the limit can be taken to zero.

The renormalized soft mode frequencies are the frequencies where ε has poles. Notice that these frequencies form two branches as shown in Fig. 2. Namely, the tunneling frequency and the $KTaO_3$ soft mode frequency anti-cross each other as a function of temperature. This result is in contrast to the Kugel et al. [32] Raman results that show that the soft mode Raman line does not split and its frequency continuously decreases as the transition temperature is approached.

The reason for this discrepancy is a slight spread in the off-center ionic displacements and the exponentially strong dependence of the tunneling frequency Ω on the ionic off-center displacements b. We assume a Gaussian distribution of the ionic off-center displacements:

$$P(b) = p_0 \exp(-(b - b_0)^2 / \sigma^2) \qquad (17)$$

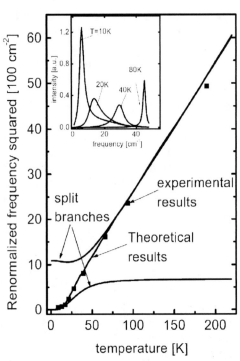

FIGURE 2. a) Temperature dependence of renormalized soft-mode frequency. The split branches and the theoretical results are obtained with and without spread in the Nb off-center displacements, respectively. Inset: Theoretical Raman lines.

Here p_0 is a normalization factor and σ is the distribution width. We have obtained the tunneling frequency dependence on the off-center displacements from the tunneling barrier ($\sim 0.096\text{eV}$) measured by Sokoloff et al. [33] using the Gomez et al. [34] calculation.

$$\Omega(b) = A\exp(-b^2 / B) \tag{18}$$

Here, $A = 0.066$ eV and $B = 6.5\times 10^{-3}\text{Å}^2$.
The dielectric function in this case can be expressed as follows:

$$\varepsilon = C/\{\tilde{\omega}_{0t}^2(T) - \omega^2 - xF(\omega,T)\}; \tag{19}$$

$$F(\omega,T) = \int \frac{|f_0|^2\, P(b)\Omega(b)\tanh(\Omega(b)/T)}{4\Omega^2(b) - \omega^2 - i\gamma\omega}\,db \tag{20}$$

The renormalized soft-mode frequencies are obtained from the poles of the dielectric function, namely:

$$\tilde{\omega}_{0t}^2(T) - \hat{\omega}^2 - F(T,\hat{\omega}) = 0 \tag{21}$$

Now if σ is large enough, this equation has a single rather than a double solution namely the Raman peak does not split. However, the Raman lines are broadened even if $\gamma \to 0$. This is seen in Fig. 2 inset.
The crystal undergoes a ferroelectric phase transition when $\varepsilon^{-1}(T_c,0) = 0$. Namely:

$$\tilde{\omega}^2(T_c) - xF(0,T_c) = 0 \tag{22}$$

and the concentration x^* that gives rise to $T_c = 0$ is given by:

$$x^* = \frac{\tilde{\omega}^2(0)}{F(0,0)}. \tag{23}$$

To compare these results with experiment we assume that the coupling coefficient between the off-center displaced ions and the soft mode is the same as in KNbO$_3$. b_0 is known from XAFS measurements. Thus the only unknown parameter in Eq. 17-20 is σ. We have adjusted this parameter so that $\hat{\omega}(T)$ is a single valued function. We found that $\sigma = 0.017\text{Å}$. This value is much smaller than the average displacement b = 0.16Å and is consistent with the static disorder value found in XAFS measurements. The renormalized value of the soft mode frequency is shown in Fig. 2 and as seen it is in very good agreement with the Raman experimental results. Using Eq. 23 we find that $x^* = 0.0068$ which is in good agreement with the value $x^* = 0.008$ estimated from experiments. The concentration dependence of T_c is obtained from Eq. 22 and is shown in Fig. 3a. As seen it is in very good agreement with experiment. Finally the inverse dielectric function was calculated using Eq. 19 as a function of temperature for a number of concentrations and is shown in Fig. 3b and it too is in very good agreement with experiment.

We have extended the calculation of the dielectric function beyondthe mean field approximation for small $T - T_c$. These calculations are beyond the scope of this paper and will be published separately. The main result of these calculations is that a central peak develops and becomes critical as T approaches T_c, while the soft-mode frequency saturates at a value of a few cm^{-1}. This central peak is different from the one in pure $KNbO_3$. The central peak in pure $KNbO_3$ is due to a relaxor associated with the correlated hoping of the Nb ions, while the central peak in $KTa_{1-x}Nb_xO_3$ for small Nb concentrations is associated with the correlated tunneling of the Nb ions. These results are in good agreement with the experimental results of Lyons et al. [35] on the central peak of $KTa_{0.001}Nb_{0.009}O_3$.

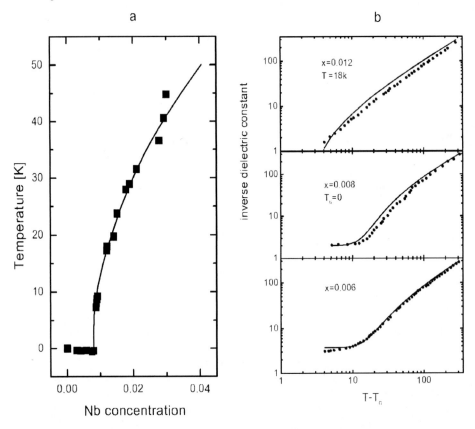

FIGURE 3. a) Nb concentration dependence of the ferroelectric transition temperature. Dots - experimental results; solid line - theory. b) Temperature dependence of the inverse dielectric constant. Dots - experimental results; solid line - theory.

MICROSCOPIC MODEL [36]

The existence of spontaneous off-center displacements in the paraelectric phase of many perovskites is puzzling because in many cases the ionic radii of the ions involved fit the space available to them in the crystal rather well. The microscopic model we developed is applicable to oxygen perovskite ferroelectrics but may not be applicable to other crystal families.

We know that the soft mode in the perovskites and in other crystal families is a result of the strong interband electron transverse phonon coupling [37-40]. Here we assume that in addition the systems have a strong intraband electron longitudinal phonon coupling. In the presence of free electrons, this coupling is responsible for the creation of small polarons.

The Hamiltonian is as follows:

$$H = H_e + H_{ph} + H_{e-ph} .$$
(24)

The first term is a one electron two band Hamiltonian. It includes the usual kinetic and potential energy terms but excludes any electron phonon coupling. The second is the usual phonon Hamiltonian. We include in this term the lowest transverse and longitudinal branches and disregard all others since we believe that they do not play an essential role in the ferroelectric transition. The third term has two components: the interband electron transverse phonon and the intraband electron longitudinal phonon coupling. Solving the problem with this Hamiltonian is rather involved because the interaction components are large and can not be treated by perturbation. We have treated the problem using the Lang-Firsov transformation [41,42,36].

Qualitatively, due to the interband e-p coupling a finite number of electrons resides on the cations. The number of these electrons relative to the number of unit cells is:

$$\frac{N_e}{N} = \frac{\omega_{0t}^2}{2\overline{\omega}\overline{E}} \coth(\frac{\overline{\omega}}{2T})$$
(25)

Here ω_{0t} is the bare transverse phonon frequency at $q = 0$; $\overline{E}$ is the average valence conduction interband separation; and $\overline{\omega}$ is the average soft phonon branch energy.

The interaction of these electrons with the longitudinal phonons leads to the creation of ground state polarons. One manifestation of these polarons are the local distortions:

$$(\frac{b}{a_0})^2 = \frac{E_p N_e}{E_{al} N}; \quad E_{al} = M^* \omega_{0t}^2 a_0^2 / 2$$
(26)

Here, b is the local off-center displacement a_0 is the unit cell dimension and E_p is the polaron energy shift.

The procedure outlined above treats the soft mode and the local off-center displacements in a unified way. Thus the interaction between the two is naturally obtained:

$$\alpha_0 = \omega_{0t}^2 E_{al} (\frac{b}{a_0})^2 \tag{27}$$

This brings us back to our phenomenological model except that now we have explicit expressions for the local spontaneous off-center displacements and the spin phonon coupling coefficient. We have compared the theoretical results with the experimental values of three parameters in $KNbO_3$ and $PbTiO_3$. The values of these parameters were calculated using parameters obtained from other independent experiments. So the calculated values were determined with no adjustable parameters. From XAFS, the off-center displacements in $KNbO_3$ and $PbTiO_3$ are 0.23Å and 0.3Å, respectively. The corresponding theoretical values are 0.2Å and 0.27Å. The experimental phase transition temperatures are 650K and 760K. The corresponding theoretical values are 690K and 845K. The experimental Curie-Weiss constants in units of 10^5K are 2.8 and 0.85 and the corresponding theoretical values are 1.74 and 1.34.

The microscopic theory also explains the puzzling fact observed in XAFS measurements that Ti atoms in $PbTiO_3$ and $BaTiO_3$ are displaced to off-center positions while in $SrTiO_3$ they are not. Similarly the Nb ions in $KNbO_3$ are displaced to off center positions while the isoelectronic Ta atoms are not. Many experiments show that $BaTiO_3$, $PbTiO_3$ and $KNbO_3$ have polaronic mobilities [43-46] while $KTaO_3$ and $SrTiO_3$ have ordinary band mobilities [47,48]. This means that the intraband e-p coupling that is essential in creating the local distortions, is large in the first group and is small in the second. Namely, the polaronic properties and the spontaneous local off-center displacements are indeed correlated as suggested by our model.

SUMMARY AND CONCLUSIONS

Structural phase transitions in perovskites and other crystal families have both displacive like and order-disorder like properties. We have demonstrated that a phenomenological model containing a soft mode, disordered spontaneous local distortions in the high symmetry phase and an interaction between them accounts quantitatively for both displacive like and order-disorder like properties. Strictly speaking the transition takes place when the renormalized relaxor mode intensity and relaxation time diverge. In fact this relaxor mode gives rise to the central peak observed in Raman and neutron experiments. Specifically, the model explains why the phase transition takes place at a temperature higher than the temperature at which the soft mode is expected to vanish yet it predicts in agreement with experiment a displacive like Curie-Weiss constant. The model also accounts quantitatively for the detailed hyper-Raman results that provide the temperature and frequency dependence of the imaginary part of the dielectric constant.

A similar model was developed for impure incipient ferroelectrics. Here the impurity ions do not hop but tunnel among the off-center equivalent positions. We have shown, that taking into account the finite spread in the size of the off-center displacements, we obtain in agreement with experiment a single renormalized soft-mode frequency that goes to zero as $T \rightarrow T_c$. Furthermore we have obtained the critical impurity concentration for $T_c = 0$ and the temperature dependence of the dielectric function with no additional parameters. The results are in excellent agreement with experiment.

We finally presented a microscopic model showing, that in some materials, the spontaneous local distortions observed in the paraelectric phase are in fact manifestations of ground state polarons. The local distortions, the phase transition temperature and the Curie-Weiss constant calculated from this model using experimentally obtained parameters are in good agreement with the experimental values.

ACKNOWLEDGEMENT

The authors gratefully acknowledge very interesting discussions with professor Sonin and the support of this work by the German-Israel Science Foundation.

REFERENCES

1. Lines, M.E., and Glass, A.M., *Principles and Applications of Ferroelectrics and Related Materials* Claredon Press, Oxford, 1977.
2. Vaks, V.G., *Introduction in Microscopic Theory of Ferroelectricity*, Nauka, Moscow in Russian, 1973.
3. Fontana, M.D., Metrat, G., Servoin, J.L., and Gervais, F., *J. Phys. C: Solid State Phys.* 17, 483-514 (1984).
4. Scott, J.F., *Rev. Mod. Phys.* 46, 83-128 (1974).
5. Axe, J.D., Harada, J., and Shirane, G., *Phys. Rev. B* 1, 1227-1233 (1970).
6. Comes, R., Lambert, M., and Guinier, A., *Acta Crystallographica A* 26, 244-254 (1970).
7. Comes, R., Currat, R., Denoyer, F., Lambert, M., and Quittet, A.M., *Ferroelectrics* 12, 3-8 (1976).
8. Comes, R., and Shirane, G., *Phys. Rev. B* 5, 1886-1991 (1972).
9. Quittet, A.M., and Lambert, M., *Solid State Commun.* 12, 1053-1055 (1973).
10. Yacoby, Y., and Just, S., *Solid State Commun.* 15, 715-718 (1974).
11. Yacoby, Y., *Z. Physik B* 31, 275-282 (1978).
12. Halperin, B.I., and Varma, C.M., *Phys. Rev. B* 14, 4030-4044 (1976).
13. Sicron, N., Ravel, B., Yacoby, Y., Stern, E.A., Dogan, F., and Rehr, J.J., *Phys. Rev. B* 50, 13168-13180 (1994).
14. Kim, K.H., Elam, W.T., and Skelton, E.F., *Mater. Res. Soc. Symp. Proc.* 172, 291-294 (1990).
15. de Mathau, N., Prouzet, E., Hasson, E., and Dexpert, H., *J. Phys. Condensed Matter* 5, 1261-1270 (1993).
16. Nishihata, Y., Kamishima, O., Ojima, K.K., Sawada, A., Maeda, H., and Terauchi, H., *J. Phys. Condensed Matter* 6, 9317-9328 (1994).

17. Hanske-Petitierre, O., Yacoby, Y., Mustre de Leon, J., Stern, E.A., and Rehr, J.J., *Phys. Rev. B* **44**, 6700-6701 (1991).

18. Rechav, B., Yacoby, Y., Stern, E.A., Rehr, J.J., and Newville, M., *Phys. Rev. Lett.* **72**, 1352-1355 (1994).

19. Sicron, N., Yacoby, Y., Stern, E.A., and Dogan, F., *J. de Physique IV* **7**, c2 1047-1049 (1997).

20. Girshberg, Ya., and Yacoby, Y., *Solid State Commun.* **103**, 425-30 (1997).

21. Salje, E.K.H., Wruck, B., and Thomas, H., *Z. Phys. B - Condensed Matter* **82**, 399-404 (1991).

22. Fontana, M.D., Idrissi, H., and Wojcik, K., *Europhys. Lett.* **11**, 419-424 (1990).

23. Vogt, H., Fontana, M.D., Kugel, G.F., and Gunter, P., *Phys. Rev. B* **34**, 410-415 (1986).

24. Kugel, G., and Fontana, M., *Phys. Rev .B* **35**, 813-820 (1987)

25. Vugmeister, B., and Glinchuk, M., *Rev. Mod. Phys.* **62**, 993-1026 (1990).

26. Hochli, U., Weibel, H., and Boatner, L., *Phys. Rev. Lett.* **39**,1158-1161 (1977).

27. Rytz, D., Chatelain, A., and Hochli, U., *Phys. Rev. B* **27**, 6830-6840 (1983).

28. Kleemann, W., Dec, J., Wang, Y.G., Lehnen, P., and Prosandeev, S.A., *J. of Physics and Chemistry of Solids* **61**, 167-176 (2000).

29. Schneider, T., Beck, H., and Stoll, E., *Phys. Rev. B* **13**, 1123-1130 (1976).

30. Oppermann, R., and Thomas, H., *Z. Phys. B* **22**, 387-396 (1975)

31. Morf, R., Schneider, T., and Stoll, E., *Phys. Rev. B* **16**, 462-469 (1977).

32. Kugel, G., Vogt, H., Kress, W., and Rytz, D., *Phys. Rev. B* **30**, 985-991 (1984).

33. Sokoloff, J.P., Chase, L.L., and Rytz, D., *Phys. Rev. B* **38**, 597-605 (1988).

34. Gomez, M., Bowen, S.P., and Krumhansl, J.A., *Phys. Rev. B* **153**, 1009-1024 (1967).

35. Lyons, K.B., Fleury, P.A., and Rytz, D., *Phys. Rev. Let.* **57**, 2207-2210 (1986).

36. Girshberg, Ya., and Yacoby, Y., *J. Phys.: Condens. Matter* **11**, 9807-9822 (1999).

37. Bersuker, I.B., "Vibronic Interaction in Molecules and Crystals" in *Springer Series in Chemical Physics*, 1989.

38. Kristoffel, N.N., and Konsin, P., *Phys. Stat. Sol. B* **149**, 11-22 (1988) and References therein.

39. Girshberg, Ya.G., and Tamarchenko, V.I., *Fiz. Tverd. Tela* **18**, 1066-1072 (1976); *Sov. Phys. Solid State* **18**, 609 (1976); *Fiz. Tverd. Tela* **18**, 3340 (1976); *Sov. Phys. Solid State* **18**, 1946-1950 (1976).

40. Girshberg, Ya.G., Bursian, E.V., and Tamarchenko, V.I., *Ferroelectrics* **18**, 39-43 (1978).

41. Lang, I.G., Firsov, Yu.A., *Zh. Eksp. Teor. Fiz.* **43**, 1843-1860 (1962); *Sov. Phys.-JETP* **16**, 1301-1312 (1963).

42. Alexandrov, A., and Capellmann, H., *Phys. Rev. B* **43**, 2042-2049 (1991).

43. Bursian, E.V., Girshberg, Ya.G., and Starov, E.N., *Phys. Stat. Sol. B* **46**, 529-533 (1971).

44. Mahgereften, D., Kirilov, D., Cudney, R.S., Bacher, G.D., Pierce, R.M., and Feinberg, J., *Phys. Rev. B* **53**, 7094-7098 (1996).

45. Bursian, E.V., Girshberg, Ya.G., and Ruzhnikov, A.V., *Phys. Stat. Sol. B* **74**, 689-693 (1976).

46. Bernasconi, P., Biaggio, I., Zgonik, M., and Gunter, P., *Phys. Rev. Letters* **78**, 106-109 (1997).

47. Wemple, S.H., DiDomenico, M., and Jayaraman, Jr., A., *Phys. Rev.* **80**, 547-556 (1969).

48. Wemple, S.H., *Phys. Rev.* **137**, A1575-1582 (1965).

Structural Instability in $Ba_{1-x}K_xBiO_3$

V. Gusakov[*], S. Zherlitsyn[†,‡], B. Lüthi[†], B. Wolf[†], F. Ritter[†], D.Wichert[†],
G. Bruls[†], S. Barilo[*], S. Shiryaev[*], C. Escribe-Filippini[§] and
J. L. Tholence[§]

[*]*Institute of of Solids State & Semiconductor Physics, 220072 Minsk, Belarus*
[†]*Physikalisches Institut, Universität Frankfurt, D-60054 Frankfurt, Germany*
[‡]*Institute for Low Temperature Physics & Engineering, 310164 Kharkov, Ukraine*
[§]*Laboratoire d'Etudes des Propriétés Electroniques des Solides, CNRS, 38042 Grenoble, France*

Abstract. We report results of the ultrasonic investigation of $Ba_{1-x}K_xBiO_3$ single crystals for two potassium concentrations $x \approx 0.35$ and $x \approx 0.47$ in a wide temperature range. The softening of both the transverse c_{44} and the longitudinal c_{11} modes have been observed at temperatures between 200 K and 50 K. In the case of $Ba_{0.65}K_{0.35}BiO_3$ a pronounced hysteresis was discovered. We proposed a model in which the softening of the elastic moduli, the hysteresis, and the maximum in the attenuation of sound can be explained by assuming a coupling of the acoustic modes with the anharmonic oscillations of O_6 octahedra.

We investigated $Ba_{1-x}K_xBiO_3$ single crystals (electro-deposition growth) $x \approx 0.35$ and $x \approx 0.47$ with onset of the superconducting phase transition at $T_c \cong 32$ K and 24 K respectively. A temperature behavior of the longitudinal c_{11} ($\mathbf{k}//\mathbf{u}//[100]$) and transverse c_{44} ($\mathbf{k}//[100]$, $\mathbf{u}//[001]$) acoustic modes have been studied. Fig. 1a shows the temperature dependence of the sound velocity of the longitudinal c_{11} mode, measured both with cooling the sample down and heating it up. There is a large softening below 150 K with a minimum at 70 K. The anomaly in the sound velocity is accompanied by a maximum in the attenuation of sound. A hysteresis occurs in the temperature range between 50 K and 250 K for both the sound velocity and the attenuation of sound. Although the c_{44} mode shows approximately the same total softening and also the temperature hysteresis exists, there is not any more step-like behavior at 100 K neither for the temperature sweep up nor for the temperature sweep down. The sound velocity of the c_{11} mode in $Ba_{0.53}K_{0.47}BiO_3$ demonstrates a rather smooth decrease with temperature (see Fig. 1a). The hysteresis has almost completely disappeared in $Ba_{0.53}K_{0.47}BiO_3$. There is no sharp anomaly in the sound velocity at the superconducting critical temperature T_c but there is a kink in the vicinity of T_c in $Ba_{0.65}K_{0.35}BiO_3$ without a visible discontinuity in magnitude of the longitudinal sound velocity at the transition. We can conclude from our experimental data that if a discontinuity exists it should be less than 5×10^{-5}. Our preliminary low temperature x-ray investigation did not show any structural change from cubic symmetry within a resolution of our x-ray technique. The observed peculiarities in the sound velocity (see Fig. 1a) and the attenuation of sound can be associated with the particular dynamics of oxygen octahedra. As it was found in EXAFS study of $Ba_{1-x}K_xBiO_3$ [1]

CP554, *Physics in Local Lattice Distortions*, edited by H. Oyanagi and A. Bianconi
© 2001 American Institute of Physics 1-56396-984-X/01/$18.00

some of the oxygen octahedra move in an anharmonic double-well potential originating from an existence of two nonequivalent bismuth positions Bi(I) and Bi(II) (a dynamic interchange $Bi(I)O_6 \leftrightarrow Bi(II)O_6$ between the positions is also possible). Movement of oxygen atoms in double-well potential results in a rearrangement of short range ordering of the octahedra with the variation of temperature. That leads to the renormalization of the elastic constants and to the observed anomalies in the sound velocity and the attenuation of sound.

The lattice Hamiltonian can be represented as:

$$H = H_h + H_{anh} + H_{int} \tag{1}$$

Here H_h is the phonon Hamiltonian of the lattice which can be written as a sum over the harmonic oscillators with the parameters normalized by empirical values of the lattice constants of crystal:

$$H_h = \sum_k \left(\frac{p_k^2}{2\mu_k} + \frac{\mu_k \omega_k^2}{2} x_k^2 \right), \tag{2}$$

where x_k, p_k, μ_k, ω_k are respectively displacement, momentum, mass and frequency of the oscillator which has a number k. The anharmonic Hamiltonian H_{anh} describes dynamic of oxygen atoms oscillated in double well potential and for the given oxygen atom can be written in the following form:

$$H_{anh} = \frac{p^2}{2m} + \frac{\alpha}{2} q^2 - \frac{\beta}{3} q^3 + \frac{\gamma}{4} q^4 \tag{3}$$

where m is mass, q and p are the configuration coordinate and canonical conjugated momentum along selected direction. H_{int} takes into account an interaction of harmonic modes of crystal with selected strong anharmonic oscillations. In general this term becomes rather complicated but for our purposes it is enough to consider a simple expression consisting of cubic and forth power contributions:

$$H_{int} = H_{int}^{(3)} + H_{int}^{(4)} \tag{4}$$

where

$$H_{int}^{(3)} = q^2 \sum_k \lambda_k x_k, \tag{5}$$

$$H_{int}^{(4)} = q^2 \sum_{kk'} \lambda_{kk'} x_k x_{k'} \tag{6}$$

and λ_k, $\lambda_{kk'}$ are coupling constants.

In order to find renormalized frequencies of the harmonic modes we use the equations of motion for the retarded two time Green functions [2,3]. This technique leads to the following expression for the renormalized frequencies:

$$\tilde{\omega}_k(T) = \omega_k(T) + \omega_k(T)\left[\lambda\left(\sigma + <q>^2\right) - \frac{\lambda_k^2}{2\mu_k\omega_k^2}\frac{\left(\sigma + 4<q>^2\right)}{m\Omega^2}\right] \equiv \omega_k + \omega_k \times f(T), \quad (7)$$

where $<q>$ is an average displacement of an atom in the anharmonic potential (in case of the harmonic approximation $<q> = 0$), $\sigma = <q^2> - <q>^2$, Ω is an effective frequency of oscillations in the anharmonic potential, $\lambda_{kk} = \lambda\mu_k\omega_k^2$, μ is a reduced mass; $\tilde{\omega}$ is a frequency of acoustic wave in the crystal with the selected sublattice moving in anharmonic potential. $\omega_k(T)$ is a temperature dependence of the ultrasonic frequency without renormalization by anharmonicity. Thus a temperature dependence of the sound velocity ($v = const \times \tilde{\omega}$) is determined by the renormalization function $f(T)$. In order to calculate the temperature dependencies of $<q>(T)$, $\sigma(T)$, and $\Omega(T)$ we use the variation inequality

$$F \leq F_0 + <H - H_0>_0, \quad (8)$$

here H is the full Hamiltonian of the system; H_0 is an approximated Hamiltonian; F is the free energy of the system, described by the Hamiltonian H; F_0 is the free energy of the system with the Hamiltonian H_0; $<\ >_0$ means averaging over H_0. We represent here sublattice oscillations in the anharmonic potential at fixed temperature as oscillations in effective shifted harmonic potential. Moreover the effective frequency Ω and displacement $<q>$ are functions of temperature. Using this approach H_0 can be written as following:

$$H_0 = \frac{p^2}{2m} + \frac{m\Omega^2(q - <q>)^2}{2}, \quad (9)$$

and H have a form of Eq. (3). Then Eq. (8) can be represented as:

$$F \leq k_BT \ln\left(2sh\left(\frac{\hbar\Omega}{2k_BT}\right)\right) + \left(\frac{\alpha}{2}<q>^2 - \frac{\beta}{3}<q>^3 + \frac{\gamma}{4}<q>^4\right) + \left(\frac{\alpha}{2} - \beta<q> + \frac{3}{2}<q>^2\gamma - \frac{m\Omega^2}{2}\right)\sigma + \frac{3}{4}\gamma\sigma^2 \quad (10)$$

where $\sigma = <(q - <q>)^2>_0 = \frac{\hbar}{2m\Omega}cth\left(\frac{\hbar\Omega}{2k_BT}\right)$.

Minimization of the right side of the Eq. (10) gives the system of equations for temperature dependence of the average displacement $<q>(T)$ and the effective frequency $\Omega(T)$:

$$-(3<q>\gamma-\beta)\frac{\hbar}{2m\Omega}cth\left(\frac{\hbar\Omega}{2k_BT}\right)=\alpha<q>-\beta<q>^2+\gamma<q>^3$$

, (11)

$$m\Omega=\alpha-2\beta<q>+3\gamma\left[<q>^2+\frac{\hbar}{2m\Omega}cth\left(\frac{\hbar\Omega}{2k_BT}\right)\right]$$

Temperature dependence of the attenuation of ultrasound can be calculated in actual case when $\omega\tau_{th}\ll 1$ using the expression:

$$\alpha_s=\frac{\gamma_G^2\tau_{th}}{3\rho}\frac{CT\omega_s^2}{v_s^3},$$ (12)

where C is the specific heat; v_s is the sound velocity, and ω_s is a frequency of ultrasound wave. An expression for the specific heat $C(T)$ can be easily derived from the free energy defined by Eq. (10).

Fig. 1b shows the calculated temperature dependence of the sound velocity in the frame of our model. The parameters of the anharmonic potential were chosen in the way to describe correctly both softening of the sound velocity and the temperature region of the hysteresis. We would like to point out that our fitting values of the energy barrier $U_0 = 0.035$ eV and the distance between two positions of the minimum $q_2 = 0.15$ Å for the $U(q)$ agree well with the values determined from an analysis of the x-ray absorption spectra [1] $U_0 = 0.04$ eV and $q_2 = 0.16$ Å.

Relation between the coupling constants λ and λ_k determines the direction of path tracing of the hysteresis loop. In case of c_{11}-mode $\lambda/\lambda_k \approx 6.5$ and the cubic interaction prevails in Eq. (4). Opposite path tracing takes place for the c_{44}-mode below 120 K. In our model this fact is explained by another ratio between the coupling constants $\lambda/\lambda_k \approx 3.5$. Calculated curves for the sound velocity and the attenuation of the c_{44}-mode are quite similar (except a direction of the tracing the hysteresis loop) to behavior of the c_{11}-mode (see Fig. 1) and are not shown here.

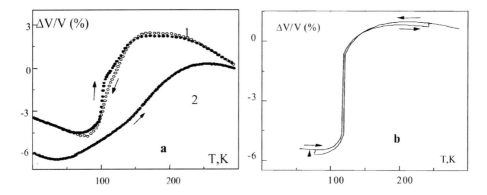

FIGURE 1. Temperature dependence of ultrasonic longitudinal wave velocity (c_{11}) for single crystal $Ba_{1-x}K_xBiO_3$; a–experiment, $1 - x \approx 0.35$, $2 - x \approx 0.47$; b–theory ; U(q) for $x \approx 0.35$; $\lambda_3/\lambda_4 \approx 6.5$.

Thus our model proceeding from the interaction of acoustic modes with movements of oxygen octahedra in double well potential gives good agreement with the experimental observations. In the context of all the statements made above we would like to emphasize that it is important to take structural instability into account in order to describe correctly the physical properties of $Ba_{1-x}K_xBiO_3$, including the origin of the SC state. Evidently the anharmonic phonon play a significant role in these compounds.

This work was supported in part by SFB 252 and by INTAS organization under grant No. 97-1371 (work in Minsk). One of the authors (S. Z.) would like to thank the Alexander von Humboldt foundation for the financial support and members of the Physikalisches Institut der J.W.Goethe Universität for hospitality.

REFERENCES

1. Menushenkov, A.P., Klement'ev, K.V., Konarev, P.V., and Meshkov, A.A., *JETP Letters* **67**, 1034–1039 (1998).

2. Zubarev, D.N., *Nonequilibrium Statistical Thermodynamics*, Nauka, Moscow, 1971, pp. 125-171.

3. Saiko, A.P., Gusakov, V.E., and Kuz'min, V.S., *Low Temp. Phys.* **20**, 764-766 (1994).

Polarized XAFS Study of the Local Structure and the Phase Transitions in Niobate Perovskites

V. A. Shuvaeva[*], Y. Azuma, K. Yagi, K. Sakaue and H. Terauchi

Advanced Research Center of Science, Kwansei-Gakuin University,
Gakuen 2-1, Sanda, Japan, 6691337
[*] *present address: Institute of Physics, Rostov State University, Stachki 194, Rostov-on-Don, 344090, Russia*

INTRODUCTION

Recent intensive XAFS investigations of perovskite crystals allowed to make considerable progress in understanding the role of disorder in phase transitions in perovskite compounds [1-4]. However there is still lack of precise information on the atomic displacements, especially in low-temperature phases. Unique opportunities for structural investigations of perovskite crystals are given by polarized XAFS. The method can provide structural details, which cannot be obtained by any other experimental technique, including XAFS obtained from powder samples. By coinciding the vector of X-ray polarization with each of the main perovskite directions in single-crystal sample, one can obtain from EXAFS spectra precise information on the direction and value of B atom displacement. Additional possibilities are provided by analysis of the pre-edge XAFS region of transition metal elements. The approach has been developed in detail for the analysis of Ti displacements in perovskite compounds [2].

Using these experimental techniques we performed a temperature-dependent polarized Nb-K XAFS study of the Nb positions in two perovskite compounds with different physical properties: ferroelectric $KNbO_3$ and antiferroelectric $NaNbO_3$. Both of them undergo a number of temperature phase transitions. In $KNbO_3$ the following sequence of the phase transitions occurs: cubic (above 425° C), tetragonal (225 - 425° C), orthorhombic (5 - 225° C) and rhombohedral (below 5° C). Many evidences have been accumulated that in $KNbO_3$ phase transitions the disorder plays an important role. In $NaNbO_3$ six phase transitions, involving different types of tilts of the octahedra and displacements of the cations, were reported to occur at - 105, 370, 480, 520, 580, 640° C [5]. The room temperature structure is characterized by antiparallel off-center displacements of Na and Nb within (010) plane and oxygen octahedra tilting. At temperatures higher than 480° C the material is paraelectric. In the paraelectric phases Nb atoms were believed to occupy positions at the centers of oxygen octahedra. Atomic displacements in $KNbO_3$ and $NaNbO_3$ were determined previously using X-ray diffraction assuming macroscopic symmetry. However locally atomic positions may deviate from the averaged ones, if atoms are statistically distributed among several symmetry equivalent positions. The value of this deviation,

CP554, *Physics in Local Lattice Distortions*, edited by H. Oyanagi and A. Bianconi
© 2001 American Institute of Physics 1-56396-984-X/01/$18.00

which can be called a 'disorder component' of the total atomic displacement, is a subject of the present study.

EXPERIMENTAL

Spectra were measured from single crystals of about $1.5 \times 1.5 \times 0.02$ mm in size in a transmission mode at BL-10B and 7C Photon Factory. The data were obtained in 16 temperature points in the range from 20 to 460° C for KNbO$_3$ and in 25 temperature points in the range from 20 to 800° C for NaNbO$_3$. The measurements were performed for two orthogonal orientations of the crystals relative to the vector of X-ray polarization, so that vector of X-ray polarization was coincided with one of the main perovskite axes and almost parallel to two of the Nb-O bonds. Data analysis has been performed using UWXAFS and FEFF7 software [6,7].

RESULTS AND DISCUSSION

As it has been reported previously, EXAFS spectra, measured from powder sample of KNbO$_3$, displayed no temperature changes [4]. However, we observed strong temperature changes of both Fourier transforms (FT) and near-edge structure of KNbO$_3$ polarized XAFS spectra, which appeared to be much more sensitive to the changes of the direction of the Nb displacement. At room temperature the 'disorder component' of Nb displacement determined by fitting to the experimental spectra is found to be quite small - not more than 0.06 Å, so that it cannot be even distinguished. These estimations are in perfect agreement with those, made on the basis of X-ray diffraction data. So in orthorhombic phase the disorder is not essential. Up to 225° C, which is the phase transition point into tetragonal phase, no significant changes of XAFS spectra are observed. In the phase transition point the 'disorder component' increases abruptly up to 0.08 Å. It should be noted that this value is still much less than the displacement along the polar axis, which is about 0.16 Å according to the X-ray diffraction data. At higher temperatures further increase of the 'disorder component' has been revealed. It was especially rapid near the phase transition into the cubic phase. Nb disorder in the cubic appeared to be essential and the 'disorder component' was close to 0.11 Å. So polarized XAFS data indicated mixed nature of the phase transitions in KNbO$_3$ and increase of the structural disorder with temperature.

In contrast to KNbO$_3$ no any step-like temperature changes of NaNbO$_3$ spectra are observed. At room temperature Nb is displaced within (010) plane, and the value of the displacement is constant up to 700° C. The displacement in the orthogonal direction at room temperature is rather small and is consistent with X-ray diffraction data. However it displays gradual growth with temperature. At temperatures higher than 480° C, when the crystal becomes paraelectric, the Nb displacements in the both directions become equal to 0.1 Å. These displacements preserve in all high-temperature phases. It shows that disorder of Nb atoms is important at least in cubic phase. The symmetry of all NaNbO$_3$ phases except the cubic phase is quite low, so

Nb displacements are not symmetry forbidden. However paraelectricity of high-temperature phases strongly suggests that there should be no ordering in any Nb off-center displacements at temperatures higher than 460° C, and that existence of such displacements may be considered as indication of a disorder.

The temperature dependence of the Nb displacements in NaNbO₃ and KNbO₃ can be illustrated by temperature dependence of the integrated intensity of the pre-edge peak of Nb K-XAFS, which is directly related to the Nb displacement. It is shown at the Figure 1. One can see the step-like changes of the dependence for KNbO₃ in contrast to gradual slow changes at the NaNbO₃ spectra.

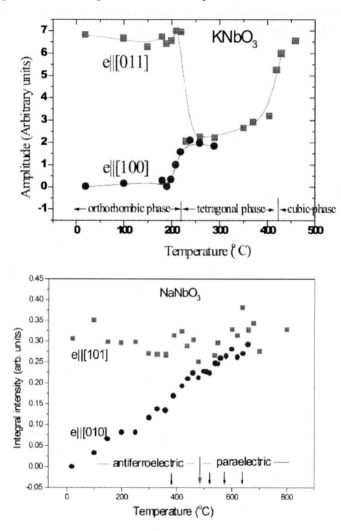

FIGURE 1. Temperature dependence of the integral intensity of the pre-edge peak

In summary, analysis of both polarized EXAFS and near-edge structure indicate that in NaNbO₃ and KNbO₃ at room temperature the disorder of Nb atoms is quite subtle, however it increases at higher temperatures. In ferroelectric KNbO₃ the 'disorder component' increases abruptly in the phase transition points, while in NaNbO₃ the changes of the Nb position are gradual and insensitive to the macroscopic symmetry. High temperature cubic phases of both compounds are essentially disordered.

REFERENCES

1. Ravel, B., Stern, E.A., Vedrinskii, R.V., and Kraizman, V.L., *Ferroelectrics* **206**, 407 (1998).
2. Vedrinskii, R.V., Kraizman, V.L., Novakovich, A. A., Demekhin, Ph. V., Urazhdin, S. V., Ravel, B., and Stern, E.A., *J. Phys. (Paris), IV,* **7**, C2-107 (1997).
3. Stern, E.A., and Yacoby, Y., *J. Phys. Chem. Solids* **57**, 1449 (1996).
4. Bell, M.I., Kim, K.H. and Elam, W.T., *Ferroelectrics* **120**, 103 (1991).
5. Glazer, A.M., and Megaw, H.D., *Acta Ctyst.* **A29**, 489 (1973).
6. Stern, E.A., Newville, M., Ravel, B., Yacoby, Y., and Haskel, D., *Physica B* **208 & 209**, 117 (1995).
7. Zabinskii, S.I., Rehr, J.J., Ankudinov, A., Albers, R.C., and Eller, M.J., *Phys. Rev. B* **52**, 2995 (1995).

Observation of Anomalous EPR Lines with Intermediate g-Factor in BaKBiO and BaPbBiO Systems

A. I. Golovashkin[a], S. V. Gudenko[b], A. P. Rusakov[c] and A. Yu.Yakubovsky[b]

[a]*P.N.Lebedev Physical Institute RAS, Moscow 117924, Russia*
[b]*Russian Research Center "Kurchatov Institute", Moscow 123182, Russia*
[c]*Moscow Steel and Alloys Institute, Moscow 117049, Russia*

Abstract. The $Ba_{1-x}K_xBiO_3$ and $BaPb_yBi_{1-y}O_3$ systems were studied by the EPR method in wide ranges of doping levels. Besides the signals with factors $g \approx 2.1$ and $g \approx 4.2$ the anomalous EPR lines with intermediate g-factors ($g \approx 3$) were also found in the both systems. The reason of such lines is local lattice distortions ("lattice defects") with appropriate magnetic moments resulting due to disorderings ("electron defects") in ordered covalent bonds Bi-O.

INTRODUCTION

The bismuthate oxide systems $Ba_{1-x}K_xBiO_3$ (BKBO) and $BaPb_yBi_{1-y}O_3$ (BPBO) have similar electronic, lattice, and superconducting properties with cuprate oxide HTS systems. Recently it was found that cuprate and bismuthate oxide systems have also similar EPR spectra. In the spectra of both groups the lines with g-factors 2.1 and ("forbidden") 4.2 were found [1–3] although Bi and Cu ions have different magnetic moments. But in $La_{1-x}Sr_xCuO_4$ (LSCO) the anomalous EPR signals with intermediate g-factor about 3 and unusual temperature dependence were also observed [1]. We present the new experimental results for the BKBO and BPBO systems in which the very similar resonance lines were observed.

EXPERIMENT

The EPR spectra of the samples were recorded in the temperature range 3 - 300K using X-band Bruker ESP-300 spectrometer equipped with Oxford Instruments helium flow-through cryostat ESR-900. Samples were synthesised by the nitrate method [4] and specially for EPR measurements protected by pure paraffin just after preparation. Laser mass-spectrometer LAMMA-1000 (sensitivity better than 10^{17} cm^{-3}) has not shown even traces of magnetic impurities or Cu in the samples. The measurements were done on the BKBO and BPBO single phase samples of different content as for dielectric and metal phases. In addition some samples were annealed in Ar

CP554, *Physics in Local Lattice Distortions*, edited by H. Oyanagi and A. Bianconi
© 2001 American Institute of Physics 1-56396-984-X/01/$18.00

atmosphere to produce oxygen vacancies. We measured also the "ageing" effect on the EPR spectra. The standard magnetic, electric, and structural measurements were done for the sample certification. For the samples the EPR signals were measured at T > T_c (T_c: critical temperature).

RESULTS

In Fig. 1 the measured resonance lines with g = 2.1, g = 4.2 ("half-field transition"), and line with g ≈ 3 are shown for BKBO with different content. Similar lines were observed also in other BKBO samples as dielectric and metal.

The results for BaBiO₃ and BPBO are shown in Fig. 2. For all measured samples besides the relatively narrow EPR lines with g = 2.1 and g = 4.2 one can see the additional wider line with intermediate g ≈ 3. The value of g changes somewhat with doping, temperature, and ageing.

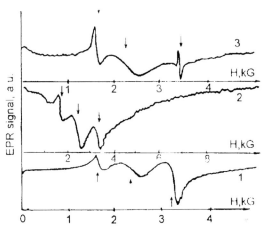

FIGURE 1. The EPR spectra of $Ba_{1-x}K_xBiO_3$: 1) x = 0.13, T = 20K; 2) x = 0.4, T = 34K; 3) x = 0.55, T = 15K (T: temperature of measurements). The observed EPR lines are indicated by arrows.

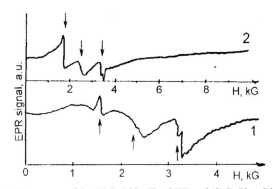

FIGURE 2. The EPR spectra of the (1) BaBiO₃ (T = 25K) and (2) BaPb₀.₇₅Bi₀.₂₅O₃ (T = 13K).

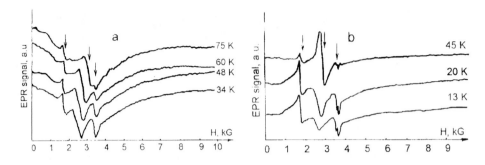

FIGURE 3. The EPR spectra for different T: (a)$Ba_{0.6}K_{0.4}BiO_3$; (b) $BaPb_{0.75}Bi_{0.25}O_3$.

In Fig. 3 the EPR signals for $BaPb_{0.75}Bi_{0.25}O_3$ and $Ba_{0.6}K_{0.4}BiO_3$ for different temperatures are shown as an example of temperature dependence. More detailed data on temperature dependencies will be published elsewhere.

There are the qualitative differences in temperature dependencies of the EPR lines intensities for lines with $g = 2.1$ and $g = 4.2$ on the one side and with $g \approx 3$ on the other side. The signal intensities of lines with $g = 2.1$ and $g = 4.2$ decrease with T but the line intensity with $g \approx 3$ increases up to $T \approx 50 - 100K$ and then decrease with T. Such a temperature dependence in general is very similar to one observed for $La_{1-x}Sr_xCuO_4$ (x = 0.075) in [1].

DISCUSSION

The nature of the observed signals with $g = 2.1$ is connected with disordering ("electron defects") of ordered covalent bonds M-O (M = Cu, Bi) [3]. Such defects have local magnetic moments. Localized pairs of such defects in triplet states give the EPR signals with $g = 4.2$ [3]. With the "electron defect" appearance the surrounding ion lattice locally deforms, *i.e.* "lattice" defects arise with the consequent magnetic moments. "Lattice" defects have much larger effective mass as compared with "electron" defects and so their g-factor will be large then 2.1. From Fig. 1 and 2 it follows that for "lattice" defects $g \approx 3$. The temperature dependence of the EPR signals intensity with $g = 2.1$ and $g = 4.2$ is defined by the activation energy (2 - 4K) for the "electron" defect creation and by the energy difference (5 - 10K) between the singlet and triplet states of such defects pairs. With temperature the intensity of signals will decrease because of saturation of the excited levels. The combination of activation and saturation mechanisms will produce the temperature dependence with maximum. The activation energy for "lattice" defects is much higher (50 - 100K), so the maximum of their signal intensity will be achieved at higher temperatures.

Thus the observed EPR signals indicate on the existence the CDW in oxygen sublattice of the HTS systems. The heating will produce the disorder in the ideally ordered covalent bonds M-O with consequent local lattice deformation.

ACKNOWLEDGMENTS

Authors thank for support the Scientific Committee on HTS Problems.

REFERENCES

1. Kochelaev, B.I., Sichelschmidt, J., Elschner, B., Lemor, W., and Loidl, A., *Phys. Rev. Lett.* **79**, 4274 (1997).
2. Eremina, R. *et al.*, *Physica B* (in print).
3. Yakubovsky, A.Yu., Gudenko, S.V., Anshukova, N.V., Golovashkin, A.I., Ivanova, L.I., and Rusakov, A.P., *JETP* **88**, 732 (1999).
4. Anshukova, N.V., Golovashkin, A.I., Ivanova, L.I., and Rusakov, A.P., *Physica C* **273**, 151 (1996).

XAFS Study on the Structural Phase Transition in PbTiO₃

T. Miyanaga[*], D. Diop[†], S. Ikeda[*], and H. Kon[*]

[*]Department of Materials Science and Technology, Faculty of Science and Technology, Hirosaki University, Aomori 036-8561, Japan
Department of Physics, Faculty of Sciences and Techniques, University Cheikh Anta Diop of Dakar, Dakar, Senegal

Abstract. The Pb L_{III}-edge EXAFS for PbTiO₃ was measured in the temperature range from 20K to 873K. Three types of atomic pairs, shorter Pb-O(1), medium Pb-O(2) and longer Pb-O(3) were analyzed independently. The former two Pb-O pairs show the "order-disorder" type phase transition; Pb-O distance does not change through T_c. On the other hand the last one shows the "displacive" type; the interatomic distance decreases discontinuously at T_c and the mean square relative displacement shows the soft mode behavior.

INTRODUCTION

Some of the perovskite crystals with an ABO₃-type molecular structure are well known to be ferroelectric compounds and undergo the structural phase transitions at specific temperatures [1-5]. Recently, X-ray absorption fine structure (XAFS) [6,7] has become a powerful technique to study the local structure of the materials based on the short-range order and is considered as a complementary technique to the diffraction methods which are based on the long-range order.

In materials undergoing structural phase transition, the average (long-range) distortion is represented by the order parameter. However, the actual local distortion can be different. If the local distortions and the order parameter have similar temperature dependencies in the range including the transition temperature, T_c, the type of the phase transition is "displacive". In this case the orientation of the local distortion is uniform over the crystal. On the other hand, if the local distortion does not change with temperature across the T_c the transition is "order-disorder" type. In this case, only the orientations of the distortion within a small domain change with temperature; the orientation is arranged along to preferable direction below T_c, but it is disordered above T_c. In this paper, we study the local structural change around Pb atoms in PbTiO₃ crystal for three types of Pb-O pairs independently and make clear the type of phase transition.

CP554, *Physics in Local Lattice Distortions*, edited by H. Oyanagi and A. Bianconi

EXPERIMENTAL AND DATA ANALYSIS

The powder sample of PbTiO$_3$ (99.99%) was checked by X-ray powder diffraction method to assure its crystal structure.

X-ray absorption spectra of Pb L_{III}-edge (13.04 keV) were measured at BL7C of Photon Factory (KEK) with transmission mode. Si(111) double crystal monochrometer was used. The temperature range for measurement was from 20 to 873 K. The EXAFS analysis was performed according to XANADU code [8]. The k-range for the Fourier transform was from 2 to 14 Å^{-1}. The non-linear least square (curve-fitting) method was applied to optimize the structural parameters in which theoretical parameters of EXAFS function obtained from FEFF 6 code [9] were used.

RESULTS AND DISCUSSION

Figure 1 (a) shows the EXAFS oscillation functions, $k^2\chi(k)$, of Pb L_{III} EXAFS for PbTiO$_3$ crystal at various temperatures. As the temperature increases, the amplitude of the EXAFS function decreases gradually. Figure 1 (b) shows the Fourier transforms of Pb L_{III} EXAFS of the PbTiO$_3$ crystal at various temperatures. The peaks at r-range from 2 to 4 Å include three Pb-O pairs and two Pb-Ti pairs. The peak intensity decreases when the temperature increases and a change of its fine structure is observed.

Three kinds of distances have been obtained for Pb-O: shorter Pb-O(1), medium Pb-O(2) and longer Pb-O(3) and two kinds of Pb-Ti; shorter Pb-Ti(1) and longer Pb-Ti(2) since the structure is distorted along to c axis. We tried to perform the 5 shells-

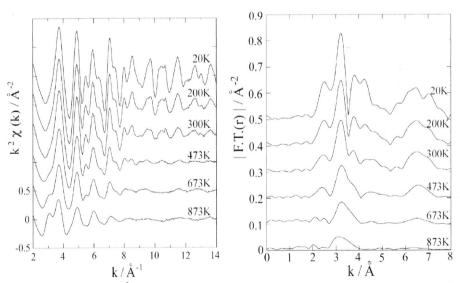

FIGURE 1. (a) EXAFS $k^2\chi(k)$ spectra of Pb L_{III}-edge for PbTiO$_3$ at various temperatures. (b) Fourier transforms of them.

fitting using eq.(1) (three Pb-O and two Pb-Ti) at temperatures higher than 300 K in order to obtain the structural parameters and obtained the reasonable results. Sicron *et al.* [7] also performed 5 shells-fitting for Pb L_{III} edge but the fitting conditions are limited in their fittings: (1) the lattice constants of a and c axis were fixed to the value at 150K; (2) and they assumed that σ^2 values for the second and the third nearest Pb-O are the same, and the ratio between σ^2 for first nearest Pb-O and that for the second and third nearest Pb-O are the same, and the ratio between σ^2 for first nearest Pb-O and that for the second and third nearest Pb-O was fixed to be constant. We improved these points: the fitting was performed on the five shells independently.

Temperature Dependence of Interatomic Distances

Figure 2 shows the variations of the interatomic distances of the three Pb-O. We can easily find the tetragonal distortion below T_c. For example in the tetragonal phase at r.t., the distance for Pb-O(1) is about 2.44 Å, that for Pb-O(2) is about 2.74 Å and that for Pb-O(3) is 3.54 Å. We can see an interesting phenomenon for the change of the interatomic distance through T_c. The existence of the local distortions for three Pb-O pairs can be found in TiO$_6$ unit above T_c as observed in the neutron diffraction [5] and in EXAFS reported by Sicron *et al.* [7]. Especially the fact that the interatomic distances of Pb-O(1,2) do not change across T_c indicates the local distortion remains in the cubic phase. Combing the general fact that the spontaneous polarization disappears at the cubic phase, we can state that the phase transition belongs to the "order-disorder" type. The most interesting phenomenon is that the interatomic distance of the longest Pb-O(3) decreases discontinuously on T_c; from 3.3 Å to 3.0 Å. Such a discontinuous change has not been reported in the EXAFS study by Sicron *et al.* [7]. The behavior of temperature dependence of the interatomic distance for Pb-O(3) suggests the "displacive type" of the structural phase transition.

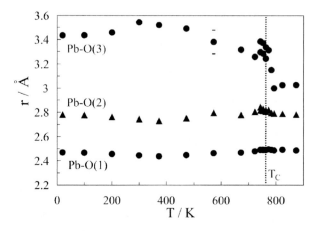

FIGURE 2. Temperature dependence of the interatomic distances for Pb-O atomic pairs for the first nearest Pb-O(1), the second nearest Pb-O(2) and the longest Pb-O(3).

Temperature Dependence of the Debye-Waller factor, σ^2

The mean square thermal amplitude of the neutron diffraction peaks has been studied for $PbTiO_3$ [5]. However, the relative mean square thermal displacement (Debye-Waller factor, σ^2) of EXAFS refers to the fluctuation in the interatomic distance between an atomic pair and it does not give the same information as that one in the diffraction methods.

The variations of σ^2 of the Pb-O pair with temperature are shown in Fig. 3. We focus our discuss to the temperature dependence of σ^2 for the longest Pb-O(3) atomic pair. The σ^2 for Pb-O(1,2) are monotonically changed as temperature [10]. However, it is quite interesting and different from Pb-O(1,2) atomic pairs that σ^2 for Pb-O(3) increases steeply (or diverge) at T_c, and keep the large value over T_c. The big jump of the σ^2 for Pb-O(3) at T_c can be interpreted as the O(3) atoms located at the one potential minimum in the double-well potential shift to the wide range of the whole potential width as thermally exited. The σ^2 is kept at the value for the wide potential at $T > T_c$. Such behavior is considered to be a direct evidence that O(3) atom exists in the double-well potential and corresponds to so called "soft mode" behavior. In the present study, we find that the Pb-O(3) atomic pair plays an important role for the soft mode behavior in $PbTiO_3$.

Summarizing these results: 1) Pb-O(1) and Pb-O(2) atomic pairs show the "order-disorder" feature, that is, the interatomic distances of these pairs do not change and σ^2 values change continuously over T_c. 2) On the other hand, the pair of Pb-O(3) shows "displacive" feature, that is, the interatomic distance decreases and σ^2 increases (or diverges) discontinuously at T_c. This phenomenon is considered as the "soft mode" behavior.

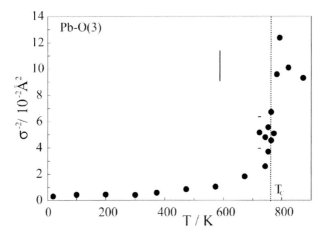

FIGURE 3. Temperature dependence of σ^2 (Debye-Waller factor) for longest Pb-O(3) atomic pair.

ACKNOWLEDGMENTS

This work was performed under the approval of Proposal No. 97G042 of Photon Factory. This work was partially supported by the Research Grants of Hirosaki University.

REFERENCES

1. Jona, F., and Shirane, G., *Ferroelectrics,* Pergamon Press, Oxford, 1962
2. Fujimori, M., *Physics of the Structural Phase Transitions,* Springer, 1997
3. Bruce, A.D., and Cowley, R.A., *Structural Phase Transitions,* Taylor & Franics LTD., 1981
4. Mabud, S.A., and Glazer, A.M., *Acta Cryst.*, **12**, 49 (1979).
5. Nelmes, R.J., Pilitz, R.O., Kuhs, W.F., Tun, Z., and Restori, R., *Ferroelectrics*, **108**, 165 (1990)
6. Koningsberger, D.C., and Prins, R., *X-ray Absorption: Principles, Application Techniques of EXAFS, SEXAFS and XANES,* John Wiley, 1988
7. Sicron, N., Ravel, B., Yacoby, Y., Stern, E.A., Dogan, F., Rehr, J.J., *Phys. Rev. B* **50**, 13168 (1994)
8. Sakane, H., Miyanaga, T., Matsubayashi, N., Watanabe, I., Ikeda, S., Yokoyama, Y., *Jpn. J. Appl. Phys. 32*, 4641 (1994)
9. Zabinsky, S.I., Rehr, J.J., Ankudinov, A., Albers, R.C., and Eller, M.J., *Phys. Rev. B* **52**, 2995 (1995).
10. Miyanaga, T., Diop, D., Ikeda, S., Kon, H., unpublished.

Local Lattice Instability of Cuprous Ions in NaBr and NaCl

S. Emura[1], K. Mutaguchi[2], Y. Ito[2], Y. Takabayashi[3], and Y. Kubozono[3]

[1] *ISIR, Osaka University, Mihogaoaka 8-1, Ibaraki Osaka 567-0047, Japan*
[2] *ICR, Kyoto University, Uji, Kyoto 667, Japan*
[3] *Faculty of Science, Okuyana University, Tsushima-naka 3-1-1, Okayama 700, Japan*

Abstract. The local lattice instability around the imperfection doped in the host matrix is invetigated with ultra-violet absorption spectra and x-ray excitation spectra. The local environment of a cuprous ion in NaCl shrinks, and in NaBr cuprous ion is lacated away from the regular lattice site. This phenomenon is discussed with the pseudo Jahn-Teller effect.

INTRODUCTION

Silver halides and cuprous halides are well known as "covalent-ionic" crystals. Several authors [1,2] had proposed for some of those to have average structure in which metal ions occupied not only central stable position but also several metastable positions. A number of anomalies of the physical properties are found [3,4]. Most of them appear to be attributed to a delicate balance of the ionicity and covalency. The central local lattice instability has been also observed in some ionic crystals doped with silver or cuprous ions as a substitutional imperfection [5-7]. The cuprous halide crystals take symmetry of zinc blende type (T_d) and alkali halides O_h. Here, the substitutional Cu^+ ion liganded with six halide ions in alkali halides will feel to be in an exotic space. Such the inconsonant crystal structure of the dopant to the matrix may give rise to an explicit change on the essential delicate balance of the ionicity and covalency. As an example, Fig. 1 schematically shows a Cu^+ ion located away from a regular lattice site to [111] direction. This situation has generally been known as "Off-center" phenomenon [8].

On origin of the lattice instability, (cooperative) Jahn-Teller effects are well known at present. It occurs in mono-materials and others on the balance with other interaction when the electronic state is degenerated. On the other hand, mixed materials with the different crystal symmetries such as in the solid solution of O_h symmetry and T_d symmetry show (local) lattice instability in some cases. In such a case, the electronic degeneracy is not necessarily condition. Furthermore, it appears to be more general. Its few cases are tentatively treated as pseudo Jahn-Teller distortion.

The optical measurements have revealed many features on these distortions. For the off-center phenomenon, no temperature dependence of the oscillator strength observed in KCl: Cu^+, KBr: Cu^+, and KI: Cu^+ gives the clear evidence. By the results of the optical observation, this instability is classified into four types [5], following the

CP554, *Physics in Local Lattice Distortions*, edited by H. Oyanagi and A. Bianconi
© 2001 American Institute of Physics 1-56396-984-X/01/$18.00

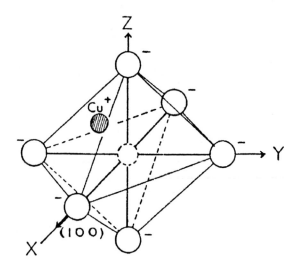

FIGURE 1. Schematic representation of the displacement of Cu^+ in $[CuX]^{5-}$.

characteristic temperature dependence of the parity-forbidden absorption band intensity. The classification also describes the corresponding adiabatic potential types. In the case of NaCl: Ag^+, it shows a clear harmonic on-center potential from the temperature dependence of the oscillator strength proportional to T. RbCl: Ag^+ shows shallow off-center potential owing to its peculiar temperature dependence of the oscillator strength.

XAFS is a direct tool to determine the exact position of the cuprous ion in its exotic space. We have observed XAFS by a fluorescent mode on the cuprous ion in NaCl [9] and NaBr. In NaCl [10], the cuprous ions locate the lattice position, but the equal six distances from the cuprous ion to the six neighbors are shorter than that of Na-Cl. This fact implies loose binding of the defect ion and the ligand ions. In NaBr, the cuprous ions are not on center, but shifts to (111) or (110) direction. This local environment of the cuprous ions is rather suitable to cuprous halides. These observations may more generally mean that the local rearrangement around the guest ions is distorted to be rather closer to the structure of the crystal formed by guest ion and its neighbors. This distortion is described by pseudo Jahn-Teller effect [11, 12].

EXPERIMENTS

The single crystals were grown from ultrapure powder by Kyropoulous method in air. The cuprous halides were added in melts of alkali halides. The concentration of Cu^+ ion was determined by atomic absorption analysis. The trapped concentration of cuprous halides in host halides is about 10% of the nominal doping one. The powder was held for a day at 400 C° in order to eliminate the hydroxyl ions before melting.

The cleaved crystals are annealed for about 2 minutes at the temperature of 50 K below the each melting point, and then rapidly quenched to the room temperature to avoid the aggregation of the impurities. The fundamental absorption measurements in the ultra-violet region were carried out at 1.2K. A light source is a high-regulated 200 W deuterium lamp. A monochromator used here is a CT-50 type grating one.

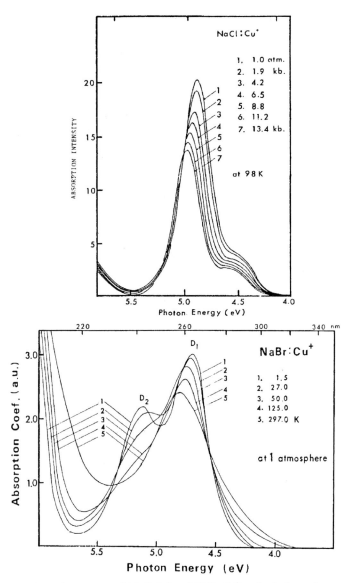

FIGURE 2. UV-absorption spectra of NaCl:Cu$^+$ and NaBr:Cu$^+$ under hydrostatic pressures and at several temperatures, respectively.

The XAFS spectra were measured at XAFS station installed BL7C in Photon Factory (Tsukuba, KEK). The x-rays were focused on a sample crystal by means of sagittal focusing of the second monochromator-Si-crystal. A Ni filter was employed to reduce the contamination of the detecting x-ray photons. The energy of the x-rays was calibrated at a pre-edge peak of the Cu metal to be 8.9788 keV. The low temperatures were achieved with a closed-cycle cryostat.

RESULTS AND DISCUSSION

Fig. 2 shows absorption spectra by Cu^+ ions in NaCl and NaBr under hydrostatic pressures and at several temperatures, respectively. The absorption band is assigned to the transition from $d^{10}(A_{1g})$ to $d^9s(E_g+T_{2g})$, which is essentially dipole-forbidden one, and is partially allowed by interaction with the lattice vibration of odd modes or static distortion of odd types. The energy diagram in the excited state is represented in Fig. 3.

The spectrum from NaBr:Cu^+ shows a different feature. The absorption intensity peculiarly depends on the temperature. That in NaCl:Cu^+ shows the standard behavior of parity forbidden transitions allowed partially by odd-mode vibrations. The temperature dependence of the oscillator strength in some spectra observed earlier is summarized in Ref. (5). This is a good monitor of the instability features. Thus, it is expected that cuprous ions in NaCl occupy the regular lattice site and that those in NaBr locate at the off-center position. However, the explicit environment around the cuprous ions cannot be analyzed by UV absorption spectroscopy.

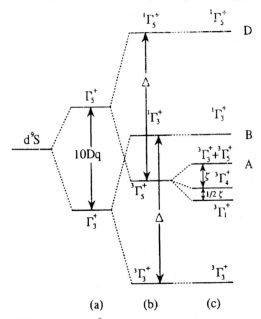

(a)   (b)   (c)

FIGURE 3. Energy splitting sheme of d^9s electron configuration in the cubic field. (a) shows the ligand field splitting, (b) the electron-electron interaction, and (c) the spin-orbital splitting.

For this goal, we try to record XAFS spectra around Cu K-edge with fluorescent mode. Fig. 4 shows XAFS spectra around Cu K-edge of NaCl:Cu$^+$ and CuCl as reference. The analyzed results are

$$
\begin{array}{ll}
\text{Cu - Cl in CuCl} & 2.34 \text{ Å,} \\
\text{Na - Cl in NaCl} & 2.82 \text{ Å,} \\
\text{and Cu - Cl in NaCl:Cu}^+ & 2.77 \text{ Å.}
\end{array}
$$

The cage constructed by Cu$^+$ and six Cl$^-$ somewhat shrinks. Based upon the empirical ionic radii, it seems to be reasonable. However, taking into consideration of covalent tendency of cuprous ion and physical interaction with six ligands, the distance appears to be very long. Therefore, cuprous ion will be in loose chemical combining. In NaBr:Cu$^+$, the bond distance between cuprous ion and the ligands Br$^-$ is found to be contrast to the case of NaCl:Cu$^+$, that is, 2.44 ± 0.05 Å under one shell analysis. This is very close to the bond distance of Cu$^+$-Br$^-$ in CuBr (2.464 Å). If it is precise, the cuprous ions in NaBr surely locate the off-center position and the direction of the off-center is preferable to [111] direction. Three-shell analysis is tried, but we do not have reasonable values at present because of noisy data. In any way, both analyses from the temperature dependence of the absorption intensity and XAFS spectrum indicate that the cuprous ions in NaBr do not locate the regular lattice site. On the degree of the off-center, both show the contrast results. The former suggests being a slight off-center, and the later to be a deep off-center.

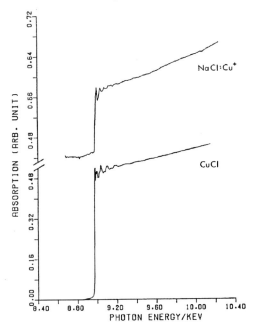

FIGURE 4. Excitation spectra around Cu K-edge from NaCl:Cu$^+$ and CuCl at about 30K.

These phenomena can be treated as pseudo Jahn-Teller problem in the ground state of cuprous ion. Taking rather strong covalency of CuBr into consideration, it is not so largely out of discussion that the ground state of cuprous ion in NaBr mixes partially with a p-like orbital, which is surely low-lying charge transfer state, through T_{lu} vibration modes. There are three T_{lu} modes for the octahedral cluster AX_6. One of them gives the vibration of only A atom, and this mode is observed as a resonant mode with low frequency [13]. This interaction is just pseudo Jahn-Teller effects. Thus, this instability is described as pseudo Jahn-Teller type instability.

REFERENCES

1. Matsubara, T., *J. Phys. Soc. Jpn.*, **38** 1076 (1975).
2. Miyake, S., and Hoshino, S., *Rev. Mod. Phys.*, **30** 172 (1958).
3. Hanson, R.C., and Helliwell, K., *Phys. Rev.* **139** 2649 (1974).
4. Potts, J.E., Hanson, R.C., and Walker, C.T., *Phys. Rev.*, **139** 2711 (1974).
5. Holland, U., and Luty, F., *Phys. Rev.* **B19** 4298 (1979).
6. Emura, S., and Ishiguro, M., *Phys. Rev.* **B38** (1988) 2851.
7. Emura, S., Maeda, H., and Kubozono, Y., *J. Phys.* (Paris), 7 C2-585 (1997).
8. Seiz, F., *Rev. Mod. Phys.*, **23** 328 (1951).
9. Emura, S., Murata, T., Maeda, H., Nomura, M., Moriga, T., and Koizumi, A., in *X-ray Absorption Fine Structure*, edited by Hasnain, S.S., Horwood, London, 1991, pp. 432.
10. Emura, S., Murata, T., Maeda, H., Nomura, M., and Moriga, T., *Jpn. J. Appl. Phys.*, Suppl. **32**-2 737 (1993).
11. Nagasaka, S., *J. Phys. Soc. Jpn.* **50** 1570 (1981).
12. Nagasaka, S., *J. Phys. Soc. Jpn.* **52** 898 (1982).
13. Weber, R., and Siebert, A.F., *Z. Phys.*, **213** 273 (1968).

On the Possibility of Spontaneous Breaking the Ideal Crystal Symmetry

E. N. Myasnikov and A. E. Myasnikova

Rostov State Pedagogical University, 344082 Rostov-on-Don, Russia

Abstract. On the base of the single-particle distribution function obtained for fermions in systems where their autolocalization is possible it is shown that there exist some critical temperature T_c such that macroscopic number of carriers from partially or completely filled band merge into autolocalized states when temperature decreases lower than T_c and return back into the band states when the temperature increases above T_c. Simultaneously with such phase transition in the system of coexisting autolocalized and delocalized fermions a window (*i.e.* a region where band states are absent) appears/disappears in the coresponding band.

WINDOWS IN THE PARTIALLY FILLED ELECTRONIC BAND DUE TO ELECTRON – PHONON COUPLING

It is ordinarily considered that autolocalization of electrons from some band with formation of large radius polarons does not introduce changes into structure of the band but only changes its filling. In the present article we intend to demonstrate that the structure of the band will also change. We shall consider mediums where there exist completely or partially filled bands of free electrons limited from above and from below by gaps. Such situation takes place in semiconductors and following accustom terminology we shall call these bands valence band and conduction band accordingly. We shall suppose that filling of the conduction band can be sufficiently high to provide existence of Fermi surface in it. Such electronic structure occurs in high-temperature superconductors so that our conclusions will be valid for them.

Ordinarily band theory considers that each of states in the valence and conduction bands can be filled by electron. However, it has been shown [1, 2] that if electrons in the conduction band with the bottom in Γ-point of the first Brillouin zone (FBZ) are autolocalized then some set of states with momentums from the FBZ becomes inaccessible, *i.e.* not existing. Absence of some set of states in any (valence or conduction) band we shall call a window in this band. Against a background of neighboring filled band states the boundary of the window will manifest itself similarly to Fermi surface (FS). Possibility of such windows existence should be kept in mind at interpretation of angle-resolved photoemission spectroscopy (ARPES) data from semiconductors including high-T_c superconductors which are reported in many recent publications [3].

Our previous works [1, 2] discussed a problem of the carrier autolocalization in a partially filled band. Two mechanisms of windows rise due to Frohlich electron-

CP554, *Physics in Local Lattice Distortions*, edited by H. Oyanagi and A. Bianconi
© 2001 American Institute of Physics 1-56396-984-X/01/$18.00

phonon interaction leading to the large radius polaron formation were in fact found there. (It should be noted that both these mechanisms are valid for any electron-phonon interaction resulting in electron autolocalization.) First mechanism is a direct consequence of Landau's bose-liquid theory [4] applied to the phonon subsystem of the polaron [1]. In accordance with Refs. [1, 4] motion of the autolocalized electron with the velocity higher than maximum group velocity u of phonons participating in the polaron formation results in destruction of the formed state of the phonon vacuum and electron will turn into delocalized state. Moreover, effective mass of the autolocalized electron demonstrates quasirelativistic dependence on the polaron velocity where u plays the role of the speed of light [5] For this reason calculation of the polaron energy dispersion or density of states in a model with dispersionless phonons (*i.e.* u = 0) (for example, [6]) can yield correct result only for one point corresponding to zero total momentum of the polaron. Thus, the first mechanism of the autolocalization influence on the conduction band structure leads to formation of a window of the size m^*u in the band of delocalized states (as it is depicted in Fig. 1). As $m^*u \ll \hbar\pi/a$ where a is the lattice constant to observe this window in the conduction band is very difficult.

Second mechanism able to generate a window in the conduction band results from Pauli exclusion rule applied to the system of fermions where their autolocalization is possible [2]. In accordance with quantum mechanics each state of a particle occupies in the single-particle phase space the volume $(2\pi\hbar)^3$. Let a carrier in autolocalized state is localized in a volume V_0. Then available electron momentums in this state are

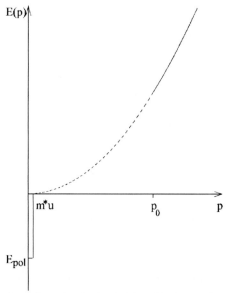

FIGURE 1. Conduction band dispersion in the vicinity of its bottom in a semiconductor with strong electron-phonon coupling.

restricted from above by a value p_0 such that

$$4\pi p_0^3 V_0 / 3 = (2\pi\hbar)^3 \qquad (1)$$

For the small radius polarons p_0 is of the order of $\hbar\pi/a$ but even for the large radius polarons $p_0 \gg m^*u$. As it can be seen from Exp. (1), if in a crystal of the volume V there are V/V_0 autolocalized conduction electrons (it is their maximum number [2]) then they occupy in 6D phase space the same region (of the volume $4\pi p_0^3 V/3$) that falls on all delocalized electrons with momentums $p < p_0$. In such a case in accordance with Pauli exclusion rule delocalized states with $p < p_0$ can not be occupied by electrons, *i.e.* a window of the radius p_0 appears in the conduction band. If autolocalized states are not completely occupied by electrons (for example if a part of them is destroyed by the heat motion) delocalized electrons with $m^*u < p < p_0$ can exist in the crystal. For example, if the polaron number in the crystal is $0.5\ V/V_0$ then in accordance with Pauli exclusion rule the number of delocalized electrons with momentums $p < p_0$ can be not higher than $0.5\ V/V_0$. For this reason the part of the conduction band dispersion with $m^*u < p < p_0$ is shown in Fig. 1 with dashed line. If this part of states in the conduction band is completely forbidden we shall speak about it as about "opened" window in the conduction band. Such opened window is demonstrated by Fig. 2 where the FBZ section by the Γ - X - M plane is shown for the case when the band bottom lies in the Γ-point of the FBZ. It can be easily shown that there will be no difference of principle if the band bottom lies in other FBZ point.

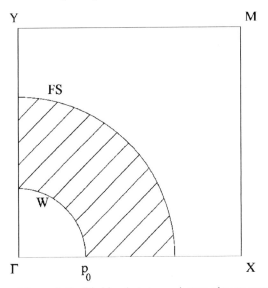

FIGURE 2. Structure of the partially filled band at strong electron-phonon coupling. Shading shows the cross-section of the FBZ region filled by delocalized carriers by the plane Γ-X-M. This region is limited by Fermi surface marked FS and spherical (with the radius p_0) boundary of the window marked W. The bottom of the band is supposed to be in Γ-point of the FBZ.

Unfortunately, Pauli exclusion rule in a system of coexisting autolocalized and delocalized electrons was not taken into account in numerous works on the polaron conductivity of substances and on bipolaron theory of the high-temperature superconductivity. Only in Ref. [2] a single-particle distribution function for the system of coexisting autolocalized and delocalized fermions has been constructed with exclusion rule kept in mind. Fig. 3 demonstrates temperature dependence of the concentration of polarons, electrons with $p < p_0$ and electrons with $p > p_0$ calculated in accordance with this distribution function for the system with fixed overall concentration of the conduction electrons. This dependence clearly demonstrates "opening and closing the window" in the conduction band at changing temperature. Since both these processes are changes of state of a macroscopic quantity of conduction electrons they can be considered as phase transitions in a system of these electrons. As it can be seen from Fig. 3 the window in the conduction band exists at temperatures lower than the polaron thermal destruction temperature T_c which as it was shown in Ref. 2 can be much lower than the polaron binding energy.

Absence of delocalized electron states in the window is caused by formation of down-polarons (d-polarons), *i.e.* polarons in which average momentum of electrons corresponds to the bottom of the band or to FBZ points in its vicinity. The boundary of this window marked in Fig. 2 by W will manifest itself in angle-resolved photoemission spectroscopy (ARPES) like additional (nested) Fermi surface. W-surface location is determined by the electron-phonon coupling strength: when it increases the value of V_0 decreases and in accordance with Eq. (1) p_0 increases inversely proportional to the autolocalized state radius. W-surface does not shift with changing the carrier concentration while it is higher than $1/V_0$. At lowering the carrier concentration FS approaches to W. At the carrier concentration lower than $1/V_0$ and T $< T_c$ delocalized carriers are absent in the band. At increasing temperature above T_c delocalized carriers appear in the band simultaneously with disappearance of the window.

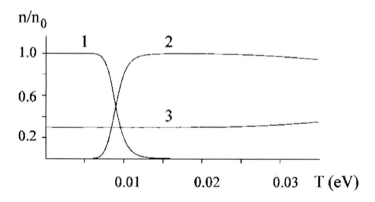

FIGURE 3. Temperature dependence of the polaron concentration (curve 1), concentration of conduction electrons with $p < p_0$ (curve 2) and concentration of electrons with $p > p_0$ (curve 3) at fixed overall concentration of the conduction electrons $n = 1.3 \, n_0$. All concentrations are expressed in units of the maximum polaron concentration $n_0 = 1/V_0$. The polaron binding energy is $0.11 \, \text{eV}$.

318

Speaking about the conditions of rise and manifestation of these features it should be recalled that electron in polarons of the large and small radius formed due to strong electron-phonon coupling is quickly moving subsystem demonstrating only adiabatic reaction to the phonon subsystem motion. In the case of weak electron-phonon interaction such adiabatic approach is not valid, autolocalized states of an electron are impossible and the window in the conduction band bottom is absent.

WINDOWS IN THE COMPLETELY FILLED BAND

Every electron in the valence band, being a charge carrier, undoubtedly participates in Frohlich interaction with optic dipole lattice vibrations. Let us consider the case when this interaction is strong and adiabatic condition [7] is satisfied. Can in such a system one valence electron with the momentum near the bottom of the completely filled band merge into a polaron state? No, it can not in accordance with Pauli exclusion rule. Indeed, free electron and autolocalized electron occupy in their 6D phase space regions quiet different in their form. Autolocalized electron has limited volume V_0 and fills partially each of states with momentums inside the sphere of the radius p_0. Hence, transition of a valence band electron from a state with certain momentum into autolocalized state is possible only if other electrons would be slightly restricted in the momentum space and clear a region of the volume V_0 in the coordinate space. Thus, such transition must be many-electron. Many-electron transition of valence electrons with momentums lower than p_0 into autolocalized states will occur if for all the system of such electrons this transition is energetically profitable. Thus, the possibility of autolocalization of carriers from the completely filled band is also determined by the strength of their interaction with phonons as it takes place in the case of conduction electrons.

If in a semiconductor there is strong coupling of valence electrons with phonons (*i.e.* coupling which can result in formation of polarons with the energy under the valence band bottom) then all the electrons with momentums lower than p_0 turn out to be in autolocalized states, i.e. form d-polarons. The number of such autolocalized electrons is $4\pi p_0^3 V / 3(2\pi\hbar)^3$ and the number of polaron states in a crystal of the volume V is V/V_0. Hence, with taking into account Eq. (1) all possible states of d-polarons of the valence band turn out to be filled. The number of free electrons in the valence band remains equal to the number of available delocalized states in it because valence band states with $p < p_0$ in accordance with Pauli exclusion rule are forbidden. Thus, carriers of the completely filled band in this case as usually will not participate in the conductivity, plasma vibrations and metallic-type screening of electric fields. So, a window opening and closing with changing temperature, as in the case of the conduction band, can be formed near the valence band bottom.

Let us now turn to the question whether polarons can exist not only in the bottom of the valence and conduction bands? In accordance with Ref. 1 at the polaron formation group velocity grad E(p) (where E(p) is the electron energy dispersion in a band) of the electron wave packet must be not exceeding in the absolute value the minimum phase velocity u of phonons participating in the electron autolocalization. It is well known that at the electron band width about 1eV the group velocity of electrons in the middle of this band ordinarily exceeds u in the order of magnitude. Therefore, autolocalization of the electron wave packet is possible only if its average momentum lies in the region of the FBZ where grad E(p) is small, *i.e.* in the vicinity of points where $v_{gr} = gradE(p) = 0$. This condition is satisfied, for example, in the band bottom and top.

In the partially filled band the top is usually not filled by carriers. Therefore, let us consider the possibility of autolocalization of electron with average momentum corresponding to the top of the valence band (resulting polaron we shall call u-polaron in contrast with d-polaron corresponding to the bottom of the band). Let in the vicinity of the valence band top momentum dependence of the electron energy can be considered as quadratic with constant effective mass. Then with increasing degree of the electron localization (*i.e.* with increasing limiting momentum p_0 of the electron in the wave packet) average kinetic energy of electron in the polaron will change proportionally to p_0^2 with a negative coefficient. Energy of Frohlich interaction of such localized electron with phonons is also negative and proportional to p_0. Thus, complete energy of the localized electron is a function of p_0 of the form

$$-p_0^2 - \gamma p_0 \tag{2}$$

where positive coefficient γ is determined by the strength of the electron-phonon coupling. Hence, in the region of momentums where quadratic dependence of energy of the band electron on its momentum is valid autolocalized states are more profitable than band ones, that must lead to collective many-electron transition from band states near the band top into autolocalized ones. This results in formation in the valence band of a window centered at the band top. Since both items in Eq. (2) are negative u-polarons formation will take place even at weak electron-phonon interaction, that is quiet different from the d-polaron case. However, at sufficiently high momentums of electrons their kinetic energy ceases to be quadratic function of the momentum with a negative coefficient and, besides, the character of their coupling with phonons changes.

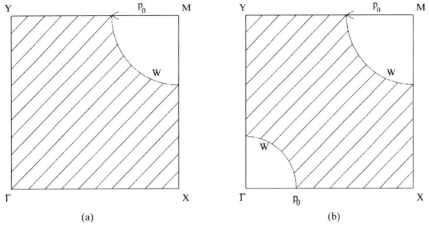

FIGURE 4. Structure of the completely filled band with bottom in Γ-point of the FBZ and top in M-point of the FBZ at weak (panel a) and strong (panel b) electron-phonon coupling. Shading shows the cross-section of the FBZ region filled by delocalized carriers by the plane Γ-X-M.

Therefore, one can hope that only for a part of valence electrons with momentums $p < p_0$ many-electron transition into autolocalized state will be energetically profitable.

Thus, at weak electron-phonon coupling in the valence band there is one window caused by u-polarons formation and centered at the band top. At strong electron-phonon coupling the band structure in the vicinity of the valence band top does not differ qualitatively from the weak-coupling case. But in the vicinity of the band bottom the window caused by d-polarons formation appears. Fig. 4 demonstrates the valence band structure for the case of weak (panel a) and strong (panel b) coupling. As Fig. 4(a) shows ARPES spectra in the former case will demonstrate presence of one window around the valence band top. In the latter case there will be two windows generally of different sizes as it is illustrated by Fig. 4(b). Besides, polarons of the valence band will manifest themselves as quasiparticle maximums, and the maximums number will be equal to the number of windows observed, *i.e.* one (due to u-polarons) in the weak-coupling case and two (caused by u- and d-polarons) in the strong-coupling case. With the temperature increase above T_c all these features will disappear.

CONCLUSION

On the base of the said above electronic structure of perfect semiconductor crystal (crystal without impurities and defects) in the ground state can be represented in a way quiet different from one based on theoretical calculation of electron states of such crystal with static symmetric arrangement of ions. At corresponding electron-phonon coupling not only electrons of the crystal ionic skeleton but also a part of valence band electrons turns out to be localized, and momentums of delocalized valence electrons can not occupy all the FBZ. At presence of autolocalized valence electrons in the crystal phonon vacuum will also change, *i.e.* lattice deformation will rise. This can result in a superstructure formation or in change of the crystal symmetry. Charge density waves caused by autolocalized valence electrons can generate incommensurate structure. Such electronic and ionic structure of the crystal will correspond to the minimum of its energy. Since this conclusion is based on calculations of autolocalization using as zero approximation ideal symmetry of semiconductor crystal and band structure of its electronic states the statement about dissymmetrical structure of the perfect crystal in the ground state appears to be well-founded. At increasing temperature in such dissymmetrical crystal a phase transition occurs due to thermal destruction of a macroscopic part of valence polarons (see Fig. 3), complete restoration of band structure of electronic states and return of ideal structure (with heat motion of ions, of course). Such a behavior is essentially similar to one observed in high-temperature superconductors, in particular, to characteristic, resembling Fig. 3 temperature dependence of their physical properties measured by X-ray and optical methods [8, 9]. In ARPES data from the ground state of such crystal absence of valence electrons in the regions (windows) restricted by some surfaces (looking therefore like additional Fermi surfaces) can be registered. These windows must disappear at temperature increase due to thermal destruction of polarons.

REFERENCES

1. Myasnikova, A.E., *Phys. Rev. B* **52**, 10457 (1995).
2. Myasnikov, E.N., and Myasnikova, A.E., *Zh. Eksp. Teor. Fiz.* **116**, 1386 (1999).
3. Ding, H. *et al.*, *Phys. Rev. Lett.* **78**, 2628 (1997); Saini, N.L. *et al.*, *Phys. Rev. Lett.* **79**, 3467 (1997); Gweon, G.-H. *et al.*, *Phys. Rev. Lett.* **81**, 886 (1998).
4. Landau, L.D., *Zh. Eksp. Teor. Fiz.* **11**, 592 (1941); *J. of Phys.* **11**, 91 (1947).
5. Myasnikova, A.E., and Myasnikov, E.N., *Phys. Rev. B* **56**, 5316 (1997); *Zh. Eksp. Teor. Fiz.* **112**, 278 (1997).
6. Alexandrov, A.S., and Dent, C.J., *Phys. Rev. B* **60**, 15414 (1999); Kornilovich, P.E., *Phys. Rev. B* **60**, 3237 (1999); Prokof'ev, N.V., and Svistunov, B.V., *Phys. Rev. Lett.* **81**, 2514 (1998).
7. Pecar, S.I., *Studies in Electron Crystal Theory*, Gostehizdat, Moscow, 1951.
8. Bianconi, A. *et al.*, *Phys. Rev. Lett.* **76**, 3412 (1996); Lanzara, A *et al.*, *Phys. Rev B* **55**, 9120 (1997).
9. Mihailovic, D. *et al.*, *J. Supercond.* **12**, 33 (1999).

Electronic Structure of BaTiO$_3$ by X-ray Absorption Spectroscopy

K. Asokan[a], J. C. Jan[a], J. W. Chiou[a], W. F. Pong[a], P. K. Tseng[a] and I. N. Lin[b]

[a]Department of Physics, Tamkang University, Tamsui, Taiwan.
[b]Department of Materials Science and Engineering, Materials Science Center, National Tsing-Hua University, Hsinchu, Taiwan

Abstract. BaTiO$_3$ is a prototype perovskite having many properties common to cuprates, manganites and nickelates. To understand the electronic structure of these perovskites, we have measured O K-edge of TiO$_2$, BaTiO$_3$ and CaTiO$_3$. Spectral features starting from 530 to 550 eV in O K-edge provides significant information to understand the role of alkaline metals like Ca, Ba. Electronic polarizability may be an important parameter to understand many properties of these perovskite materials.

INTRODUCTION

Barium titanate (BaTiO$_3$) has been an important ferroelectric prototype pervoskite investigated thoroughly using different techniques. The characteristic feature of its unit cell is the existence of TiO$_6$ octahedra [1-3]. Between ~ 10 and 120°C barium titanate is a ferroelectric insulator, with highly anisotropic dielectric and electro-optical properties. The mobility of holes in BaTiO$_3$ crystals is highly anisotropic at room temperature [4]. Presence of other 3d transition metals in similar environment leads to many interesting phenomena like high Tc superconductivity, colossal magnetoresistance, ferromagnetic and charge ordering [5]. Interestingly existence of anisotropic normal state properties is common to all these perovskites and lattice dynamics is considered to play an important role. The microscopic origin of these phenomena is still a mystery to scientific community. To understand the macroscopic properties of a material it is necessary to have a firm grasp on the electronic structure of constituent atoms which ultimately determines all of the observable macroscopic properties like phase stability, ferroelectric and magnetic behavior.

In this paper, we report x-ray absorption near edge spectra of O K-edge of BaTiO$_3$, CaTiO$_3$ and TiO$_2$ and understand the spectral features with reference to TiO$_2$. A brief discussion on the role of electronic polarizability of ions is presented in the context to understand various properties of perovskites.

CP554, *Physics in Local Lattice Distortions*, edited by H. Oyanagi and A. Bianconi
© 2001 American Institute of Physics 1-56396-984-X/01/$18.00

EXPERIMENTAL

BaTiO₃ and CaTiO₃ samples were prepared from reagent grade $CaCO_3$, $BaCO_3$ and TiO_2 powders (Merck Co, Darmstat, Germany) via standard ceramic routes. These samples were characterized by x-ray diffraction [6]. Room temperature x-ray absorption spectra were obtained for O K-edge in fluorescence mode using seven-element Ge detector. These measurements were carried out using the high-energy spherical grating monochromator beamline with electron beam energy of 1.5 GeV and a maximum stored current of 200 mA at the Synchrotron Radiation Research Centre, Hsinchu, Taiwan.

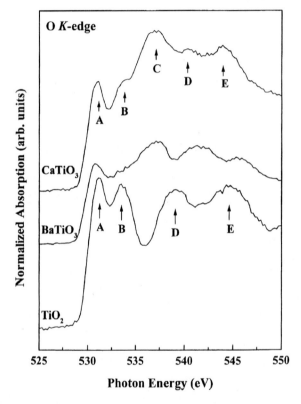

FIGURE 1. Normalized x-ray absorption spectra of O K-edge measured in fluorescence mode using seven-element Ge detector for TiO_2, $BaTiO_3$, and $CaTiO_3$ compounds. Note the spectral feature that appears at 537 eV in $BaTiO_3$ and $CaTiO_3$ and the feature at 534 eV attributed to e_g band is significantly different in all these systems.

RESULTS AND DISCUSSION

Fig. 1. shows the normalized x-ray absorption spectra of O K-edge of $BaTiO_3$, $CaTiO_3$ and TiO_2 samples in the energy range of 525 to 550 eV. All distinct features are marked from A to E and energy positions are centered at ~531, 534, 537, 540 and 545 eV respectively. These spectra are consistent with the data existing in literature especially for TiO_2 where considerable amount of work exists in literature [7-11]. The molecular orbitals of O $2p$ states can be observed in the O K-edge absorption spectrum and depends upon the energy and symmetry.

In a purely ionic model, for oxygen, the $1s \rightarrow 2p$ channel would be closed in the x-ray absorption process [10, 12, 13]. However, presence of considerable amount of covalent contribution reduces the number of filled states with O $2p$ character. Hence the strength of the O $1s$ signal at the threshold is related to the degree of covalency [12, 13]. The information on unoccupied oxygen p states in these compositions are of great importance to understand the hybridization of Ti and O states in TiO_6 octahedra which forms the basic structural unit in these compounds [7-9].

To understand the spectral features of O K-edge, it is necessary to recall the environment of the O. In titanates, oxygen ions are surrounded by six cations: four Ba cations separated a distance of $\mathbf{a}/\sqrt{2}$ from them (where $\mathbf{a}$ is the parameter of the cubic lattice) on the corners of a square whose centre is an anion and two Ti cations separated a distance of $\mathbf{a}/2$ at right angles to the square comprised of $\mathbf{a}$ sites [2].

Calculations showed that the double peak structure in the low-energy part of the spectra (region under peaks A and B) could be assigned to the Ti $3d$ derived states whereas the high-energy part of the spectra (region under peaks D and E) is formed by the delocalised states derived from antibonding O $2p$ and Ti $4sp$ bands [9, 14]. The first two peaks A and B are similar to TiO_2 with a splitting of ~2.5 eV. These are attributed to the unoccupied O $2p$ states which are covalent mixed with the Ti t_{2g} and e_g states [7, 8]. The e_g states are directed towards the O and therefore have a higher energy than the t_{2g} states. The distortion of the Ti site from octahedral to lower symmetry results in an asymmetric broadening of the peaks assigned to e_g. For TiO_2, the broadening is mainly caused by a tetragonal distortion of the octahedra which cannot be counteracted by the core-hole potential. In $BaTiO_3$ and $CaTiO_3$, the effect of core-hole potential is more prominent for the e_g band due to the overlap of first Ba and Ca related bands causing large dispersional broadening [12, 13].

The spectral feature at C is seen only in $BaTiO_3$ and $CaTiO_3$ which is likely to arise due to O $2p$ states hybridized with Ba $5d$ and Ca $3d$ states. The structure marked as D and E are seen in all samples and may be attributed to Ti $4sp$ states [8]. Note that these peaks are shifted towards lower energy side for E and higer energy side for D with reference to TiO_2. It is evident that there is finite hybridization of O $2p$ states with Ti $3d$ wave functions in the ground state for all compounds apart from admixture of sp states of Ca or Ba. Eventhough all these compounds posses TiO_6 octahedra in their unit cell, local distortions of the TiO_6 octahedra appear after the introduction of Ca and Ba. These alkaline cations play an important role in determining the complex ferroelectric and structural phase behavior. As evident from the Fig.1 that new spectral feature at 537 eV which is totally absent in TiO_2 compounds appears. It is concluded that the large energy spread (~15eV) of O $2p$ states is an indication of strong covalency in the TiO_2 compounds [14]. This indicates addition of Ca or Ba considerably changes the covalency of O $2p$ character. Covalency strongly depends on the electronic structure of the materials. Considerable evidence exists to prove that the presence of alkaline earth metals is essential for many interesting properties of perovskites [4].

Ferroelectric and dielectric properties derive from covalent as well as ionic contributions of the constituent atoms. Ferroelectric nature depends on the total polarizability of the system comprising of three components namely electronic, ionic and dipolar polarizability [16, 17]. Basically, the electronic polarizability (EP) arises from the displacement of the electron shell relative to a nucleus. In case of a compound, EP depends on the environment in which it is placed and it varies from element to element and critically depends on the electronic structure. Recently Ishihara *et al.* have studied the covalency contributions to the EP in some dielectric compounds and related these two concepts [18]. Adding Ba or Ca to TiO_2 compound leads to ferrolectric nature and increases the total electronic polarizability of cations of the compound. Since x-ray absorption measures the electronic environments around ion, one can treat EP can also related to the unoccupied states. As evident from our measurements, the O $2p$ character is mixed into many different unoccupied states. It is also known that O has variable EP depending upon the chemical environment. In fact these changes are evident from absorption spectra. From this discussion, EP of oxygen ion is an important in perovskites. Its value varies considerably from 0.5 to 3.2 $\mathring{A}^3$ depending upon the crystal structure and the cations involved [19]. The EP of oxygen and alkaline metals determine the lattice dynamics of TiO_6 octahedra. Temperature or any thermodynamical parameter can change the electronic structure and which is again can be related to change in EP. Such a change is likely to cause clustering and ordering to compete due to electric dipole. This quantity may determine the *order and disorder type* transitions arising due to permanent dipoles or dipole moment induced phase transitions also called as *displacement type* transitions. Other properties like the off-stoichiometry, which is also the characteristic of these compounds expected to originate due to variable oxygen. Many properties appear to originate due to the EP contribution of ions and it may be responsible for the basic perovskite framework.

CONCLUSIONS

We have reported x-ray absorption near edge spectra of O K-edge of $BaTiO_3$, $CaTiO_3$ and TiO_2. The spectral features were compared with TiO_2. A brief discussion on the role of electronic polarizability of ions was presented in the context to understand various properties of perovskites.

ACKNOWLEDGMENTS

One of us (K. A.) is thankful to Prof. G. K. Mehta, Director, Nuclear Science Centre, New Delhi for granting leave and encouragement. The authors (K. A. and W. F. P.) wish to acknowledge support by the National Science Council of the Republic of China under Contract No. NSC-89-2112-M-032-008. The excellent cooperation during the beamtime from SRRC staff is highly appreciated.

REFERENCES

1. Kwei, G.H., Lawson, A.C., Billinge, S.J.L., Cheong, S.W., *J. Phys. Chem.* **97**, 2368-2377 (1993).
2. *Encyclopedia of Chemical Technology* **10**, edited by Howe-Grant, M., John Wiley & Sons Inc, New York, 1980, pp.1-30.
3. Cohen, R.E., *Nature* **359**, 136-138 (1992); Cohen, R.E., and Krakauer, H., *Phys. Rev. B* **42**, 6416-6423 (1990).
4. Mahgerefteh, D., Kirillov, D., Cudney, R.S., Bacher, G.D., Pierce, R.M., and Feinberg, J., *Phys. Rev. B* **53**, 7094-7098 (1996).
5. Rao, C.N.R., and Raveau, B., *Colossal Magnetoresistence, Charge ordering and related properties of Manganese oxides*, World Scientific, Singapore, 1998.
6. Lin, T.F., Hu, C.T., and Lin, I.N., *J. Appl. Phys.* **67**, 1042-1047 (1990).
7. Brydson, R., Sauer, H., Engel, W., Thomas, J.M., Zeitler, E., Kosugi, N., and Kuroda, H., *J.Phys. Condens. Matter* **1**, 797-812 (1989).
8. Abbate, M., Potze, R., Sawatzky, G.A., Schlenker, C., Lin, H.J., Tjeng, L.H., Chen, C.T., Teehan, D., and Turner, T.S., *Phys. Rev. B* **51**, 10150-10153 (1995).
9. de Groot, F.M.F., Faber, J., Michiels, J.J., Czyzyk, M.T., Abbate, M., and Fuggle, J.C., *Phys. Rev. B* **48**, 2074-2080 (1993).
10. de Groot, F.M.F., Fuggle, J.C., Thole, B.T., and Sawatzky, G.A., *Phys. Rev. B* **41**, 928-937 (1990).
11. Wu, Z.Y., Ouvard, G., Gressier, P., and Natoli, C.R., *Phys. Rev. B* **55**, 10382-10391 (1997).
12. Van der Laan, G., *Phys. Rev. B* **41** 12366-12368 (1990).
13. Pedio, M., Fuggle, J.C., Somers, J., Umbach, E., Haase, J., Linder, Th., Hofer, U., Grioni, M., de Groot, F.M.F., Hillert, B., Becker, L., and Robinson, A., *Phys. Rev. B* **40** 7924-7927 (1989).
14. Ruus, R., Aarik, A., Aidla, A., Uustare, T., Kikas, A., *J. Electron Spectrosc. Relat. Phenom* **92**, 193-199 (1998) and references therein.
15. Matsubara, T., *The structure and properties of matter*, Springer, Berlin, 1982.
16. Kittel, C., *Introduction to solid state physics*, Wiley, New York, 1985, pp361-386.
17. Asokan, K., *Materials Lett.* **14**, 251-254 (1992).
18. Ishihara, S., Tachiki, M., and Egami, T., *Phys. Rev. B* **53**, 15563-15570 (1996); *ibid. Phys. Rev B* **49**, 16123-16128 (1994).
19. Dimitrov, V., and Sakka, S., *J. Appl. Phys* **79**, 1736-1745 (1996).

Theory for Photoinduced Structural Phase Transitions

H. Mizouchi, H. Ping and K. Nasu

Institute of Materials Structure Science, The Graduate University for Advanced Study,
1-1,Oho,Tukuba,Ibaraki,305 Japan

Abstract. For the photoinduced ionic → neutral structural phase transition in the organic charge transfer crystal TTF-CA, we, at first, theoretically show the adiabatic path of this transition, which starts from a single charge transfer (CT) exciton in the ionic phase, but finally results in a macroscopic neutral domain. In the next, using a more simplified model, we theoretically study nonlinear nonequilibrium quantum time-evolution dynamics of the photoinduced phase transition, with a special emphasis on the multistability of the ground state and the "initial condition sensitivity" of the exciton proliferation, both characterize the nonlinearity of this system.

INTRODUCTION

As is already well known, an electron in an insulating crystal induces a local lattice distortion around itself, when it is excited by a photon. This phenomenon called "lattice relaxation" of an optical excitation, has been studied in various kinds of insulating crystals for these fifty years. According to the original concept of this lattice relaxation, however, it is tacitly assumed to be a microscopic phenomenon, in which only few atoms and electrons are involved [1].

In the recent years, on the other hand, there have been appeared many unconventionally photo-active solids, where the relaxation of optical excited states results in various collective motions involving a large number of atoms and electrons. In some cases, it results in a macroscopic excited domain with new structural and electronic orders quite different from the original ones. This situation can be called "photoinduced phase transition" [2], and the purpose of this paper is to review our recent theoretical studies on this phenomenon.

This problem is closely related with the hidden multistability intrinsic to each solid. If the ground state of a solid is degenerate or pseudo-degenerate, being composed of several true and false ground states with each structural and electronic orders different from others, we call it multistable. In this case, the photo-absorption, being initially a single-electron excitation from one of the true ground states, can triggers local but macroscopic instabilities. The photo-absorption can induce low-lying collective excitations during the lattice relaxation, and can finally produce a hidden ground state at the expense to create boundaries between the two states. Thus a local but

CP554, *Physics in Local Lattice Distortions*, edited by H. Oyanagi and A. Bianconi
© 2001 American Institute of Physics 1-56396-984-X/01/$18.00

macroscopic excited domain appears. In other words, the initially created single-electron excitation proliferates during the relaxation.

ADIABATIC PATH OF PHOTOINDUCED IONIC→NEUTRAL PHASE TRANSITION IN TTF-CA CRYSTAL

As one of the typical examples for photoinduced structural phase transitions, here, we will be concerned with the photoinduced ionic→neutral transition in a molecular crystal TTF-CA. Both TTF and CA are planar organic molecules as schematically shown at the top of Fig. 1, and this crystal has a quasi 1-d structure, in which these two molecules are alternately stacked along this 1-d chain axis. In the true ground state of this crystal at absolute zero of temperature, both TTF and CA become a cation and an anion, respectively, and make a dimer with each other as shown in Fig. 1(a). This is the ionic phase. On the other hand, we also have the neutral phase, in which neutral TTF and CA are stacked alternately without dimerization, as shown in Fig.1(b). This is the accidentally pseudo-degenerate false ground state. At absolute zero of temperature, it is just above the aforementioned ionic true ground state.

Keeping this material in the low enough temperature, but shining a strong laser light of about $0.6 \sim 2.2\text{eV}$ onto it, we can generate a neutral domain even in the ionic ground state as shown in Fig. 1(d), and this domain is composed of $200 \sim 1000$ neutral pairs [3]. This is nothing else but the photoinduced structural phase transition.

We can think of a simple intuitive scenario for this transition in such a way as schematically shown in Fig. 1(c) and (d). That is, a single photon can make a single neutral pair, and after that, the number of this pair will increase, just like a domino-game. By the recent experimental studies shown in Fig. 2, however, this simple scenario is proved wrong [4].

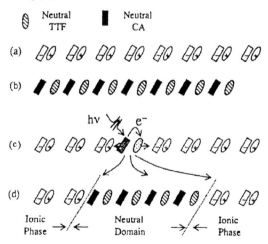

FIGURE 1. The schematic view of TTF-CA crystal; (a) the ionic phase, (b) the neutral phase, (c) the CT exciton, and (d) the neutral domain.

In this figure, we have shown the photo-absorption spectrum of TTF-CA, as a small inset. It has two peaks at 0.6eV and 2.2eV. The first one, being the elementary optical excitation of this crystal, corresponds to the intermolecular charge transfer (CT) exciton. Among a macroscopic number of ionic pairs (TTF$^+$ and CA$^-$) in Fig.1(a), only a single pair returns back to a neutral one (TTF^0 and CA^0) by this excitation, as shown in Fig. 1(c). While, the second peak at 2.2eV corresponds to an intramolecular electronic transition of TTF$^+$. The thick solid line in Fig. 2 denotes the efficiency of neutral phase generation as function of the photon intensity, when the photon energy is fixed at the CT exciton (0.6eV). We can clearly see that there is a threshold in the intensity, below which the neutral phase can never be generated. It means that a single CT exciton alone can never result in the neutral phase, but only through a nonlinear cooperation between several photo-excited CT excitons, the new phase can be attained. This is the first note-worthy characteristic.

The second note-worthy characteristic, which we can see from Fig. 2, is the difference between the dashed line and the aforementioned thick solid line. Exactly speaking, the horizontal axis of this figure does not simply denote the photon intensity itself, but denotes the total photon energy which is absorbed in the unit volume of the TTF-CA crystal. This total energy is calculated by taking the three quantities into account; the absorption coefficient, the energy of the photon and its intensity. Thus, we can compare the 0.6eV excitation and the 2.2eV excitation (the dashed line), on an equal footing. For example, if we focus on the point with value 0.25 (10^{19}eV/cm^3) of the horizontal axis in Fig. 2, we can find that the efficiency changes very high or

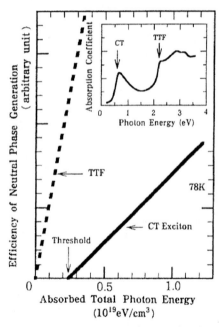

FIGURE 2. The efficiency of the neutral phase generation as a function of the absorbed total photon energy [4].

almost zero, depending sensitively on the photon energy, (that is, on the way of the excitation), although the total absorbed photonic energies are the same each other. This is the so-called "initial condition sensitivity", peculiar to the dynamics of nonlinear systems [5].

Keeping this in mind, very recently, we have theoretically clarified the adiabatic path, which starts from a Franck-Condon type optical excitation in the ionic ground state, and terminates up to the large neutral domain formation. This theory is based on the extended Peierls-Hubbard model for valence electrons of TTF and CA molecules, and can nicely describe the dimerized ground state, the first order nature of the ionic-neutral phase transition, as well as the presence of the CT exciton at around 0.6eV. In Fig. 3, we have shown the adiabatic path obtained by this theory [6].

The horizontal axis of this figure denotes the total number of neutral pairs ($\equiv$ N) created in the ionic ground state at absolute zero of temperature. The adiabatic potential energy surface of the excited state ($\equiv E_{x1}$) has two local minima, at the region of small N ($\approx$ 0) and also at the region of large N ($\approx$ 40). The first one corresponds to the CT exction, while the second one is the neutral domain. They are separated by a high potential barrier. The energy increase due to the two ionic-neutral phase boundaries, shown in Fig. 1(d) by dashed lines, makes small size neutral domains too unstable. They can become stable only when its size exceeds a critical value. As N increase further, the energy of this neutral domain again turns to increases because of the weak inter-chain interactions. In the adiabatic potential energy surface of the ground state ($\equiv E_g$), this neutral domain also appears in the region of large N ($\approx$ 50), and we can get many shallow local minima, as shown in the small inset of Fig. 3.

Thus, the lowest state of a single CT exciton can not relax down to the neutral domain, but an excess energy is necessary, as schematically shown by the upper

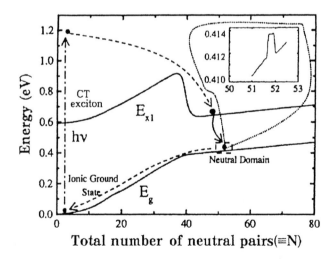

FIGURE 3. The adiabatic potentials as functions of N [6].

dashed line in Fig. 3. This excess energy will be converted into excess vibronic energy and (or) kinetic energy of exciton, and may induce various nonlinear processes during the relaxation.

ITERATIVE EQUATION FOR PROLIFERATION AND INITIAL CONDITION SENSITIVITY

Let us now proceed to the quantum time-evolution dymanics of the photoinduced structural phase transition, and clarify how excitons with the large excess energy will proliferate and grow up into a macroscopic domain. In the previous section, we are concerned with the TTF-CA crystal and could get various interesting characteristics even within the adiabatic approximation. In order to proceed to the quantum dynamics, however, the situation realized in this crystal is too complicated to clarify it theoretically. Hence, we extract only the following four points as its essence.

1. There must be a hidden multistability. A false ground state has to be just above the true one, and the energy difference between them per unit volume should be larger than the thermal energy, but should be much smaller than the visible photon energy.

2. There must be a nonlinear mechanism, through which the exciton can proliferate. This mechanism should be efficient enough to overcome various radiative and noradiative decay channels of the excitons. These decay channels will act in the every intermediate stages of the relaxation, to hinder the proliferation, as shown in Fig. 3 by dashed lines.

3. At the final state of the Franck-Condon transition, a large excess energy should be given to the excitons. By using these excess energies, the initially created excitons can proliferate, under the energy conservation low.

4. The resultant domain should be a local minimum in the adiabatic potential surface of the ground state, so that this domain can have a sufficiently long life time, and is worth to be called the photoinduced phase transition. Thus, the intially created excitons have to go through various intersections of the potential surfaces diabatically, so that it can finally reach to the local energy minimum, not in the excited state potential surface, but in the ground state one, as shown in Fig. 3.

In order to clarify these points from a unified point of view, we will be concerned with the real-time quantum-dynamics of the following model for a many-exciton system coupling strongly with Einstein phonons in a one-dimensional crystal [7]. Its Hamiltonian ($\equiv H$) is given as,

$$H = E_0 \sum_l B_l^+ B_l - T \sum_{<l,l'>} B_l^+ B_{l'} - V \sum_{<l,l'>} B_l^+ B_l B_{l'}^+ B_{l'} + \sum_l \sum_{l'(\neq l)} G(|l - l'|) B_l^+ B_l (B_{l'}^+ + B_{l'})$$

$$+ \omega \sum_l b_l^+ b_l - (\omega S)^{1/2} \sum_l B_l^+ B_l (b_l^+ + b_l), \tag{1}$$

where, B_l^+ is the creation operator of an exciton at a lattice site l with its energy E_0. T is the exciton transfer, that operates from site l to other sites l', only when these two sites are in the nearest neighbour relation, which is denoted by $< l, l' >$. V is an effective attraction between two excitons at neighbouring sites. Occupations of a single site by more than one excitons are excluded from the beginning. $G(|l - l'|)$ is the inter-site third order anharmonicity of the excitons. This anharmonicity originally comes from the long range nature of the Coulomb interaction between electrons and holes constituting these excitons. Thus, the appearance of G in the present many-electron system seems to be quite natural or rather trivial. However, excitons can proliferate through this fourth term. The fifth term of Eq. (1) is the Einstein phonon at site l with the creation operator b_l^+ and the frequency ω. The last term is the exciton-phonon coupling with a coupling constant S.

This exciton-phonon system is our relevant system, and we also assume that there are bosonic reservoir modes and the radiation field linearly coupling with these exciton and phonon fields. Consequently, in our relevant system, various relaxation channels occur, such as the vibrational (or phonon) relaxations, radiative and nonradiative decays of excitons. Hence, the main point of our present theoretical research is to find realistic but specific conditions, under which a large size domain is created, in spite of the abovementioned hindrances.

In order to find the bistable condition mentioned in the point 1, we have set parameters of the relevant system so that the state with n excitons ($n = 10 \sim 30$) has a higher energy than the ground state ($n = 0$), only a little. When this condition is fulfilled, we will get a new macroscopic state just above the ground state. If we neglect T in Eq. (1), this condition is realized when ($E_0 - S$) and V are well balanced. While, when S is sufficiently large, the large vibronic excess energy is prepared at the Franck-Condon state, as required by the point 3. As for the diabatic nature mentioned in the point 4, we will perform full quantum mechanical calculations for a finite ω.

On the other hand, the proliferation shown in the point 2, can occur through the aforementioned anharmonic term, using the excess energy as its energy source. We can also think of various other mechanisms for the proliferation. However, it should be noted that the Coulomb type nonlinearity is an instantaneous force with no retardation effect, as compared with various phonon-mediated mechanisms.

For practical calculations, the following parameters are used, $\omega = 0.1$eV, $E_0/\omega = 9.5$, $T/\omega = 1$, $S/\omega = 8$, $V/\omega = 1.7$, $G(1)/\omega = 0.2$, $G(2)/\omega = 0.1$, $G(3)/\omega = 0.067$. The coupling constants of the reservoir to the phonon and exciton fields are assumed to be 0.16ω, so that relaxations can occur within one period of the phonon. While, the coupling between the exciton and the photon fields are determined so that the radiative damping time becomes 10^{-9} sec at the Franck-Condon State. During the relaxation, on the other hand, this damping is assumed to decrease in proportion to the third power of the photon energy.

By the parameters thus selected, we have solved the time evolution of the density matrix of our relevant system within the Markov approximation for the reservoir and the radiation field. Since we are interested in the early time dynamics of the proliferation in connection with the initial condition sensitivity, we will be concerned with the case where a pair of excitons are created by various photons. While, depending on the way of this photo-excitation, the distance ($\equiv d$) between these two excitons at the Franck-Condon state changes from $d = 1$ to $d = \infty$ (the unit of length is the lattice constant). The total photon energy necessary for this excitation is almost independent of d, provide that T in Eq. (1) is small. While the proliferation which is brought about through G in Eq. (1) will be different depending on d. In any case, however, we have to solve complicated quantum eigenstates realized in this exciton-phonon coupled system.

In order to simplify this complexity, we focus only one exciton with the excess vibronic energy. We call it "mother exciton". As for the other excitons, we approximate them by frozen ones localized at their lattice sites with only the lowest vibronic energies. Even in this case, the mother exciton can proliferate using her excess energy. While the other excitons help it, since, through G in Eq. (1), they can induce mixing between the quantum eigenstates with different exciton numbers, although they are frozen. Thus, the mother exciton can make her daughter exciton. Calculating this process as a function of time t after the excitation, we can see the excess energy gradually decreases, while the total number of the exciton ($\equiv n$) will increase from $n = 2$ toward $n = 3$, due to the birth and subsequent growth of the daughter exciton. Hence our system again becomes a complicated many-exciton-phonon coupled system. To overcome this difficulty, we again approximate the mother exciton as the frozen one just at the time when her daughter has grown up to be an adult or a new mother, that is, $n = 3$. At this time, the excess energy has already decreased from its original value because of the dissipation, and we assume that the new mother inherits only this reduced excess energy. However, this new mother can make her own daughter, using this residual excess energy.

This is nothing but the generation crossover, and we will repeat this procedure, step by step, iteratively, until the original excess energy will be used up. The resultant numerical solutions of this iterative equation are shown in Fig. 4, as functions of d and time t (in the unit of the phonon period $2\pi/\omega$). We can see that the occurrence or nonoccurrence of the proliferation sensitively depends on d, that is, how close the initial two excitons are created. The infinitely distant excitation case ($d = \infty$) can result in almost no proliferation, since the nonlinearity can not act in this case. On the other hand, the most closely excited case ($d = 1$) also results in no net proliferation, since, in the fourth term of Eq. (1), the nonlinearity becomes too strong, and the exciton annihilation becomes more dominant than the creation. Only the moderately distant excitation case ($d = 3 \sim 4$) can successfully result in the vital proliferation. Thus, our system is highly sensitive to the initial condition, or the way of the photo-excitation, even if the total photon energy is the same.

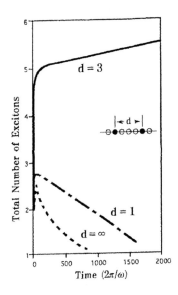

FIGURE 4. The total number of excitons as functions of time and the initial distance d.

ACKNOWLEDGMENTS

The authors are very grateful to professor K. Tanimura for presenting us his experimental results before publication. This work is supported by the Grant-in-aid of Japanese Ministry of Education, Science, Sport and Culture, the priority area (B) "Photoinduced Phase Transitions and their Dynamics".

REFERENCES

1. Ueta, M., Kanzaki, H., Kobayashi, K., Toyozawa, Y., and Hanamura, E., *Excitonic Processes in Solids*, Springer-Verlag, Berlin, 1984
2. Nasu, K., *Relaxations of Excited States and Photoinduced Structural Phase Transitions*, Springer-Verlag, Berlin, 1997, pp. 3-16.
3. Koshihara, S., Takahashi, Y., Sakai, H., Tokura, Y., and Luty, T., *J. Chem. Phys. B* **103**, 2592-2600 (1999).
4. Suzuki, T., Sakamaki, T., Tanimura, K., Koshihara, S., and Tokura, Y., *Phys. Rev. B* **60**, 6191-6193 (1999).
5. Thompson, J., and Stewart, H., *Nonlinear Dynamics and Chaos*, John Wiley & Sons, New York, 1986.
6. Huai, P., and Nasu, K., *J. Phys. Soc. Jpn.* **69**, 1788-1800 (2000).
7. Mizouchi, H., and Nasu, K., *J. Phys. Soc. Jpn.* **69**, 1543-1552 (2000).

Photoinduced Local Lattice Distortions in II-VI Semiconductors

J. Mustre de León and F. J. Espinosa

Departamento de Física Aplicada, Cinvestav-Mérida, AP73, Mérida, Yucatán, 97310, MEXICO

Abstract. We present XAFS results in CdTe heavily doped with In which show the presence of a reversible photoinduced local structural change at low temperatures. Results of ab-initio calculations, which show the effect in the local electronic structure caused by the observed local structural change, are also presented. These results imply the presence of a strong local electron-lattice coupling in II-VI semiconductors. This characteristic, not previously recognized, is likely to be the missing component in explaining carrier passivation and other novel electrical properties.

INTRODUCTION

The change in electrical properties of crystalline compound semiconductors, induced by photoexcitation, has been recognized since the early 60's [1], and already several applications make use of such changes [2]. A particular striking example of such changes is photoinduced persistent photoconductivity (PPC). In PPC the conductivity of the material changes by several orders of magnitude after photoexcitacion. More important, it remains altered for several hours o days after the photoexcitation [1,3]. The microscopic origin of PPC is still not understood. However, it appears that its explanation lies beyond the 1-electron models, which have been successfully used to explain other electrical properties.

In contrast, just very recently reports of possible structural changes induced by photoexcitation have appeared in the literature [4,5]. Under photoexcitation the electrons access excited states, and the structure of the lattice could be modified due to a change in Coulomb interaction between excited electrons and ions. Hence, the existence of structural changes under illumination, at optical frequencies, would imply the presence of a strong coupling between electronic states and the lattice. Consequently, the description of such changes, strictly speaking, lies beyond the realm of 1-electron calculations, which assume that the lattice remains stationary when there are transitions between different electronic states [1]. Such calculations have been used to predict the electrical and optical behavior of compound semiconductors with great success [6]. However, these materials have common characteristics with systems, in which the electron-lattice coupling is known to play an important role in determining their physical properties. Examples of such systems are C_{60} [7], amorphous semiconductors [8], manganese oxides [9], and high temperature superconductors [10]. Moreover, photoinduced structural changes have been reported

CP554, *Physics in Local Lattice Distortions*, edited by H. Oyanagi and A. Bianconi

in these materials [7-10]. Among the common characteristics one can cite the presence of reduced numbers of free charge carriers and the tendency to exhibit charge localization around defect sites.

Specifically in the case of II-VI semiconductors, during the last decade, several ab-initio 1-electron calculations have shown that defects in these materials can be in metastable configurations which depend on the local charge around the defect [11-13]. Hence, these calculations suggest the possibility of photoinduced structural changes. Also, explanations of PPC imply a concurrent change in the lattice structure [1,3]. Experimental techniques commonly used to study the structure of II-VI semiconductors, like diffraction, are sensitive to the average crystalline structure. However, calculations [11-13] suggest that structural changes are related to the charge state of defects, and thus are local. Consequently, there are few experimental techniques that can provide direct evidence of photoinduced structural changes in these materials. X-ray absorption fine structure (XAFS) has proven to be a useful technique characterize the local atomic structure in semiconductors [14,15]. Also this technique has provided direct evidence of local structural changes under photoexcitation in high-Tc superconductors [10] and amorphous semiconductors [8]. Consequently, it would be a choice technique to determine the existence of photoinduced structural changes in II-VI semiconductors.

In this manuscript we review recent XAFS results in CdTe heavily doped with In which show the presence of a reversible photoinduced local structural change centered around the In [5]. We also present results of ab-initio calculations, which show the effect in the local electronic structure caused by the observed local structural change.

EXPERIMENTAL SETUP

Details of sample preparation and characterization of the CdTe:In sample, with In concentration of 6.7 at.% are given in Refs. [5], and [16]. Here we only note that this sample was a powder sample, which only exhibited the zincblende phase characteristic of II-VI semiconductors [15]. This is an important point, since it has been commonly assumed that at high dopant concentrations phase separation will occur in these materials [17]. Although the observation of a photoinduced structural change is independent of the presence of a single phase, the specific interpretation of the phenomenon does depend on having a single crystallographic phase.

The XAFS sample was mounted in a cryostat, whose windows were sealed with aluminized mylar to prevent photoexcitation. Spectra were measured at the Stanford Synchrotron Radiation Laboratory (SSRL) beam line 4-1 with a Si(220) monochromator. Data was taken in transmission and fluorescence modes, using a 13-element Ge detector for the fluorescence measurements. A first set of measurements was done in at T ~ 80 K. Two scans were taken for the K-edge XAFS spectra of Cd and Te and three scans for In. A second set of measurements was done for the same edges but now after illuminating the sample for five minutes with the white spectra from a halogen lamp. Dark conditions were restored sealing the windows of the cryostat with aluminized mylar before data acquisition [5]. Data reduction to obtain the appropriate XAFS spectra was performed using standard prescriptions and details are given in Refs. [5,16]. These set of measurements was repeated after raising the temperature of the sample to room temperature and repeating the cycle, finding the same results described below.

RESULTS

After photoexcitation, from the normalized absorption of In K-edge we observe an increase in the amplitude up to about 10 eV above the absorption edge concurrent with a small decrease in intensity in the region below the absorption edge (Fig. 1a). A decrease in the intensity of the Cd spectrum is observed from the edge up to 10 eV above (Fig.1b). While no detectable changes were observed for the Te K-edge (Fig. 1c). These results show a change in the local electronic structure around In and Cd atoms. Since the local electronic structure is intimately related with the local atomic structure such results already suggest a local structural change induced by the photoexcitation process. However, observed changes in the absorption edge of a particular atom do not necessarily imply a structural change centered at that particular atom.

Structural changes are directly seen in the Fourier transform magnitude (FTM). In the case of In, we observe an increase in the amplitude of the main peak at $R - \Delta = 2.5$ Å (Fig. 2a). We note that this result *directly* shows a change in the structure of the nearest neighbor shell of atoms around In dopants caused by the photoexcitation. It is also notable that there is no clear indication of further shells from the In FTM, indicating a very broad distribution of farther out neighbors. From the Cd FTM, the first peak shows a decrease in amplitude (Fig. 2b), but its magnitude is within the experimental uncertainty. As it was observed in the absorption edge, from the Te FTM we do not observe any change in the main peak (Fig. 2c). Changes present in farther out shells are all within experimental error. We note that consecutive scans in the dark, before illumination did not exhibit any change. This fact indicates that x-rays do not produce changes as those produced by optical wavelengths. Consequently, the photoinduced structural change seems to be of a different nature to that reported in manganese oxide materials [9].

Quantitative local structure information was obtained fitting the XAFS spectra in over the region $2.6 < k < 13.4$ Å^{-1}. Photoelectron scattering factors were derived from a standard ab-initio calculation [18]. For the In XAFS fit, the distance to neighbors, R, the Debye-Waller factor, σ, and the number of neighbors, N, were left as floating parameters. An energy shift E_0 of 6.8 eV above the edge was used and the multielectron-excitation scale factor, $S^2_0 = 1.2$. For Cd (Te) the number of neighbors was fixed to the crystallographic value of CdTe, E_0 was set to 5.0 (5.3) eV above the edge and S^2_0 was set to 0.94 (0.85). The results of the fits for In are shown in Ref. [5]. In this case we obtained a fit that reproduced the spectra using only a single shell of neighbors, in agreement with the FT magnitude shown in Fig. 1.

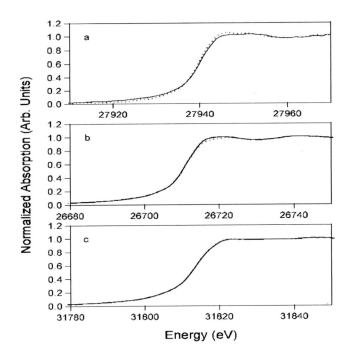

FIGURE 1.. Absorption edge (K) before illumination (solid line) and after illumination (dotted line) of (a) In, (b) Cd, and (c) Te (c.f. Ref. 5).

From the Cd and Te XAFS fits one can identify the first three shells of neighbors. The distances obtained are in agreement with the zincblende structure of CdTe. The structural parameters derived from the Cd and Te K-edges do not exhibit significant changes under photoexcitation. We note the smaller distances obtained in Te edge compared with those in Cd. We have observed this trend even in undoped CdTe, and might reflect the presence of intrinsic defects.

Given the fact that the In K–edge exhibits only structural information about the nearest neighbors, it is not possible to construct a detailed model of the lattice structural change induced by the photoexcitation. However, the observation of changes in the In nearest neighbor environment together with the result of negligible changes on the Cd and Te nearest neighbor environments rule out several model of defect which have been proposed in similar systems [12]. Thus, the simplest model consistent with the observations above is the single breaking bond (SBB) model proposed in Ref. [11]. In this model the dopant atom (In) enters substitutionally replacing the cation (Cd). However, one of its anion neighbors (Te) moves from its ideal crystallographic position leaving the In atom 3-coordinated instead of having its normal tetrahedral coordination (Fig. 3a). In this configuration the In dopant retains

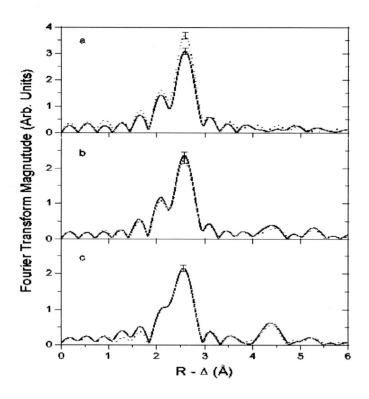

FIGURE 2. Fourier transform magnitude of the XAFS before illumination (solid line) and after illumination (dotted line) for (a) In,(b) Cd, and (c) Te. (c.f. Ref. 5).

one of the electrons which normally would enter as a free carrier. Hence, this model has been used to explain the observed carrier compensation in n-type doped II-VI semiconductors, after certain threshold of carrier concentration is achieved.

After photoexcitation the traped electron is promoted to the conduction band and the In atom relaxes to its normal crystallographic position, where it is tetrahedrally coordinated to the Te atoms (Fig. 3b). This explains the observed increase in coordination number after photoexcitation (Fig. 2a).

TABLE 1. Cd and Te K-edge XAFS fits results

	Cd(B.I)	Cd (A.I.)	Te (B.I.)	Te (A.I.)
R_1(Å)	2.80(1)	2.80(1)	2.79(1)	2.79(1)
σ_1 (Å)	0.039(3)	0.043(3)	0.050(2)	0.051(2)
R_2 (Å)	4.57(1)	4.58(2)	4.54(2)	4.53(2)
σ_2 (Å)	0.10(1)	0.12(1)	0.09(1)	0.09(1)
R_3(Å)	5.36(2)	5.34(2)	5.29(2)	5.31(3)
σ_3 (Å)	0.11(1)	0.12(1)	0.11(1)	0.12(1)

We modeled the XAFS and the X-ray absorption near edge spectra (XANES) produced by SBB centers. We used the functional form of the structure for an SBB center proposed in Ref. [19] for CdTe:In, but we used distances derived from the fits obtained in this work. Details of the XAFS simulation are presented in Ref. [5].

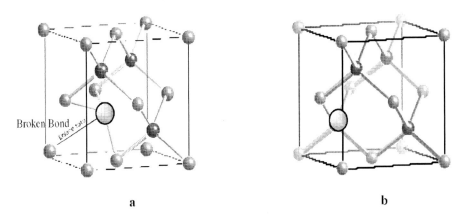

FIGURE 3. (a) In (big sphere) in a SBB center showing a broken In-Te bond (assumed configuration before photoexcitation), (b) tetrahedrally coordinated In (big sphere) entering substitutionally for Cd (assumed configuration after photoexcitation).

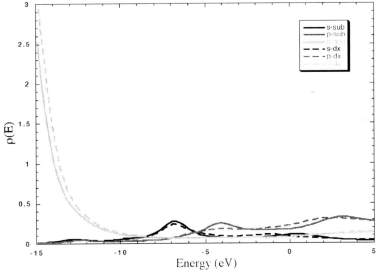

FIGURE 4. Projected In-local density of states calculated for In entering substitutionally for Cd (solid line) and for In in a SBB center (dashed line, assumed configuration after photoexcitation). The Fermi level is placed at –6.1 eV.

For the simulation of the XANES we performed a full multiple scattering calculation with spherically symmetric self consistent scattering potentials, using the program of Ref. [20] in a cluster of 50 atoms centered around an In atom. The comparison of the XANES simulations using a SBB center or In entering substitutionally for Cd shows the same trend as that observed in the experiment, namely an increase in the 'white line' and a small loss of spectral weight below the edge after photoexcitation (Fig. 1a). To interpret these changes we calculated the projected angular momentum local density of states around In. As shown in Fig. 4 the main change in the spectrum is a change in the density of p-states of the In. Also there are changes in s- and d-states not reflected in the K-edge absorption. We note that for a direct comparison between the calculated local density of states with the XANES spectra it is necessary to calculate the projected local density of states in the presence of a core hole. However, the general trend shown by the calculation of the ground state projected local density of states (Fig. 4) is consistent with the experimental result (Fig. 1a). Also, consistent with the experimental observations the Cd and Te local density of states do not exhibit appreciable changes under photoexcitation. It should be possible to detect these changes with techniques, which can probe electronic states near the Fermi level.

SUMMARY AND CONCLUSIONS

In summary, we have observed a change in the nearest neighbor environment of In, in CdTe:In, after photoexcitation at low temperatures. The photoinduced structural change implies a strong coupling between the charge state and the structure around indium with a barrier that prevents indium to return to its original position at low temperatures. It is important to note that in this study in spite of the high In concentration, the estimated probability of finding two defect centers adjacent to each other is ~ 2%. Consequently, our results are relevant for the regime in which these centers can be considered as nonoverlapping and are expected to control the passivation of free carriers. Recently we have observed a photoinduced structural change in ZnSe:Cl, another n-type doped II-VI semiconductor [21]. However, the nature of the change in that system appears to be different from that found in CdTe:In. In the case of ZnSe:Cl we have observed changes in the local structure around the host atoms and the change seems to be irreversible. These results suggest that photoinduced structural changes are a general characteristic of doped II-VI semiconductors and that strong local electron-lattice interaction is present in these materials. The inclusion of previously ignored strong local electron-lattice interaction in the modeling of these materials is likely to be the missing element necessary to explain observed phenomena like carrier passivation and persistent photoconductivity. It would be also important to determine whether the observed local lattice distortions tend to form ordered structures, as those found in manganese oxides or high temperature superconductors, and that plays a role in determining the properties of these materials.

ACKNOWLEDGMENTS

We thank M. Zapata-Torres for providing samples used in this study. We also would like to thank S. D. Conradson for the Ge detector used in the experiments performed at the Stanford Synchrotron Radiation Laboratory (SSRL), which is operated by the Department of Energy, Office of Basic Energy Sciences. We acknowledge A. Ankudinov and J. J. Rehr for providing us the code used in Ref. [20]. This work was supported by CONACyT, Mexico.

REFERENCES

1. Redfield, D., Bube, R.H., *Photoinduced Defects in Semiconductors*, Cambridge University Press, New York: 1996.

2. Linke, R.A., Redmond, I., Thio, T., and Chadi, D.J., *Appl. Phys. Lett.* **83**, 661-667 (1998).

3. Khachaturyan, K., Kaminska, M., Weber, E.R., Becla, P., and Street, R.A., *Phys. Rev. B* **40**, 6304-6310 (1989).

4. Nissila, J., Saarinen, K., Hautojarvi, P., Suchocki, A., and Langer, J., *Phys. Rev. Lett.* **82**, 3276-3279 (1999).

5. Espinosa, F.J., Mustre de Leon, J., Conradson, S.D., Pena, J.L., and Zapata-Torres, M., *Phys. Rev. Lett.* **83**, 3446-3499 (1999).

6. Yu, P., Cardona, M., *Fundamentals of Semiconductors : Physics and Materials Properties*, Springer, New York, 1999.

7. Suzuki, M., *J. of Luminisensce* **87**,661-663 (2000).

8. Kolobov, A.V., Oyanagi, H., Kondo, M., Matsuda, A., Roy, A., and Tanaka, K., *J. of Luminisensce* **83**, 205-208 (1999); Kolobov, A.V., Oyanagi, H., Tanaka, K, and Tanaka, K., *Phys. Rev. B* **55**, 726-734 (1997).

9. Cox, D.E., Radaelli, P.G., Marezio, M., and Cheong, S.W., *Phys. Rev. B* **57**, 3305-3314 (1998).

10. Mustre de Leon, J., Conradson, S.D., Batistic, I., and Bishop, A.R., *Phys. Rev. Lett.* **65**, 1675-1678 (1991); Tyson, T.A., Mustre de Leon, J., Conradson, S.D., Bishop, A.R., Neumeier, J.J., Roder, H., and Zang, J., *Phys. Rev. B* **53**, 13985-13988 (1996).

11. Chadi, D.J., *Phys. Rev. Lett.* **72**, 534-537 (1994).

12. Park, C.H., and Chadi, D.J., *Phys. Rev. Lett.* **75**, 1134-1137 (1995).

13. Garcia, A., and Northrup, J.E., *Phys. Rev. Lett.* **74**, 1131-1134 (1995).

14. Sette, F., Pearton, S.J., Poate, J., Rowe, J.E., and Stohr, J., *Phys. Rev. Lett.* **56**, 2637-2640 (1986).

15. Espinosa, F.J., Mustre de Leon, J., Zapata-Torres, M., Castro-Rodriguez, R., Pena, J.L., Conradson, S.D., and Hess, N.J., *Phys. Rev. B* **55**, 7629-7632 (1997).

16. Espinosa, F.J., Mustre de Leon, J., Zapata-Torres, M., Castro-Rodriguez, R., Pena, J.L., and Conradson, S.D., *Phys. Rev. B* **61**, 7428-7432 (2000).

17. Desnica, U.V., *Prog. Cryst. Growth and Charac.* **36**, 291-357 (1998).

18. Mustre de Leon, J., Rehr, J.J., Zabinski, S.I., and Albers, R.C., *Phys. Rev. B* **44**, 4146-4156 (1991).

19. Park, C.H. and Chadi, D.J., *Phys. Rev B* **52**, 11884-11890 (1995).

20. Ankudinov, A.L., Ravel, B., Rehr, J.J., and Conradson, S.D., *Phys. Rev. B* **58**, 7565-7576 (1998).

21. Espinosa, F.J., Mustre de Leon, J., and Hernandez-Calderon, I., *Rev. Mex. Fis.* **45S1**, 167-170 (1999); Espinosa, F.J., Mustre de Leon, J., and Hernandez-Calderon, I., *unpublished.*

Photoinduced Paramagnetism of Group III Impurities in $A_{IV}B_{VI}$ Narrow-Gap Semiconductors.

A. N. Vasil'ev[*], T. N. Voloshok[*], J. K. Warchulska[†] and H. Kageyama[‡]

[*]*Low Temperature Physics Department, Moscow State University, Moscow 119899, Russia*
[†]*International Laboratory of High Magnetic Fields and Low Temperature, Wroclaw 53-421, Poland*
[‡]*Institute for Solid State Physics, University of Tokyo, Kashiwa, Chiba 277-8581, Japan*

Abstract. At low temperatures some ionic, covalent and mixed bonding compounds, as well as semiconducting heterostructures and quantum wells exhibit persistent photoconductivity. This term is used to describe the striking observation that the conductivity of these compounds and/or structures is greatly enhanced by visible or infrared illumination and that the low resistance state is maintained for a long time after switching off the illumination. To describe this effect in variously doped ionic-covalent semiconductors the models of repulsive barrier for both electron emission and capture were introduced based primarily on the assumption of dopant's displacement in the host's crystal lattice. Here we report on the magnetic counterpart of this phenomenon, which however does not exactly meet the expectations based on transport measurements. It was found that the magnetic response of $A_{IV}B_{VI}$ narrow-gap semiconductors doped with C_{III} impurities possesses features of both relaxation phenomena and light-induced phase transition. Exposure of PbTe:Ga, PbTe:In and $Pb_{0.75}Sn_{0.25}Te$:In single crystals to white-light illumination at low temperatures resulted initially in an increase of the diamagnetic response and then in the appearance of a sharp paramagnetic peak at heating.

The objects of the present study belong to the unique family of $A_{IV}B_{VI}$ semiconductors constituted by nine binary compounds (Ge, Sn, Pb) - (S, Se, Te) and their solid solutions. All lead chalcogenides as well as high-temperature phases of germanium telluride and tin telluride possess the rocksalt structure [1]. On lowering the temperature GeTe at 670 K and SnTe at $T \leq 100$ K transform into the rhombohedral phase through the softening of transverse optical mode in [111] direction, while lead telluride remains a virtual ferroelectric with anomalously high dielectric constants [2]. While doped with C_{III} elements the inherently unstable $A_{IV}B_{VI}$ semiconductors show persistent photoconductivity, which is most pronounced if the impurity levels fall into the forbidden gap [3, 4]. In contemporary understanding [5], C_{III} impurities in $A_{IV}B_{VI}$ matrix dissociate, $2C_{III}^{0} = C_{III}^{-} + C_{III}^{+}$, into negatively charged C_{III}^{-} and positively charged C_{III}^{+} centers, creating therefore deep and shallow levels in the energy spectrum of the semiconductor. The transitions between these levels lead to the appearance of a resonant peak at the edge of fundamental adsorption in the $A_{IV}B_{VI}$:C_{III} spectra of photoconductivity. However, the microscopic picture for

CP554, *Physics in Local Lattice Distortions*, edited by H. Oyanagi and A. Bianconi
© 2001 American Institute of Physics 1-56396-984-X/01/$18.00

this deep-shallow transformation is far from being clear. Both C_{III}^- and C_{III}^+ being in singlet states are diamagnetic, but the intermediate metastable C_{III}^0 state may carry a magnetic moment. To get additional insight into the processes of photoexcitation of C_{III} dopants in $A_{IV}B_{VI}$ semiconductors the magnetic properties of PbTe:Ga, PbTe:In and $Pb_{0.75}Sn_{0.25}Te$:In single crystals were studied under various regimes of white-light illumination along with the measurements of transport properties.

The resistivity of the bulk samples ($1 \times 1 \times 5$ mm^3) was measured with the help of the four-probe technique in a brass chamber equipped with a miniature incandescence lamp. Its radiation parameters were chosen to fit the resonant peak of photoconductivity in PbTe:Ga. At cooling in darkness the resistivity of $Pb_{0.75}Sn_{0.25}$Te:0.5at.%In and PbTe:0.3at.%Ga show the semiconducting behaviour as expected for the activation of electrons from deep levels into the conduction band. Illumination within a few seconds (~ 20 mW/cm^2) at low temperatures resulted in a decrease of resistivity by a few orders of magnitude persisting for many hours. At heating, as shown in Fig. 1, the resistivity of the samples increased reaching maximums at T ~ 20 K in $Pb_{0.75}Sn_{0.25}$Te:0.5at.%In and at T ~ 70 K in PbTe:0.3at.%Ga. The heating of the samples under permanent illumination (~ 2 mW/cm^2) resulted in a decrease of the resistivity peaks, but their positions remain practically unchanged. The effect of persistent photoconductivity was less pronounced in PbTe:1at.%In, since the In level in PbTe falls into the conduction band.

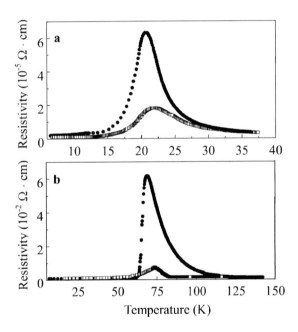

FIGURE 1. Temperature dependencies of resistivity at heating after brief illumination at low temperature (solid circles) and under permanent illumination (open circles). **a**, $Pb_{0.75}Sn_{0.25}$Te:0.5at.%In; **b**, PbTe:0.3at.%Ga.

The influence of illumination on the magnetic properties of $A_{IV}B_{VI}$:C_{III} compounds was studied mostly by a Faraday balance technique. The construction of this apparatus did not allow for the complete screening of the sample from the daylight. Additional illumination of the sample was provided by the incandescense lamp through the channel of measuring device. The sample mounted on Kevlar thread was cooled down to liquid helium temperature in the absence of a magnetic field and then it was briefly illuminated ($\sim$ 20 mW/cm^2). After that the source of additional illumination was switched off and measurements were performed at heating with the rate of about 1K/min at H = 4.28 kOe. Surprisingly, the results of magnetic measurements did not meet the expectations proceeding from the transport measurements.

The temperature dependencies of magnetic susceptibility of PbTe:1at.%In, $Pb_{0.75}Sn_{0.25}$Te:0.5at.%In and $Pb_{0.75}Sn_{0.25}$Te:2at.%In are shown in Fig. 2. At low temperatures the illumination of samples resulted in an increase of diamagnetic response compared with the reference data of undoped sample of PbTe (see the inset in Fig. 2a). At heating an additional diamagnetism was lifted off and at approximately 50 K strong paramagnetic peak with sharp high temperature edge appeared. As can be seen from the comparison of curves for $Pb_{0.75}Sn_{0.25}$Te:0.5at.%In and $Pb_{0.75}Sn_{0.25}$Te:2at.%In (Fig. 2b), an increase of impurity content resulted in an increase of both paramagnetic background and paramagnetic peak.

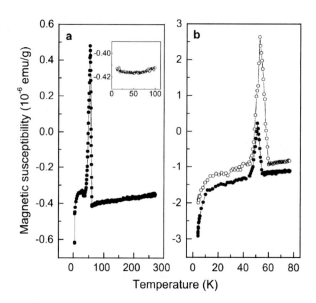

FIGURE 2. Temperature dependencies of magnetic susceptibility at heating after brief illumination at low temperature. **a**, PbTe:1at.%In; **b**, $Pb_{0.75}Sn_{0.25}$Te:0.5at.%In (solid circles) and $Pb_{0.75}Sn_{0.25}$Te:2at.%In (open circles). The inset to panel **a** shows the temperature dependence of the magnetic susceptibility of the undoped sample of PbTe.

The magnetic properties of PbTe:0.3at.%Ga were studied by both Faraday balance and SQUID magnetometer techniques. In Faraday balance measurements this sample shows mainly the same behaviour as other $A_{IV}B_{VI}:C_{III}$ compounds. The SQUID measurements revealed another unexpected feature of paramagnetic singularity. In this case no illumination of the sample by incandescence lamp was employed, but prior to measurements the sample was exposed to daylight at room temperature and then was cooled to liquid helium temperature. Once again the behaviour similar to that described above was present, however at subsequent cooling in dark chamber of SQUID magnetometer the paramagnetic peak was not observed at all and the diamagnetic downturn at low temperatures was substituted by a paramagnetic upturn. The increase of the exposure to white-light illumination at room temperature resulted finally in the appearance of both diamagnetic downturn and paramagnetic peak, as shown in Fig. 3. In general, the magnetic response of a PbTe:0.3at.%Ga single crystal resembles the behaviour of a spin glass system, showing a paramagnetic peak in the zero field cooled regime and a smooth paramagnetic upturn in the field cooled regime. At any temperature the magnetic response of the sample consisted of field dependent paramagnetic and field independent diamagnetic parts, but the pronounced hysteresis of the magnetization could be observed (Fig. 4) only below the paramagnetic peak temperature.

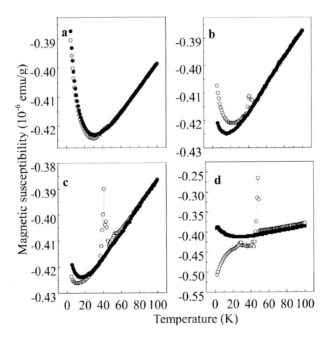

FIGURE 3. Temperature dependencies of magnetic susceptibility of PbTe:0.3at.%Ga both at heating (open circles) and cooling (solid circles) measured in the dark chamber of SQUID magnetometer after different exposures to white-light illumination (t_{exp}) at room temperature. **a**, $t_{exp} = 5 \times 10^2$ sec; **b**, $t_{exp} = 5 \times 10^3$ sec; **c**, $t_{exp} = 1.5 \times 10^4$ sec; **d**, $t_{exp} \geq 10^5$ sec.

In our opinion, the difference in critical temperatures for singularities in transport and magnetic properties reflects a previously hidden aspect of the interaction of light with $A_{IV}B_{VI}{:}C_{III}$ compounds. At cooling in darkness the electrons leaving the conduction band occupy dopant centers, C_{III}^- being a substitute for metal in octahedral surrounding of chalcogene ions and C_{III}^+ being shifted probably in the [111] direction. Illumination of the sample resulted in the recharging of C_{III}^- and C_{III}^+ centers leading to a metastable situation where the charges of some impurities do not correspond to their local surroundings. Different light-induced displacements of dopants could result in the formation of an energy level either in the conduction band or in the gap. In the former case it would result in persistent photoconductivity, in the latter case it would lead to variations in magnetic properties. The recharging of C_{III}^- and C_{III}^+ centers always requires redistribution of two electrons, otherwise the illumination of sample at low temperatures would be accompanied by a paramagnetic upturn and not by a diamagnetic downturn. The electrons at shallow impurity levels in the gap presumably possess higher radii of localization, providing therefore the increase of diamagnetic response at low temperatures. At heating, the probability of overcoming the potential barrier separating stable and metastable states of impurities increases. The relaxation of elastic stresses at any center overcoming the potential barrier is followed at the paramagnetic peak temperature by an avalanche-like process of recharging due to the proximity of the system to a phase transition. This simulation of a phase transition is accompanied by a sharp increase of paramagnetic susceptibility in

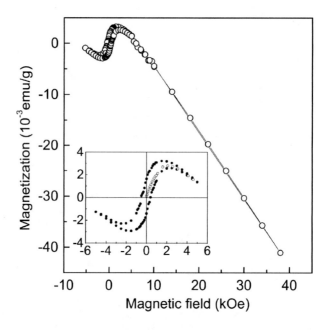

FIGURE 4. Field dependence of PbTe:0.3at.%Ga magnetization at T = 5K. The inset shows the hysteresis in low magnetic field.

the same way as it occurs at phase transitions in $A_{IV}B_{VI}$ semiconductors [6]. A remarkable feature of the obtained results is the spin glass behaviour of a system, nominally not containing any magnetic ingredients. It means that in some crystal environments atoms of non-transition metals may carry a magnetic moment as was predicted in [7, 8] and, moreover, may even interact, a small number of electrons being present in the conduction band [9].

REFERENCES

1. Littlewood, P.B., *J. Phys. C* **13**, 4855-4873 (1980).
2. Kawamura, H., *The Lecture Notes on Physics*, **133**, 470-494 (1980).
3. Akimov, B.A., Bogoyavlenskiy, V.A., and Zimon, S.P., *Semicond. Sci. Techn.* **14**, 679-683 (1999).
4. Martinez, A., Abbundi, R.J., Houston, B., Davis, J.L., and Allgaier, R.S., *J. Appl. Phys.* **57**, 1165-1170 (1985).
5. Drabkin, I.A., and Moizhes, B.Ya., "The spontaneous dissociation of impurity neutral states on positevely and negatively charged states" in *Sov. Phys.: Semiconductors* **15**, 625-630 (1981).
6. Litvinov, V.I., and Dugaev, V.K., *Sov. Phys.: JETP*, **77**, 335-343 (1979).
7. Anderson, P.W., *Phys. Rev.* **124**, 41-53 (1961).
8. Kondo, J., *Solid State Physics* **23**, New York: Academic Press, 1969, pp.184-281.
9. Bastard, G., and Lewiner, C., *Phys. Rev. B* **20**, 4256-4265 (1979).

Local Mechanism of the Reversible Photo-induced Phase Transition in Co-Fe Prussian Blue Analogues

T. Kawamoto, Y. Asai and S. Abe

Electrotechnical Laboratory, Agency of Industrial Science and Technology (AIST),
1-1-4 Umezono, Tsukuba 305-8568, Japan.

Abstract. Local mechanism is proposed for the reversible photo-induced magnetic phase transition observed in Prussian blue analogues, $A_y Co[Fe(CN)_6]_x$ (A = alkali metal). We determined '*cluster potentials*' with ab-initio quantum chemical calculations. These potentials represent the local tendency, *e.g.*, which spin state is preferred at each local cluster. From the obtained potentials, it is suggested that the photo-excited clusters are different between the forward and backward transitions. Within our model, it is expected that the absorption spectra is sensitive to x, due to the variation of cobalt atoms perfectly surrounded by six cyano-group. We propose the pressure dependence of the photo-induced phase transition to confirm our model.

INTRODUCTION

A bi-directional photo-induced magnetization was first discovered in a cobalt-iron Prussian blue analogue, $K_{0.31}Co[Fe(CN)_6]_{0.77} \cdot 3.85H_2O$ [1-3]. Illumination of visible light (500 - 700 nm) at low temperature induces a bulk magnetization (presumably, ferrimagnetism), which can be eliminated by illumination of near-IR light (~ 1300 nm). This magnetization is believed to be accompanied with the cooperative effect [4, 5]. We propose a local model to describe the primary relaxation paths of electronic excited states in local regions (the local process), which play the role of a seed for nucleation and growth of new phase domains (the global process) [4].

The crystal structure of the parent compound $ACoFe(CN)_6$ (case of $x = 1$) is shown in Fig. 1(a). Fe and Co constitute a lattice of NaCl structure, while cyano-groups are located between Fe and Co. In the case of $x < 1$, there are vacant Fe sites, for which the surrounding six cyano-groups are replaced by water molecules (Fig. 1(b)). This implies that the cyano groups octahedrally surrounding a Co atom are partially substituted by water molecules depending on the number of nearest neighbor Fe vacancies.

The low spin (LS) configuration of the ground nonmagnetic state and the high spin (HS) configuration of the meta-stable magnetic state are most likely $Co^{III}(d\varepsilon^6, S = 0)$ $Fe^{II}(d\varepsilon^6, S = 0)$ and $Co^{II}(d\varepsilon^5 d\gamma^2, S = 3/2)$ $Fe^{III}(d\varepsilon^5, S = 1/2)$, respectively [1, 2]. The LS and HS states are converted to the intermediate states by photo-induced charge transfer (CT) between iron and cobalt atoms, and then to the final states HS and LS by intersystem crossing due to spin-orbit coupling at cobalt sites. The major structural

CP554, *Physics in Local Lattice Distortions*, edited by H. Oyanagi and A. Bianconi
© 2001 American Institute of Physics 1-56396-984-X/01/$18.00

difference between the LS and HS states lies in the Co-N bond length d(Co-N), which is longer by about 0.2 Å in HS than in LS [6, 7]. This implies that the transition between LS and HS states requires a volume change, which may be a major origin of a large energy barrier between them.

In the nonstoichiometric crystal, the ligand substitution will stabilize the HS state of cobalt due to the suppression of energy splitting between Co-$d\gamma$ and $d\varepsilon$ orbitals [8, 9]. The dependence of the stability of each spin state on the number of cyano-groups can be represented with 'cluster potential' [4] derived for a model cluster consisting of a cobalt ion and their six ligands. To take into account the partial substitution of ligands by water molecules, we consider with various numbers N_W of H_2O substitutions, for example, (a) Co(NC)$_6$ ($N_W = 0$), and (b) Co(NC)$_5 \cdot H_2O$ ($N_W = 1$). The energies are calculated for each value of d(Co-N) common to all the Co-N bonds in the cluster as well as in the environmental lattice. This choice diminishes the interfacial strain energy due to local volume change, being a suitable way to extract the tendency of the local cluster.

RESULT

Fig. 2 schematically shows calculated cluster potentials as functions of d(Co-N) for $N_W = 0$ and 1 [4]. It indicates that the LS state is more stable than the HS state in Co($N_W = 0$). On the other hand, the HS state is preferable in Co($N_W = 1$). We also obtained the Frank-Condon excitation energies of the CT excitations. The calculated excitation energy ~ 2.3 eV of LS → LS1 in Co($N_W = 1$) reasonably corresponds to the

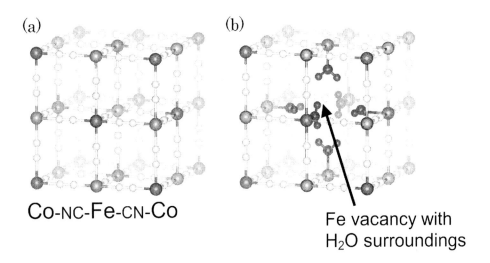

(a)

Co-NC-Fe-CN-Co

(b)

Fe vacancy with H_2O surroundings

FIGURE 1. Crystal structure of A$_y$Co[Fe(CN)$_6$]$_x$ neglecting alkali metals: (a) $x = 1$; (b) $x < 1$. In (b), there is a vacancy at an iron site surrounded by six water molecules substituting cyano anions.

observed absorption energy 2.4 eV inducing the magnetization. This implies that the photo-induced LS → HS transition is triggered by CT excitation mainly between a Co atom with $N_W = 1$ and a nearby Fe atom. The reverse (HS → LS) transition experimentally required near-IR irradiation ($hv = 0.9$ eV). It is assigned to the excitation energy of Co($N_W = 0$) at ~ 0.8 eV. This implies that the HS → LS transition is triggered mainly by the excitation of Co atoms with $N_W = 0$, in contrast to the forward transition. This is reasonable, because the final LS state is much lower in energy than the initial HS state in the case of $N_W = 0$ shown in Fig. 2(a).

The suggested main relaxation paths in the reversible transitions are indicated by arrows in Fig. 2. The path for the LS → HS transition is shown in the case of $N_W = 1$, and the path for the reverse transition is indicated in the case of $N_W = 0$. The Frank-Condon photo-excitations induce CT between cobalt and iron cations. The intermediate excited states are subsequently transformed into local meta-stable states via intersystem crossing.

DISCUSSION

Here we propose some experiments to make further confirmation of our model. First, x dependence of the photo-absorption spectra should be observed. The fraction

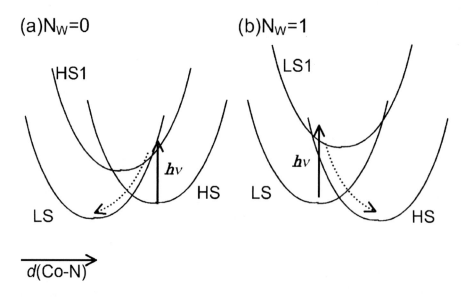

FIGURE 2. Schematic potentials of the cobalt-centered clusters with (a) $N_W = 0$ and (b) $N_W = 1$. Possible primary relaxation paths for the LS → HS and HS → LS transitions are indicated by arrows in (b) and (a), respectively. LS1 and HS1 represent the intermediate state in each transition. Solid and dotted arrows represent photo-excitation and relaxation with intersystem crossing.

of Co(N_W) for N_W = 0, 1 and 2 can strongly depend on x, if iron vacancies are randomly distributed. In the case of $x = 0.9$, the fraction of Co(N_W = 1) is 37%, which is very comparable to 34% in the case of $x = 0.77$. On the other hand, the fraction of Co(N_W = 0) largely increases to 56% when $x = 0.9$ which should be contrasted with 21% when $x = 0.77$. Therefore, it is expected that the oscillator strength of the 3.9 eV peak is much enhanced in the LS state while the peak around 2.3 eV is unchanged.

A pressure effect should also be studied. The cobalt high spin state which is inevitably accompanied by the volume expansion will become unstable under high pressure. The cluster potential energy surface of Co(N_W = 1)-HS state is shifted upward by applying pressure. The situation becomes optimal due to increasing of the relative stability of Co(N_W = 1)-LS state. Therefore we expect under pressure that the HS-LS transition can be induced by the light illumination with a different photon energy for the excitation of Co(N_W = 1) clusters.

ACKNOWLEDGMENTS

The authors acknowledge Prof. K. Hashimoto at Tokyo University and Dr. O. Sato at Kanagawa Academy of Science and Technology for useful discussions. Computations have been done partly using the facilities of AIST Tsukuba Advanced Computing Center.

REFERENCES

1. Sato, O., Einaga, Y., Iyoda, T., Fujishima, A., and Hashimoto, K., *J Electronchem. Soc.* **144**, L11 (1997).
2. Sato, O., Iyoda, T., Fujishima, A., and Hashimoto, K., *Science* **272**, 704 (1996).
3. Verdaguer, M., *Science* **272**, 698 (1996).
4. Kawamoto, T., Asai, Y., and Abe, S., *Phys. Rev. Lett.* to be submitted.
5. Goujon, A., Roubeau, O., Varret, F., Dolbecq, A., Bleuzen, A., and Verdaguer, M., *Eur. Phys. J. B* **14**, 115 (2000).
6. Yokoyama, T., Ohta, T., Sato, O., and Hashimoto, K., *Phys. Rev. B* **58**, 8257 (1998).
7. Yokoyama, T., Kiguchi, M., Ohta, T., Sato, O., Einaga, Y., and Hashimoto, K., *Phys. Rev. B* **60**, 9340 (1999).
8. Kawamoto, T., Asai, Y., and Abe, S., *Phys. Rev. B* **60**, 12990 (1999).
9. Kawamoto, T., Asai, Y., and Abe, S., *J. Lumin.* **87-89**, 658 (2000).

3. COLOSSAL MAGNETO RESISTANCE (CMR)

Valence State and Local Structure in Magnetite and Ca-doped Manganites

J. García and G. Subías[*]

[*]Instituto de Ciencia de Materiales de Aragón, C.S.I.C.-Universidad de Zaragoza,
Pedro Cerbuna 12, 50009 - Zaragoza, Spain

Abstract. X-ray resonant scattering experiments at the Fe K-edge of magnetite and X-ray absorption spectroscopy at the Mn K-edge of Ca-doped manganites show the lack of charge localization on the transition metal atom, neither temporal nor spatially. First, the evolution of the intensity of (002) and (006) forbidden reflections in magnetite, as a function of the incoming X-ray beam energy, the azimuthal angle and the X-ray polarization, shows that the atomic anomalous scattering factors for all the octahedral Fe atoms are identical. This result demostrates both, the absence of charge fluctuation above the Verwey transition and the absence of charge ordering below it. Second, high-resolution XANES spectra of $RE_{1-x}Ca_xMnO_3$ series can not be reproduced by a linear combination of $REMnO_3$ and $CaMnO_3$ spectra. Accordingly, the electronic state of Mn atoms can not be considered as a mixture of Mn^{3+} and Mn^{4+} pure ionic states in the intermediate compounds. Finally, the experimental evidence of charge ordering in $Mn^{3+}/Mn^{4+} = 1$ manganites, given by resonant scattering experiments, has been also critically re-examined. We show that these experiments prove the existence of two different kinds of Mn atoms, although they can not be identify as pure Mn^{3+} and Mn^{4+} ions.

INTRODUCTION

The description of the electronic state of transition metal atoms in transition metal oxides has been usually made on the basis of the ionic model, which implies an electronic transfer of one d-electron to the neighbouring oxygen atoms, being the remaining electrons fixed at the transition metal d-orbitals. However, this model can not be apply for the so-called mixed valence oxides. In these compounds, the formal valence state of the transition metal ion is non-integer and it is not clear where the extra-electron is localized. The usual way to attempt this problem is to assume that the electron is spatially localized, i.e. part of the atoms have a n^+ valence state and the others have a $n–1^+$ valence state. Sometimes, this localization is periodic and gives rise to the so-called charge ordering phases. Another possibility is considering that the extra-electron is temporally localized, so it can hope among different atoms in the lattice giving rise to electrical conduction. Generally, the lack of metallic conduction in mixed valence oxides has been considered as a proof of spatial atomic charge localization, for example, the Verwey transition in magnetite [1]. A similar description has been given for the charge localization process in the so-called charge ordering manganites [2]. However, after 60 years, there is not consensus about the kind of charge ordering that takes place [3].

CP554, *Physics in Local Lattice Distortions*, edited by H. Oyanagi and A. Bianconi
© 2001 American Institute of Physics 1-56396-984-X/01/$18.00

The electronic state of an atom in a solid can be investigated by means of X-ray absorption spectroscopy. The XANES spectrum gives information about the electronic and geometrical local structure of the absorbing atom. The energy position of the absorption edge is intimately correlated with the valence state of the atom, i.e. the chemical shift. Moreover, the experimental spectrum is the result of the incoherent sum of each individual electronic processes, so different local electronic configurations give rise to different XANES spectra and, in principle, we could determine the existence of different types of atoms from an electronic point of view, by deconvolution of the experimental spectrum. As an example, if we consider two different valence states for an atom in the sample, the XANES spectrum would be obtained by addition of the representative spectra of each kind of atom. This property has been mainly used to study the localization of f states in lanthanide compounds for both, spatial charge localization and fluctuating systems [4].

Directly related to the X-ray absorption coeficient is the anomalous scattering factor of an atom. The involved electronic transitions are the same: the incoming photon is virtually absorbed to promote a core electron to an empty excited state and subsequently, decays to the same core hole, emitting a second photon. The advantage of X-ray anomalous diffraction is the coherence of the diffraction process which adds the site selection to the chemical and moment selections of the absorption process. Anomalous diffraction experiments are particularly interesting in the case of reflections forbidden by symmetry where the structure factor is given by the difference of atomic scattering factors of crystallographic equivalent atoms. In particular, for charge ordering reflections, the anomalous scattering factors for atoms with different valence state are different at energies close to the absorption edge. Consequently, we observe scattered intensity at the absorption edge due to the chemical shift between both valence states. The breakdown of the forbidden reflections can also be associated to the presence of anisotropy of the anomalous scattering factor as a function of the relative direction between the X-ray polarization vector respect to the orientation of the anisotropic axis in the unit cell (ATS reflections) [5]. This kind of reflections, sometimes considered as a proof of orbital ordering, also shows azimuthal periodicity, i.e. the intensity depends on the azimuthal angle of the diffraction plane.

In conclusion, X-ray absorption spectroscopy gives us direct information in order to resolve the problem of electronic localization. Moreover, the appearing of resonant forbidden reflections is a direct proof of the presence of electronic anisotropy at the atoms and can also help us to determine the local electronic state. Taking into account that the interaction time is very short for the photoabsorption process, these experiments can also resolve the temporal charge localization.

In this paper, we present our last experimental results in two important systems, magnetite and Ca-doped manganites, which demostrate the absence of charge localization either spatial or temporally. Magnetite has been investigated by means of X-ray resonant scattering at the Fe K-edge [6,7] and Ca-doped manganites have been studied by X-ray absorption spectroscopy at the Mn K-edge, conventional [8] and high resolution by recording the intensity of the K_β emission line[9]. Moreover, we have also re-interprated the X-ray resonant scattering experiments in the so-called charge ordered manganites [10].

MAGNETITE

Magnetite belongs to the family of oxides which crystallizes in the spinel cubic structure, AB_2O_4, with a space group $Fd\bar{3}m$. The chemical formula can be written as $Fe^{3+}(Fe^{3+}, Fe^{2+})O_4$, where brackets indicate atoms located at the B sites (16d) of the spinel lattice. Magnetite is metallic at room temperature and presents a phase transition to a semiconductor state at about 125 K. This metal-insulator transition, known as Verwey transition, was postulated to be an order-disorder transformation of the octahedral Fe^{3+} and Fe^{2+} ions [1]. Then, the electrical conductivity in the high temperature phase is explained by an electron hopping between both octahedral Fe^{2+} and Fe^{3+} sites, giving rise to a charge fluctuation. Moreover, the insulator behaviour, below the Verwey transition, is explained by the localization of the mobile electron at the octahedral B sites, giving rise to a charge ordering phase.

We report here the observation of (002) and (006) forbidden reflections in a single crystal of magnetite by means of X-ray resonant scattering experiments at the Fe K-edge as a function of temperature [6,7]. The X-ray experiments were performed at the beam line D2AM at the European Synchrotron Radiation Facility (E.S.R.F.) in Grenoble. σ-incident geometry was used and the polarization analysis (σ - σ component) was performed by a MgO(222) crystal analyser.

The energy dependence of the top intensity of (002) and (006) reflections at room temperature is given in Fig. 1. No intensity has been observed below the Fe K-edge, as expected for a forbidden Thompson reflection. Three main features can be distinguished as a function of the energy: i) a resonance in the pre-peak region of the fluorescence spectrum, ii) a main resonance at the Fe absorption K-edge and iii) an

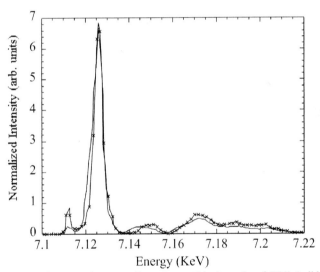

FIGURE 1. Comparison between the energy dependence of the intensity of (002) (solid line) and (006) (crosses) reflections in magnetite at room temperature, self-absorption corrected.

oscillatory behaviour at energies above the absorption edge (EXAFS region). It is noticed that the pre-peak resonance is higher for the (006) reflection than for the (002) one. The dependence of the intensity on the azimuthal angle is shown in Fig. 2. It is observed a $\pi/2$ periodicity on φ. The energy dependence of the intensity is also proportional among the different azimuthal angles, so the intensity for these reflections can be factorized as $I = f(E) g(\theta, \varphi)$.

The structure factor for these forbidden reflections is given by $\hat{F}_{002} = 4(\hat{f}_{B_1} + \hat{f}_{B_2} - \hat{f}_{B_3} - \hat{f}_{B_4})$, where B_i represent the octahedral iron atoms within an octant of the unit cell and f_i are the corresponding atomic anomalous scattering factors. The site symmetry for the A atoms is T_d, so the pure dipolar and quadrupolar scattering factors are isotropic and we can not expect any contribution from these terms. Only the mixed dipolar-quadrupolar term is anisotropic and it could give a contribution. For the B atoms, the real site symmetry is trigonal ($\bar{3}m$). In this case, the dipolar scattering factor is anisotropic, being the anisotropy axis in the direction of the threefold axis, and the contribution to the structure factor is different from zero. This dipolar contribution of the B atoms is given, in the trigonal reference frame of each atom, by a diagonal tensor with two degenerate components $f_\perp$ (i.e. directions perpendicular to the trigonal axis) and a different component $f_\parallel$ in the direction of the trigonal axis. Therefore, the appearance of these reflections comes from the different orientation of the trigonal axis in the octant of the unit cell.

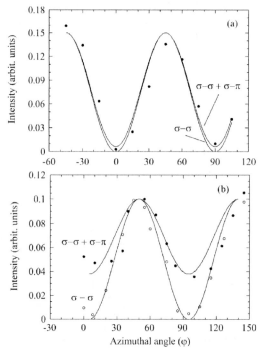

FIGURE 2. Azimuthal dependence of the intensity of (a) (002) and (b) (006) reflections in magnetite at room temperature at the energy of the absorption K-edge.

The intensity of the (0,0,4n+2) forbidden reflections as a function of the azimuthal angle (φ) for both, σ - σ and σ - π channels, is the following [6]:

$$I^{\sigma\sigma'}_{(004n+2)} = |\frac{16}{3}(f_\parallel - f_\perp)|^2 \sin^2(2\varphi) \tag{1}$$

$$I^{\sigma\pi'}_{(004n+2)} = |\frac{16}{3}(f_\parallel - f_\perp)|^2 \sin^2(2\theta)\cos^2(2\varphi) \tag{2}$$

The experimental azimuthal dependence of the intensity agrees nicely with the dipolar model in the whole energy range. In the other hand, the difference ratio between the intensity of (002) and (006) reflections at the pre-peak resonance can not be explained by the dipolar model and it would depend on the transfer moment $Q = \vec{k} - \vec{k}'$. "Ab initio", multiple scattering calculations performed for a cluster around the tetrahedral (A) and octahedral (B) atoms have shown that only a mixed dipole-quadrupole term, coming from the A atoms, can give contribution to the anomaly at the pre-peak energy. On the other hand, the main resonance and the extended part have its origin in the dipolar term associated to the B atoms [6]. Moreover, the oscillatory signal of the extended part comes from the addition of cosine terms whose frecuencies are twice the difference between interatomic distances of different coordination shells. It is important to point out that the anomalous scattering factor is the same for all the iron atoms, tetrahedral and octahedral, and the only difference is due to the distinct orientation of each anisotropy axis in the unit cell.

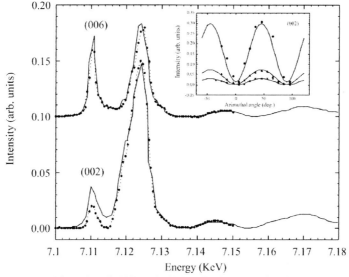

FIGURE 3. Integrated intensity of (002) and (006) reflections as a function of the energy at room temperature (solid line) and at T=30 K (full dots). Inset: Azimuthal dependence of the intensity of the (002) reflection at T=30 K at the following energies; absorption K-edge (circles), pre-peak (triangles) and EXAFS (crosses).

The same experiment has been performed at temperatures below the Verwey transition. The energy and the azimuthal dependences of the intensity of (002) and (006) reflections are identical above and below the Verwey transition, as it is shown in Fig. 3. As a conclusion, we can say that no essential modification of the p and mixed p-d orbital ordering has been observed in magnetite due to the metal-insulator transition.

In summary, our results show that only one kind of tetrahedral and octahedral iron ions exits in magnetite, above and below the Verwey transition. As the anomalous scattering factor depends on the valence state of the atom, the charge must fluctuates in a time scale lower than 10^{-16} sec. Therefore, in the high temperaure phase, the incident photon would see a random distribution of different scattering factors, due to the Fe^{3+} and Fe^{2+} states, and no coherence for diffraction would be fulfilled (see ref. [6] for details). So, we can conclude that the charge does not fluctuate in the high temperature phase of magnetite and the mechanism for the conduction will be based in a band transport. In the insulating phase, at temperatures below the Verwey transition, the ordering of the charge would give an isotropic contribution at the resonance. The identity of the spectra above and below the phase transition and the same azimuthal behaviour as in the high temperature phase demostrate that no charge ordering occurs below the phase transition. In ref. [7], we have discussed in detail the different proposed charge ordering models.

CA-DOPED MANGANITES

A systematic study of manganese perovskites, $RE_{1-x}A_xMnO_3$ (RE:La, Tb, Pr, Nd: A:Ca, Sr), has been performed by means of XANES spectroscopy at the Mn K-edge. High resolution spectra have been measured at the beam line ID26 at E.S.R.F. in Grenoble by recording the intensity of the Mn K_β fluorescence spectrum [9].

Normalized XANES spectra for the $La_{1-x}Ca_xMnO_3$ series, obtained by measuring both, total yield fluorescence and Mn K_β line intensities, are compared in Fig. 4a at room temperature. The several physical features are clearly best resolved by measuring the Mn K_β line intensity. The spectral line shapes are similar for all the samples with a pre-edge structure (A) and a main resonance at the absorption edge (B).

The main difference arises from the chemical shift, which is nearly linear with the Mn formal valence state and it reaches a value of about 4.5 eV between $REMnO_3$ and $CaMnO_3$. Moreover, the spectral differences beyond the absorption edge are due to small changes in the local structure among the different samples.

We have also measured the XANES spectra of different $RE_{0.5}Ca(Sr)_{0.5}MnO_3$ and $La_{0.5}Sr_{1.5}MnO_4$ samples where the small variations in the local structure are negligible. All the samples exhibit a main resonance at nearly the same energy position and a similar slope at the edge, even the $La_{0.5}Sr_{1.5}MnO_4$ sample with a different crystallographic structure. Therefore, the presence of the main peak is common for all these manganites and it can be used as a mark of the Mn electronic state. If we describe the Mn electronic state in the manganites as a mixture of Mn^{3+} and Mn^{4+} ions, the experimental XANES spectrum would be obtained by the addition of the characteristic spectra of the Mn^{3+} and Mn^{4+} ions. We have analysed the high-resolution XANES spectra for the whole series, using as references those of $LaMnO_3$ and $CaMnO_3$ compounds. The comparison between the added and the experimental high-resolution spectra for the $La_{1-x}Ca_xMnO_3$ series is shown in Fig 4b. The edge position is reasonably well reproduced, but the slope of the experimental edge is sharper and the two features displayed at the absorption edge in the simulated spectra are absent in the experimental ones.

Therefore, our results show that the valence state of the Mn atom for the intermediate composition compounds can not be described as a mixture of the 3+ and 4+ pure valence states of the end members of the series. This conclusion is in conflict with the presence of charge and orbital ordering, claimed by X-ray resonant sca\-ttering experiments at the Mn K-edge of the so-called charge ordering manganites, in spite of the same electronic process involved in both kinds of experiments.

We have compared the X-ray resonant scattering experiments with our XANES data and a geometrical model has been proposed to take into account all the experimental results [10]. The main conclusions are the following: i) the anisotropy of the so-called orbital ordered Mn atoms is enough to observe simultaneously both, charge $(0,k,0)$ and orbital $(0,k/2,0)$ reflections and ii) the energy and azimuthal dependences for the $(0,k,0)$ charge ordering reflection, obtained from the atomic scattering factors provided by the experimental XANES spectra, are incompatible with the published data. Moreover, the energy position of the simulated resonances of charge and orbital ordering reflections is shifted by about 2 eV, in contrast with the published data where both resonances occurs at the same energy. The anomalous scattering factor for the so-called "Mn^{3+}", atoms has been calculated considering a

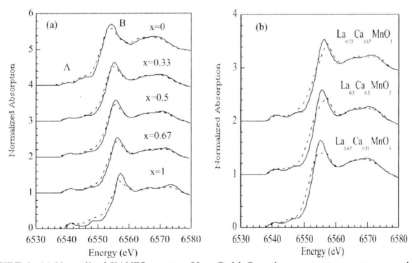

FIGURE 4. (a) Normalized XANES spectra of $La_{1-x}Ca_xMnO_3$ series at room temperature recorded by Mn K_β emission (solid line) and total yield fluorescence (dashed line). (b) Comparison between the normalized high-resolution XANES spectra of $La_{1-x}Ca_xMnO_3$ series (solid line) and the spectra obtained by the average addition of the end members compounds ones (dashed line).

tetragonal distorted octahedron of oxygens whereas for the so-called "Mn^{4+}" atoms, a octahedron of oxygen atoms with C_{2v} symmetry has been considered. The calculated intensity for the (0,k,0) reflection at $\varphi = 90$, as a function of the energy, is shown in Fig. 5a for several values of C (Thompson scattering). It is clear that the intensity of the (0,k,0) charge ordering reflection changes by one order of magnitude depending on the C value. This fact could explain the remarkable differences for the intensity of the (0,k,0) and (0,k/2,0) reflections among the different samples. The azimuthal behaviour at the resonant energy also depends on the C value (see Fig. 5b).

We note that the experimental energy and azimuthal dependences found for the $Nd_{0.5}Sr_{0.5}MnO_3$ sample [11] would correspond to the low values of C while the $\sin^2(2\varphi)$ dependence reported for the $Pr_{0.5}Ca_{0.5}MnO_3$ sample [12] would correspond to high C values. Summarinzing, this geometrical model explain the simultaneous appearance of "charge" (0,k,0) and "orbital" (0,k/2,0) reflections and also reproduces nicely either the energy or the azimuthal dependences for both reflections.

CONCLUSIONS

X-ray absorption and X-ray resonant scattering experiments in mixed valence oxides, magnetite and doped manganites, have shown that no electronic localization occurs on the atomic scale. In the case of magnetite, the mechanism for the electrical conduction above the Verwey transition should be described in terms of band transport

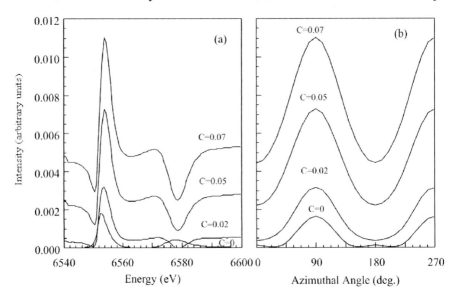

FIGURE 5. (a) Calculated intensity of the $(0,k,0)$ reflection at $\varphi = 90$ for different values of the isotropic atomic scattering factor C, as a function of the photon energy. (b) Azimuthal dependence of the $(0,k,0)$ reflection at E = 6554 eV (Mn K-edge) for the different values of C.

instead of the electronic hopping among the octahedral iron atoms, postulated by Verwey. Below the Verwey transition, the abrupt change in the electrical conductivity should be due to an electronic localization in a spatial scale larger than the atomic one. Similar conclusions have been obtained for the manganites. The charge ordering transition can be well described as a structural phase transition, where a phonon mode condenses, giving rise to a new periodicity on the oxygen atoms. This oxygen motion differenciates two kinds of Mn atoms from a local point of view, but it does not allow us to distinguish two different valence states. As a concluded remark, we can say that charge localization is not an extended phenomenum in mixed valence transition metal oxides, contrary to the conventional believings.

REFERENCES

1. Verwey, E. J. W., *Nature* (London) **144**, 327-328 (1939).
2. Kuwahara, H., Tomioka, T., Asamitsu, A., Moritomo, Y., and Tokura, Y., *Science* **270**, 961-963 (1995).
3. Brabers, V. A. M., *Handbook of Magnetic Materials*, edited by K. H. J. Buschow, Elsevier Science, Vol. **8**, 189-324 (1995).
4. Kanamori, J., and Kotani, A., *Core Level Spectroscopy in Condensed Systems*, Springer, Heidelberg, 1988.
5. Dmitrienko, V. E., *Acta Cryst. A* **39**, 29-35 (1983); Dmitrienko, V. E., Acta Cryst. A **40**, 89-95 (1984); Templeton, D. H., and Templeton, L. K., *Acta Cryst. A* **41**, 133-142 (1985); Templeton, D. H., and Templeton, L. K., *Acta Cryst. A* **42**, 478-481 (1986).
6. García, J., Subías, G., Proietti, M. G., Renevier, H., Joly, Y., Hodeau, J. L., Blasco, J., Sánchez, M. C., and Berar, J. F., *Phys. Rev. Lett.* **85** 578-581 (2000).
7. García, J., Subías, G., Proietti, M. G., Blasco, J., Renevier, H., Hodeau, J., and Joly, Y., *Phys. Rev. B*, to be published.
8. Subías, G., García, J., Proietti, M. G., and Blasco, J., *Phys. Rev. B* **56**, 8183-8191 (1997).
9. García, J., Sánchez, M. C., Subías, G., and Blasco, J., *Phys. Rev. B*, submitted.
10. García, J., Sánchez, M. C., Blasco, J., Subías, G., and Proietti, M. G., *Phys. Rev. B*, submitted.
11. Nakamura, K., Arima, T., Nakazawa, A., Wakabayashi, Y., and Murakami, Y., *Phys. Rev. B* **60**, 2425-2428 (1999).
12. Zimmerman, M. V., Hill, J. P., Gibbs, D., Blume, M., Casa, C., Keimer, B., Murakami, Y., Tomioka, Y., and Tokura, Y., *Phys. Rev. Lett.* **83**, 4872-4875 (1999).

THz Time-Domain Spectroscopic Studies on Magnetoresistive Manganites

N. Kida and M. Tonouchi

Research Center for Superconductor Photonics, Osaka University
2-1 Yamadaoka, Suita, Osaka 565-0871, Japan

Abstract. Low-energy (1.5 - 10 meV) charge dynamics of the magnetoresistive manganites $La_{0.7}Ca_{0.3}MnO_{3-\delta}$ ($\delta \sim 0$ and ~ 0.04) thin films have been studied by using terahertz time-domain spectroscopy. We directly obtained complex optical spectra in the ferromagnetic metallic phase and analyzed by a simple Drude model. On the basis of the present results, low-energy charge dynamics are briefly discussed.

INTRODUCTION

Since the discovery of the colossal magnetoresistive phenomenon in hole-doped manganites, wide variety methods have been applied to study their unique features and revealed new exotic properties of the manganites [1]. Among them, optical spectroscopy is a powerful tool for investigations of the electronic structure of the manganites [2, 3]. Up to date, various spectroscopic studies concerning the large spectral reconstruction in the typical order of several eV with temperature and magnetic field have been observed, nevertheless the low-energy charge dynamics reflecting the coherent motion of the charge carriers are not well clarified.

In this study, we describe a THz time-domain spectroscopic (TDS) studies of $La_{0.7}Ca_{0.3}MnO_{3-\delta}$ (LCMO). Using THz-TDS technique, we directly measured complex optical spectra without the Kramers-Kronig transformation, which is widely used in the optical spectroscopy. In addition to the above advantage, the THz beam (several meV) is suitable for the observation of the Drude term in the metallic phase of the manganites.

EXPERIMENTAL SETUP

LCMO thin films were deposited on MgO(100) substrates by a pulsed laser deposition technique. We prepared two compounds by changing the oxygen pressure. Room temperature X-ray diffraction profile indicates that obtained films were of single phase. The temperature-dependent resistivity measurements show that obtained two films have the insulator-metal (IM) transition temperature (T_{IM}) around 230 K and 190 K, respectively. According to the phase diagram [4], we deduced δ of the

CP554, *Physics in Local Lattice Distortions*, edited by H. Oyanagi and A. Bianconi
© 2001 American Institute of Physics 1-56396-984-X/01/$18.00

obtained films; $\delta \sim 0$ (230 K) and ~ 0.04 (190 K). It is noticed that $\delta \sim 0.04$ is a critical doping level.

An experimental configuration of our system is shown in Fig. 1. The source is THz radiation from a dipole-type low temperature grown GaAs (LT-GaAs) photoconductive switch under a voltage bias of 15 V or the surface of InAs, excited by femtosecond optical pulses from a mode-locked Ti:sapphire laser emitting 80 fs pulses at 800 nm. The THz beam was collimated by an off-axial paraboloidal mirror. The transmitted THz beam through the sample is detected by another bow-tie-type LT-GaAs photoconductive switch. A part of the laser light was split by a beam splitter and introduced to the detector photoconductive switch after a time delay. The detector device becomes instantaneously conductive when hit by the light pulse and detects a photocurrent caused by the electric field of the radiation. The waveform of the radiation is obtained by scanning the time delay. The time resolution is ~ 300 fs, determined by the detection system. The detailed transformation procedure from the experimentally observed transmittance to complex optical spectra, is published in Ref. [5].

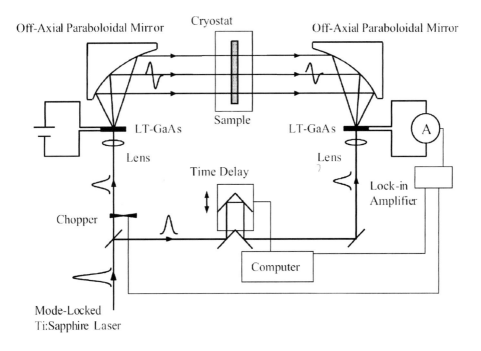

FIGURE 1. Experimental configuration for THz time-domain spectroscopy.

RESULTS AND DISCUSSION

Fig. 2(a) shows the imaginary part (ε_2) of dielectric spectra below 10 meV for $\delta \sim 0$ compound in the ferromagnetic metallic phase [5]. At low-temperatures, ε_2 clearly shows a steep increase with decreasing photon energy, forming the Drude-like spectra. With increasing temperature, a Drude term decreases and finally disappears well below T_{IM} determined by the resistivity measurement. For detailed discussions, we applied a simple Drude model including the scattering rate (Γ) and the plasma frequency (ω_p) as parameters. The solid curves are least-squares fits to experimental data using a simple Drude model; $\varepsilon_2(\omega) = \Gamma\omega_p^2 /[\omega(\omega^2 + \Gamma)]$. We obtained $\omega_p \sim 1.6$ eV and $\Gamma \sim 100$ meV for $\delta \sim 0$. Although not shown here, we also measured ε_2 spectra of $\delta \sim 0.04$ compound and derived $\omega_p \sim 1.46$ eV and $\Gamma \sim 102$ meV at 20 K from the same fitting procedure [see, Fig. 2(b) and 2(c)].

To quantify the spectral change of a Drude term with temperature (T), we show the T/T_{IM} dependence of Γ and ω_p as shown in Fig. 2(b) and 2(c), respectively. The data for $\delta \sim 0$ and ~ 0.04 compounds are represented by open and close circles, respectively. With increasing temperature, both Γ increase in proportion to T^2 as indicated by solid lines in Fig. 2(b). As clearly seen, the ferromagnetic metallic phase can be divided into two phases at the characteristic temperature (T'); one is a metallic

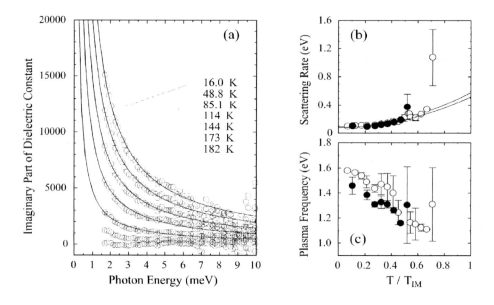

FIGURE 2. (a) Imaginary part of dielectric constant spectra as a function of temperature in the ferromagnetic metallic phase of $La_{0.7}Ca_{0.3}MnO_{3-\delta}$ for $\delta \sim 0$ compound. Scattering rate (b) and the plasma frequency (c) as a function of temperature (T) normalized by the insulator-metal transition temperature (T_{IM}) for $\delta \sim 0$ ($\circ$) and ~ 0.04 ($\bullet$) compounds.

phase with a Drude peak and another without a Drude peak [3, 5]. The obtained T' of $\delta \sim 0$ and ~ 0.04 compounds is $\sim 0.7\, T_{IM}$ and $\sim 0.5\, T_{IM}$, respectively. Around T', clear deviations of the experimental data from the T^2 dependence are observed for both compounds.

With increasing δ toward to the IM phase boundary, the magnitude of $\Gamma(T)$ does not change dramatically, while that of $\omega_p(T)$ decreases. In conventional metals, Γ is mainly affected by the impurity scattering at low temperature. Therefore, the magnitude of Γ is related to that of the residual resistivity (ρ_0). The observed magnitude of Γ is comparable between two compounds, but ρ_0 is quite different, indicating that the impurity scattering is not a major factor of great influence in the motion of the charge carriers in the ferromagnetic metallic phase of the manganites. In fact, the obtained effective mass (m^*) is few times larger than that estimated from specific heat measurements [4]; we derived $m^* = 15 - 20 m_0$ (m_0: bare electron mass) for both compounds assuming the carrier density of 1.6 holes per Mn-site according to Hall effect measurements by Chun et al. [6]. These results simply indicate that the motion of the charge carriers has an incoherent nature even in the ferromagnetic metallic ground state. Previously, Okuda et al. have revealed that ρ_0 of LCMO single crystals shows the strong pressure-dependent behavior and concluded that the anomalous scattering channel of the charge carriers exists in the ferromagnetic metallic phase [4], which is consistent with our results.

We also estimated the mean free path (l), and Fermi wavelength (λ_F) of the conduction carriers [5]. In $\delta \sim 0$ compound, the condition $\lambda_F > l > a$ (a: lattice constant) is obtained even in the ferromagnetic metallic ground state, which is contrast to the relation of the conventional metal. We found that this condition is also hold good in the $\delta \sim 0.04$ compound. With increasing T toward T', Γ increases while ω_p decreases. Namely, the hopping amplitude of the charge carriers decreases. In the vicinity of T', conditions $l < a$ and $l < \lambda_F$ are obtained in both compounds.

Recently, Sawaki et al. have systematically investigated the Al-substitution effect on transport properties in $La_{1-x}Sr_xMnO_3$ (LSMO) single crystals [7]. They revealed clearly that the coherent-incoherent crossover occurs when l reaches λ_F at T^* [7] and categorized the metallic state into the low-temperature coherent metallic phase (LCM) and the high-temperature incoherent metallic phase (HIM) [3, 7]. In their classification, the condition $l < \lambda_F$ is violated in LCM but in HCM. In LCMO as investigated here, characteristics of the HIM is clearly observed by our spectroscopic studies. On the contrary, the LCM is not realized in the measured T range. Sawaki et al. also reported that the critical resistivity (ρ_c) of LSMO is about 300 $\mu\Omega$ cm at T^*. The magnitude of ρ_c is comparable with that of ρ_0 for LCMO with $\delta \sim 0$ compound [5]. This may be the reason of the missing of the LCM in LCMO. In order to confirm this assumption, THz-TDS studies on LSMO are now in progress.

ACKNOWLEDGMENTS

We thank Mr. M. Yamashita for the experimental support and Prof. M. Hangyo for fruitful discussions.

REFERENCES

1. *Colossal Magnetoresistive Oxides*, edited by Y. Tokura, Gordon and Breach Science, New York, 2000.
2. Okimoto, Y. *et al.*, *Phys. Rev. B* **55**, 4206-4214 (1997).
3. Takenaka, K., Sawaki, Y., and Sugai, S., *Phys. Rev. B* **60**, 13011-13015 (1999).
4. Okuda, T. *et al.*, *Phys. Rev. B* **61**, 8009-8015 (2000).
5. Kida, N., Hangyo, M., and Tonouchi, M., *Phys. Rev. B* **62** Nov. 1 Issue (2000) in press; J. Magn. Magn. Mater. in press.
6. Chun, S. H. *et al.*, *Phys. Rev. B* **61**, R9225-R9228 (2000).
7. Sawaki, Y. *et al.*, *Phys. Rev. B* **61**, 11588-11593 (2000).

Crucial Role of Jahn-Teller Distortions in Stabilizing Magnetic Ordering in Insulating Manganite Phases

S. Fratini[1,2], M. Capone[3], M. Grilli[2] and D. Feinberg[1]

[1] *LEPES, CNRS, BP 166, 38042 Grenoble, France*
[2] *INFM and Dip. di Fisica, Univ. di Roma, Piazz. Aldo Moro 2, 00185 Roma, Italy*
[3] *INFM and SISSA-ISAS, Via Beirut 2-4, 34013 Trieste, Italy*

Abstract. The phase diagram of manganites combines magnetic, orbital, lattice and charge ordering. We underline the essential role of strong cooperative Jahn-Teller interactions in selecting the magnetic structure and fine tuning by external parameters. First, the realistic Jahn-Teller distortions is $LaMnO_3$ controls the orbital order and consistently explain the layered antiferromagnetic ordering. Secondly, the phase diagram of half-doped manganites of the family $La_{0.5}Ca_{0.5}MnO_3$ can be reproduced, with the observed ferromagnetic, charge-ordered (CE) antiferromagnetic and paramagnetic insulating phases.

INTRODUCTION

The prominent feature of manganese perovskites is their colossal magnetoresistive properties. In between $LaMnO_3$(trivalent Mn) and $CaMnO_3$(tetravalent Mn), an extremely rich phase diagram has emerged. Understanding the interplay between spin, charge and orbital ordering requires at least; i) strong on-site Coulomb interactions; ii) the "double exchange" mechanism due to e_g electron itineracy and Hund's exchange with the more localized t_{2g} electron spins, favouring ferromagnetism [1-3]; iii) superexchange between localized electrons; iv) large electron-lattice interactions due to Jahn-Teller (JT) effect on Mn^{3+} ions [4,5]. In principle, long-range Coulomb interactions should also be included, although most models are limited to nearest-neighbour *Mn-Mn* repulsion.

Although Goodenough [6] proposed long time ago a qualitative understanding of the phase diagram of the (La,Ca)MnO₃ family, a full microscopic description is still lacking, with a clear identification of the dominant mechanism for every doping. We show in Section I that understanding the insulating antiferromagnet $LaMnO_3$ demands that the strong Jahn-Teller deformations are taken into account. Following the same viewpoint, we show in Section II that the charge-ordered CE chain antiferromagnetic structure of $(A,A')_{0.5}MnO_3$ can be explained when considering; i) the Jahn-Teller modes on Mn^{3+} ions; ii) the breathing mode on Mn^{4+} ions; iii) the shear mode inducing shortening and lengthening of Mn-O bonds. The strong first order character of the metal-insulator transitions, varying temperature or magnetic field, is reproduced thanks to a self-consistent screening mechanism.

CP554, *Physics in Local Lattice Distortions*, edited by H. Oyanagi and A. Bianconi
© 2001 American Institute of Physics 1-56396-984-X/01/$18.00

A-TYPE ANTIFERROMAGNETISM IN LaMnO$_3$

Stoechiometric LaMnO$_3$ is a layered antiferromagnet, with ferromagnetic couplings (F) in two directions and antiferromagnetic (AF) coupling in the other [7]. The AF directions are associated to a shortening of the Mn-O bonds, while in the F directions long and short bonds alternate. We hereafter neglect other deformations such as tilting and concentrate only on the Mn-O bond length deformations, which can be understood in terms of cooperative JT effect. The corresponding lifting of e_g degeneracy results in orbital ordering, with occupied d orbitals pointing preferentially in the directions of long Mn-O bonds. Long ago, Kugel and Khomskii [8] proposed that superexchange (SE) in the presence of e_g orbital degeneracy is the cause of the observed structure. Hund's rule tends to favour different orbitals on neighbouring sites thus ferromagnetic coupling. Starting with degenerate e_g orbitals, intraorbital SE dominates in the c-direction (defined as the z-axis), leading to AF coupling, while interorbital SE dominates in the ab-directions, yielding F coupling. JT couplings are neglected, hence orbital and magnetic ordering result from superexchange (SE) only. Occupied orbitals are dominantly $d_{z^2-x^2}$, $d_{z^2-y^2}$, therefore, for Cu^{2+} in KCuF$_3$ (hole orbital), a small JT coupling would imply a shortening as experimentally observed ($c/a < 1$). However, for Mn^{3+} ions with one *electron* in the e_g levels, JT coupling would trigger a lengthening of the c-axis ($c/a > 1$), in contradiction with the actual structure.

This sets the limits of the Kugel-Khomskii model for LaMnO$_3$. Actually, the assumption that the e_g degeneracy is lifted principally by superexchange is not correct in this system. The typical JT splitting ε is indeed much larger than the SE splitting, of order t^2/U, thus a few meV: the deformations of Mn-O bonds are extremely large, more than ten per cent. Photoemission and optical conductivity [9,10] indicate that JT splittings are as large as a few tenths of eV, comparable to the electronic hopping integrals.

The microscopic model is detailed in ref. [11]. It includes hopping between $d_{x^2-y^2}$ (x) and d_{z^2} (z) orbitals, the Coulomb interactions U, $U' = U - 2J_H$ and J_H, the AF superexchange J_t between neighboring t_{2g} spins, and the electron-lattice interaction. The JT modes are defined in terms of the short (s), medium (m) and long (l) Mn-O bonds by $Q_2 = \sqrt{2}\,(l - s)$ and $Q_3 = \sqrt{2/3}\,(2m - l - s)$, the m bonds lying in the z direction and the s, l ones in the x, y planes. With this convention, both KCuF$_3$ and LaMnO$_3$ show compression along the c-axis, i.e. $Q_3 < 0$. On the other hand, due to the respectively hole-like and electron-like character of the relevant e_g states, the coupling constant g is positive in KCuF$_3$, and negative in LaMnO$_3$ where the $d_{x^2-y^2}$ orbital is favoured ($\tau_z = +1$). In what follows, we redefine Q_3 as Q_3 sgn(g), e.g. $Q_3 > 0$ in KCuF3 and $Q_3 < 0$ in LaMnO$_3$.

We believe that no simple short-range interaction model is able to account for the observed lattice distortions. Thus we treat the JT deformations $\mathbf{Q} = (Q_2, Q_3)$ as external fields, which in turn determine the various magnetic couplings. The Hamiltonian is diagonalized for a system of two sites located either on the same xy plane or on adjacent planes.

The JT energy splitting $\varepsilon = g\sqrt{Q_2^2 + Q_3^2}$ and the deformation anisotropy ratio $r \equiv Q_2/Q_3$ are given external parameters. We determine the ground states of a two-site cluster with total spin $S_T = 4$, $M_{S_T} = 4$ and $S_T = 3$, $M_{S_T} = 3$. Then the magnetic coupling is given by the energy difference $E(S_T = 4, M_{S_t} = 4) - E(S_T = 3, M_{S_t} = 3) = 2J$ for each lattice direction, $J < 0$ indicating a ferromagnetic coupling.

Our calculation [11] shows that generically, for small JT splitting and $Q_3 < 0$, the interorbital superexchange dominates in all directions and yields ferromagnetic coupling. On the contrary, Fig. 1 shows that for large JT coupling, the correct A-type ordering is stabilized for sufficient anisotropy r, typically $|r| \sim 2\text{-}3$, in a rather large window of the Hund coupling. A too large J_H always favours ferromagnetism.

The results are summarized in a phase diagram at zero temperature, in terms of the JT energy splitting and the deformation ratio, for $Q_3 < 0$ ($Q_3 > 0$ in the inset): ferromagnet FFF, Néel antiferromagnet AAA, AAF (C) phase, and the observed layered FFA phase. At small values of the JT splitting, the phase diagram is prominently occupied by a FFF phase. On the other hand, in the positive Q_3 case like in KCuF$_3$ the FFA phase is realized over a much broader range of ε.

One can also attempt at precise estimates of J_{xy} and of J_z. Taking $Q_2/|Q_3|=3.2$ as observed [12], $t = 0.124$ eV, $U = 5.81$ eV, $J_H = 1.2$ eV, $J_t = 2.1$ meV, and $\varepsilon = 0.325$ eV, one obtains the observed $J_{xy} = -0.83$ meV and $J_z = 0.58$ meV [13]. The "anomalous" trend $|J_{xy}| > |J_z|$ is correctly reproduced, and our fit is quite flexible concerning parameters U, J_H or t, provided ε is large enough.

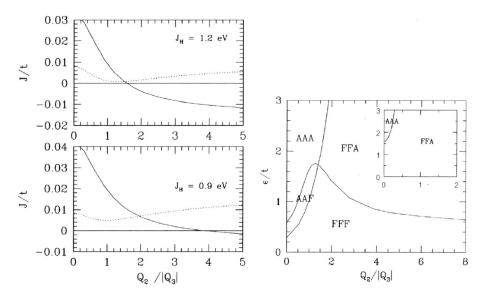

FIGURE 1. Left panel: Magnetic couplings J_{xy} (solid line) and J_z (dotted line) vs. the deformation ratio $r = Q_2/|Q_3|$ for negative values of Q_3 for $t = 0.14$eV, $U = 6$eV, $\varepsilon = 0.3$eV, $J_t = 2.1$meV and $J_H = 1.2$eV ($J_H = 0.9$eV) in the upper (lower) panel. Right panel: Zero-temperature phase diagram for $Q_3 < 0$, $t = 0.14$eV, $U = 6$eV, $J_t = 2.1$mV and $H_H = 1.2$eV.

THE *CE* STRUCTURE OF (A,A')$_{0.5}$MnO$_3$

In the doped compounds, charge ordering tends to build in for sufficient doping, bringing together orbital and magnetic ordering. The competition with double exchange yields a very large magnetoresistance, obtained when the applied magnetic field is able to align t_{2g} spins, thereby favouring the metallic phase. As proposed by Millis *et al.* [4], a large electron-lattice coupling is involved, with the formation of Jahn-Teller polarons in the insulating phase. Large cooperative Jahn-Teller distortions consist of Jahn-Teller deformations around Mn^{3+} ions, and "breathing mode" deformations with shorter Mn-O bonds around Mn^{4+} ions. Charge ordering in (LaCa)$_{0.5}$MnO$_3$ is accompanied by CE-type antiferromagnetic order involving ferromagnetic and antiferromagnetic zigzag chains crossing each other. Mizokawa *et al.* [14], and Yunoki and coworkers [15] have underlined the prominent role of Jahn-Teller deformations.

The Jahn-Teller splitting, although smaller than in LaMnO$_3$, is still large in (LaCa)$_{0.5}$MnO$_3$. On the other hand, the magnetic couplings (which in a cubic lattice give rise to critical temperatures T_c between 100K and 400K) are in the range of a few meV. This holds as well for the superexchange (antiferromagnetic) couplings as, for the (ferromagnetic) double-exchange ones. Moreover, the energy scale set by the external magnetic field needed to turn the FM phase into the CE (AFCO) phase ranges from a few Teslas to 20 Teslas or more, thus of the order of 0.4-4 meV per atom. This points out that the free energies of the above phases are very close, in the range of a few meV per atom. Owing to the much larger electron-phonon and Coulomb interactions, the latter play a dominant role in stabilizing the low-temperature CE phase. It follows that CE and FM phases are (meta)stable minima of the free energy, separated by rather high barriers. In fact, the phase transitions (with temperature or magnetic field) between charge ordered and charge disordered phases are strongly first-order. Tendencies to phase separation between FM and CO phases have been demonstrated in La$_{0.5}$Ca$_{0.5}$MnO$_3$, Pr$_{0.7}$Sr$_{0.3}$MnO$_3$ and other compositions [16]. One also notices that charge ordering is always strong when it exists : fine tuning of the chemical composition between CO and FM low temperature phases [17] does not allow to stabilize "weak" charge ordering. Moreover there does not seem to be any evidence of a ferromagnetic charge ordered phase, as would be expected from any coupling constant owing to the perfect nesting of the Fermi surface on the half-doped phase. This points towards strong interactions (electron-phonon or Coulomb) in the insulating phase, while they are screened in the metallic phase. This feature is overlooked by mean-field treatments, but can be recovered by taking into account exchange-correlation corrections to the intersite Coulomb repulsion, as shown by Sheng and Ting [18]. Since the lattice distortions here also come from Coulomb interactions (between Mn and O ions), we have generalized the screening idea to electron-phonon interactions and used for this purpose a phenomenological approach.

We succeed in i) obtaining, for realistic values of the parameters, FM, CE and paramagnetic phases; ii) exploring by small variations of those parameters the different kinds of phase diagrams, with temperature and magnetic fields: of the type of La$_{0.5}$Sr$_{0.5}$MnO$_3$ (no charge ordering, FM-PM transition with increasing T); of the type of Nd$_{0.5}$Sr$_{0.5}$MnO$_3$ (CE-FM-PM transitions with T, CE-FM with H); of the type of

$Pr_{0.5}Ca_{0.5}MnO_3$ (CE-PMCO-PM transitions with T, CE-FM with H) iii) obtaining first-order transitions between CE and FM phases.

Here we restrict ourselves to a single-orbital model in two dimensions, which quantitatively reproduces the various phase diagrams and their tuning by subtle variations of the bandwidth [19]. We assume an infinite repulsion ($U = \infty$) between electrons on the same lattice site, and an infinite Hund coupling ($J_H = \infty$) between the localized t_{2g} spins and the itinerant e_g spins. We consider a two-dimensional plane of the structure, with a half-filled band made of a single e_g orbital. $\tilde{t}_{ij} = t\langle\cos(\theta_{ij}/2)\rangle$ is the transfer integral between neighboring Mn sites whose ionic spins $\mathbf{S_i}$ and $\mathbf{S_j}$ make an angle θ_{ij} [2].

The nearest-neighbour Coulomb repulsion V is taken into account, as well as three elastic lattice modes, the electrons coupling to a Jahn-Teller (JT) mode Q_2 and holes to a "breathing" mode Q_b (g_2 and g_b are the coupling strengths, K_2 and K_b the spring constants). In the planar geometry, the other Jahn-Teller mode Q_l is not relevant. We also introduce a shear mode Q_s, driven by Q_2, and essential to reconcile the alternating Mn^{4+} breathing and Mn^{3+} JT distortions which develop in the ordered phases. A substantial shear deformation is indeed observed in $La_{0.5}Ca_{0.5}MnO_3$ [20]. It results in half the Mn-O-Mn bonds being shorter and the others larger than average.

A term $H_{SE} = \sum_{<ij>}[J_1 - J_2 Q_s]\vec{S}_i \cdot \vec{S}_j$ represents the antiferromagnetic (AF) superexchange interaction J_1 between the ionic spins on neighboring sites, which are treated as classical. The coupling $J_2 Q_s$ reflects the Goodenough rule: it can either enhance or reduce the AF coupling depending on the sign of the shear deformation, which accounts for the fact that longer (shorter) Mn-Mn bonds have a more (less) antiferromagnetic character [6].

We use a mean-field approximation, using de Gennes' procedure [3] to treat the magnetic part. Exchange and correlation corrections related to the Coulomb repulsion and to the electron-lattice interactions are accounted for by interpolating between the weak and strong coupling regimes [18]. We consider the following magnetic phases: ferromagnetic (F), paramagnetic (P), Néel anti-ferromagnetic (not stabilized in the physical range of parameters), and CE-type ordering. The unit cell is made of 8 nonequivalent Mn sites in a plane.

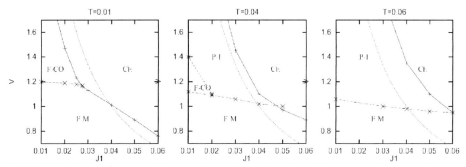

FIGURE 2. Phase diagrams as function of the AF coupling J_1 and the electron-phonon or Coulomb parameter V, for different temperatures (see text).

In Fig. 2 we report typical phase diagrams at various temperatures as a function of the magnetic coupling $J = J_1 S^2/t$ with $J_1 = J_2$ ($S = 3/2$) and of the JT coupling and repulsive e-e interaction, arbitrarily choosing $V = 0.5(g^2/K)$. At low temperature a metallic (without charge-ordering) ferromagnetic phase (FM) appears for small V and g^2/K. It is naturally suppressed by the increase of the antiferromagnetic (AF) superexchange coupling J_1. When the charge mobility is suppressed by the CO terms, one finds two distinct possible phases. At low values of the AF coupling the pure CO effects dominate and a ferromagnetic (F) CO phase occurs at sufficiently large values of V (F-CO). The transition is first order, as found in Ref [18], due to the exchange-correlation terms. On the other hand, by increasing the AF coupling, the CO ferromagnetism is destabilized and a CE phase takes place. This latter phase indeed realizes the best compromise between the electron mobility, favored by the ferromagnetic bonds, on one hand, and the charge ordering in the other hand, with the AF interactions increasing with J_1. The CE ordering here arises due to competing lattice displacements.

The temperature and field evolution is represented in Fig. 3, corresponding to very small variations of the bandwidth. It has been suggested [21] that this parameter plays a primary role in determining the competition between the FM and the insulating phases [22]. To be consistent, the superexchange couplings J_1 and J_2 also vary, and the dotted curves $V/t = A/J$ in Fig.2 correspond to similar physical systems, where the only nearest neighbour electronic hopping amplitudes have been varied. The results summarized in Fig. 3 show that our simple approach allows for a unified qualitative description of different half-doped materials. Transitions CE-FM and FM-PI are strongly first order. In particular the most insulating behavior in the first panel of Fig. 3 is consistent with the generic features of $Pr_{0.5}Ca_{0.5}MnO_3$. On the other hand the second one shows the same qualitative behavior of $La_{0.5}Ca_{0.5}MnO_3$ or

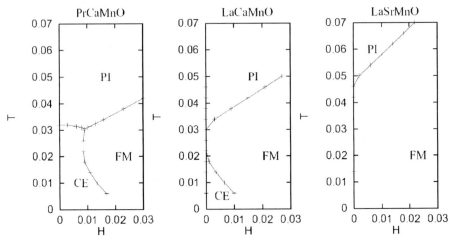

FIGURE 3. Phase diagrams as function of temperature and magnetic field, for three typical bandwidth parameters

$(Nd_{1-y}Sm_y)_{0.5}MnSr_{0.5}O_3$. Finally the most metallic system in the third panel is a good qualitative description of $La_{0.5}Sr_{0.5}MnO_3$. A semi-quantitative agreement is an indication that the right physics is captured. In fact, the pure double-exchange ferromagnetic critical temperature ($J_1/t = 0$) is, from our 2D mean-field calculation, $T_c^{DE} \simeq 0.085t$, thus $\simeq 0.13t$ in 3D. For an average value $t = 0.3$ eV one obtains a transition temperature $\simeq 450K$. It is reduced by the presence of the antiferromagnetic coupling, for instance in the third panel of Fig. 3, in zero field $T_c \simeq 0.05t$ thus $\simeq 270K$ in 3D. Then one obtains in the second one the value $T_c^{CE} \simeq 180K$, and in the first one $T_c^{CE} \simeq 170K$. These values are reasonable, as compared with experimental ones, in particular one notices that T_c^{CE} is strongly reduced compared to T_c^{DE}.

Concerning the role of magnetic field, by taking a typical value of $\simeq 0.3$ eV, one finds that $H/t = 0.015$ (where $H = g\mu_b SH$) roughly corresponds to 10 T. This value agrees well with the typical values experimentally used to investigate the (T,H) dependence of the low-temperature CE insulating phase and the intermediate-temperature uniform FM phase [17,21].

Nevertheless, in the present mean-field scheme, the CO paramagnetic phase is the only possible description of an insulating non-magnetic phase at moderate temperatures. In strong coupling, a more refined treatment would allow for the disordering of the charges, thus producing the PM polaronic phase which is observed in all manganites above a few hundreds of kelvins, and in particular the sequence CE-(PM)CO-(PM)CD in PrCaMnO.

CONCLUSION

On the basis of microscopic models, we have shown the crucial role of lattice distortions in stabilizing the magnetic structure of insulating manganites. In $LaMnO_3$ the layered AF ordering cannot be understood without a sizeable JT coupling, and the experimental values of the magnetic couplings are reconciled with the typical microscopic parameters. For half-doped systems, the cooperative lattice distortions stabilize the CE phase with charge ordering, and self-consistent screening is a feature which makes the metal-insulator transition first order. Considering explicitly orbital ordering should improve the present picture.

REFERENCES

1. Zener, C., *Phys. Rev.* **82**, 403 (1951).
2. Anderson, P.W., Hasegawa, H., *Phys. Rev.* **100**, 675 (1955).
3. De Gennes, P.G., *Phys. Rev.* **118**, 141 (1960).
4. Millis, A.J., Littlewood, P.B., and Shraiman, B.I., *Phys. Rev. Lett.* **74**, 5144 (1995).
5. Zang, J., Bishop, A.R., and Röder, H., *Phys. Rev. B* **53**, 8840 (1996).
6. Goodenough, J. B., *Phys. Rev.* **100**, 564 (1955).
7. Wollan, E.O., and Koehler, W.C., *Phys. Rev.* **100**, 545 (1955).
8. Kugel', K.I., and Khomskii, D.I., *Sov. Phys. JETP* **37**, 725 (1973); *Zh. Eksp. Teor. Fiz.* **64**, 1429 (1973).

9. Dessau, D.S., and Shen, Z.-X., in *Colossal Magnetoresistance Oxides*, edited by Tokura, Y., Gordon & Breach, Monographs in Cond. Matt. Science, 1999.

10. Jung, J.H., Kim, K.H., Eom, D.J., Noh, T.W., Choi, E.J., Yu, J., Kwon, Y.S., and Chung, Y., *Phys. Rev. B* **55**, 15489 (1997).

11. Feinberg, D., Germain, P., Grilli, M., and Seibold, G., *Phys. Rev. B* **57**, 5583 (1998); Capone, M., Feinberg, D., and Grilli, M., *Eur. J. Phys. B* **17**, 103 (2000).

12. Rodríguez-Carjaval, J., Hennion, M., Moussa, F., Moudden, H., Pinsard, L., and Revcolevschi, A., *Phys. Rev. B* **57**, 3189 (1998).

13. Moussa, F., Hennion, M., Rodríguez-Carjaval, J., Moudden, H., Pinsard, L., and Revcolevschi, A., *Phys. Rev. B* **54**, 15149 (1996).

14. Mizokawa, T., and Fujimori, A., *Phys. Rev. B* **56**, R493 (1997).

15. Yunoki, S., Hotta, T., and Dagotto, E., *Phys. Rev. Lett.* **84**, 3714 (2000).

16. Mori, S., Chen, C.H., and Cheong, S.-W., *Phys. Rev. Lett.* **81**, 3972 (1998).

17. Tokura, Y., Kuwahara, H., Moritomo, Y., Tomioka, Y., and Asamitsu, A., *Phys. Rev. Lett.* **76**, 3184 (1996).

18. Sheng, L., and Ting, C.S., *Phys. Rev. B* **57**, 5265 (1998).

19. Fratini, S., Grilli, M., and Feinberg, D., in preparation.

20. Radaelli, P.G. *et al.*, *Phys. Rev. B* **55**, 3015 (1997)

21. Tokura, Y. *et al.*, *J. Appl. Phys.* **79**, 5288 (1996).

22. It has been pointed out (Egami, T., and Louca, Despina, *J. of Supercond.* **13**, 247 (2000)) that a rapid change in lattice constant K, rather than necessarily small changes of t, could be the clue for the very different behaviours of the systems $(A,A')_{0.5}MnO_3$. In our case this would correspond to an abrupt change along a vertical line in Figs. 2, and would enhance the first-order character of the transitions.

Jahn-Teller Distortion and Magnetoresistance in Sr$_{1-x}$Ce$_x$MnO$_3$

A. Sundaresan[a,*], J. L. Tholence[a], A. Maignan[b], C. Martin[b], M. Hervieu[b], B. Raveau[b], E. Suard[c], T. Eto[d], F. Honda[d] and G. Oomi[d]

[a]LEPES-CNRS, BP 166, 38042 Grenoble Cedex 09, France
[b]Laboratoire CRISMAT, ISMRA, 6 Bd du Marechal Juin 14050 Caen Cedex, France
[c]Institut Laue-Langaevin, BP 156X, 38042 Grenoble Cedex 9, France
[d]Department of Physics, Kyushu University, 4-2-1 Ropponmatsu, Fukuoka 810-8560, Japan
*Present address: Electrotechnical Laboratory, 1-1-4 Umezono, Tsukuba 305 8568, Japan

Abstract. Structural, magnetic and electrical transport properties of Sr$_{1-x}$Ce$_x$MnO$_3$ (x = 0, 0.1, 0.2, 0.3 and 0.4) have been investigated. For x = 0.1, it undergoes a first-order structural and metal-insulator transitions due to Jahn-Teller ordering (T$_{JT}$ ~ 315 K) which stabilizes C-type antiferromagnetic ground state (T$_N$ = 290 K) that is insensitive to an applied magnetic field of 7 T. With increasing x, the Jahn-Teller distortion decreases and a small magnetoresistance appear. However, there is no ferromagnetism or colossal mangnetoresistance in any of these materials due to the stability of the JT distortion and C-type antiferromangetism.

INTRODUCTION

Hole doped manganites R$_{1-x}$A$_x$MnO$_3$ (where R is rare-earths and A is alkaline earths or Pb) exhibit various interesting physical properties including colossal magnetoresistance. The origin of this phenomenon is believed to be due to double exchange ferromagnetic interaction and Jahn-Teller polaron. Recently, it has been reported that electron doping in the antiferromagnetic (G-type) insulator CaMnO$_3$ (with Jahn-Teller inactive Mn(IV):t$_{2g}^3$ electronic configuration) exhibits large negative magnetoresistance [1]. However, the magnitude of this mangnetoresistance remains smaller than that observed for hole doped manganites [2]. Therefore, it is important to investigate the structural and the nature of magnetic interactions those are primarily responsible for the mangnetoresistance in these materials. For this reason, we have investigated the structural, magnetic and magnetotransport properties of a newly synthesized electron doped system Sr$_{1-x}$Ce$_x$MnO$_3$.

EXPERIMENTS

Polycrystalline samples Sr$_{1-x}$Ce$_x$MnO$_3$ with various values of x were prepared by the solid state reaction of the stoichiometric mixture of SrCO$_3$, CeO$_2$, and Mn$_2$O$_3$ at 1600°C in air. The purity and the composition of the samples were checked by electron diffraction using a JEOL 200 Cx and a JEOL 2010 electron microscopes. The

CP554, *Physics in Local Lattice Distortions*, edited by H. Oyanagi and A. Bianconi
© 2001 American Institute of Physics 1-56396-984-X/01/$18.00

ED patterns at various temperatures were recorded keeping a constant electron current density. Powder neutron diffraction data were collected at room temperature and at 200 K (only for x = 0.1) on the high-resolution powder diffractometer D2B at the Institut Laue-Langevin, Grenoble, France with a wavelength of 1.594 Å. Additional neutron data for x = 0.1 at 350 K were also collected on the D1A instrument with λ = 1.911 Å. The data were analyzed by the Rietveld analysis method. Magnetization and resistivity measurements were carried out with a commercial (Quantum Design, USA) SQUID magnetometer and a Physical Property Measuring System, respectively. Temperature dependence of thermal expansion, $\Delta L/L$, was also measured for x = 0.1 under various hydrostatic pressures.

RESULTS AND DISCUSSIONS

For x = 0.1, the energy dispersive spectroscopy (EDS) analyses performed on numerous grains showed the homogeneity of the crystallites which exhibit a composition close to the nominal one, $Sr_{0.9}Ce_{0.1}MnO_3$. At room temperature, the ED patterns are characterized by intense Bragg reflections which are those of the perovskite subcell (a_p = 3.85Å) and weak extra reflections indicating an I-type lattice with a $\approx$ b $\approx a_p\sqrt{2}$ and c $\approx 2a_p$. The reconstruction of the reciprocal space at 360 K showed a cubic P-type lattice (a_p). For x = 0.4, the ED investigation showed an orthorhombic cell with a ~ $a_p\sqrt{2}$, b ~ $2a_p$ and c ~ $a_p\sqrt{2}$.

Rietveld analysis of the room temperature neutron diffraction data revealed that the samples with x = 0.1, 0.2 and 0.3 have tetragonal structure with the space group I4/mcm. The lattice parameters, cell volume and various bond lengths are given in Table 1. It can be seen that there is a large increase in the lattice parameter 'a', whereas the parameter 'c' decreases with increasing x. The observed increase of lattice volume is due to the larger difference in the ionic radii between the Mn^{3+} and Mn^{4+} ions than between Sr^{2+} and Ce^{3+} ions. For all x there are two different bond lengths and angles. For x = 0.1 there are two longer and four shorter bonds and the difference between them is ~ 0.04 Å. This indicates that there is a strong Jahn-Teller distortion of the MnO_6 octahedron and the doping electrons occupy the d_z^2 symmetry of the Jahn-Teller split state (elongated octahedron). With increase of x, the Jahn-Teller distortion decreases. For x = 0.4, a tendency of the octahedra to be compressed seems to appear (two shorter and four longer bonds). This indicates that the doping electrons occupy the $d_{x^2-y^2}$ symmetry (compressed octahedron) which is consistent

TABLE 1. Lattice parameters, volume and selected bond lengths.

$Sr_{1-x}Ce_xMnO_3$	Lattice Parameters and Volume	Bond lengths
X = 0.1	a = b = 5.3637(2); c = 7.7481(3) Å V = 222.90(1) Å^3	Mn-O(1)x2 = 1.937(1) Å Mn-O(2)x4 = 1.900(2) Å
X = 0.2	a = b = 5.3971(1); c = 7.7454(1) Å V = 225.33(5) Å^3	Mn-O(1)x2 = 1.936(1) Å Mn-O(2)x4 = 1.916(3) Å
X = 0.3	a = b = 5.42881(1); c = 7.7256(1) Å V = 227.68(6) Å^3	Mn-O(1)x2 = 1.931(1) Å Mn-O(2)x4 = 1.929(3) Å
X = 0.4	a = 5.4522(3); b = 7.6813(3); c = 5.4712(2) Å ;V = 229.12 Å^3	Mn-O(1)x2 = 1.937(1) Å Mn-O(2)x4 = 1.940(1) Å

380

with the macroscopic orthorhombic symmetry (space group Imma). The Mn-O(1)-Mn bond angle in x = 0.1, 0.2 and 0.3 remains 180 degree and the Mn-O(2)-Mn angle is smaller than 180 degree. At 350 K, the structure is ideal cubic perovskite with the lattice parameter a = 3.8217(6) Å which is in agreement with the electron diffraction analysis as discussed above. In contrast to the low temperature tetragonal phase, there is no static Jahn-Teller distortion in the cubic phase and the MnO_6 octahedra become regular with six equal Mn-O bond lengths (1.911 Å).

For x = 0.1, the temperature dependence of resistivity and susceptibility between 4.5 K and 400 K is shown in Fig. 1. It can be seen that the resistivity at high temperature is low and is independent of temperature down to 315 K. This is in agreement with the fact that the nuclear structure is cubic at 350 K where the doped electrons occupy the degenerate e_g band and is responsible for metallic conductivity. At 315 K there is a jump in resistivity or metal-insulator transition below which the resistivity increases logarithmically with decrease of temperature down to 120 K. Below this temperature it increases more rapidly. Although the exact temperature at which the structural or JT transition occurs is not known due to lack of neutron data between 300 and 350 K, it is conceivable from the metal-insulator transition that it occurs at T_{JT} = 315 K and is responsible for the observed metal-insulator transition. The insulating behavior in the tetragonal phase is consistent with the strong Jahn-

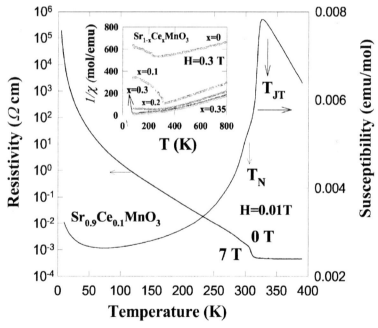

FIGURE 1. Temperature dependence of magnetic susceptibility and resistivity for x = 0.1. T_{JT} is Jahn-Teller Transition and T_N = Néel Temperature.

Teller distortion and the localization of doping electrons on the d_z^2 orbital as revealed by the different Mn-O bond lengths as discussed above. The resistivity showed a hysteritic thermal behavior near the metal-insulator transition indicating the first order nature of JT and associated metal-insulator transitions.

While cooling, the susceptibility for x = 0.1 increases and follows a Curie-Weiss behavior with a positive θ_P and peaks at about 325 K resembling an antiferromagnetic transition (Fig. 2). But, the neutron diffraction data at 300 K reveal no contribution due to magnetic ordering. Therefore, the peak at 325 K in susceptibility results from the suppression of the short range ferromagnetic interactions as indicated by positive θ_P in the metallic cubic phase by the Jahn-Teller ordered insulator transition. The presence of ferromagnetic interactions is further evidenced from the high temperature inverse susceptibility (H = 0.3 T) data (as shown in the inset), where the paramagnetic intercept is positive (θ_P = 68 K). At about 290 K, a shoulder in our low field (H = 0.01 T) susceptibility and a small anomaly in the resistivity are related to the long-range antiferromagnetic ordering. Rietveld analysis of the neutron diffraction data at 200 K confirms the antiferromagnetic ordering of manganese moments with a chain-like or C-type spins aligned along the c-direction with an ordered manganese moment of 2.17(1) μ_B. Below 50 K, the increase of susceptibility for x = 0.1 is due to paramagnetic contribution from cerium Ce^{3+} ions. Unlike the antiferromagnetic (A and CE-type) insulator state driven by Coulomb repulsion between the e_g electrons, the C-type antiferromagnetic insulator state caused by the Jahn-Teller ordering is almost insensitive to applied magnetic field (7 T) and therefore there is no CMR effect. Similar behavior has also been observed for $Sr_{0.85}Pr_{0.15}MnO_3$. However, this behavior is in contrast with the calcium rich system $Ca_{0.85}Sm_{0.15}MnO_3$ where the Jahn-Teller and antiferromagnetic ordering is suppressed by the application of magnetic field and there is a large negative magnetoresistance. It should be noticed that the antiferromagnetic ordering temperature for calcium rich system is much lower (T_N = 125 K) than that in strontium rich systems (T_N = 295 K) and $Sr_{0.85}Pr_{0.15}MnO_3$ (T_N = 260 K). This behavior indicates that the Jahn-Teller ordering associated with C-type antiferrromagnetic ordering in systems with large A-site cation is more stable than that in systems with smaller A-site cation which explains the absence of CMR effect in the strontium rich electron doped systems.

For x > 0.1, all the samples show a spin glass behavior and the temperature dependence of resistivity showed a semiconducting behavior. Magnetoresistance measurements showed a small negative magnetoresistance. A detailed study on this system is reported elsewhere [4].

Temperature dependence of thermal expansion under various pressures showed that the T_{JT} decreases with increase of pressure at a rate of dT_{JT}/dp = −20(1) KGPa^{-1}. The large suppression of T_{JT} indicates the instability of the J-T tranision due to a widening of the conduction electron band under high pressure. This may in turn results in the weakening of the antiferrromagnetic interactions. A hysteretic thermal behavior in $\Delta L/L$ was observed near T_{JT}, which further support that the J-T transition is first-order.

In conclusion, for low level of electron doping (x = 0.1) in $Sr_{1-}Ce_xMnO_3$, a static Jahn-Teller distortion and associated metal insulator transition and an antiferromagnetic transition occur at ~ 315 K and 295 K, respectively. In contrast to hole doped or Ca rich electron doped system, there is no ferromagnetism or CMR effect in any of these materials.

REFERENCES

1. Bokov, V.A., Grigoryan, N.A., and Bryzhina, M.F., *Phys. State. Solidi.*, **20**, 745 (1991).
2. Martin, C., Maignan, A., Damay, F., Hervieu, M., and Raveau, B., *J. Solid State Chem.*, **134**, 198 (1997).
3. Sundaresan, A., Tholence, J.L., Maignan, A., Martin, C., Hervieu, M., Raveau., B., and Suard, E., *Eur. Phys. J. B* **14**, 431 (2000).

Spin-lattice Correlations and Temperature Dependent Mn *K*-edge Structure in the La₁₋ₓCaₓMnO₃ Systems

F. Bridges, D. Cao and M. Anderson

Physics Dept. University of California, Santa Cruz CA, 95064 USA

Abstract. The $La_{1-x}Ca_xMnO_3$ system exhibits novel transport properties and unusual correlations between local distortions, magnetism, and the electronic structure. A large distortion, parameterized by the width of the pair distribution function, σ, develops as T is increased through T_c for values of x roughly from 0.2-0.5 (the colossal magnetoresistance (CMR) regime); we associate this change of the local distortion with the formation of polarons. Changes in σ^2 as T is reduced below T_c depend on the magnetization, M, and can be described by the equation $\ln \Delta\sigma^2 = AM/M_o + B$, where A and B are constants and M_o is the saturation magnetization. Here we show that σ^2 is reduced when a magnetic field is applied and that the above equation holds even when the peak in the magnetoresistance occurs well below T_c. We also report on temperature dependent structure in the absorption edge, some of which again correlates with changes in the local structure.

INTRODUCTION

The substituted lanthanum manganites ($La_{1-x}A_xMnO_3$, where A is a divalent atom - Ca, Ba, Pb, etc) are excellent examples of systems[1-3] in which there is a strong interplay between the electronic bands, magnetism and the local structure. As such they are important test cases for developing a better understanding of these interesting correlations. In this paper we report on several of these effects and outline some of the outstanding questions that need to be answered.

For x roughly in the range 0.2-0.5, the $La_{1-x}Ca_xMnO_3$ systems exhibit a large magnetoresistance - hence the name colossal magnetoresistors (CMR). The spin-charge coupling is usually described in terms of the double exchange model (DE), originally proposed by Zenner [4], however, to explain the large magnitude of the magnetoresistance a significant electron-lattice coupling is also needed [5,6]. Experimentally, a large local distortion is observed in these materials; it develops as T is increased through the ferromagnetic (FM) transition temperature, T_c, and is associated with the formation of small polarons [7-9]. The slightly increased lengths of some of the Mn-O bonds near and above T_c, appear to suppress the conductivity and thus enhance the magnetoresistance.

We have parameterized the local order/distortion by the width of the pair distribution function, σ, for each atom pair in the system [7,10]; here we focus mainly

CP554, *Physics in Local Lattice Distortions*, edited by H. Oyanagi and A. Bianconi
© 2001 American Institute of Physics 1-56396-984-X/01/$18.00

on σ for the Mn-O bond which we obtain from XAFS measurements. The different contributions to the broadening of the pair distribution function add in quadrature; i.e.

$$\sigma^2 = \sigma^2_{static} + \sigma^2_{phonons} + \sigma^2_{polaron} \qquad (1)$$

where σ^2_{static} is any static contribution, $\sigma^2_{phonons}$ arises from lattice vibrations, and $\sigma^2_{polaron}$ is the distortion associated with polaron formation; it reaches its maximum value σ^2_{pm} just above T$_c$.

In recent work [10] we have shown that there is a well defined relationship between changes in the local distortions and the sample magnetization, M. We first define the decrease in the polaron contribution (from its maximum value, σ^2_{pm}) by:

$$\Delta\sigma^2 = \sigma^2_{static} + \sigma^2_{phonons}(T) + \sigma^2_{pm} - \sigma^2_{data}(T). \qquad (2)$$

Then the relationship between $\Delta\sigma^2$ and M is given by:

$$\ln(\Delta\sigma^2) = AM / M_o + B, \qquad (3)$$

where A and B are constants and M_o is the saturation value of M. Note however, that each data point is at a different temperature. This relationship suggests that σ^2 should depend on the applied magnetic field. Here we show directly that the local structure changes when a magnetic field is applied at a fixed temperature near T$_c$. In addition we find that Eq. 3 persists even when the magnetic transition is broad or when the magnetoresistance peak-temperature, T$_{MR}$, occurs well below T$_c$, as is the case reported here for some co-doped samples containing Ca and Ti or Ga.

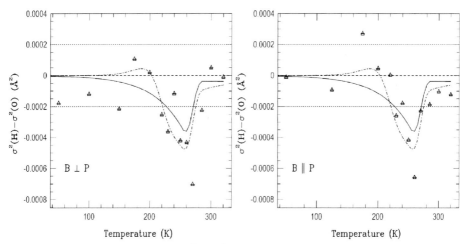

FIGURE 1. A plot of the difference σ^2(B)-σ^2(0) with the magnetic field perpendicular (left) and parallel (right) to the X-ray polarization vector. From reference 11. The dotted lines indicate the 2σ error level.

In addition, the absorption edge structure provides information about the energy bands; the pre-edge structure provides constraints on parameters such as U and J_H, while small features in the main edge also appear to reflect changes in the local structure.

MAGNETIC FIELD EFFECTS

The magnetic transition in the CMR systems moves to a slightly higher temperature with increasing applied magnetic field; consequently if σ^2 decreases with increasing magnetization as implied from earlier measurements, then it should also decrease when a static magnetic field, B, is applied at a fixed temperature near T_c. In Fig. 1 we plot the change in σ^2 ($\sigma^2(B) - \sigma^2(0)$) as a function of T for a 30% Ca sample. The difference has a maximum amplitude near $T_c = 260K$. The difference function can be roughly modeled as a rigid shift of the transition to higher T (solid line Fig. 1); however, since the FM transition broadens in a magnetic field, we have also included a broadening of the structural transition which changes the shape of the curve (See dot-dash line) and improves the quality of the fit [11]. Similar results are obtained for a 21% sample, with the peak in the difference occurring near 190K, just below the transition temperature of 210K. These results confirm that the local distortions are determined in part by the sample magnetization.

CO-DOPED SAMPLES

To further explore the changes in the local structure and the role of the Mn valence in these systems, we have co-doped several 30% Ca samples with Ti or Ga on the Mn site. Ti has a valence of +4 while Ga is +3; consequently adding these dopants should change the ratio of Mn^{+4} to Mn^{+3}. However at relatively low concentrations (4-6%) these secondary dopants already play a significant role, with the peak in resistivity moving rapidly to lower temperatures; at 10% Ti there is no longer a metal/semiconductor transition and the sample is very resistive at low T. The effects for Ga are smaller.

The changes in the magnetic behavior are less pronounced. T_c does decrease a little, but the two main effects are 1) the transition broadens and 2) the saturation magnetization decreases with increasing concentration, (down $\sim$ 15% for Ga and 50 % for Ti at 10% concentration). Thus for these systems the metal/semiconductor transition at T_{MS} and the magnetic transition at T_c are decoupled. Consequently this provides an excellent test case for extending the investigation of correlations between σ^2, magnetization, and transport, to a new regime. For these systems we again find essentially the same relationship between $\Delta\sigma^2$ and M as given by Eq. 3, although there is a little curvature for $M/M_o < 0.3$ because of the rounding of the transition (See Fig. 2). Note that the resistivity and magnetoresistance peaks occur in the top half of this magnetization range (i.e. for M roughly 70-90% of M_o), but there is no feature in Fig. 2 at the magnetization that corresponds to T_{MS}.

The physical basis for the above relationship is still not clear and there may be other functional forms that can also describe the data. We have therefore plotted $\ln\Delta\sigma^2$ vs $\ln(1-T/T_c)$ to see if a power law in reduced temperature would model the data, but such plots yield strongly curved lines. The observed dependence of $\Delta\sigma^2$ on M clearly needs to be understood before a complete model can be developed for these systems.

Although there is no simple connection between $\Delta\sigma^2$ and transport we have found an unusual correlation: for a given co-dopant, the peak in resistivity occurs when σ^2 is reduced to a given value. When σ^2 does not reach this level as occurs for the 10% Ti sample, there is no resistivity peak. For the Ti doped samples, a lower value of σ^2 is required than for Ga, which suggests that Ti inhibits the electronic transport more than Ga. Further, there appears to be a trade-off between a static distortion that is not removed (at least not down to ~ 30K) and a polaron distortion that changes with T. At 300K, there is only a small difference in σ^2 for the different samples. The main difference in the temperature dependence of the different samples is primarily the amount of distortion that can be removed at low T. Thus the addition of the second dopant appears to introduce static distortions that impede charge transport, with the larger static distortions and higher resistivity occurring for Ti. A more detailed discussion will be given in a longer paper [12].

NEAR EDGE STRUCTURE

The Mn K-edge provides additional information and constraints on the electronic bands in these systems. Small changes in the edge and pre-edge correlate well with T_c, and the driving force again appears to be changes in the local structure about the Mn atom.

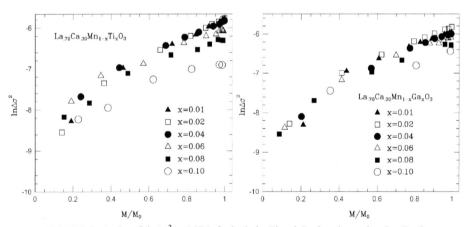

FIGURE 2. A plot of $\ln\Delta\sigma^2$ vs M/M_o for both the Ti and Ga doped samples. See Eq. 3.

In Fig. 3 we plot the pre-edge structure for LaMnO₃ and the 30% Ca sample at several temperatures. There are two peaks, A_1 and A_2 for all of the substituted LaMnO₃ systems (also a third A_3, for high Ca concentrations), plus the peak labeled B. For the CMR systems, the amplitudes of the A_1 and A_2 change in opposite ways (A_1 increases while A_2 decreases) as the sample becomes magnetic, with most of the change occurring in a 60K range just below T_c. The A_1-A_2 splitting is small about 2.1 eV at 300K, and decreases about 0.4 eV in the FM state. Based on the calculations by Elfimov *et al.* [13] we attribute these peaks to weak dipole transitions made allowed by hybridization of the Mn 4p states with an odd symmetry combination of Mn 3d states (e_g) on *neighboring* Mn atoms [14]. To match the small splitting observed experimentally, Elfimov reduced U and J_H to 4 eV and 0.7 eV respectively. These values indicate considerable covalency in this system. The reduction of the A_1-A_2 splitting in the magnetic state may indicate a further change in covalency.

The main edge has very little structure and shifts roughly uniformly with Ca concentration [9,10]. The lack of significant structure is at first surprising if the material is viewed as a mixture of ionic Mn^{+3} and Mn^{+4} sites. However the Mn 4p band is very broad (about 15 eV) and hence the 4p states are extended and will overlap with neighboring Mn atoms (a necessary condition for the hybridization described above). This partially explains the lack of significant structure in the edge; however, increased covalency likely also plays a role.

To investigate the structure more carefully, we have taken the difference between a data file at temperature T and one at 300K which we use as a fiducial trace. To do so, the energy scale for each plot must first be corrected (to about 0.02 eV) using the reference data and carefully normalized above the edge. The resulting difference files for the 30% Ca sample are compared with those for the LaMnO₃ sample in Fig. 4; there is clearly a weak but reproducible temperature dependence. Here two aspects are clear – 1) the opposite dependencies of the amplitudes for A_1 and A_2 and 2) a new

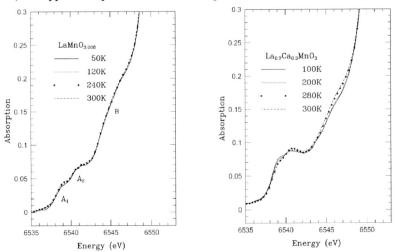

FIGURE 3. A comparison of the pre-edge structure for 30% Ca with that of LaMnO₃ for several different temperatures.

structure for the 30% sample in the main edge (a dip/peak structure, see vertical lines on Fig. 4) that begins to be observed near T_c and grows rapidly just below T_c. The dip/peak structure has a separation of about 2 eV; we associate this splitting with the difference in the positions of the partial density of states (PDOS) for the Mn $4p_x$ (long Mn-O bond) and $4p_{y,z}$ (short Mn-O bond) states, as calculated by Elfimov it *et al.* [13].

The Mn K-edge is a sum over all the Mn 4p states and thus will be a weighted sum over only two configurations, $4p_x$ and $p_{y,z}$, irrespective of any averaging caused by the extended nature of the 4p states. The short core hole lifetime will of course broaden the entire 4p PDOS. If one thinks in terms of valence (a mixture of Mn^{+3} and Mn^{+4}) and considers an average over only three Mn atoms as a result of the extended 4p states, then there will be a broad distribution of average charge which would not yield the observed structure. Thus we interpret the dip/peak structure in the difference data as arising from local distortions in the system. An extensive paper on the XANES will appear elsewhere [15].

CONCLUSIONS

We have shown that changes in the local structure are correlated with magnetism and transport in the substituted manganites and play an important role in determining their properties. First, the application of a magnetic field at T_c decreases the broadening of the Mn-O pair distribution function. This was anticipated from earlier experiments, but no model has yet been developed to understand this result in detail. On the experimental side, higher fields will be needed to determine the field dependence of this effect.

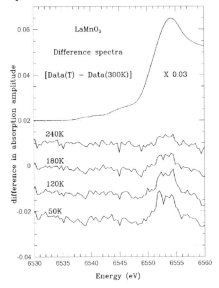 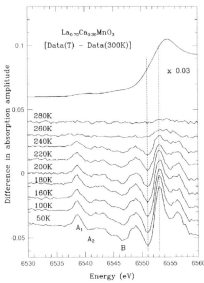

FIGURE 4. A comparison of the difference files for 30% Ca with that for $LaMnO_3$. Note the scale - the absorption edge at the top has been multiplied by 0.03 to fit on this scale.

For a wide number of systems we have found that the decrease in the polaron contribution to σ^2, $\Delta\sigma^2$, is a linear function of M, even when the magnetoresistance peak occurs far below T_c as occurs for co-doped samples containing Ca and Ti (Ga). Thus in these systems the local structure and magnetism appear more correlated than the local structure and charge transport. However there is some correlation between σ^2 and resistivity - the peak in the resistivity occurs when σ^2 has been reduced to a specific value, with a larger reduction required for Ti than for Ga. Neither of these results is understood in terms of a well defined model.

Finally the Mn pre-edge is temperature dependent as is a weak structure in the main K-edge. Both of these features are correlated with T_c for CMR samples. It is likely that they are related to changes in the local structure but again a model is needed.

ACKNOWLEDGEMENTS

The experiments were performed at the Stanford Synchrotron Radiation Laboratory, which is operated by the U.S. Department of Energy, Division of Chemical Sciences, and by the NIH, Biomedical Resource Technology Program, Division of Research Resources. Some experiments were carried out on UC/National Laboratories PRT beam time. The work is supported in part by NSF grant DMR-97-05117.

REFERENCES

1. Jonker, G.H., and van Santen, J.H., *Physica* (Amsterdam) **16**, 337 (1950).
2. Wollan, E.O., and Koehler, W.C., *Phys. Rev.* **100**, 545 (1955).
3. Schiffer, P., Ramirez, A., Bao, W., and Cheong, S-W., *Phys. Rev. Lett.* **75**, 3336 (1995).
4. Zener, C., *Phys. Rev.* **82**, 403 (1951).
5. Millis, A.J., Littlewood, P.B., and Shraiman, B.I., *Phys. Rev. Lett.* **74**, 5144 (1995).
6. Röder, H., Zang, J., and Bishop, A.R., *Phys. Rev. Lett.* **76**, 1356 (1996).
7. Booth, C.H., Bridges, F., Snyder, G. J., and Geballe, T.H., *Phys. Rev. B* **54**, R15606 (1996).
8. Billinge, S.J.L., DiFrancesco, R.G., Kwei, G.H., Neumeier, J.J., and Thompson, J.D., *Phys. Rev. Lett.* **77**, 715 (1996).
9. Subías, G., García, J., Proietti, M.G., and Blasco, J., *Phys. Rev. B* **56**, 8183 (1997).
10. Booth, C.H., Bridges, F., Kwei, G.H., Lawrence, J.M., Cornelius, A.L., and Neumeier, J.J., *Phys. Rev. B* **57**, 10440 (1998).
11. Cao, D., Bridges, F., Booth, C.H., and Neumeier, J.J., *Phys. Rev. B* (2000).
12. Cao, D., Bridges, F., Ramirez, A.P., Olapinski, M.P., Subramanian, M.A., Booth, C.H., and Kwei, G., (unpublished).
13. Elfimov, I.S., Anisimov, V.I., and Sawatzky, G.A., *Phys. Rev. Lett.* **82**, 4264 (1999).
14. Bridges, F., Booth, C.H., Kwei, G.H., Neumeier, J., and Sawatzky, G.A., *Phys. Rev. B* **61**, R9237 (2000).
15. Bridges, F. *et al.* (unpublished).

Lattice Distortion and Isotope Effect in Thin Films LaPrCaMnO Manganites

N. A. Babushkina[*], L. M. Belova[*], E. A. Chistotina[*],
O. Yu. Gorbenko[†], A. R. Kaul[†], and K. I. Kugel[‡]

[*]Russian Research Cente "Kurchatov Institute", Kurchatov sqr. 1, Moscow 123182, Russia
[†]Department of Chemistry, Moscow State University, Vorobievy Gory, Moscow119899, Russia
[‡]Institute for Theoretical and Applied Electrodynamics, RAS, Izhorskaya str. 13/19,
Moscow 127412, Russia

Abstract. The effect of $^{16}O \rightarrow ^{18}O$ isotope substitution on electrical resistivity and magnetoresistance, was studied for $La_{0.35}Pr_{0.35}Ca_{0.3}MnO_3$ epitaxial thin films deposited onto $LaAlO_3$ and $SrTiO_3$ substrates. For the films on $LaAlO_3$, the isotope substitution resulted in the reversible transition from a metal-like to insulating state. The films on $SrTiO_3$ remained metallic at low temperatures for both ^{16}O and ^{18}O, but the shift of the resistivity peak corresponding to onset of metallic state exceeded 63 K after $^{16}O \rightarrow ^{16}O$ substitution. This drastic difference in transport properties can be related to the substrate-induced strains having opposite signs for the films on $LaAlO_3$ and $SrTiO_3$.

INTRODUCTION

The electron-phonon interaction is widely known as a key factor in the physics of manganites [1]. The isotope substitution is a direct experimental method to study the effect of lattice vibrations on physical properties. An unusually large isotope shift of the Curie temperature T_c for La-based manganites [2]. The isotope effect turns out to be even more pronounced in the system with competing ferromagnetic (FM), antiferromagnetic (AF), and charge ordering (CO), such as La-Pr and La-Nd manganites, where the isotope driven metal-insulator transition (MIT) can be observed near the FM-AF phase boundary [3,4].

Until now the isotope effect has been studied mostly in the bulk samples of manganites. Epitaxial thin films are very interesting objects for the study of isotope effect. First, their good crystallinity comparable to that of high-quality single crystals provides an opportunity to study the intrinsic properties of the material and eliminate spurious effects related to grain boundaries. Second, the film-substrate lattice mismatch producing epitaxial strain in the film adds new features to the isotope effect in the perovskite materials. In this paper, we discuss the isotope effect in $(La_{1-y}Pr_y)_{0.7}Ca_{0.3}MnO_3$ perovskite thin films at $y = 0.5$ on $LaAlO_3$ (LAO) and on $SrTiO_3$ (STO) substrates.

CP554, *Physics in Local Lattice Distortions*, edited by H. Oyanagi and A. Bianconi
© 2001 American Institute of Physics 1-56396-984-X/01/$18.00

RESULTS

Aerosol MOCVD used to prepare thin films includes deposition from the vapor produced by evaporation of precursor solution nebula in the carrier gas flow [5]. The method used for oxygen isotope substitution $^{16}O \rightarrow {}^{18}O$ in thin films was similar to that [3]. The film thickness was 60 nm.

XRD study demonstrates that the difference of the substrate lattice constants results in the tetragonal strain of the film lattice in the interface plane. The lattice constant of STO is larger, consequently the tensile strain in the film demands an in-plane expansion of the perovskite cube. For the films on LAO substrates the lattice constant

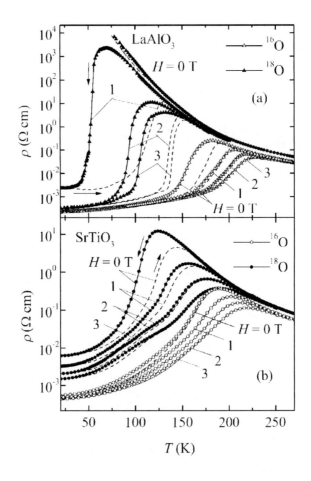

FIGURE 1. Temperature dependence of electrical resistivity for $La_{0.35}Pr_{0.35}Ca_{0.3}MnO_3$ films at various magnetic fields: 0 T, 1 T, 2 T, 3 T: (a) films on LAO substrate, with ^{16}O (open triangles) and ^{18}O (solid triangles); (b) films on STO substrate, with ^{16}O (open circles) and ^{18}O (solid circles).

of the substrate is smaller than that in the manganite film, and this causes in-plane contraction of the films.

Fig. 1 are shown that the $^{16}O \rightarrow {}^{18}O$ isotope substitution produces different effect on the $\rho(T)$ behavior of the films deposited on LAO and STO. This difference stems from additional strains induced by substrates. At zero magnetic field, the film with ^{16}O on LAO exhibits the resistivity peak at $T_p \approx 181$ K. Resistivity of the film with ^{18}O on LAO behaves in a qualitatively different way: it is growing monotonously when the temperature is lowered. As a result, the $^{16}O \rightarrow {}^{18}O$ isotope substitution induces the transition from metallic to insulator-like resistivity at low temperatures. We have already observed such a behavior for ceramic $(La_{1-y}Pr_y)_{0.7}Ca_{0.3}MnO_3$ samples at $y = 0.75$ [3]. The behavior of the films on STO remains metal-like at low temperatures even after the isotope substitution, but the resistivity peak shifts from 186 K (^{16}O) to 123 K (^{18}O) at $H = 0$, that is $\Delta T_p = 63$ K.

The electrical resistivity decreases significantly under effect of magnetic field. Magnetic field $H = 1$ T transforms the sample with ^{18}O on LAO from insulating to metallic state with $T_p = 70$ K owing to the suppression of charge ordering. For the samples with ^{16}O on LAO and the samples with ^{16}O and ^{18}O on STO, the applied magnetic field causes a pronounced shift of $\rho(T)$ curves and of T_p toward high temperatures as implied by the double exchange model.

The $\rho(T)$ curves for all films measured on heating and on cooling exhibit hysteresis behavior. At $H = 0$, it was found that the hysteresis is rather small for the samples on STO annealed in ^{16}O, but it is more pronounced for the samples on LAO. The $^{16}O \rightarrow {}^{18}O$ isotope substitution causes broadening of the hysteresis loops for films on both LAO and STO. Existence of hysteresis suggests possibility of a phase separation (coexistence of (FM) and (AF) regions). Such a coexistence was observed by the neutron powder diffraction experiments [6] for samples annealed in ^{16}O and ^{18}O.

DISCUSSION

The study of $^{16}O \rightarrow {}^{18}O$ isotope substitution in $(La_{0.5}Pr_{0.5})_{0.7}Ca_{0.3}MnO_3$ thin films revealed a significant difference in the electrical properties of the films deposited on LAO and STO substrates. This difference can be explained by additional strains induced by substrates. In fact, the localization of itinerant electrons implies a narrowing of the conduction band. The value of the conduction bandwidth W depends on the cosine of the Mn-O-Mn bond angle. At a certain critical value of the bond angle, the charge carriers become localized. Because of the strain, the film on LAO is actually closer to the critical state for the localization of itinerant carries than the film on STO. The strain causes the decrease in the Mn-O-Mn bond angles in the plane of the film on LAO. On the contrary, in the plane of the film on STO the Mn-O-Mn bond angles are increased. The in-plane contraction of the thin film deposited onto the LAO substrate shifts the crossover from the metal-like to the charge ordered state toward smaller y values resulting in the metal-insulator transition at $y = 0.5$. For ceramic samples this transition takes place at $y = 0.75$. For thin $(La_{1-y}Pr_y)_{0.7}Ca_{0.3}MnO_3$ films on STO the Mn-O-Mn bond angle grows, distortions become smaller and the films remain metallic after the $^{16}O \rightarrow {}^{18}O$ substitution, although they exhibit a pronounced isotope shift of the resistivity peak.

The vicinity of this phase boundary underlies also a pronounced temperature hysteresis of resistivity for the film on LAO, which is a general signature of the first order phase transitions and of the phase separation accompanying them. Thus, the significant isotope effect in the width of hysteresis loops demonstrates an important role of the electron-lattice coupling in manganites, which is additionally enhanced in the vicinity of the FM-CO phase boundary.

CONCLUSION

The isotope effect for thin films on different substrates is a unique tool to study the role of magnetic ordering, charge ordering, and phase separation in manganites near the crossover between ferromagnetism and charge ordering. The system turns out to be most sensitive to isotope substitution. $(La_{0.35}Pr_{0.35})Ca_{0.3}MnO_3$ epitaxial thin films deposited onto both $LaAlO_3$ and $SrTiO_3$ substrates are in the vicinity of this crossover and demonstrate a huge isotope effect. However, the films on $LaAlO_3$, appear to be so close to the phase boundary that $^{16}O \rightarrow {}^{18}O$ isotope substitution leads to the metal-insulator transition, phase separation phenomena, and a unique sensitivity of transport properties to the applied magnetic field.

ACKNOWLEDGMENT

We are grateful to A. N. Taldenkov, A. V. Inyushkin, A. A. Nikonov, and S. Yu. Shabanov for helpful assistance. The work was supported by INTAS (Projects 97-0963 and 97-11954) and by the Russian-Dutch Program for Scientific Cooperation.

REFERENCES

1. Millis, A.J., *Nature* (London) **392**, 147-149 (1998).
2. Zhao, G.M., Conder, K., Keller, H., and Müller K.A., *Nature* (London) **381**, 676-678 (1996).
3. Babushkina, N.A., Belova, L.M., Gorbenko, O.Yu., Kaul, A.R., Bosak, A.A., Ozhogin, V.I., and Kugel, K.I., *Nature* (London) **391**, 159-161 (1998).
4. Zhao, G.-M., Keller, H., Hofer, J., Shengelaya, A., and Müller, K.A., *Solid State Commun.* **104**, 57-61 (1997).
5. Gorbenko, O.Yu., Fuflyigin, V.N., Erokhin ,Y.Y., Graboy, I.E., Kaul, A.R., Tretyakov, Yu.D., Wahl, G., and Klippe, L., *J. Mater. Chem.* **4**, 1585-1589 (1994).
6. Balagurov A.M., Pomjakushin, V.Yu., Sheptyakov, D.V., Aksenov, V.L., Babushkina, N.A., Belova, L.M., Taldenkov, A.N., Inyushkin, A.V., Fischer, P., Gutmann, M., Keller, L., Gorbenko, O.Yu., and Kaul, A.R., *Phys. Rev. B* **60**, 383-387 (1999).

Crucial Role of Jahn-Teller Distortions in LaMnO$_3$

M. Capone[1], D. Feinberg[2] and M. Grilli[3]

[1] Istituto Nazionale di Fisica della Materia and International School for Advanced Studies (SISSA-ISAS), Via Beirut 2-4, 34013 Trieste, Italy
[2] Laboratoire d'Etudes des Propriétés Electroniques des Solides, Centre National de la Recherche Scientifique, associated with Université Joseph Fourier, BP 166, 38042 Grenoble Cedex 9, France
[3] Istituto Nazionale di Fisica della Materia and Dipartimento di Fisica, Università di Roma "La Sapienza", Piazzale Aldo Moro 2, 00185 Roma, Italy

Abstract. The crucial role of the cooperative Jahn-Teller deformations in determining the magnetic properties of stoechiometric LaMnO$_3$ is studied by means of exact diagonalization. The layered antiferromagnetic order mainly results from the strong cooperative Jahn-Teller deformations, stabilizing a certain type of orbital ordering. The main result is that antiferromagnetic (ferromagnetic) coupling along the c-direction (ab-planes) can be understood only if the Jahn-Teller energy is much larger than the superexchange couplings, which is consistent with experiments. The crucial role of the deformation anisotropy Q_2/Q_3 is also emphasized: realistic values of Q_2/Q_3 are needed to stabilize the layered order. The experimental values of the superexchange couplings can be recovered using reasonable sets of parameters.

The complex phase diagram of the Mn perovskites requires at least the following ingredients to be understood: i) strong on-site electron-electron correlation; ii) the "double exchange" mechanism due to the interplay between e_g electron itineracy and Hund's exchange with the more localized t_{2g} electron spins, which favours ferromagnetism [1-3]; iii) superexchange between t_{2g} electrons as well as between e_g electrons on neighbouring sites; iv) sizeable electron-lattice interactions, in particular due to Jahn-Teller (JT) effect on Mn^{3+} ions [4, 5]. All these elements are necessary to understand the interplay between spin, charge and orbital ordering. The latter lifts the degeneracy of the e_g orbitals by a cooperative Jahn-Teller lattice deformation and leads to tetragonal or orthorhombic deformations of the cubic structure.

Quite Surprisingly, even the understanding of the insulating antiferromagnet LaMnO$_3$ is still lacking, although it represents an essential starting point towards the doped phases. This phase, when fully stoechiometric, presents a layered antiferromagnetic order, with ferromagnetic couplings (F) in two directions and antiferromagnetic (AF) coupling in the other [6]. The AF directions present shortened Mn-O bonds, leading to tetragonal distortion, while in the F directions long and short bonds alternate, yielding the overall orthorhombic structure. These deformations can be understood in terms of cooperative JT effect. The corresponding lifting of e_g degeneracy can be viewed as an orbital ordering, with occupied d orbitals pointing preferentially in the directions of long Mn-O bonds.

CP554, *Physics in Local Lattice Distortions*, edited by H. Oyanagi and A. Bianconi
© 2001 American Institute of Physics 1-56396-984-X/01/$18.00

In this work we present a phase diagram for this compound obtained by exact-diagonalization, considering both e_g orbitals and the quantum $\frac{3}{2}$-spins coming from localized t_{2g} electrons. As described in details in Ref. [7,8], we considered a model that takes into account both Coulomb and lattice (namely JT) interactions. For simplicity we will not distinguish between the Hund exchange energy between electrons in the e_g and t_{2g} orbitals, both indicated by J_H. The antiferromagnetic superexchange coupling between neighboring t_{2g} spins J_t is considered, while the JT interaction between the e_g electrons and the (cooperative) lattice deformation is parametrized by the JT splitting $\varepsilon = g\sqrt{Q_1^2 + Q_2^2}$ and the anisotropy ratio $r = Q_2 / Q_3$.

In this work we treat the JT deformations $\mathbf{Q} = (Q_2, Q_3)$ as external fields imposed by a lattice ordering involving a much higher energy scale than the magnetic ones. Therefore in the following the various magnetic couplings will be determined in terms of *assigned* lattice deformations. This assumption is justified by the experimental observation that the JT energy splitting is much larger than all magnetic couplings.

We exactly solve a system of two sites located either on the same xy plane or on adjacent planes and the suitable hopping matrix elements between the various orbitals have been considered.

We studied the values of in-plane (J_{xy}) and z-axis (J_z) superexchange constants for various values of the electronic parameters at fixed lattice parameters. The occurrence of the various magnetic phases can be cast in a phase diagram at zero temperature illustrating the stability region of these phases in terms of the JT energy splitting and the deformation ratio. Fig. 1 reports the phase diagram for a set of parameters (see caption).

At moderate and large values of ε a Néel AAA phase is found for weak planar distortions (small r). In the very-small-r region, J_{xy} is naturally positive, while the superexchange between e_g levels along z, although ferromagnetic, is small so that the direct superexchange between t_{2g} spins may easily dominate and gives rise to the AAA phase. The AAA phase is replaced by the so-called C-like antiferromagnetic AAF phase in the $J_t = 0$ case. At small-to-intermediate values of ε, a progressive increase of $|r|$ drives the system towards the phase AAF. In this phase J_{xy} keeps its AF character, while the negative superexchange between e_g levels along z is small, but no longer is overcome by J_t. At larger values of ε the AAF phase is not present, but the physical nature of the AAA phase changes upon increasing $|r|$. In particular while at low $|r|$ the AF along z is determined by J_t, at larger $|r|$, the superexchange between e_g levels along z is itself AF and therefore the t_{2g} superexchange contributes, but it is not strictly necessary to the AF coupling along z. On the other hand, a further increase of $|r|$ promotes a F coupling along the planes and leads to the A-type antiferromagnetism FFA experimentally observed in undoped LMO.

At small values of the JT splitting, the phase diagram is prominently occupied by a FFF phase. The FFF phase at low and moderate ε's is greatly stabilized by the increase of the Hund coupling J_H, as expected.

Within the present exact numerical treatment, it is also possible to give quantitative estimates of J_{xy} and of J_z. As an example, we report here a realistic sets of parameters (among many others) providing the values $J_{xy} = -0.83$ meV and $J_z = 0.58$ meV experimentally observed with inelastic neutron scattering [9]. Assuming $Q_2/|Q_3| = 3.2$, a value largely confirmed by many groups [10], we take $t = 0.124\ eV$, $U = 5.81\ eV$, J_H

= 1.2 eV, J_t = 2.1 meV, and ε = 0.325 eV. The quite reasonable values of the model parameters needed to reproduce the measured magnetic couplings is an indirect test of the validity of the considered model. We emphasize that the "anomalous" trend $|J_{xy}| > |J_z|$ is correctly reproduced, and that our fit is relatively flexible concerning parameters U, J_H or t, provided ε is large enough.

One further relevant result is that, when the Mn-O octahedron is compressed along z, a FFA phase is only obtained for a sizable (staggered) Q_2 deformation of the planar unit cell. This finding agrees with the *ab initio* calculations of Ref. [11].

Our analysis also points out the relevant role played by the Hund coupling, which enhances the ferromagnetic component of the superexchange processes. Quite relevant turns out to be also the Hund coupling between the e_g electrons and the t_{2g} spins. In this latter regard, we explicitly checked that, keeping J_H finite between the e_g electrons, but decoupling them from the t_{2g} spins no longer gives rise to the FFF phase at low values of the JT splitting. Instead at $\varepsilon \sim 0$ a FFA phase is found in agreement with the results of Ref. [12] for a model, which only considered e_g electrons and no JT splitting. This indicates that the determination of the stable phase (at least) at small values of the JT energy must take in due account the Hund coupling thereby

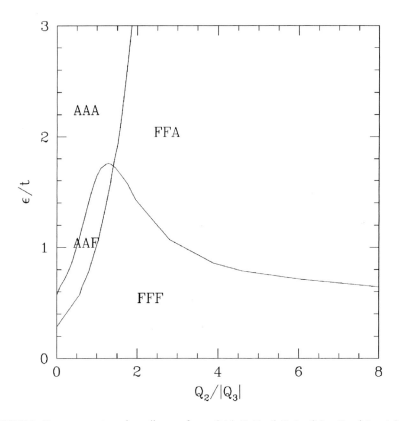

FIGURE 1. Zero-temperature phase diagram for $t = 0.14eV$, $U = 6eV$, $J_t = 2.1meV$ and $J_H = 1.2eV$.

including the t_{2g} levels. Secondly a quantitative determination of the stability region for the FFA phase and of the value of the magnetic couplings is subordinate to the consideration of the J_H term.

Finally we showed that using reasonable parameters the experimental values of the magnetic couplings can easily be reproduced. Of course precise estimates depend on the knowledge of the various couplings entering the model, which are not always available neither from experiments nor from reliable first principle calculations. However calculating the magnetic couplings for various parameters and matching the numerical results with the experimentally obtained values provides useful connections between the involved parameters and set limits to the poorly known physical quantities.

REFERENCES

1. Zener, C., *Phys. Rev.* **82**, 403-405 (1951).
2. Anderson, P.W., and Hasegawa, H., *Phys. Rev.* **100**, 675-681 (1955).
3. de Gennes, P.G., *Phys. Rev.* **118**, 141-154 (1960).
4. Millis, A.J., Littlewood, P.B., and Shraiman, B.I., *Phys. Rev. Lett.* **74**, 5144-5147 (1995).
5. Zang, J., Bishop, A.R., and Röder, H., *Phys. Rev. B* **53**, 8840-8843 (1996).
6. Wollan, E.O., and Koehler, W.C., *Phys. Rev.* **100**, 545-546 (1955).
7. Capone, M., Feinberg, D., and Grilli, M., *Eur. Phys. J. B* **17**, 103-109 (2000).
8. Feinberg, D., Germain, P., Grilli, M., and Seibold, G., *Phys. Rev. B* **57**, 5583-5586 (1998).
9. Moussa, F., Hennion, M., Rodríguez-Carjaval, J., Moudden, H., Pinsard, L., and Revcolevschi, A., *Phys. Rev. B* **54**, 15149-15155 (1996).
10. Rodríguez-Carjaval, J., Hennion, M., Moussa, F., Moudden, H., Pinsard, L., and Revcolevschi, A., *Phys. Rev. B* **57**, 3189-3192 (1998).
11. Solovyev, I., Hamada, N., and Terakura, K., *Phys. Rev. Lett.* **76**, 4825-4828 (1996).
12. Kugel', K.I., and Khomskii, D.I., *Sov. Phys. JETP* **37**, 725-732 (1973); *Zh. Eksp. Teor. Fiz.* **64**, 1429-1435 (1973).

Effects of Lattice Distortion on the Stability of CE-phase of Perovskite Manganites

T. Ohsawa and J. Inoue

Department of Applied Physics, Nagoya University, Nagoya 464-8603, Japan

Abstract. The two-orbital double-exchange model with effects of lattice distortion and Coulomb repulsion is studied for quarter-filled perovskite manganites using a mean field approximation. A phase diagram of the ground state (GS) of two-dimensional systems is depicted in terms of the lattice distortion and t_{2g} spins exchange. The calculated results show that the ferromagnetic state with $x^2 - y^2$ orbital ordering is stabilized at finite lattice distortion. The GS becomes CE phase with increasing superexchange coupling between t_{2g} localized spins.

Perovskite manganites $A_{1-x}B_xMnO_3$ have unique characteristics due to strong coupling between fundamental degrees of freedom of electrons: charge, spin, and orbital [1]. Recent experiments [2] have shown a rich variety of the magnetic phase diagram of perovskite manganites. In particular, the ground state (GS) of the quarter-filled manganites forms ferromagnetic (F) zigzag chains which are coupled antiferromagnetically. The phase is called CE type antiferromagnetic (AF) state. This phase can be transformed into F state by an external magnetic field of a few Tesla being accompanied with a lattice distortion and the colossal magnetresistance. Recent studies using a mean field approximation, Monte Carlo method [3] and the exact diagonalization method [4] for the two-orbital double exchange (DE) [5, 6] model have shown that a charge-ordered (CO) state of the CE phase occurs due to the on-site Coulomb repulsion and isotropic electron hopping. In these studies, however, the spin degree of freedom is ignored, namely, electrons can hop only between sites with ferromagnetically aligned localized t_{2g} spins. Effects of lattice distortion are also neglected. In this paper, we study the stability of the CE phase at the quarter-filling taking into account the coupling between spins and the effect of the lattice distortion.

We adopt a model of two-orbital DE model with the effects of lattice distortion, intra-, inter-orbital and inter-site Coulomb repulsion and superexchange interaction between localized spins. The effect of the lattice distortion is taken into consideration by introducing an e_g level splitting. The Hamiltonian which we adopt is written as;

CP554, *Physics in Local Lattice Distortions*, edited by H. Oyanagi and A. Bianconi
© 2001 American Institute of Physics 1-56396-984-X/01/$18.00

$$H = \sum_{<ij>\mu\nu} t_{ij}^{\mu\nu} c_{i\mu\sigma}^{\dagger} c_{j\mu\sigma} - K \sum_{i} \mathbf{S}_i \cdot \mathbf{s}_i + U \sum_{i\mu} n_{i\mu\uparrow} n_{i\mu\downarrow} + U' \sum_{i\sigma\sigma'} n_{i\mu\sigma} n_{i\mu\sigma'}$$

$$+ V \sum_{<ij>} n_i n_j + j \sum_{<ij>} \mathbf{S}_i \cdot \mathbf{S}_j \tag{1}$$

$$+ g \sum_{i\sigma} (c_{ia\sigma}^{\dagger} \ \ c_{ib\sigma}^{\dagger}) \begin{pmatrix} \cos\theta_i^o & \sin\theta_i^o \\ \sin\theta_i^o & -\cos\theta_i^o \end{pmatrix} \begin{pmatrix} c_{ia\sigma} \\ c_{ib\sigma} \end{pmatrix} .$$

Here $t_{ij}^{\mu\nu}$ represents the hopping of e_g electrons between orbital μ on site i and orbital ν on site j, K, U, U' and V stand for the Hund coupling between e_g electrons and localized t_{2g} spins, intra-, inter-orbital Coulomb repulsion and nearest neighbor (n. n.) Coulomb repulsion, respectively, J is the antiferromagnetic interaction between localized spins, and g characterizes the effect of Jahn-Teller distortion on the orbitals. The value of g is defined such that the energy level of $x^2 - y^2$ orbital is lower than that of $3z^2-r^2$ orbital by $2g$ when $g > 0$ and $\theta_i^o = 0$. **S** and **s** stand for the spins of localized t_{2g} and itinerant e_g electrons, respectively, $n_{i\mu\sigma}$ is the number operator on site i for orbital ν and spin σ, and $n_i = \sum_{\nu\sigma} n_{i\nu\sigma}$. Orbitals μ and ν are given by linear combination of orbital a, b which are $3z^2 - r^2$ and $x^2 - y^2$, respectively. Here θ_i^o represents the weight of combination of orbitals a and b. Since $La_{0.5}Sr_{1.5}MnO_4$ [7] shows the CE phase, we treat a two dimensional (2D) system.

In contrast to the previous works [3, 4], we take finite value of $K = 4t_0$, where t_0 is the hopping between $x^2 - y^2$ orbitals on n. n. sites. In this case, hopping between sites with AF spin alignment is allowed. We take $U = U' = 10t_0$, $V = 0 \sim t_0/4$, and apply the mean field approximation for U, U' and V terms. The value of $< n_{i\mu\sigma} >$ is determined self-consistently. We treat the localized spin as classical spins and assume ferromagnetic alignment of the localized spins along zigzag chains and spin-spiral structure between zigzag chains, the canting angle of the latter is denoted as θ_{spin}. We take 4-sites unit cell along the zigzag chain as shown in the inset of Fig. 1. In order to characterize the orbital ordering within the unit cell, the value of θ_i^o of B and D sites are fixed to be zero, while those of A and C sites are determined self-consistently to minimize the GS energy for given values of g and J. At the same time, the GS energy is also minimized with respect to θ_{spin}. The procedure can determine the various spin, orbital and charge orderings. For example, when $\theta_{spin} = \pi$, $\theta_A^o = - \pi/3$ and $\theta_C^o = \pi/3$, the phase is of CE type.

The phase diagram calculated for the GS at the quarter-filling, $n = 0.5$, is shown in Fig. 1. When both J and g are small, we obtain a charge-ordered F (CO-F) state. The charge difference between A and B (or C and D) is $< n_A > - < n_B > \sim 0.3$. With increasing g, metallic F state with $x^2 - y^2$ orbital ordering is stabilized, which can be identified as the A-type AF phase in the 3D structure. The transition between the CO-F and M-F is of first - order. With increasing J, both CO-F and M-F phase are transformed into the CO-AF phase around $J = 0.025t_0$. The transition is also of first-order and no spin-spiral state appears. We see that the transition is almost independent of g. The orbital ordering of the CO-AF phase is depicted in the inset, and we can identify this CO-AF phase as the CE phase. The charge difference between A and B sites in the CE phase has been calculated to be ~ 0.1 irrespective of the value of g.

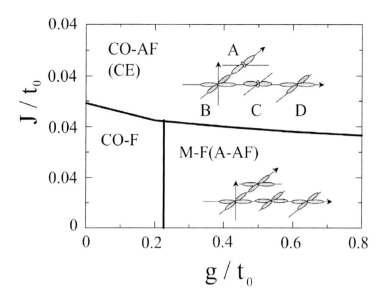

FIGURE 1. A phase diagram of the ground state in J-g plane with $K = 4t_0$, $U = U' = 10t_0$ and $V = 0$. CO and M stand for the charge ordered and metallic states, respectively. The CO-AF and M-F states correspond the CE and A-AF phases, respectively. The insets show the orbital orderings within the unit cell adopted in the calculation.

When J is small, the x^2 - y^2 orbital is naturally stabilized with increasing g and, at the same time, the F spin ordering appears to increase the kinetic energy. It is also reasonable that the AF state appears with increasing J. The CO-AF, that is, the CE phase is realized because the additional periodicity within the unit cell due to the orbital ordering produces an energy gap at the Fermi energy and thereby lowers the GS energy. Because of this periodicity a CO appears even when $V = 0$. When both J and g are small, the orbital ordering is of $\theta_A^o = -\pi/2$ and $\theta_C^o = \pi/2$ type. Because this type of orbital ordering also gives rise to the additional periodicity within the unit cell, an energy gap at the Fermi level opens to produce the charge ordering.

Now, let us compare our theoretical results with experimental ones. Moritomo *et al.* [9] studied the phase diagram of $(Nd_{1-z}La_z)_{1/2}Sr_{1/2}MnO_3$ and showed that the A-AF phase changes to CE phase with decreasing z. As the replacement of Nd ions by La ions gives rise to a change in the lattice distortion, we may consider the change of z corresponds the change of J/t_0. Suppose that the decrease of z corresponds to increase of J/t_0, our results are qualitatively in agreement with the experimental ones. The present results indicate that the A-type AF phase can be altered to the CE phase by changing J/t_0 by static or chemical pressure. The CO-F phase in Fig. 1 has not been observed. The reason may be that the three-dimensionality washes out this phase in 3D systems or that the lowering of the symmetry in 2D systems brings about non-zero value of g to stabilize z^2 - y^2 orbitals. We have shown that CO phase appears even if $V = 0$. However the degree of CO in the CE phase is rather small as $< n_A > - < n_B > \sim$

0.1. The degree of CO may be increased by a non-zero values of V, for example, $V = t_0/4$ gives that $< n_A > - < n_B > \sim 0.2$. The precise value of V, however, is unable to determine in the present calculation.

In conclusion, the calculated results for small J show that a CO-F phase is stabilized when there is no lattice distortion, and the F phase with $x^2 - y^2$ orbital ordering is stabilized at finite distortion. The ground state becomes the CE phase with increasing superexchange coupling between localized spins. The results suggest that the A-type AF phase is transformed into the CE phase when the electron hopping is suppressed by static and chemical pressure.

The authors would like to thank Y. Moritomo for valuable discussions.

REFERENCES

1. Wollan, E. O., and Koehler, W. C., *Phys. Rev.* **100**, 545 (1955).
2. Tokura, Y., Tomioka, Y., Kuwahara, H., Asamitsu, A., Moritomo, Y., and Kasai, M., *Physca C* **263**, 544 (1996).
3. Yunoki, S., Hotta, T., and Dagotto, E., *Phys. Rev. Lett* **84**, 3714 (2000).
4. van den Brink, J., Khaliullin, G., and Khomskii, D., *Fhys. Rev. Lett.* **83**, (1999) 5118.
5. Zener, C., *Phys. Rev.* **82**, 403 (1951), Anderson, P. W., and Hasegawa, H., *Phys. Rev.* **100**, 675 (1955), de Gennes, P.G., *Phys. Rev.* **118**, 141 (1960), Kugel', K. I., and Khomskii, D. I., *JETP Lett.* **15**, 446 (1972).
6. Ishihara, S., Inoue, J., and Maekawa, S., *Physica C* **263**, 130 (1996), *Phys. Rev. B* **55**, 8280 (1997).
7. Sternlieb, B. J., Hill, J. P., Wildgruber, U. C., Luke, G. M., Nachumi, B., Moritomo, Y., and Tokura, Y., *Phys. Rev. Lett.* **76**, 2169 (1996).
8. Inoue, J., and Maekawa, S., *Phys. Rev. Lett.* **74**, 3407 (1995).
9. Moritomo, Y., Kuwahara, H., Tomioka, Y., and Tokura, Y., *Phys. Rev. B* **55**, 7549 (1997).

The Crucial Role of Jahn-Teller Effect in Forming of Orbital and Magnetic Structures of Charge Ordered Manganite $R_{0.5}A_{0.5}MnO_3$

L. E. Gontchar, A. E. Nikiforov and S. E. Popov

Department of Physics, Urals State University, 620083, 51 Lenin ave., Ekaterinburg, Russia

Abstract. This work is devoted to the description of the orbital and magnetic structures of charge ordered manganites, caused by crystal and charge structures and the Jahn-Teller effect. The model under consideration is based on the orbital dependency of the magnetic interactions in Jahn-Teller crystal. The orbital structure is considered dependent upon the Jahn-Teller distortions of the environment of the Mn^{3+} ions and upon the crystal field. Therefore, it is shown that the orbital structure of the charge ordered phase is formed mainly by cooperative Jahn-Teller distortions and, taking into account the charge superstructure, causes the non-trivial magnetic structure.

INTRODUCTION

A lot of attention was paid to crystals of rare-earth manganites with the general formula $R_{1-x}A_xMnO_3$ (R = La, Pr, Nd, Y, Eu, Sm, A = Ca, Sr) after discovery of the effect of colossal magnetoresistance (CMR) at some concentration x. These compounds have also a number of nontrivial properties shown at various concentration of an alkaline-earth impurity. One of such properties is the charge ordering (CO).

The CO phase arises at doping of a crystal by non-isovalent impurity at some temperature. In this phase, free charge carriers are localized and form the superstructure. In manganites, this localization causes an alternation of magnetic ions Mn^{3+}/Mn^{4+}.

The greatest attention is usually paid to mutual influence of magnetic and transport properties, less attention is paid to interplay between structural and magnetic characteristics. The phenomenon of CO is interesting because there is a necessity of taking into account all aspects of manganites' physics (localization of charges, structural distortions, orbital ordering, and magnetic interactions).

The model of the charge ordered crystal is following:

- The hole that has appeared because of non-isovalent doping, is localized on the manganese ion;
- The Mn^{3+} and Mn^{4+} ions form the ordered structure;
- The CO is accompanied by the lattice distortions: rotations and e_g-distortions (JT distortion) of oxygen octahedra;
- The orbital structure is determined by the linear JT interaction of nearest oxygen ions and by the low-symmetric crystal fields;

CP554, *Physics in Local Lattice Distortions*, edited by H. Oyanagi and A. Bianconi
© 2001 American Institute of Physics 1-56396-984-X/01/$18.00

- The Pmmn space group can describe the crystal structure of all CO compounds of the type $R_{0.5}A_{0.5}MnO_3$.

CRYSTAL AND ORBITAL STRUCTURES OF MANGANITES

The space symmetry is described in work [1] by D_{2h}^{13} (*Pmmn*) space group. This description takes into account CO and cooperative Jahn-Teller distortions of oxygen octahedra around a trivalent manganese ion. In previous works [2–4], the symmetry of the CO crystal was considered to coincide with the structure of pure manganite — $D_{2h}^{16}I$ (*Pnma*) — or monoclinic (taking into account charge-inequivalent manganese positions).

Complex crystal structure and charge one form orbital and magnetic structures. They are represented on Figure 1.

As there are charge inequivalent atoms in the crystal, there is an additional to JT interaction low-symmetric crystal field, which acts upon the trivalent manganese

$$V_n = A_n X_\theta + B_n X_\varepsilon,\qquad(1)$$

where $X_{\theta,\varepsilon}$ are orbital operators.

After the establishment of the distorted structure, the ground state wave function of these ions has a form

$$\Psi_n = \sin\frac{\Theta_n}{2}\,\varphi_{n\theta} + \cos\frac{\Theta_n}{2}\,\varphi_{n\varepsilon},\qquad(2)$$

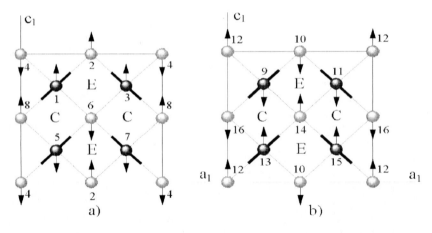

FIGURE 1. Charge (C-type), orbital (A'-type) and magnetic (CE-type) structures of the CO compound in two next planes along the Y-axis (part a and b). Dark spheres designate Mn^{3+}, light spheres designate Mn^{4+}; the bold lines designate the direction of the e-orbitals. The arrows show the directions of the magnetic moments. The numbers enumerate magnetic sublattices.

where

$$\sin\Theta_n = \frac{V_e Q_{\varepsilon n} + B_n}{\sqrt{(V_e Q_{\theta n} + A_n)^2 + (V_e Q_{\varepsilon n} + B_n)^2}}, \cos\Theta_n = \frac{V_e Q_{\theta n} + A_n}{\sqrt{(V_e Q_{\theta n} + A_n)^2 + (V_e Q_{\varepsilon n} + B_n)^2}} \quad (3)$$

if we consider that $V_e = -0.025$ a. u., as for pure manganite.

The magnitude of the Θ angle alters its sign from ion to ion:

$$\Theta_1 = -\Theta_3 = \Theta_5 = -\Theta_7 = \Theta_9 = -\Theta_{11} = \Theta_{13} = -\Theta_{15} = \Theta. \quad (4)$$

ISOTROPIC EXCHANGE INTERACTION AND SINGLE-ION ANISOTROPY

If there are the charge inequivalent magnetic ions in the crystal lattice, various types of the exchange are possible. For this charge and orbital structure (Figure 1), the exchange interactions are of four types.

1. $Mn^{3+}- Mn^{3+}$ along Y-axis

$$J_b^1 = J_0 \frac{\cos^2 \varphi_b}{r_b^{10}} F_b(\Theta), \text{ where } F_b(\Theta) = 1 + 2\alpha\cos\Theta + \beta\cos^2\Theta, \quad (5)$$

J_0, α, β are the parameters, which could be related with the experimental data, Θ is the angle that characterize the orbital structure, r is the length of Mn-O bond, φ is the angle of Mn-O-Mn bond. The values of the parameters are $J_0 = 1.69 \cdot 10^4$ K·Å^{10}, $\alpha = 1.0$, $\beta = 4.5$.

2. $Mn^{4+} - Mn^{4+}$ along Y-axis

$$J_b^2 = J_0' \frac{\cos^2 \varphi_b'}{(r_b')^{10}}. \quad (6)$$

3. $Mn^{3+} - Mn^{4+}$ in the basic plane along the pseudo-cubic x_p axis for $\Theta > 0$ and along y_p for $\Theta < 0$

$$J_{ac}^1 = J_0'' \frac{\cos^2 \varphi_{ac}'}{r_{ac}^{10}} F_{ac}''(\Theta), \text{ where } F_{ac}''(\Theta) = 1 - \frac{\alpha''}{2}(\cos\Theta - \sqrt{3}\sin\Theta). \quad (7)$$

4. $Mn^{3+} - Mn^{4+}$ in the basic plane along the pseudo-cubic x_p axis for $\Theta < 0$ and along y_p for $\Theta > 0$.

The parameters J_0', J_0'', α'' are estimated from available experimental data. The values of the parameters are $J_0' = 1.12 \cdot 10^4 \, \text{K} \cdot \text{Å}^{10}$, $J_0'' = -4.9 \cdot 10^3 \, \text{K} \cdot \text{Å}^{10}$, $\alpha'' = 5.1$.

In the case of the high-spin state of Mn^{3+} ion, the single-ion anisotropy plays the significant role. We obtain an expression of single-ion anisotropy dependent on the angle of orbital ordering in quasi-cubic axes:

$$\hat{H}_{an}^{(n)} = D_n S_{nz_p}^2 + E_n (S_{nx_p}^2 - S_{ny_p}^2), \text{ where } D_n = 3P\cos\Theta_n, \ E_n = \sqrt{3}P\sin\Theta_n \quad (8)$$

RESULTS AND DISCUSSION

The cooperative Jahn-Teller effect leads to the establishment of orbital structure. It means the certain ordering of e-orbitals of Mn^{3+} ions in the crystal. Because of dependency of isotropic exchange interaction and single-ion anisotropy on filling of orbital states of interacting ions, it leads also to the certain magnetic structure.

For calculation of this magnetic structure, we use the model of sixteen magnetic sublattices (eight on each charge state of manganese). Let us write down the following Hamiltonian, which is derived from Eqs. (5), (6), (7), (8):

$$\hat{H} = \sum_{n>m} J_{nm}(\mathbf{S}_n \cdot \mathbf{S}_m) + \sum_{n=2k-1} \hat{H}_{an}^{(n)} \quad (9)$$

where

$J_{12} = J_{23} = J_{38} = J_{18} = J_{56} = J_{67} = J_{47} = J_{45} = J_{9\,10} = J_{10\,11} = J_{11\,16} = J_{9\,16} = J_{12\,13} = J_{13\,14} = J_{14\,15} = J_{12\,15} = J_{ac}^1$,

$J_{14} = J_{16} = J_{36} = J_{34} = J_{78} = J_{58} = J_{27} = J_{25} = J_{9\,12} = J_{9\,14} = J_{11\,12} = J_{11\,14} = J_{10\,13} = J_{10\,15} = J_{15\,16} = J_{13\,16} = J_{ac}^2$,

$J_{19} = J_{3\,11} = J_{5\,13} = J_{7\,15} = J_b^1$,

$J_{2\,10} = J_{4\,12} = J_{6\,14} = J_{8\,16} = J_b^2$.

The term of the single-ion anisotropy for all trivalent manganese ions in a cell is derived by appropriate rotations of local axes to the general orthorhombic reference frame X, Y, Z in Eq. (8). For trivalent manganese ions in the magnetic cell, the angles of the orbital structure both the angles α_1, α_2 correspond as follows:

$$Mn_1(\alpha_1, \alpha_2, \Theta), \ Mn_3(-\alpha_1, \alpha_2, -\Theta), \ Mn_5(\alpha_1, \alpha_2, \Theta), \ Mn_7(-\alpha_1, \alpha_2, -\Theta),$$
$$Mn_9(\alpha_1, -\alpha_2, \Theta), \ Mn_{11}(-\alpha_1, -\alpha_2, -\Theta), \ Mn_{13}(\alpha_1, -\alpha_2, \Theta), \ Mn_{15}(-\alpha_1, -\alpha_2, -\Theta). \quad (10)$$

For $Pr_{0.5}Sr_{0.41}Ca_{0.09}MnO_3$, the calculations of the magnetic structure based on the experimental data [1] were carried out.

The minimization of magnetic energy (9) results to multi-sublattice magnetic structure, which has a non-collinear form. In the *Pmmn* space group, the classification of possible structures determines such structures as (Gx, 0, C'z) for Mn^{3+} and (0 0 Gz) for Mn^{4+} without the account of rotary distortions. In another case, one determines it as (G_X, A'_Y, C_Z) for Mn^{3+} and $(0\ C_Y\ G_Z)$ for Mn^{4+}. On the basic directions of the magnetic moments, such structure also refers to CE_Z. The values of the components of magnetic moments of Mn^{3+} ions are $\mu_X = 0.21\mu_B$, $\mu_Y \approx 0$, $\mu_Z = 3.17\mu_B$ (the magnetic moment of Mn^{3+} ion are equal to $3.18\mu_B$ [1]). The values of the components of magnetic moments of Mn^{4+} ions are $\mu_X = 0$, $\mu_Y \approx 0$, $\mu_Z = 2.83\mu_B$ (the magnetic moment of Mn^{4+} ion are equal to $2.83\mu_B$ [1]). Because of absence of Y-component in the magnetic structure, there are six various directions of the spins.

CONCLUSION

Thus, we put forward the convenient (for consideration and calculations) semi-empirical approach, in which the magnetic properties of pure and doped ($x = 0.5$) insulating manganites could be described. In this model crystal, charge and orbital structures are taken into account.

In our model, the crucial role belongs to the electron-lattice interaction. The results of a calculation have shown, what exactly this interaction determines the formation of the orbital structure of examined crystals. The orbital structure, in its turn, strongly influences the magnetic interactions and, hence, is the reason of observable magnetic structure.

We have taken into account following magnetic interactions: the superexchange and the single-ion anisotropy of Jahn-Teller ions (Mn^{3+}). We have neglected by the double exchange, by the anisotropic exchange interaction, and by the anisotropy of the Mn^{4+} ion.

In our model, it is possible to estimate the dependencies of the exchange parameters for regular and doped manganites in terms of the orbital structure.

ACKNOWLEDGMENTS

The research described in this paper was made possible in part by Award N REC-005 of the US Civil Research Development Foundation for the Independent States of Former Soviet Union (CRDF).

REFERENCES

1. Damay, F. et al., *JMMM* **184**, p.71 (1998).
2. Jirak, Z. et al., *JMMM* **53**, p. 153 (1985).
3. Kawano, H. et al., *Phys. Rev. Lett.* **78**, p. 4253 (1997).
4. Tomioka, Y. et al., *Phys. Rev. Lett.* **74**, p. 5108 (1995).

Magnetic Ordering and Spin-state of Cobalt in $R_{0.67}Sr_{0.33}CoO_3$

M. Paraskevopoulos, J. Hemberger and A. Loidl

Experimentalphysik V, Elektronische Korrelationen und Magnetismus, Institut fuer Physik, Universitaet Augsburg, D-86159 Augsburg, Germany

Abstract. We report on detailed magnetization and specific heat studies in $R_{0.67}Sr_{0.33}CoO_3$. All samples reveal ferromagnetic order with increasing T_c for increasing average ionic radius $\langle r_A \rangle$. The magnetic measurements in $R_{0.67}Sr_{0.33}CoO_3$ clearly indicate a spin-state transition as a function of $\langle r_A \rangle$. For some samples a further magnetic transition accompanied with a reduction in magnetization has been observed at low temperatures. Whether a temperature induced spin-state transition for the cobalt ions or crystal field effects for the rare earth ions hold for the reduced magnetization is still open.

INTRODUCTION

The subtle balance between crystal-field energy Δ_{cf} and intraatomic exchange energy J_{ex} (Hund's coupling) for trivalent cobalt in $LaCoO_3$ is the precursor of a peculiar spin state transition observed around 90 K. At the lowest temperatures, Co^{3+} is in a low-spin configuration (LS) $t_{2g}^6 e_g^0$ revealing a nonmagnetic ground state with $S = 0$. Upon heating thermal excitation leads to gradual population of intermediate spin states (IS) $t_{2g}^5 e_g^1$ with $S = 1$. At even higher temperatures a metal-insulator transition occurs and the high spin-state (HS) $t_{2g}^4 e_g^2$ with $S = 2$ probably becomes relevant.

In general the magnetic and electronic properties of ABO_3 perovskite oxides can easily be tuned by modifying the lattice parameters. The crystal symmetry changes from cubic to rhombohedral and subsequently to orthorhombic, depending on the size of the A and B ions. The different lattice distortions are determined by the tolerance factor t, which is defined as $t = (\langle r_A \rangle - \langle r_O \rangle)/(\sqrt{2}(\langle r_B \rangle - \langle r_O \rangle))$. Here $\langle r_i \rangle$ represents the ionic size of each element in the ABO_3 perovskites. When t is close to 1, the cubic perovskite structure is obtained (*e.g.* $SrTiO_3$). As t decreases the lattice symmetry transforms to rhombohedral and subsequently to orthorhombic, in which the B-O-B bond angle more and more deviates from 180°. Concomitantly the crystal field (CF) slightly changes at the B site. In most cases this effect can be neglected (*e.g.* in the manganites). However in cobalt based perovskites the aforementioned balance between Δ_{cf} and J_{ex} can easily be disturbed yielding transitions between different electronic configurations. This possibility of inducing and controlling a spin-state transition through structural tuning motivated us to reinvestigate the $R_{1-x}A_xCoO_3$ series (R = rare earth, A = divalent ion) series. In this paper we report on systematic magnetic measurements in $R_{0.67}Sr_{0.33}CoO_3$ which clearly indicate a spin-state

CP554, *Physics in Local Lattice Distortions*, edited by H. Oyanagi and A. Bianconi
© 2001 American Institute of Physics 1-56396-984-X/01/$18.00

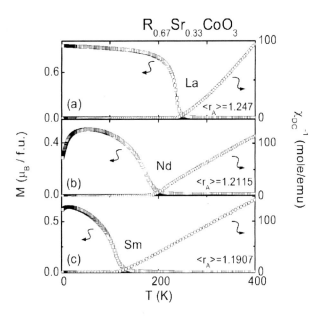

FIGURE 1. Temperature dependence of the dc-magnetization (left scale) and inverse magnetic susceptibility χ (right scale) for the samples R = La, Nd and Sm.

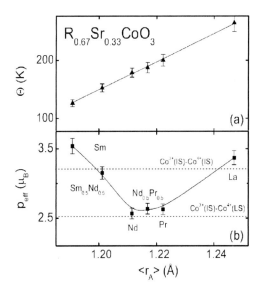

FIGURE 2. The paramagnetic Curie temperature Θ (part a) and the effective magnetic moment p_{eff} (part b) as a function of average ionic radius $\langle r_A \rangle$.

transition as a function of $\langle r_A \rangle$. For some samples a further magnetic transition has been observed at low temperatures.

MEASUREMENTS AND RESULTS

In Fig. 1 we show the temperature dependence of the magnetization (left scale) for R = La, Nd and Sm. Within the same frame the inverse susceptibility $1/\chi$ is shown *vs.* temperature (right scale). In the paramagnetic regime all samples obey the Curie-Weiss (CW) law. The paramagnetic CW temperature Θ steadily decreases on decreasing ionic radii of the R ions (Fig. 2a). All samples show the evolution of spontaneous magnetization below the transition temperature T_c, which decreases with decreasing ionic radius of the rare-earth ion (Fig. 2a-b). The suppression of T_c is closely related to the change in the lattice distortion induced through the reduced average radius of the A ion. As $\langle r_A \rangle$ decreases from La to Sm, the hexagonal lattice structure transforms to orthorhombic, in which the Co-O-Co bond angle deviates significantly from 180°, yielding a reduced effective d - d electron transfer via the O $2p$ states.

The results of the analysis of susceptibility and magnetization are summarized in Fig. 2a-b. The paramagnetic Curie temperature decreases linearly on decreasing average ionic radii $\langle r_A \rangle$ driven by the same reasons as discussed in the case of T_c. To elucidate the spin states of the cobalt ions in the various samples we calculated the effective paramagnetic moments p_{eff} by subtracting the contribution of the rare-earth ion (Fig. 2b). A minimum of p_{eff} can be observed for $\langle r_A \rangle \sim 1.21$ indicating the structurally triggered spin-state transition.

Hysteresis measurements taken for various samples at T = 5 K are plotted in Fig. 3. For clarity the rare earth contributions have been subtracted. The value of the spontaneous magnetization M_s at 5 K decreases from La to Nd, show a minimum for R = Nd and becomes almost constant for smaller ionic radii. The ordered magnetic moments deduced from the M(H) curves are significantly smaller than predicted from an simple ionic picture. This behaviour is opposite to what has been observed in manganites systems, where the ionic picture worked quite well. In the cobalt compounds under consideration the formation of a narrow σ^* band evolving the Co e_g states, calls for an itinerant electron description. The origin of ferromagnetism in cobaltates is still controversial although a itinerant magnetism picture seems to be favored. For smaller ionic radii larger coercitivity fields can be seen (inset b). This effect most probably is driven by an enhanced magnetic anisotropy induced through the L-S coupling of the rare earth ions. If the spin-orbit coupling of the cobalt ions also contributes to the anisotropy is not clear.

We now focus on the low temperature properties of the Nd sample. At about 50 K the magnetization begins to decrease with decreasing temperature. The same behavior but significantly weaker has also been observed for R = $Pr_{0.5}Nd_{0.5}$, $Nd_{0.5}Sm_{0.5}$ and Sm. In order to study the role of the rare-earth ion we prepared samples with identical $\langle r_A \rangle$ as for the Nd compound. The temperature dependence of the magnetization M for this samples is shown in Fig. 4. It can be clearly seen that the ferromagnetic transition temperature T_c remains almost constant however the low temperature transition

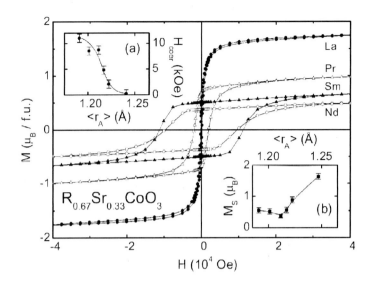

FIGURE 3. Field dependence of the magnetization at 5 K for the $R_{0.67}Sr_{0.33}CoO_3$ compounds (R=La, Pr, Nd and Sm). In the insets the coercive field H_{coer} and the spontaneous magnetization M_s as a function of $\langle r_A \rangle$ are shown.

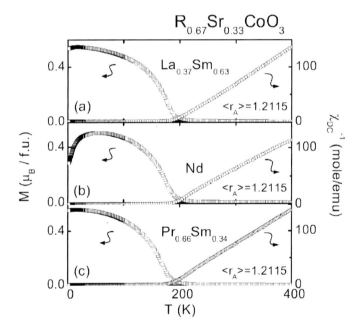

FIGURE 4. Temperature dependence of the dc-magnetization (left scale) and inverse magnetic susceptibility χ (right scale) for the samples $R = La_{0.37}Sm_{0.63}$, Nd and $Pr_{0.66}Sm_{0.34}$.

accompanied with a reduction in M clearly observed in the Nd compound, can hardly be seen in $La_{0.37}Sm_{0.63}$ and is almost absent for $Pr_{0.66}Sm_{0.34}$. This points out that this feature is not primarily related to structural changes. A possible explanation could be given if one considers the multiplet splitting of the rare earth ions. The schottky anomalies for the rare earth ions as deduced from specific heat measurements (Fig. 5a, b) indicate population of higher lying terms with increasing temperature. It is known that on reducing temperature a reduction in the magnetic moment of rare earth ions can occur. Additionally we cannot exclude partially magnetic ordering of the rare earth ion driven by the internal magnetic field caused by the ferromagnetic ordering of the cobalt spins. When a antiferromagnetic ordering in respect with the cobalt spin subsystem sets in, a reduction of the magnetization is expected. Finally a temperature induced spin-state transition for the cobalt ions as a result of the changing crystal field could also hold for the reduced magnetization. Further investigations are now on the way to clarify this issue.

CONCLUSIONS

The main aim was to investigate the ground state properties of a system with competing crystal-field and intraatomic (Hund's rule coupling) energies as a function of tolerance factor. All compounds reveal ferromagnetic order, with both T_c and the magnitude of the ordered moment depend sensitively on the ionic radius of the A-site ion: as the tolerance factor decreases, yielding increasing deviations from a 180° bond angle, T_C and M_S decrease significantly. A spin-state transition as a function of $\langle r_A \rangle$ has been found. For the origin of the low temperature magnetic transition observed for some samples, no final explanation could be given.

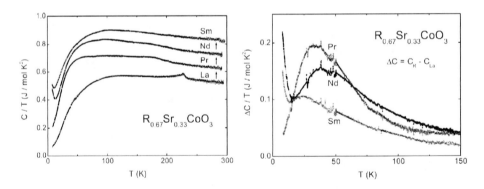

FIFURE 5. a) Specific heat for the samples R = La, Pr, Nd and Sm. The curves are shifted in steps of 0.1 J/molK^2. A sharp transition associated with a pronounced peak has only been observed for R= La. For the other samples the anomaly at T_c are smeared out. b) Schottky anomalies for R = Pr, Nd and Sm obtained by taking the difference between the R curve and the La one.

Investigation of the Microscopic Mechanisms of $La_{1.5}Sr_{0.5}NiO_4$ Charge-ordered Phase Stability

A. E. Nikiforov, V. A. Smorkalov and A. Yu. Zaharov

Department of Physics, Urals State University, Lenin av. 51, 620083 Ekaterinburg, Russia

Abstract. Within the framework of pair potential approximation and shell model the we have studied the charge-ordering phenomenon in $La_{1.5}Sr_{0.5}NiO_4$. The charge-ordered phase is shown to be stable. According to the results of our calculations the charge redistribution plays the main role decreasing the total energy of crystal. On the other hand the lattice distortions accompanying the charge redistribution is shown to be associated with lattice relaxation within NiO_2 layers.

INTRODUCTION

The crystal $La_{2-x}Sr_xNiO_4$ has received considerable attention during the last few years. Initially it was associated with the assumption that the Sr-doped nickelates can become superconductor. These hopes ware based mainly on the fact that $La_{2-x}Sr_xNiO_4$ is isostructural to well-known $La_{2-x}Sr_xCuO_4$ cuprate which is high-temperature superconductive material within the wide range of composition. However these expectations have not been justified: $La_{2-x}Sr_xNiO_4$ is isolator up to $x \sim 1$ since holes is localized within NiO_2 layers. The hole localization is connected with formation of the polarons consisting of nickel ion and of four neighbor planar oxygen ions. Moreover it turns out that the experimental data can be explained under assumption that these polarons arrange to periodical structure below some critical temperature T_{CO}. This phenomenon was named charge ordering.

Present work is devoted to investigation of charge ordered phase in $La_{2-x}Sr_xNiO_4$ crystal. Particularly our research has three aims. First of all, we would like to confirm theoretically the possibility of charge ordering for layered perovskites and to prove its stability. We also would like to show that such charge redistribution results in (directly or indirectly) the total crystal energy decrease. And we tried to find microscopic mechanisms responsible for charge ordered phenomenon in nickelates.

MODEL

The equilibrium structural parameters can be found as a result of the minimization of the total crystal energy. The crystal energy E is presented within the framework of pair potential approximation and shell model as follows:

CP554, *Physics in Local Lattice Distortions*, edited by H. Oyanagi and A. Bianconi
© 2001 American Institute of Physics 1-56396-984-X/01/$18.00

$$E = \frac{1}{2}\sum_{\substack{i,j \\ i \neq j}} V_{i,j} + \frac{1}{2}\sum_i k_i \, |\vec{S}_i|^2 \qquad (1)$$

where the index i numbers all ions in the unit cell, index j numbers all the crystal ions, k_i is the core-shell force constant, and s_i is the 'i'-ion shell displacement relative to its core. For the interionic pair potential $V_{i,j}$ we use the same expression as in [1]. The values of all constants needed to simulate $La_{2-x}Sr_xNiO_4$ as well as the methods used to determine them are given in our previous paper [1].

STRUCTURE FEATURES

In this work we study two structural phases of $La_{1.5}Sr_{0.5}NiO_4$ – high-temperature tetragonal phase (HTT) and charge-ordered phase (CO). On the base of the results obtained in [2,3] we consider only holes localizing on nickel ions. In HTT phase all holes are distributed uniformly over nickel ions and as a result all nickel ions have charge (+2.5) (see Figure 1). In CO phase there are two different sort of nickel ions – Ni^{2+} and Ni^{3+} (see Figure 1). These ions in CO phase are arranged periodically.

In order to simulate the presence of Sr^{2+} ions in crystal we use the following approach. We assume that strontium ions are distributed over crystal bulk chaotically in both HTT and CO phases. As a result we consider uniform sublattice of some effective ion ($La_{1.5}Sr_{0.5}$) instead of considering disorder La^{3+} sublattice in which some ions are substituted by Sr^{2+} ions. This approach gives us possibility to preserve long-

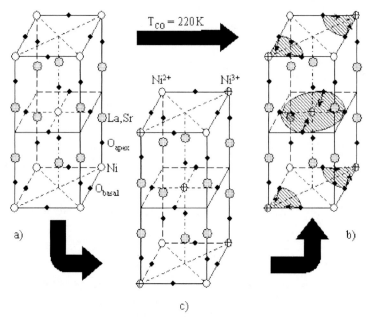

FIGURE 1. The structure of HTT (a), CO (b) and pseudo-HTT (c) phases.

range order and to use symmetry analysis of structural phases. According to symmetry analysis the space group of HTT phase is D_{4h}^{17} and the space group of CO phase is D_{2h}^{19}.

RESULTS AND DISCUSSION

In order to find the microscopical reasons responsible for charge ordering phenomenon in $La_{1.5}Sr_{0.5}NiO_4$ we have carried out the simulation of structural transition from HTT to CO phase. We use one more structural phase – pseudo-HTT (see Figure 1) – which has the same structure as HTT phase but the nickel ions charges are similarly to CO phase. Pseudo-HTT phase are used to distinguish between two different contributions to stability of CO phase: from charge redistribution itself and accompanying this redistribution lattice relaxation.

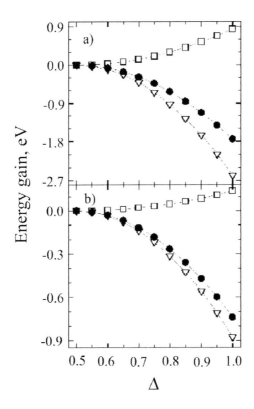

FIGURE 2. Dependence of energy gain for total crystal energy (circle), short range interactions (square), Coulomb interactions (triangle) on the share of hole for the most charged nickel ion Δ (for HTT phase $\Delta = 0.5$ and for CO phase $\Delta = 1$): (a) HTT → CO transition, (b) HTT → pseudo-HTT transition.

From the result of simulation (see Figure 2) it becomes clear that considered CO phase is stable and has lower energy than HTT phase, and that during HTT $\rightarrow$ CO phase transition structure adjustment under hole redistribution over nickel sublattice gives more energy gain than hole redistribution itself. It turned out that the relaxation of NiO_2 layer structure decrease energy more than crystal relaxation along direction perpendicular to these layers. And, at last the main interaction decreasing full energy of the crystal is Coulomb interaction Ni^{3+}-O^{2-}(*basal*) which is responsible for polaron formation.

ACKNOWLEDGMENTS

The research described in this publication was made possible in part by Award No. Rec-005 of the U.S. Civilian Research & Development Foundation for the Independent States of the Former Soviet Union (CRDF).

REFERENCES

1. Zaharov, A.Yu., Nikiforov, A.E., Smorkalov, V.A., *Phys. Solid State* **42**, 1441-1445 (2000).
2. McQueeney, R.J., Sarrao, J.L., Osborn, R, *Phys. Rev. B* **60**, 80-83 (1999).
3. Yi, Y.-S., *et al*, *Phys. Rev. B* **58**, 503-513 (1999).

Ultrasonic Investigation of Perovskite Manganese Compounds La$_{1-x}$Sr$_x$MnO$_3$

T. Goto, Y. Nemoto, H. Hazama, Y. Tomioka*, A. Asamitsu†
and Y. Tokura*‡

Graduate School of Science and Technology, Niigata University, Niigata 950-2181, Japan
**Joint Research Center for Atom Technology (JRCAT), Tsukuba 305-0046, Japan*
†*Cryogenic Center, University of Tokyo, 113-0032, Japan*
‡*Department of Applied Physics, University of Tokyo, 113-8656, Japan*

Abstract. We made the ultrasonic measurements on the perovskite manganese La$_{1-x}$Sr$_x$MnO$_3$ (x = 0.12, 0.165) to examine the electron-lattice interaction in the system with both orbital and charge degrees of freedom. The softening of $(C_{11} - C_{12})/2$ above the structural transition point T_s = 290 K for x = 0.12 and T_s = 310 K for x = 0.165 is well described in terms of the quadrupolar susceptibility for O_2^0 and O_2^2 of $d\gamma$-doublet of Mn^{3+}. Elastic softening associated with the charge ordering of Mn^{4+} ions in a sea of Mn^{3+} ions for x = 0.12 and x = 0.165 is also presented.

INTRODUCTION

The colossal magnetoresistance (CMR) in the manganese oxides La$_{1-x}$Sr$_x$MnO$_3$ with perovskite structure attracts much attention from the viewpoint of technological application as well as academic interest [1, 2]. The antiferromagnetism of the end material LaMnO$_3$ of an insulator changes into the ferromagnetism in a hole-doping compound La$_{1-x}$Sr$_x$MnO$_3$ with metallic character for $x > 0.20$. The $3d^4$ state of Mn^{3+} ion is split into the ground state of $d\varepsilon$-triplet and the excited state of $d\gamma$-doublet at about 2 eV by a crystalline electric field potential. Because of the ferromagnetic Hund-rule coupling due to the intra-atomic exchange, the spin $S = 1/2$ of $d\gamma$-orbit with E$_g$ symmetry aligns parallel to spin $S = 3/2$ of $d\varepsilon$-orbits with T$_{2g}$ symmetry. The double-exchange interaction mediated by $d\gamma$-carriers of manganese ions gives rise to the ferromagnetism and plays an important role for CMR in the hole-doping compounds. In addition to magnetic transition due to the spin degree of freedom, the orbital freedom of $d\gamma$-orbit gives rise to an orbital ordering with anti-parallel orientation. The charge freedom of Mn^{4+} ions in a sea of Mn^{3+} ions leads the charge ordered phase in lowering temperature [3]. Furthermore the coupling of the lattice to the quadrupolar moment due to orbital freedom of Mn^{3+} ion and the charge fluctuation mode of Mn^{3+} and Mn^{4+} ions gives rise to the structural change of the crystal. It has been pointed out that the Jahn-Teller coupling of $d\gamma$-orbit plays an important role for increase of $d\gamma$-carrier associated with CMR [4]. The investigation of the elastic properties of the manganese compounds by the ultrasonic measurement is an

CP554, *Physics in Local Lattice Distortions*, edited by H. Oyanagi and A. Bianconi
© 2001 American Institute of Physics 1-56396-984-X/01/$18.00

important issue for clarification of the electron-lattice interaction of $d\gamma$-orbit and its interplay for the origin of CMR.

EXPERIMENTAL

In the present paper, we show the elastic properties of the manganese compounds $La_{1-x}Sr_xMnO_3$ ($x = 0.12, 0.165$) measured by the ultrasonic method. The single crystals in the present experiment were grown by the floating zone furnace with the halogen lamp and ellipsoidal mirror. The velocity of the ultrasound was measured by a home made apparatus with the phase different detector. The piezoelectric transducers of $LiNbO_3$ were bonded on the plane parallel surfaces of the crystal for generating and detecting of the ultrasound. The $LiNbO_3$ plates of 36°Y-cut and X-cut are available for the transducers of longitudinal and transverse sound waves, respectively. The magnetic field up to 80 kOe was generated by a superconducting magnet.

QUADRUPOLAR EFFECT

In Fig. 1(a) we show the temperature dependence of the elastic constants (C_{11}–$C_{12})/2$ and C_{44} for the compound of $x = 0.12 \approx 1/8$. Here we made the ultrasonic measurements in accordance with the axes of the proto-type cubic crystal. The transverse ($C_{11} - C_{12})/2$ mode associated with the strain $\varepsilon_v = \varepsilon_{xx} - \varepsilon_{yy}$ exhibits a pronounced softening above the structural transition point $T_s = 290$ K, while the transverse C_{44} mode increases monotonously above T_s. Furthermore the longitudinal C_{11} mode and the bulk modulus C_B, which are not presented here, increase above T_s. In lowering temperature the ($C_{11} - C_{12})/2$ as well as C_{44} mode exhibits the softening again above the charge ordering point $T_{co} = 145$ K. The ultrasonic echo signal for both transverse and longitudinal modes disappears completely in the low temperature phase below T_{co}, because the sound wave is scattered intensively by the elastic domain in a triclinic structure. The ferromagnetic transition around 170 K has been reported, but the transition point is not clear in the present elastic measurements. The $d\gamma$-orbit with E_g symmetry of Mn^{3+} ion in $La_{1-x}Sr_xMnO_3$ possesses the charge monopole with A_{1g} symmetry, the electric quadrupole $O_2^0 = (2l_z^2 - l_x^2 - l_y^2) / \sqrt{3}$, $O_2^2 = l_x^2 - l_y^2$ with E_g symmetry and the octupole moment $T_{xyz} = l_x l_y l_z$ of A_{2g}, because the direct product of E_g reduces to $E_g \times E_g = A_{1g} + A_{2g} + E_g$ [5]. The quadrupole of O_2^0 is estimated as $|< \phi_{eg} | O_2^0 | \phi_{eg} >| = 2\sqrt{3}$ for the $d\gamma$-orbit of Mn^{3+} ion. The coupling of O_2^0, O_2^2 of $d\gamma$-orbit in Mn^{3+} to the tetragonal strain $\varepsilon_u = (2\varepsilon_{zz} - \varepsilon_{xx} - \varepsilon_{yy}) / \sqrt{3}$, the orthorhombic strain $\varepsilon_v = \varepsilon_{xx} - \varepsilon_{yy}$ of E_g symmetry is responsible for the softening of the ($C_{11} - C_{12})/2$ mode above $T_s = 290$ K in Fig. 1(a). We employ $C_\Gamma = C_\Gamma^0(T - T_c^0)/(T - \Theta)$ for fitting of the elastic softening above T_s. Here the characteristic temperatures are defined as $\Theta = g_\Gamma' |< \phi_{eg} | O_{\Gamma\gamma} | \phi_{eg} >|^2$ and $T_c^0 = \Theta - E_{JT}$. The Jahn-Teller coupling energy is written as $E_{JT} = Ng_\Gamma^2 |< \phi_{eg} | O_{\Gamma\gamma} | \phi_{eg} >|^2 / C_\Gamma^0$. We obtained $T_c^0 = 240$ K and $\Theta = 188$ K by the solid line of Fig. 1(a). The dashed line in Fig. 1(a) means a background C_Γ^0. The quadrupole-strain coupling constant $|g_{\Gamma3}| = 1167$ K and the ferro-type quadrupolar inter-site coupling $g_{\Gamma3}' = 16$ K > 0 were obtained. It should be noted that the absence of the quadrupole $O_{yz} = l_y l_z + l_z l_y$, $O_{zx} = l_z l_x + l_x l_z$, $O_{xy} = l_x l_y + l_y l_x$ with T_{2g} symmetry for $d\gamma$-orbit of Mn^{3+} ion consists with the monotonous increase of C_{44} above T_s in Fig. 1(a). It seems that the octupole moment T_{xyz} is irrelevant to the present systems.

In Fig. 1(b) we show the elastic constants $(C_{11} - C_{12})/2$ and C_{44} of the compound $x = 0.165$, which is characterized by CMR. The $(C_{11} - C_{12})/2$ mode exhibits a softening above the structural transition point $T_s = 310$ K, but the C_{44} mode shows an increase above T_s. This behavior of both transverse modes in $x = 0.165$ is similar to the result of $x = 0.12$ in Fig. 1(a). The solid line of $(C_{11} - C_{12})/2$ above T_s in Fig. 1(b) is obtained for the characteristic temperatures $T_c^0 = 190$ K and $\Theta = 175$ K. The quadrupole-strain interaction $|g_{\Gamma 3}| = 496$ K and the inter-site quadrupolar interaction $g_{\Gamma 3}' = 15$ K > 0. It is noticeable that the Jahn-Teller coupling energy $E_{JT} = T_c^0 - \Theta = 15$ K of $x = 0.165$ is smaller than $E_{JT} = 52$ K for $x = 0.12$. The anomalies associated with the ferromagnetic transition at $T_C = 260$ K has been clearly observed in Fig. 1(b).

CHARGE ORDERING

The charge degree of freedom associated with the hole-doped Mn^{4+} ions in a sea of Mn^{3+} ions of $La_{1-x}Sr_xMnO_3$ plays an important role for CMR. The Coulomb interaction among the Mn^{4+} ions gives rise to the charge ordering in lowering

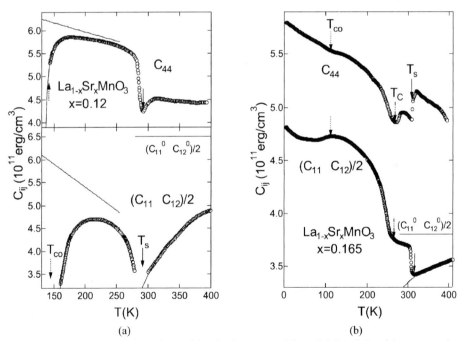

FIGURE 1. a) Temperature dependence of the elastic constants $(C_{11} - C_{12})/2$ and C_{44} of the compound $x = 0.12$ in $La_{1-x}Sr_xMnO_3$. b) Temperature dependence of the elastic constants $(C_{11} - C_{12})/2$ and C_{44} of the compound $x = 0.165$ in $La_{1-x}Sr_xMnO_3$.

temperature. The charge fluctuation among the different valences of Mn^{4+} and Mn^{3+} is assisted by thermally activated process. The freeze of the charge fluctuation mode with symmetry breaking character gives rise to the charge ordering. When the charge fluctuation mode couples to the lattice degree of freedom, the softening of the elastic constant and the structural change are expected across the charge ordering point. For example we refer the Verwey transition at $T_c = 120$ K of magnetite Fe_3O_4, where the transeverse C_{44} mode exhibits a softening around T_c [6]. The elastic soft C_{44} mode around the charge ordering point at $T_c = 292$ K of the rare-earth pnictide Yb_4As_3 is also a typical example [7]. The elastic softening of $(C_{11} - C_{12})/2$ and C_{44} above $T_{co} = 145$ K of $x = 0.12$ in Fig. 1(a) is caused by the coupling of the charge fluctuation mode and the appropriate elastic mode. Solid lines in Fig. 1(a) above T_{co} mean the fittings by $C_\Gamma = C_\Gamma^0 (T - T_c^0) / (T - \Theta)$. The characteristic temperatures $T_c^0 = 134$ K and $\Theta = 132.5$ K for C_{44} are smaller than $T_c^0 = 150$ K and $\Theta = 142$ K for $(C_{11} - C_{12})/2$. This means that the charge fluctuation mode coupled with the $(C_{11} - C_{12})/2$ mode dominates for the charge ordering in $x = 0.12$ in compared to the contribution from the mode coupled with the C_{44} mode. The interplay of the lattice coupling for the transition to the monoclinic structure is not clear at the present stage.

In the compound of $x = 0.165$, the charge ordering at $T_{co} = 100$ K has already been reported by the neutron scattering measurements [3]. Actually the elastic constants in Fig. 1(b) show round minimums around T_{co}. This behavior is reflected by the glass character of the charge ordering in $x = 0.165$, which is distinguished from the symmetry breaking character of the charge ordering in the compounds $x = 0.12$ ($\approx$ 1/8) and in Yb_4As_3. In the case of $x = 0.165$, the disaccord of the lattice periodicity with the number of the Mn^{4+} ions in the system and the increased carriers of the $d\gamma$-electrons in the ferromagnetic state favor the glass character of the charge ordering.

CONCLUSION

The elastic softening of the transverse $(C_{11} - C_{12})/2$ mode due to the interaction of the quadrupole O_2^0, O_2^2 $d\gamma$-orbit in Mn^{3+} has been commonly observed across the structural transition point T_s in the compounds $x = 0.12$ and $x = 0.165$. The increase of the transverse C_{44} mode above T_s is consistent with the absence of the quadrupole O_{yz}, O_{zx} and O_{xy} for $d\gamma$-orbit of Mn^{3+}. The elastic softening of $(C_{11} - C_{12})/2$ and C_{44} of $x = 0.12$ above the charge ordering point T_{co} indicates the symmetry breaking character of the charge fluctuation mode in the system. The round minimums around T_{co} of $x = 0.165$ indicates the glass character in the charge ordering. In conclusion the ultrasonic measurement is a powerful probe to examine the lattice effects due to both orbit and charge degrees of freedom in the manganese oxides with CMR.

REFERENCES

1. Urushibara, A., Moritomo, Y., Arima, T., Asamitsu, A., Kido, G., Tokura, Y., *Phys. Rev. B* **51**, 14103 (1995).

2. Tomioka, Y., Asamitsu, A., Kuwahara, H., Moritomo, Y., and Tokura, Y., *Phys. Rev. B* **53**, 1689 (1996).

3. Yamada, Y., Hino, O., Nohdo, S., Kanao, R., Inami, T., and Katano, S., *Phys. Rev. Lett.* **77**, 904 (1996).

4. Millis, A.J., Mueller, R., and Shraiman, B.I., *Phys. Rev. B* **54**, 5405 (1996).

5. Shiina, R., Nishitani, T., and Shiba, H., *J. Phys. Soc. Jpn.* **66**, 3159 (1997).

6. Moran, T.J., and Lüthi, B., *Phys. Rev.* **187**, 710 (1969).

7. Goto, T., Nemoto, Y., Ochiai, A., and Suzuki, T., *Phys. Rev. B* **59**, 269 (1999).

Phase Separation between Novel Bi-stripe Order and CE-type Charge Order in Half-doped Bilayer Manganite LaSr$_2$Mn$_2$O$_7$

M. Kubota[1,*], H. Yoshizawa[1], K. Shimizu[2], K. Hirota[2],
Y. Moritomo[3], and Y. Endoh[4]

[1] Neutron Scattering Laboratory, I.S.S.P., University of Tokyo, Tokai, Ibaraki, 319-1106, Japan
[2] CREST, Department of Physics, Tohoku University, Aoba-ku, Sendai, 980-8578, Japan
[3] CIRSE, Nagoya University, Nagoya, 464-8601, Japan
[4] Institute for Materials Research, Tohoku University, Aoba-ku, Sendai 980-8577, Japan
* Present address: Photon Factory, Institute of Materials Structure Science, KEK,
1-1 Oho, Tsukuba-shi, 305-0801, E-mail: mkubota@post.kek.jp.

Abstract. We have revealed that a bi-stripe order with the [1 0 0] propagation vector (*vertical* direction) exists in the A-type antiferromagnetic spin ordered phase in half-doped bilayer manganite LaSr$_2$Mn$_2$O$_7$, which also exhibits the CE-type charge/spin order in the intermediate temperature. Although the CE-type charge order (CO) diminishes at low temperatures, the bi-stripe order exists over a whole temperature. Unlike other half-doped manganites in which the CE-type CO controls the transport properties, the bi-stripe order is responsible for the insulating behavior of the resistivity in LaSr$_2$Mn$_2$O$_7$. Thereby, the picture of phase separation between the metallic region and the CE-type charge/orbital ordered insulating region is not valid in the LaSr$_2$Mn$_2$O$_7$ system.

INTRODUCTION

A phase separation between a charge order (CO) and a ferromagnetic (FM) order has been recently observed in perovskite manganese oxides [1]. Within the FM domain, carriers are mobile due to the double-exchange interaction, whereas they are localized due to the Coulomb interaction and the Jahn-Teller distortion in the CO domain. Near the 50 % hole-doping, the metallic state exhibits either the FM ordering or an A-type antiferromagnetic (AFM) ordering in which FM planes are stacked antiferromagnetically, as seen in a typical example of Pr$_{1/2}$Sr$_{1/2}$MnO$_3$ [2]. On the other hand, an insulating state exhibits a so-called CE-type charge/orbital ordering, as seen in Pr$_{1/2}$Ca$_{1/2}$MnO$_3$ [3]. In the CE-type CO, Mn^{3+} and Mn^{4+} ions are alternately arranged in a checkerboard pattern.

A bilayer manganite La$_{2-2x}$Sr$_{1+2x}$Mn$_2$O$_7$ has MnO$_2$ FM sheets separated by (La$_{1-x}$Sr$_x$)$_2$O$_2$ layers, which give rise to the two-dimensional character in this system. The crystal structure has the tetragonal symmetry ($I4/mmm$). In the half-doped sample LaSr$_2$Mn$_2$O$_7$ ($x = 0.50$), the CE-type CO with the breathing-type distortion and the A-type AFM spin order emerge simultaneously at $T_{N,CO}^{A}$ (= 210 K) and the CE-type

CP554, *Physics in Local Lattice Distortions*, edited by H. Oyanagi and A. Bianconi
© 2001 American Institute of Physics 1-56396-984-X/01/$18.00

AFM spin order appear below T_N^{CE} (= 145 K) [4]. Thus the CE-type AFM region coexists with the A-type AFM region in the intermediate temperature. Comparing the temperature dependence of the resistivity with those of other half-doped manganites, a new-type of a phase separation (PS) picture was suggested in which the system breaks into the metallic A-type AFM state and the insulating CE-type CO state [4]. Although this PS picture qualitatively explains a trend of resistivities for a number of half-doped manganites, it was puzzling that why $La_{2-2x}Sr_{1+2x}Mn_2O_7$ with $x = 0.45$ sample is insulating despite the lack of CE-type CO. Concerning this issue, Kubota et al. have demonstrated very recently that a short-ranged bi-stripe ordering with the propagation vector [1 0 0] (vertical direction) exists in the $x = 0.50$ sample by a neutron diffraction measurement [5].

In order to elucidate the interplay between the conventional CE-type CO and the novel stripe-like CO, neutron diffraction measurements were performed on a single crystal of $LaSr_2Mn_2O_7$. We have found that the correlation lengths and temperature dependences of two kinds of lattice distortions completely differ from each other. In this report, we shall argue that in the $LaSr_2Mn_2O_7$ system the short-ranged bi-stripe CO in the A-type AFM state is responsible for the insulating behavior of the resistivity over a whole temperature region.

EXPERIMENTAL

Neutron diffraction measurements were carried out on triple-axis spectrometers GPTAS and TOPAN in the JRR-3M of JAERI, Tokai, Japan. Several incident neutron momentums were utilized, depending on the necessity of the intensity and the resolution, along with various combinations of collimators with PG filters. The (h0l) and (hhl) reciprocal planes were aligned as the scattering planes, and the samples were set in aluminum capsules filled with helium gas and were attached to the cold head of a closed-cycle helium gas refrigerator.

RESULTS

Fig. 1 shows typical profiles of the diffuse scattering for the bi-stripe order (upper panel) and the CE-type CO (lower panel) observed in the tetragonal c plane at 120 K. The longitudinal lattice distortion associated with the short-ranged bi-stripe CO is observed around $Q = (2.3\ 0\ 1)$ in the (h0l) scattering plane, whereas the breathing-type distortion is observed around $Q = (3/2\ 3/2\ 1)$ in the (hhl) scattering plane. The horizontal bars in Fig. 1 give a measure of the instrumental resolution. The line width of the profile for the bi-stripe CO is wider than the instrumental resolution, while that for the CE-type CO is resolution-limited, indicating that the bi-stripe CO remains short ranged, in contrast with a quasi long range order of the CE-type CO within a limit of the resolution of neutron diffraction.

Next, we examine the temperature dependence of the two orderings and resistivity in detail. The T dependences of the peak intensity for the bi-stripe and the CE-type order are shown in Fig. 2(a) and (b), respectively. The strong intensity can be observed even at 300 K for the bi-stripe order. With decreasing T, the intensity grows up to ~ 175 K, but levels off below ~ 175 K. The CE-type CO is invisible at room temperature, emerges below $T_{N,CO}^A$, and reaches at the maximum at T_N^{CE}. It suddenly starts to diminish below T_N^{CE}, and then vanishes at low temperatures, as shown in Fig. 2(b). A distinct hysteresis is observed in the temperature dependence of the CE-type CO.

It should be noted that the two types of charge orders coexist in the LaSr$_2$Mn$_2$O$_7$ sample. The present results indicate that in LaSr$_2$Mn$_2$O$_7$ the CE-type charge ordered

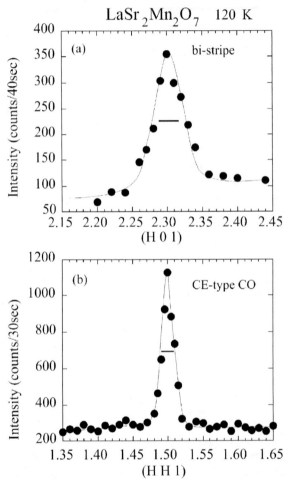

FIGURE 1. The comparison of profiles of the longitudinal distortion (a) and the breathing-type distortion (b). The upper profile was measured along the [1 0 0] direction in the ($h0l$) scattering plane. The lower profile was measured along the [1 1 0] direction in the (hhl) scattering plane. The horizontal line means the instrumental resolution.

state appears due to the phase separation in the intermediate temperature, but it is unstable at low temperatures. In contrast to this, the bi-stripe short-range CO is embedded in the A-type AFM state, and is stable within each two-dimensional FM layers at low temperatures, as was the case for the insulating $x = 0.48$ A-type AFM sample [5, 6]. As shown in Fig. 2 (c), the resistivity of the $x = 0.50$ sample shows a small anomaly between $T_{N,CO}^{A}$ and 80 K. However, an overall trend of the resistivity is a progressive increase with decreasing T, namely, the insulating behavior. We believe that the bi-stripe order embedded in the A-type AFM state is responsible for this behavior.

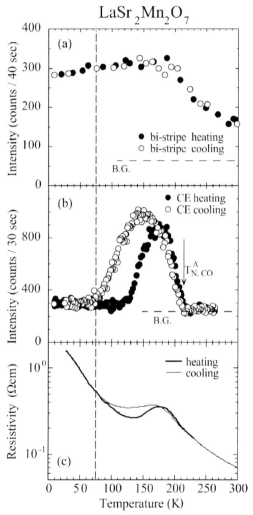

FIGURE 2. The temperature dependence of the bi-stripe order (a), the CE-type charge order (b), and the resistivity (c)(Ref.4).

Now, we focus on the dimensionality of charge ordering and magnetism in the A-type AFM phase and the CE-type CO phase. In the A-type AFM phase, the bi-stripe CO causes a longitudinal distortion with the propagation vector along the [1 0 0] (Mn-O-Mn) direction. This indicates that holes form a number of one-dimensional short segments within the background FM spin arrangement. With respect to magnetism, spins have a two-dimensional FM coupling within the *ab* plane in the A-type AFM state. On the other hand, the CE-type CO, forming a checkerboard pattern of Mn^{3+} and Mn^{4+} ions, causes a breathing-type distortion with the propagation vector along the [1 1 0] (*diagonal*) direction. The orbital degree of freedom in the CE-type CO allows one to recognize a ferromagnetically coupled one-dimensional zigzag chain of Mn^{3+} ions within the CE-type AFM spin pattern, within the framework of the double-exchange mechanism. Between such zigzag chains, spins are coupled antiferromagnetically [7]. Therefore, in the CE-type CO region, the charge ordered pattern forms two-dimensional sheets, while spins form a one-dimensional FM interaction in the *ab* plane. Accordingly, in the light of the dimensionality for the alignment of a charge and spin ordering, we can consider that the A-type phase and the CE-type phase have a *conjugate* relation in the $LaSr_2Mn_2O_7$ system.

Brink *et al.* pointed out that the Coulomb interaction between carriers on the same Mn ions also plays an important role on the formation of the CE-type CO [8]. On the other hand, in the A-type AFM phase, the stripes are stacked in phase within a bilayer despite the Coulomb repulsion. This indicates that the energy gain of superexchange interactions overwhelms the energy loss of the Coulomb interaction in the A-type AFM phase of $LaSr_2Mn_2O_7$. In fact, we evaluated that the volume fraction of the CE-type CO phase is minor even in the intermediate temperature region, being only a quarter to the A-type AFM phase [4]. The dominance of the A-type AFM phase suggests that the superexchange interaction is more significant factor in determining the alignment of holes in the $LaSr_2Mn_2O_7$ system.

CONCLUSION

We have demonstrated that the bi-stripe order exists in the A-type AFM state in the $LaSr_2Mn_2O_7$ system. Holes aligned in line form the short-ranged bi-stripe order over a whole temperature region, and dominate the transport properties. The bi-stripe order is distinct contrast to the CE-type CO in other manganite such as $Pr_{1/2}Ca_{1/2}MnO_3$ and $La_{1/2}Sr_{3/2}MnO_4$ which controls the insulating behavior.

REFERENCES

1. Moreo, A., Yunoki, S., Dagotto, E., *Science* **283**, 202 (1999), and references therein.
2. Kawano, H. *et al.*, *Phys. Rev. Lett.* **78**, 4253 (1997).
3. Tomioka, Y., Asamitsu, A., Moritomo, Y., and Tokura, Y., *J. Phys. Soc. Jpn.* **64**, 3626 (1995).
4. Kubota, M., Yoshizawa, H., Moritomo, Y., Fujioka, H., Hirota, K., and Endoh, Y., *J. Phys. Soc. Jpn.* **68**, 2202 (1999).

5. Kubota, M., Oohara, Y., Yoshizawa, H., Fujioka, H., Shimizu, K., Hirota, K., Moritomo, Y., and Endoh, Y., *J. Phys. Soc. Jpn.* **69**, 1986 (2000).

6. Kubota, M., Fujioka, H., Hirota, K., Ohoyama, K., Moritomo, Y., Yoshizawa, H., and Endoh, Y., *J. Phys. Soc. Jpn.* **69**, 1606 (2000); Kubota, M., Fujioka, H., Ohoyama, K., Hirota, K., Moritomo, Y., Yoshizawa, H., and Endoh, Y., *J. Phys. Chem. Solids* **60**, 1161 (1999); Hirota, K., Moritomo, Y., Fujioka, H., Kubota, M., Yoshizawa, H., and Endoh, Y., *J. Phys. Soc. Jpn.* **67**, 3380 (1998).

7. Goodenough, J.B., *Phys. Rev.* **100**, 564 (1955).

8. Brink, J., Khaliullin, G., Khomskii, D., *Phys. Rev. Lett.* **83**, 5118 (1999)

Microstructure Related to the Charge and Orbital Ordered State in Perovskite Manganites

S. Mori[*], N. Yamamoto[*] and T. Katsufuji[†]

[*]*Department of Physics, Tokyo Institute of Technology, Meguro-ku, Tokyo 152-8551, Japan*
[†]*Department of Advanced Materials Science, University of TokyoBunkyo-ku, Tokyo 113-8656, Japan*

Abstract. We investigated thoroughly change in the microstructure related to the charge ordered (CO) state in half-doped manganites such as $Pr_{0.5}Ca_{0.5}MnO_3$ and $Nd_{0.5}Ca_{0.5}MnO_3$ by the transmission electron microscopy. We found the incommensurate (IC) to commensurate (C) structural phase transition in $Pr_{0.5}Ca_{0.5}MnO_3$ and $Nd_{0.5}Ca_{0.5}MnO_3$. The magnitude of the wave vector characterizing the CO state changes with respect to the temperature. Real space images taken by using the superlattice spots due to the CO state revealed that the IC structure is characterized as the regular arrangement of the discommensurations with the phase slip of π along the [100] direction. Furthermore, the IC CO state is present as the microdomains with the size of 20-30nm.

INTRODUCTION

Perovskite manganites with general formula $A_{1-x}A'_xMnO_3$ ($A=La^{3+}$, Pr^{3+}, Nd^{3+} and Gd^{3+}, rare earth, $A' = Ca^{2+}$) have been investigated extensively because of their unusual physical properties such as a colossal magnetoresistance (CMR) effect and a metal-to-insulator transition due to the charge ordering [1,2]. Recent extensive studies have revealed that an spatial inhomogeneous state of magnetic, electric and/or lattice systems has closely connected with their properties of the manganites [3,4]. It is important to elucidate microscopic nature in the understanding of their properties in the manganites.

The stable charge and orbital ordered state has been found in half-doped manganites such as $La_{0.5}Ca_{0.5}MnO_3$. It is proposed that the CO state is characterized as the alternate arrangement of Mn^{3+} and Mn^{4+} species and the zigzag ordering of d_{z^2}-type orbital involved in the Mn^{3+} species [5]. Recently both electron and neutron diffraction experiments found the presence of the incommensurate (IC) to commensurate (C) structural phase transition in $La_{0.5}Ca_{0.5}MnO_3$, which strongly correlated to the ferromagnetic to antiferromagnetic transition [5,6]. However, the microstructure related to both the IC to C transition and the IC structure in half-doped manganites has not sufficiently understood to date.

In this paper we reported the microstructure related to the CO state in half-doped manganites such as $Pr_{0.5}Ca_{0.5}MnO_3$ and $Nd_{0.5}Ca_{0.5}MnO_3$ by the transmission electron microscopy. The IC CO state is found to be present in the paramagnetic insulator (PI) state in the intermediate temperature region. In addition, we found the IC to C

CP554, *Physics in Local Lattice Distortions*, edited by H. Oyanagi and A. Bianconi
© 2001 American Institute of Physics 1-56396-984-X/01/$18.00

structural phase transition in $Pr_{0.5}Ca_{0.5}MnO_3$ and $Nd_{0.5}Ca_{0.5}MnO_3$, which correlates well with the paramagnetic to antiferromagnetic transition.

EXPERIMENTAL RESULTS AND DISCUSSION

In this work both ceramic and single crystals of $Pr_{0.5}Ca_{0.5}MnO_3$ and $Nd_{0.5}Ca_{0.5}MnO_3$ were used. The observation was carried out by using the JEM-200CX transmission electron microscope (TEM), which is equipped with the low-temperature holder. The samples inside the TEM are located under the magnetic field by the object lens with the magnitude of ~ 2T.

First of all, we examined the electron diffraction (ED) patterns in the CO state of $Pr_{0.5}Ca_{0.5}MnO_3$ [7]. As a result, we found that superlattice reflection spots due to the CO structure appear at the (1/2 0 0)–type positions, in addition to the fundamental orthorhombic structure with the space group Pnma. This means that the CO state at low temperature is characterized as the superstructure with the wave vector of $\bar{\delta} = 1/2\,\bar{a}_0^*$. Thus, we examined change in the wave vector (δ) related to the CO state in the cooling and subsequent warming processes. It is found that the IC CO structure is present in the PI state of $Pr_{0.5}Ca_{0.5}MnO_3$ and $Nd_{0.5}Ca_{0.5}MnO_3$ and the incommensurability (ε) characterizing the IC structure changes in the cooling and subsequent heating processes with a small thermal hysteresis. It should be noticed that the wave vector of the IC structure can be defined as $\bar{\delta} = (1/2 - \varepsilon)\,\bar{a}_0^*$, where $\bar{a}_0^*$ is a lattice vector along the [100] direction in the orthorhombic structure and ε is defined as a deviation of the position of the superlattice spot from its commensurate position in the reciprocal space. Fig. 1 shows the temperature dependence of the incommensurability (ε) obtained in $Pr_{0.5}Ca_{0.5}MnO_3$ (a) and $Nd_{0.5}Ca_{0.5}MnO_3$ (b). As can be clearly seen in Fig. 1(a), ε has a finite value of 0.11 at the transition

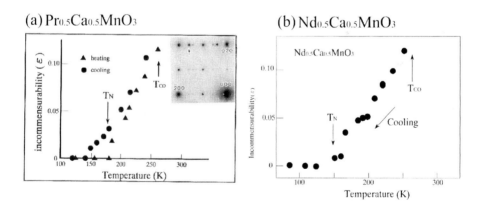

FIGURE 1. Temperature dependence of the incommensurability (ε) characterizing the IC structure in (a) $Pr_{0.5}Ca_{0.5}MnO_3$ and (b) $Nd_{0.5}Ca_{0.5}MnO_3$, respectively. (Inset in (a)) Electron diffraction pattern showing the presence of the IC structure in $Pr_{0.5}Ca_{0.5}MnO_3$.

temperature of 250K and decreases gradually with decreasing temperature. Finally ε reaches zero around 150K and the CO state becomes commensurate and is characterized by the wave vector of $\bar{\delta} = 1/2\,\vec{a}_0^*$ with $\varepsilon = 0$ at lower temperature. On the other hand, on warming from the CO state, ε remains zero until around 180K and increases drastically with increasing temperature. Finally ε reaches a value of 0.11 around the transition temperature of 250K. The transition temperature (180K) of the IC to C structural transition on warming coincides with the paramagnetic to antiferromagnetic one obtained by magnetic and resistivity measurements [8]. In the case of $Nd_{0.5}Ca_{0.5}MnO_3$, the IC to C transition is found at 150K and strongly correlated to the magnetic transition from the paramagnetic to antiferromagnetic phase, as shown in Fig. 1(b).

We examine carefully change in the microstructure related to the CO state by dark field imaging method. Fig. 2 shows typical microstructure related to the CO state in $Nd_{0.5}Ca_{0.5}MnO_3$. Note that the images shown in Fig. 2 are obtained by using the superlattice reflection spots characterizing the CO state and then bright contrast regions in Fig. 2 correspond to ones where the CO state is formed. As shown in Fig. 2(a) taken at 120K, the CO state is seen as a large domain structure with a coherent length of > 100nm. It should be noticed that dark line contrast in Fig. 2(a) should be regarded as the anti-phase boundaries in the commensurate structure. In $Nd_{0.5}Ca_{0.5}MnO_3$, we found the IC structure in the intermediate temperature range between 150K and 250K in the ED experiments, as shown in Fig. 1(b). Fig. 2(b) shows the microstructure taken at 192K in the IC phase. Bright contrast regions with the size of 20 - 30nm, which should be regarded as the CO regions, can be seen in Fig. 2(b). On the other hand, dark contrast regions should be identified as the charge disordered ones regarded as the PI state. On further warming to the transition temperature, the microdomains of the CO state has shrunk down to the size of about 10 - 20nm and finally disappear above the transition temperature of about 260K. The image in Fig. 2(b) revealed that the CO state with the IC wave vector exists as the microdomains with the size of 10 - 30nm in the PI state. That is, the state in the intermediate temperature region is characterized as fine mixture of the CO state and the charge disordered one.

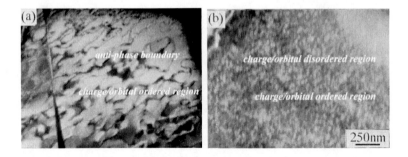

FIGURE 2. Typical microstructure related to the CO state in $Nd_{0.5}Ca_{0.5}MnO_3$. The images are obtained at (a) 120K and (b) 192 K, respectively.

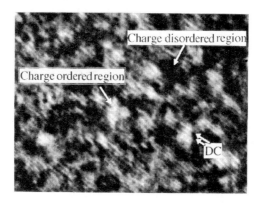

FIGURE 3. Microstructure in the IC CO phase of $Pr_{0.5}Ca_{0.5}MnO_3$ showing the regular arrangement of DC's along the [100] direction. The image is taken at 200K in the warming process.

In order to elucidate the IC structure, we carefully examined microstructure related to the CO state in the intermediate temperature region between 180K and 250K in $Pr_{0.5}Ca_{0.5}MnO_3$. We found the presence of the regular arrangement of the discommensurations (DC's) in the IC structure. Fig. 3 is a microstructure related to the CO state with the IC wave vector ($\varepsilon = 0.045$) taken at 200K in $Pr_{0.5}Ca_{0.5}MnO_3$. As indicated by arrows in Fig. 3, dark line contrasts which should be identified as the DC's with a phase slip of π can be clearly seen. It should be noticed that average distance of two neighboring dark line contrasts is about 12nm, which is almost consistent that estimated from the incommensurability (ε) obtained experimentally, Note that ε is 0.045 at 200K on warming. This image shown here revealed that the IC structure is characterized as the regular arrangement of the DC's with the phase slip of π.

In summary, we carefully examined the microstructure related to the CO state in $Pr_{0.5}Ca_{0.5}MnO_3$ and $Nd_{0.5}Ca_{0.5}MnO_3$ by the TEM. We found the IC to C structural phase transition, which strongly correlated with the paramagnetic to antiferromagnetic transition. Furthermore, the CO state with the IC wave vector exists as the microdomains with the size of 10 - 30nm, and the IC structure is characterized as the regular arrangement of the DC's with the phase slip of π.

ACKNOWLEDGMENTS

The authors are grateful to Prof. Y. Yamada at Waseda University for useful discussion. This work was supported by a Grant-in-Aid for Priority Areas of Scientific Research, "Novel properties in transition metal oxides" from the Ministry of Education, Science, Sports and Culture.

REFERENCES

1. Chahara, K. *et al.*, *Appl. Phys.Lett.* **62**, 780 (1993).
2. For a review, see *Colossal Magnetoresistance Oxides*, edited by Tokura, Y., Gorden & Breach Science Publishers, New York, (1999).
3. Uehara, M. *et al*, *Nature* **399**, 560 (1999).
4. Moreo, A. *et al.*, *Science* **283**, 2014 (1999).
5. Radaelli *et al.*, *Phys. Rev. B* **55**, 3015 (1997).
6. Chen, C.H., and Cheong, S-W., *Phys.Rev.Lett.* **76**, 4042 (1996); Mori, S., Chen, C.H., and Cheong, S-W., *Phys. Rev. Lett.* **81**, 3972 (1998).
7. Mori, S. *et al.*, *Phys. Rev. B* **59**, 13555 (1999).
8. Tomioka, Y. *et al.*, *Phys. Rev. B* **53**, 1689 (1996).

Heat Transport Anomalies around Ferromagnetic and Charge-order Transitions in La$_{1-x}$Ca$_x$MnO$_3$

H. Fujishiro and M. Ikebe

Faculty of Engineering, Iwate University, Morioka 020-8551, Japan

Abstract. The thermal conductivity κ(T) and the thermal dilatation dL(T)/L of La$_{1-x}$Ca$_x$MnO$_3$ have been measured and the characteristic anomalies of κ correlated to those of dL/L have been found. κ was substantially enhanced below the ferromagnetic (FM) transition temperature T$_c$ in FM metallic samples (0.20 ≤ X ≤ 0.45). In contrast, κ(T) was depressed below T$_{CO}$ in charge ordered (CO) samples (X ≥ 0.50). The enhancement and depression were closely connected with the local distortion of MnO$_6$ octahedra through the relaxation of the Jahn-Teller (J-T) distortion by mobile holes.

INTRODUCTION

As recent revived studies on perovskite-based manganese have confirmed, (RE$_{1-x}$AE$_x$)MnO$_3$-type crystals (RE = trivalent rare-earth ions such as La, Pr: AE = divalent alkaline-earth ions such as Ca, Sr, Ba) undergo a variety of dramatic phenomena such as the colossal magnetoresistance (CMR) and the insulator-metal (I-M) transition [1]. Recently it has been widely recognized that the local lattice distortions of the MnO$_6$ octahedra play an important role in determining the transport properties of the doped holes and the complex behaviors of magnetic and structural transitions. The thermal conductivity κ is a valuable tool to investigate the effect of the lattice dynamics near the phase transitions of the manganite system [2-5]. For the samples in which the simultaneous ferromagnetic (FM) and I-M transition occurs, κ(T) increases abruptly below the FM transition temperature T$_c$, while no such κ(T) anomaly is observed for the FM insulating samples. Cohn *et al.* proposed that the lattice thermal resistance is mainly governed by the local distortions of the MnO$_6$ octahedra [2]. The κ(T) enhancement should not be linked only to the ferromagnetism but rather to the electronic transition to metallic state which suppresses significantly the amplitude of Jahn-Teller (J-T) distortion.

In this paper, we investigate the thermal conductivity κ(T) of La$_{1-x}$Ca$_x$MnO$_3$ and discuss the κ(T) anomalies around the FM and the charge-order (CO) transitions. We discuss the origin of the κ(T) anomalies from the view point of the lattice distortions referring to the results of the thermal dilatation dL(T)/L.

CP554, *Physics in Local Lattice Distortions*, edited by H. Oyanagi and A. Bianconi
© 2001 American Institute of Physics 1-56396-984-X/01/$18.00

EXPERIMENTAL

The La$_{1-x}$Ca$_x$MnO$_3$ (LCMO) (X = 0.15 ~ 0.60) samples were prepared by a standard solid-state reaction method which was described in detail elsewhere [6,7]. The thermal conductivity κ(T) was measured by a steady-state heat flow method [8]. The dilatation dL(T)/L was measured using a strain-gauge method. An automatic measuring system was used for these measurements including a Gifford-McMahon cycle helium refrigerator as a cryostat.

RESULTS AND DISCUSSION

Figure 1 shows the temperature dependence of the thermal conductivity κ(T) for the LCMO (0.15 ≤ X ≤ 0.30) samples. The electronic contribution κ$_e$ as estimated by the Wiedemann-Franz law is quite small in the present samples and the heat conduction is overwhelmingly due to phonons. For the ferromagnetic X = 0.15 sample, whose electrical resistivity ρ(T) behaves nonmetallic, κ(T) is low and no anomaly is observed around T$_c$. κ(T) of La$_{0.85}$Ca$_{0.15}$MnO$_3$ suggests the existence of very strong phonon scattering over the entire temperature range. On the other hand, for the X ≥ 0.20 samples which show the FM metallic behavior [6], we can see that the κ(T) shows an enhancement for T < T$_c$, which increases and becomes step-like just below T$_c$ for X = 0.25 and 0.30. These results are consistent with the reported data [2]. The thermal dilatation dL(T)/L for the same samples are shown in Fig. 2. dL(T)/L of the X = 0.15 sample does not show any anomaly at T$_c$. In contrast, the samples which show the κ(T) anomaly exhibits a distinct lattice contraction below T$_c$. The lattice contraction may mean the relaxation of the J-T distortion as a result of the increasing itinerancy of holes, which in turn results in a considerable reduction in the phonon scattering [2].

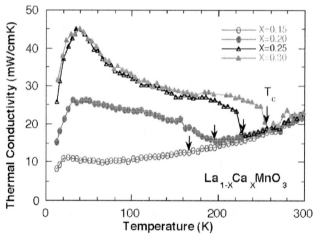

FIGURE 1. The temperature dependence of the thermal conductivity κ(T) for the LCMO (0.15 ≤ X ≤ 0.30) samples.

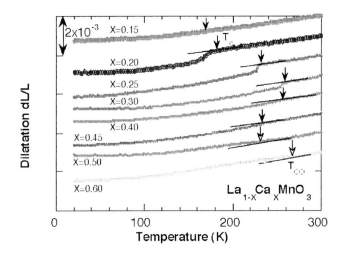

FIGURE 2. The temperature dependence of the thermal dilatation dL(T)/L for the LCMO ($0.15 \leq X \leq 0.60$) samples.

Thus the step-like enhancement in $\kappa(T)$ just below T_c may originate from the notable relaxation of the J-T distortion around T_c.

In Fig. 1, $\kappa(T)$ of $X = 0.20$, 0.25 and 0.30 LCMO takes a clear maximum at $T_m \sim$ 35K. Usually, the origin of such a maximum of the lattice thermal conductivity is ascribed to the phonon umklapp scattering. However, the umklapp process does not explain either the anomalous step-like jump of κ or $d\kappa(T)/dT > 0$ for $T > T_c$ of these samples. As well as the step-like jump at $T \approx T_c$, the maximum at T_m and $d\kappa(T)/dT < 0$ for $T_m < T < T_c$ should also originate from the gradual relaxation process of the J-T distortion associated with the gradual reduction of $\rho(T)$ of metallic samples with decreasing temperature. It is to be noticed that this effect of the gradual relaxation cannot be expected in non-metallic samples such as $La_{0.85}Ca_{0.15}MnO_3$.

Figure 3 shows the thermal conductivity $\kappa(T)$ for the $0.40 \leq X \leq 0.60$. For the $X = 0.40$ and 0.45 samples which are FM and metallic, $\kappa(T)$ shows the upturn below T_c, though the step-like jump at $T \approx T_c$ almost disappears. Because of reduction in J-T active Mn^{3+} ion concentration with increasing X, the effect of relaxation of the local lattice distortion may be somewhat weakened in these samples.

For the $X = 0.50$ sample, in which the FM state vanishes and the CO state appears, the magnitude of $\kappa(T)$ drastically decreases and the $\kappa(T)$ reduction is observable around the CO transition temperature T_{CO} (= 230K). Thus, the $\kappa(T)$ behaviors of $X = 0.45$ and 0.50 samples show a clear contrast reflecting the difference in the ordered phase. The $X = 0.60$ sample shows the similar $\kappa(T)$ behavior as $X = 0.50$. As shown in Fig. 2, for the FM samples ($X \leq 0.45$), dL(T)/L decreases below T_c with decreasing temperature, while for the CO samples ($X \geq 0.50$), the thermal dilatation shows quite a opposite character; dL(T)/L abruptly increases below T_{CO}. The increase of dL(T)/L below T_{CO} in contrast to the decrease below the Curie temperature T_c of the metallic

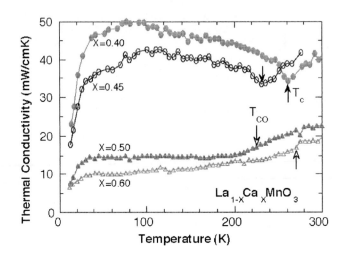

FIGURE 3. The temperature dependence of the thermal dilatation $\kappa(T)$ for the LCMO ($0.40 \leq X \leq 0.60$) samples.

samples suggests that the lattice distortion rather increases below T_{CO}. It should also be noticed that $\rho(T)$ increases below T_{CO}, which may inhibit the relaxation of the J-T distortion due to mobile carriers. Therefore the $\kappa(T)$ reduction below T_{CO} may come from the increase of the lattice distortion.

In summary, the experimental results of the thermal conductivity $\kappa(T)$ and of the thermal dilatation $dL(T)/L$ exhibited anomalies closely correlated each other. The $\kappa(T)$ anomalies characteristic of each phase, *i.e.*, the FM insulator, FM metal and charge ordered, are systematically understood on the basis of the relaxing (or enhancing) process of the local Jahn-Teller distortions by the mobile carriers. In this way, the heat transport properties of the manganite system sensitively reflect the microscopic phase transitions such as spin, charge and lattice order/disorder.

REFERENCES

1. Tomioka, Y., Asamitsu, A., Moritomo, Y., Kuwahara, H., and Tokura, Y., *Phys. Rev. Lett.* **74**, 5108-5112 (1995).
2. Cohn, J. L., Neumeier, J. J., Popoviciu, C. P., McClellan, K. J., and Leventouri, Th., *Phys. Rev. B* **56**, R8495-R8498 (1997).
3. Visser, D. W., Ramirez, A. P., and Subramanian, M. A., *Phys. Rev. Lett.* **78**, 3947-3950 (1997).
4. Hejtmanek, J., Jirak, Z., Arnold, Z., Marysko, M., Krupicka, S., Martin, C., and Damay, F., *J. Appl. Phys.* **83**, 7204-7206 (1998).
5. Ikebe, M., Fujishiro, H., and Konno, Y., *J. Phys. Soc. Jpn.* **67**, 1083-1085 (1998).
6. Fujishiro, H., Fukase, T., Ikebe, M., and Kikuchi, T., *J. Phys. Soc. Jpn.* **68**, 1469-1472 (1999).
7. Fujishiro, H., Ikebe, M., Konno, Y., and Fukase, T., *J. Phys. Soc. Jpn.* **66**, 3703-3705 (1997).
8. Ikebe, M., Fujishiro, H., Naito, T., and Noto, K., *J. Phys. Soc. Jpn.* **63**, 3107-3114 (1994).

X-ray Diffuse Scattering from (Nd$_{1-y}$Sm$_y$)$_{1-x}$Sr$_x$MnO$_3$ and Pr$_{1-x}$Ca$_x$MnO$_3$

S. Shimomura, T. Tonegawa, K. Torashima, K. Tajima, N. Wakabayashi, H. Kuwahara[†,*], Y. Tomioka[†], and Y. Tokura[†,¶]

Department of Physics, Faculty of Science and Technology, Keio University, 3-14-1 Hiyoshi, Kohoku-ku, Yokohama 223-8522, Japan
[†]*Joint Research Center for Atom Technology (JRCAT), Tsukuba 305-8562, Japan*
[¶]*Department of Applied Physics, University of Tokyo, Tokyo 113-8656, Japan*
[*]*Present address: Department of Physics, Sophia University, 7-1 Kioicho, Chiyodaku, Tokyo 102-8554, Japan, and PRESTO, JST, Japan.*

Abstract. X-ray diffuse scattering measurements on Pr$_{1-x}$Ca$_x$MnO$_3$ and (Nd$_{1-y}$Sm$_y$)$_{1-x}$Sr$_x$MnO$_3$ have been performed and characteristic patterns of the intensity distribution were observed. In Pr$_{1-x}$Ca$_x$MnO$_3$ exhibiting the charge-ordering transition, the intensity maxima were observed around $(0, \pm \zeta, 0)$ and $(\pm \zeta, 0, 0)$ with $\zeta \sim 0.4$ at 300K. The value of ζ increases with decreasing temperature and discontinuously changes to 0.5 at the transition temperature. In (Nd$_{1-y}$Sm$_y$)$_{1-x}$Sr$_x$MnO$_3$, the diffuse scattering profiles have intensity maxima around $\zeta \sim 0.3$. The intensity increases with decreasing temperature above the ferromagnetic insulator-metal transition temperature (T_C), but its intensity vanishes below T_C. The local Jahn-Teller distortion associated with an independent polaron can account for the intensity patterns observed in both systems except for the local intensity maxima.

INTRODUCTION

Manganese oxide compounds with distorted perovskite structures, $R_{1-x}A_x$MnO$_3$ (R is a trivalent rare earth ion and A is a divalent alkali earth ion), have attracted interest because of the colossal magnetoresistance (CMR). It has been pointed out that the electron-lattice interaction is significant in these compounds and the formation of lattice polarons, carriers self-localized by lattice distortion, may be relevant to the CMR effect [1]. In order to investigate the local lattice distortion around localized charges, we have measured the diffuse scattering in Pr$_{1-x}$Ca$_x$MnO$_3$ and (Nd$_{1-y}$Sm$_y$)$_{1-x}$Sr$_x$MnO$_3$ ($x \sim 0.5$) systems.

The former system of Pr$_{1-x}$Ca$_x$MnO$_3$ ($0.3 \leq x \leq 0.5$) shows the charge-ordering transition and the antiferromagnetic transition [2]. Below the charge-ordering transition temperature (T_{CO}), the Mn^{3+} and Mn^{4+} ions are ordered alternatively in the a-b plane [3]. The cooperative Jahn-Teller (JT) ordering simultaneously occurs, resulting in the doubling of the crystallographic unit cell. The purpose of the present study is to reveal the existence of the local lattice distortions and the development of the correlation between the distortion fields above T_{CO} [4].

CP554, *Physics in Local Lattice Distortions*, edited by H. Oyanagi and A. Bianconi
© 2001 American Institute of Physics 1-56396-984-X/01/$18.00

In the other system of $(Nd_{1-y}Sm_y)_{1-x}Sr_xMnO_3$ ($x \sim 0.5$), the average ionic radius of the R site, depending on y, changes the transport properties [5]. The crystal with $y = 0$ undergoes the paramagnetic insulator to ferromagnetic metal transition at T_C and then to the charge-ordering antiferromagnetic transition at T_{CO}. As y increases, a large decrease in the resistivity becomes prominent at T_C. For $0.85 \leq y \leq 0.95$, the charge-ordered phase disappears. We measured the diffuse scattering from the crystals with $y = 0.875$ [6] having T_C and those with $y = 0.75$ having both T_C and T_{CO}.

EXPERIMENTS

Single crystals of $Pr_{1-x}Ca_xMnO_3$ ($x = 0.35, 0.4, 0.5$), $(Nd_{0.875}Sm_{0.125})_{0.52}Sr_{0.48}MnO_3$, and $(Nd_{0.75}Sr_{0.25})_{0.5}Sr_{0.5}MnO_3$ were grow by the floating zone method. These compounds have orthorhombic structures with the pseudocubic relation among the lattice parameters specified by $a \sim b \sim c/\sqrt{2} \sim \sqrt{2}a_c$, where a_c is a lattice parameter of the cubic perovskite. All the samples were composed of pseudocubic domains. X-ray scattering measurements were performed using a rotating anode source equipped with a two-axis diffractometer. The incident beam with Mo-K_a radiation (50 kV, 150 mA) was monochromatized by the 002 reflection of a pyrolytic graphite crystal. The sample sizes were approximately $2 \times 2 \times 0.5$ mm^3. The sample was mounted on a cold head of a closed-cycle herium refrigerator. The intensity distribution patterns of the diffuse scattering were observed in the reciprocal ($hk0$) plane. Only the measurement of the pattern around (10, 4, 0) in $Pr_{0.6}Ca_{0.4}MnO_3$ was performed by using synchrotron radiation (SPring-8, BL02B1).

RESULTS AND DISCUSSION

The local lattice distortion induced around localized charges causes the diffuse scattering. Fig. 1(a) shows the temperature dependence of the diffuse scattering profiles along the [010] direction through (10, 0, 0) point in $Pr_{0.5}Ca_{0.5}MnO_3$. The diffuse scattering intensity with local intensity maxima around (10, $\pm \zeta$, 0) with $\zeta \sim 0.4$ are clearly seen at 300K. With decreasing temperature, the local intensity maxima grow, and ζ shifts towards 0.5. At T_{CO} of 235 K, ζ discontinuously changes to the commensurate value of 0.5 and the width decreases to an experimental resolution. Similar diffuse scattering was also observed in the $x = 0.35$ and 0.4 crystals. The temperature dependences of ζ are summarized in Fig. 1(b). All the data were collected for decreasing temperature. The wave vector ζ is incommensurate above T_{CO} even in the $x = 0.5$ crystal having the stoichiometric hole-concentration corresponding to the charge-ordered state. In contrast, below T_{CO} the wave vector locks into the commensurate value of 0.5 even in the $x = 0.35$, 0.4 crystals.

In $(Nd_{1-y}Sm_y)_{0.52}Sr_{0.48}MnO_3$ crystals with $y = 0.875$, the diffuse scattering was also observed above T_C as shown in Fig. 2(a). The profiles seem to have local intensity maxima around (10, ±0.3, 0). Its intensity increases with decreasing temperature but abruptly disappears below T_C of 133K [Fig. 2(b)]. This temperature dependence is quite similar to that of the resistivity shown in Fig. 2(c). This indicates that the local lattice distortion arises from the localized charges and it spreads out in a large spatial region in the ferromagnetic metallic phase. In the crystal with $y = 0.75$ having both T_C and T_{CO}, similar diffuse scattering was observed above T_C. In contrast to $Pr_{1-x}Ca_xMnO_3$, the apparent diffuse scattering was not detected in the temperature range $T_{CO} < T < T_C$, because the ferromagnetic metallic phase is stable above T_{CO} in the $y = 0.75$ crystal.

In order to determine the pattern of distortion fields, we measured patterns of the intensity distributions in the $(hk0)$ reciprocal plane. The characteristic patterns were observed depending on the reciprocal points. As an example, the pattern around (10, 4, 0) in $Pr_{0.6}Ca_{0.4}MnO_3$ is shown in Fig. 3(a). The pattern observed around (8, 0, 0) extends along the [010] direction and that around (8, 8, 0) shows a butterfly-shaped pattern extending along the [010] and [100] directions. Similar patterns were also

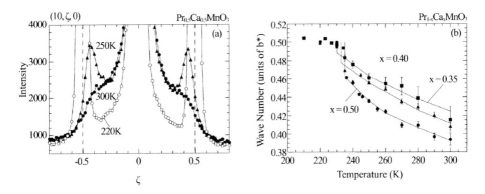

FIGURE 1. (a) Temperature dependence of diffuse scattering profiles in $Pr_{0.5}Ca_{0.5}MnO_3$.
(b) Temperature dependence of the modulation wave number in $Pr_{1-x}Ca_xMnO_3$ ($x = 0.35, 0.4, 0.5$).

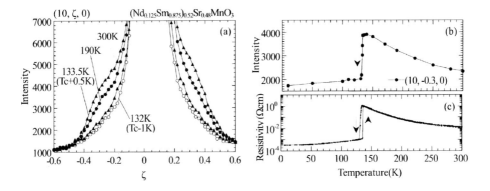

FIGURE 2. (a) Temperature dependence of diffuse scattering profiles in $(Nd_{1-y}Sm_y)_{0.52}Sr_{0.48}MnO_3$ ($y = 0.875$). Temperature dependences of (b) the diffuse scattering intensity and (c) the resistivity.

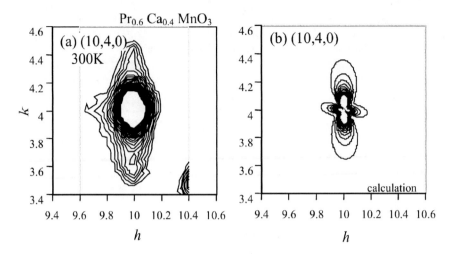

FIGURE 3. (a) Contour plot of diffuse scattering intensities around $(10, 4, 0)$ in $Pr_{0.6}Ca_{0.4}MnO_3$. (b) Calculated intensity distribution based on the assumption of the local Jahn-Teller distortion.

observed in $(Nd_{1-y}Sm_y)_{1-x}Sr_xMnO_3$ ($y = 0.875, 0.75$ and $x \sim 0.5$). Fig. 3(b) shows the calculated intensity based on the assumption that the local lattice distortion arises from the JT effect of a $Mn^{3+}O_6$ octahedron having two long Mn-O bonds and four short Mn-O bonds. The observations can be qualitatively interpreted by the calculations except for the local intensity maxima. This indicates that the local lattice distortion due to polarons is associated with the JT distortion. The intensity maxima probably arise from the correlation between the distortion fields.

ACKNOWLEDGMENTS

We would like to thank H. Ohtsuka, N. Ikeda, T. Shobu, Y. Noda for their assistance in the synchrotron radiation experiment. This work was partly supported by a Grant-in-Aid for Science Research on Priority Areas from the Ministry of Education, Science, Sports and Culture, Japan. The synchrotron radiation experiment was performed at the SPring-8 with the approval of the JASRI (Proposal No. 1999B0263-ND-np).

REFERENCES

1. Millis, A.J., *Nature* (London) **392**, 147 (1998); and references therein.
2. Tomioka, Y., Asamitsu, A., Kuwahara, H., Moritomo, Y., and Tokura, Y., *Phys. Rev. B* **53**, R1689-R1692 (1996).
3. Zimmermann, M.V., Hill, J.P, Gibbs, Doon, Blume, M., Casa, D., Keimer, B., Murakami, Y., Tomioka, Y., and Tokura, Y., *Phys. Rev. Lett.* **83**, 4872-4875 (1999).

4. Shimomura, S., Tonegawa, T., Tajima, K., Wakabayashi, N., Ikeda, N., Shobu, T., Noda, Y., Tomioka, Y., and Tokura, Y., *Phys. Rev. B* **62**, 3875-3878 (2000).

5. Kuwahara, H., Moritomo, Y., Tomioka, Y., Asamitsu, A., Kasai, M., Kumai, R., and Tokura, Y., *Phys. Rev. B* **56**, 9386-9396 (1997).

6. Shimomura, S., Wakabayashi, N., Kuwahara, H., and Tokura, Y., *Phys. Rev. Lett.* **83**, 4389-4392 (1999); similar observations of the diffuse scattering in $La_{1.2}Sr_{1.8}Mn_2O_7$ exhibiting the CMR have been reported by Vasiliu-Doloc, L., Rosenkranz, S., Osborn, R., Sinha, S.K., Lynn, J.W., Mesot, J., Seeck, O., Lee, W.-K., and Mitchell, J.F., *Phys. Rev. Lett.* **83**, 4393-4396 (1999).

Stripe Structure in the Ferromagnetic Insulating Phase of La$_{1-x}$Sr$_x$MnO$_3$

M. Arao, Y. Koyama, Y. Inoue^a and Y. Moritomo^b

Kagami Memorial Laboratory for Materials Science and Technology, Waseda Univ.,
and Dept. of Materials Science and Engineering, Waseda Univ.,
Shinjuku-ku, Tokyo 169-8555, Japan.
^{a}NISSAN ARC, Ltd., Yokosuka, Kanagawa 237-0061, Japan.
^{b}CIRSE, Nagoya University, Nagoya 464-8601, Japan.

Abstract. The atomic displacement involved in the crystal structure of the ferromagnetic insulating phase in La$_{1-x}$Sr$_x$MnO$_3$ has been reexamined by transmission electron microscopy, using a $x = 0.15$ single crystal. An analysis of the electron diffraction patterns taken from the $x = 0.15$ samples showed that the atomic displacement consists of a superposition of the Jahn-Teller displacement with $\vec{q} = <\frac{1}{2}\frac{1}{2}0>_C$ and the inverse Jahn-Teller one with $\vec{q} = <\frac{1}{2}\frac{1}{2}\frac{1}{4}>_C$, where the subscript c denotes the cubic Brillouin zone. The crystal structure involving these displacements can be identified as a stripe structure.

INTRODUCTION

It was theoretically suggested that when holes are doped into a correlated insulator, a stripe structure characterized by a charge modulation of doped holes appears in a transitional state between the antiferromagnetic insulating and metallic phases [1]. Such stripe structures have been reported in 3d-transition metal oxides such as La$_{1-x}$Ca$_x$MnO$_3$ [2]. In the case of La$_{1-x}$Sr$_x$MnO$_3$, there is the transitional state around $x = \frac{1}{8}$, which is called ferromagnetic insulating (FI) phase. Yamada *et al.* investigated the features of the crystal structure of the FI phase by means of neutron diffraction and found that superlattice reflections appear at $\vec{q} = [00\frac{1}{2}]_o$ in the reciprocal space of the FI phase, where O denotes a high-temperature P_{bnm} structure without the Jahn-Teller distortion, the O structure [3]. In addition, a recent synchrotron X-ray diffraction experiment indicated that there exists a new type of orbital ordering in the FI phase [4]. In spite of these works, the presence of the stripe structure in the transitional state of La$_{1-x}$Sr$_x$MnO$_3$ is still open to question. The reason is that the O phase exhibits a complex domain structure. Then we have reexamined the atomic displacement involved in the crystal structure of the FI phase by transmission electron microscopy, using a single variant region of a La$_{0.85}$Sr$_{0.15}$MnO$_3$ single crystal.

CP554, *Physics in Local Lattice Distortions*, edited by H. Oyanagi and A. Bianconi
© 2001 American Institute of Physics 1-56396-984-X/01/$18.00

EXPERIMENTAL PROCEDURE

The $La_{0.85}Sr_{0.15}MnO_3$ single crystal grown by a floating-zone technique was used in this work. The details of the growth procedure and its physical properties were already described in the previous paper [5]. Specimens for transmission-electron-microscopy observation were thin films obtained by Ar-ion thinning of the single crystal. The observation was carried out by means of an H-800 transmission electron microscope with the help of a cooling holder. Imaging plates were also used as a recording medium to avoid sample drift during exposure, and to measure intensities of diffraction spots.

RESULTS

Electron diffraction patterns taken from a single variant in the $La_{0.85}Sr_{0.15}MnO_3$ single crystal at 15K are shown in Fig. 1, together with an intensity profile along the [001] direction through the $\bar{1}00$ reflection. Electron incidences in Fig. 1(a) and (b) are, respectively, parallel to the $[0\bar{1}0]$, and [001] direction and diffraction spots are indexed in terms of the high-temperature O structure. In addition to the fundamental reflections due to the O structure, there are new diffraction spots indicated by arrows A and B in Fig. 1(a). An interesting feature is that the intensity of spot B is by $\sim 10^{-1}$ weaker than that of spot A. In the profile, these reflections are clearly seen. An analysis of the diffraction patterns indicated that spot A with the strong intensity is identified as a forbidden reflection of the O structure, while spot B with the weak intensity is located at $\vec{q} = [00\frac{1}{2}]_o$.

We checked the extinction rule of these new reflections at 15K. Fig. 2 is one part of the constructed reciprocal lattice of the FI phase. From the lattice, it was found that the forbidden reflections exist only in the $(0\bar{1}0)^*$ plane, and that the superlattice

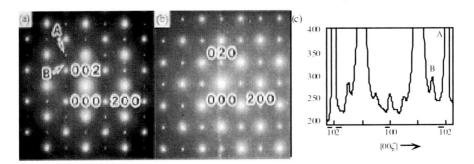

FIGURE 1. Electron diffraction patterns taken from the single-variant region of the $x = 0.15$ sample at 15K. Electron incidences in (a) and (b) are, respectively, parallel to the $[0\bar{1}0]$, and [001] directions, and diffraction spots are indexed in terms of the O structure. An intensity profile along the [001] direction through the $\bar{1}00$ reflection is also shown in (c).

reflections are present at the $\vec{q} = <00\frac{1}{2}>_o$ position, as was mentioned above. An important feature of the forbidden and superlattice reflections is that these reflections are missing along the [001] direction through the origin 000. It should here be remarked that the forbidden reflections can be divided into two groups: the $\vec{q} = <\frac{1}{2}\frac{1}{2}0>_c$ - and $\vec{q} = <\frac{1}{2}00>_c$ -type reflections in the cubic Brillouin zone. Since the $\vec{q} = <\frac{1}{2}\frac{1}{2}0>_c$ - and $\vec{q} = <\frac{1}{2}00>_c$ -type positions are equivalent in the O structure, one of the irreducible representations at $\vec{q} = <\frac{1}{2}\frac{1}{2}0>_c$ is associated with the atomic displacement, which gives rise to the forbidden reflections. In addition, the presence of the superlattice reflections at $\vec{q} = <00\frac{1}{2}>_o = <00\frac{1}{4}>_c$ indicates that the FI structure involves another atomic displacement with the wave vector of $\vec{q} - <00\frac{1}{2}>_o$. The missing of the forbidden and superlattice reflections along the [001] direction through the origin suggested that these atomic displacements have a transverse-wave character.

DISCUSSION

The obtained experimental data indicated that there exist two atomic displacements in the crystal structure of the FI phase. Among these two displacements, the atomic displacement associated with the irreducible representation at $\vec{q} = <\frac{1}{2}\frac{1}{2}0>_c$ was suggested to give rise to the forbidden reflections. Then we first discuss the displacement for the forbidden reflections. In order to determine the displacement, we took into account both the transverse-wave character of the displacement and the presence of orbital ordering in the FI phase. As a result of the checking all representation at $\vec{q} = <\frac{1}{2}\frac{1}{2}0>_c$, the M_2 irreducible representation was understood to be responsible for the displacement. That is, the (x^2-y^2)-type Jahn-Teller displacement at the $\vec{q} = <\frac{1}{2}\frac{1}{2}0>_c$ occurs in the FI phase. On the other hand, displacement giving rise to the superlattice reflections at $\vec{q} = <00\frac{1}{2}>_o = <00\frac{1}{4}>_c$ must be related to charge ordering of doped holes. Because doped holes release the Jahn-Teller distortion, an appropriate candidate is the inverse Jahn-Teller displacement at $\vec{q} = <\frac{1}{2}\frac{1}{2}\frac{1}{4}>_c$, as was suggested by Yamada et al.[3] The atomic displacement involved in the FI phase is therefore to be a superposition of the Jahn-Teller displacement with

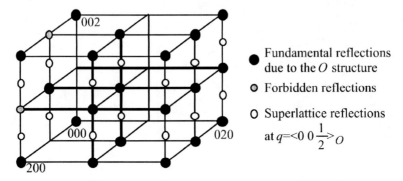

FIGURE 2. Part of the constructed reciprocal lattice of the FI phase.

$\vec{q} = <\frac{1}{2}\frac{1}{2}0>_c$ and the inverse Jahn-Teller one with $\vec{q} = <\frac{1}{2}\frac{1}{2}\frac{1}{4}>_c$. The atomic displacement model in the FI phase for $x = \frac{1}{8}$ is depicted in Fig. 3. The solid and dashed lines, respectively, correspond to the MnO_6 octahedra in the crystal structure of the FI phase and the O structure. The atomic shift of each ion is represented by an arrow. From the figure, the crystal structure of the FI phase is understood to be regarded as a stripe structure along the [001] direction, which consists of the severely, moderately, and little distorted layers. In this stripe structure, doped holes should mainly occupy the little distorted layer. From the fact that the intensities of the superlattice reflections are much weaker than those of the forbidden reflections, however, there exist a large disorder in the charge modulation.

CONCLUSION

The present work showed that the atomic displacement in the crystal structure of the FI state is characterized by a superposition of the Jahn-Teller displacement with $\vec{q} = <\frac{1}{2}\frac{1}{2}0>_c$ and the inverse Jahn-Teller displacement with $\vec{q} = <\frac{1}{2}\frac{1}{2}\frac{1}{4}>_c$, where subscript c denotes the cubic Brillouin zone. That is, the crystal structure can be regarded as a stripe structure consisting of four layers along the [001] direction. A characteristic feature of this stripe structure is that the large disorder should be involved in the charge modulation.

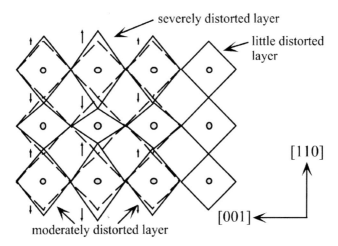

FIGURE 3. Atomic-displacement model in the ferromagnetic insulating phase: The schematic diagram is viewed from the [1$\bar{1}$0] direction. In the diagram, the atomic shift of each ion is overdrawn and the tilting of the octahedra is omitted.

REFERENCES

1. Emery, V. J., and Kivelson S. A., *Physica C* **235-240**, 189-192 (1994).
2. Mori, S., Chen, C. H., and Cheong, S. W., *Nature* **392**, 473-476 (1998).
3. Yamada, Y., Hino, S., Nohdo, S., Kanao, R., Inami, T., and Katano, S., *Phys. Rev. Lett.* **77**, 904-907 (1996).
4. Endoh, Y., Hirota, K., Ishikawa, S., Okamoto, S., Murakami, Y., Nishizawa, A., Fukuda, T., Kimura, H., Kaneko, K., and Maekawa, S., *Phys. Rev. Lett.* **82**, 4328-4331 (1999).
5. Urushibara, A., Moritomo, Y., Arima, T., Asamitsu, A., Kido, G., and Tokura, Y., *Phys. Rev. B* **51**, 14103-14109 (1995).

4. LATTICE EFFECTS ON ELECTRONIC STRUCTURE

Temperature Dependent Lattice Distortion in La$_{1-x}$Ce$_x$Ru$_2$ by Ru K-edge Absorption Spectroscopy

Z. Y. Wu[*§†], N. L. Saini[§] and A. Bianconi[§]

[*]Beijing Synchrotron Radiation Facility, Beijing, China
[§]Unitá INFM and Dipartimento di Fisica, Università di Roma "La Sapienza"
P.le Aldo Moro 2, 00185 Roma, Italy
[†]Laboratori Nazionali di Frascati, INFN, 00044, Frascati, Italy

Abstract. X-ray absorption near edge structure (XANES) spectroscopy has been used to investigate local lattice and electronic structure of the intermetallic La$_{1-x}$Ce$_x$Ru$_2$ system. Origin of different features is explored by one electron multiple scattering calculations. The experimental and theoretical analysis demonstrate that the Ce doping in La$_{1-x}$Ce$_x$Ru$_2$ has direct influence on the states at the Fermi level and hence the hybridization between the f and d states induced by doping.

INTRODUCTION

X-ray absorption near edge structure (XANES) was introduced as a tool to investigate intermediate short-range order and electronic structure in complex materials [1]. The x-ray absorption coefficient $\mu(E)$ is given by the product of the matrix element and the joint density of states for the electronic transitions from the initial to final states. The dipole matrix element from the initial state, the core level of well defined symmetry, selects the partial density of final states for the allowed electronic transitions. The XANES spectroscopy probes the final states in an energy range of about 50 eV above the chemical potential. While at the Fermi level the mean free path is large, it decreases rapidly with increasing the energy of the final state since it is strongly scattered by the many body electronic excitations and its mean free path becomes of the order of 5 Å. Therefore the XANES spectra can be solved in the real space describing the final state as an outgoing spherical wave which interferes with the waves backscattered from the neighbouring atoms within a cluster of atoms in the intermediate range of the order of 5 Å [2-4].

The hybridization of localized f electrons (magnetism) and itinerant d electrons (superconductivity) drives intermetallics towards the border line between magnetic and metallic systems. Among these intermetallic systems, LaRu$_2$ provides an interesting case [5-8] since it is formally a 4f^0 system showing a superconducting transition at 4.4 K. Systematic studies of Ce substitution in La$_{1-x}$Ce$_x$Ru$_2$ have shown an anomalous evolution of the superconducting transition temperature from T$_c$ ~ 4.4 K at x = 0.0 to T$_c$ ~ 0.3 K at x ~ 0.5. The lattice parameters are found to show a

CP554, *Physics in Local Lattice Distortions*, edited by H. Oyanagi and A. Bianconi
© 2001 American Institute of Physics 1-56396-984-X/01/$18.00

continuous evolution with substitution of Ce indicating an average compression of the unit cell [5]. The cooling of the sample introduces a further compression of the unit cell showing overall flexibility of the lattice structure. The anomalous lattice flexibility of $CeRu_2$ was shown by ultrasonic technique measuring elastic stiffness that shows significant lattice softening without structural transition in the normal state [9]. Furthermore, the importance of lattice in the electronic properties has been indicated by an abnormal pressure dependence of the superconducting transition temperature of the $La_{1-x}Ce_xRu_2$ system [10]. This unusual effect of pressure on T_c indicates interesting evolution of the electron-lattice interactions in the $La_{1-x}Ce_xRu_2$ superconductor. It is still to be understood whether the decrease of the T_c in $La_{1-x}Ce_xRu_2$ is due to increasing magnetic interaction or due to change in the local electronic density of states and/or local lattice distortions.

Here we have exploited capabilities of XANES spectroscopy and made high resolution Ru K-edge XANES measurements to study the effect of Ce in the $La_{1-x}Ce_xRu_2$ intermetallic system. The advantage of the high photon flux at the ESRF synchrotron facility and fluorescence detection by a multi-element solid state detector has been taken to measure the absorption spectra with very high signal to noise ratio in order to explore the small changes. Multiple scattering calculations are made to identify origin of different XANES features [11]. The results suggest that the Ce substitution in place of La in the $LaRu_2$ introduce large change in the density of states at the Fermi level and hence modify the hybridization between the f and d. The decrease in the superconducting transition appears to be related with change in the density of states may be due to scattering with localized f electrons injected by the mixed valent Ce.

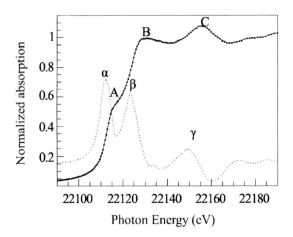

FIGURE 1. Normalized Ru K-edge XANES spectrum of the $LaRu_2$ and its derivative. The spectrum was measured at 200K on $LaRu_2$ in the fluorescence yield mode and is shown normalized to the atomic absorption.

EXPERIMENTAL

Ru K-edge absorption measurements were performed on well characterized powder samples of $La_{1-x}Ce_xRu_2$ prepared using arc melting under argon atmosphere, followed by high-temperature vacuum homogenization. The details about samples and their characterization could be found elsewhere [12]. The fluorescence yield (FY) high resolution absorption measurements were made at the beamline BM29 of European Synchrotron Radiation Facility (ESRF), Grenoble, using Si(311) double crystal monochromator. The Ru K_α fluorescence yield was collected using multi-element Ge X-ray detector array covering a large solid angle of the X-ray fluorescence emission. The temperature dependent measurements were performed on the samples mounted in a closed cycle He refrigerator. The sample temperature was controlled by a PID controller and monitored with an accuracy of ±1K. Several absorption scans at the same temperature were recorded to ascertain the reproducibility and to obtain spectra with very high signal to noise ratio.

RESULTS AND DISCUSSION

Fig. 1 shows a normalized high resolution Ru K-edge X-ray absorption near edge structure (XANES) spectrum, measured in fluorescence yield mode on $LaRu_2$ at 200K. The XANES spectrum shows at least three well-resolved features denoted by A, B and C. The peak A appears near the threshold of the Ru K-edge while the peak B and C appear at about 12 eV and about 38 eV respectively above the threshold. The first derivative of the XANES spectrum is also shown with the peaks denoted by α, β and γ.

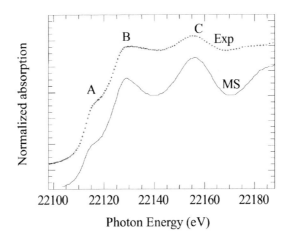

FIGURE 2. The XANES spectrum calculated by multiple scattering (MS) theory is compared with the experimental spectrum (Exp). The MS spectrum is a result of calculations made using a cluster of five shells including all atoms contains within about 5 Å from the central Ru. The agreement between the calculated and experimental spectra in respect to the intensity and the energy positions is quite good.

LaRu$_2$ is a laves phase with a face-centered cubic (MgCu$_2$-type) structure [13]. We have calculated real-space multiple scattering of the photoelectron excited from the Ru 1s-state to the atoms surrounding the central Ru for a cluster of atoms. Fig. 2 shows the calculated spectrum for a cluster of five-shells containing all atoms within 5 Å from the central Ru. The experimental spectrum is also plotted for comparison. A good agreement between the two spectra suggests that a cluster of about five shells is sufficient to describe the characteristic XANES features of the LaRu$_2$ system.

From the cluster dependence of the XANES multiple scattering peaks we have found that the peak A is associated with the unoccupied states made of mixed Ru p-d states, *i.e.*, central Ru 4p states hybridized with d orbitals of outer Ru shell. The Ru p-d states are also mixed-up with higher-shell La- d/f orbitals. Therefore the peak A contains information on the density of states at the Fermi level in the LaRu$_2$ system. On the other hand, the peak B is assigned to the 1s → 4p transition. In the real space, the feature B is attributed to multiple scattering within the Ru atomic shell while the peak C is dominantly due to single-scattering events between absorber and the shell of 6 Ru atoms. The details of the calculations are published elsewhere [11].

We have measured the Ru K-edge XANES spectra on the LaRu$_2$ at several temperatures to investigate the effect on the local lattice and local density of states at the Fermi level. Fig. 3 shows the absorption differences between the spectra measured at different temperatures. We do not observe any appreciable temperature dependence down to ~ 140 K as evident from the absorption difference between the spectra measured at 140 K and 200 K. On the contrary, there is a large change in the main absorption features by lowering the temperature down to 15K that could be seen in the absorption difference between the spectra measured at 15 K and 200 K showing a maximum of ~ 4% of the normalized absorption. The peaks in the derivative spectrum

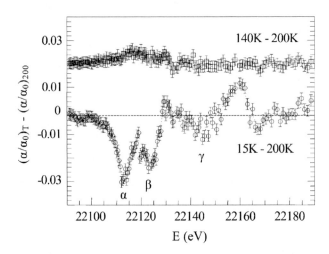

FIGURE 3. Absorption differences revealing temperature dependence (note that the upper difference curve is artificially displaced by 0.02).

(α, β, γ) appears almost at the same energy positions (see, *e.g.* Fig., 2) where the absorption difference shows peaks, suggesting that the temperature dependence is mainly due to an energy shift and it appears to be related to an overall contraction of the LaRu$_2$ lattice.

Now we turn towards the influence of Ce doping on the local electronic density of states and the local lattice displacements in the La$_{1-x}$Ce$_x$Ru$_2$. Fig. 4 shows the absorption difference between the spectrum measured on La$_{0.5}$Ce$_{0.5}$Ru$_2$ and LaRu$_2$ at 200 K (symbols). We observe a sharp decrease of the unoccupied density of states at the Fermi level (α) and a complex variation of the spectral features in energy range up to $\sim$ 50 eV. In order to understand the spectral variations we have calculated the XANES spectra taking into account the compression of the crystalline lattice (from 7.7 Å to 7.63 Å as given in [5]) and the substitution of La by Ce in the native LaRu$_2$, *i.e.*, considering the nominal composition La$_{0.5}$Ce$_{0.5}$Ru$_2$ where the compression (Ce is smaller than La) and the electronic configuration of Ce are taken into account. The absorption difference between the calculated spectra on the La$_{0.5}$Ce$_{0.5}$Ru$_2$ and the LaRu$_2$ is plotted in Fig. 4 (solid line). The agreement in the experiment and calculations is quite good and we can reproduce the variation of the peaks in the energy range 10 - 50 eV, however, we are not able reproduce the peak at the Fermi level (see arrow). This suggests that the substitution of Ce in place of La does not change the one-electron density of states at the Fermi level as shown by the multiple scattering calculations. Therefore the disagreement between the calculated and experimental differences is presumed to be related to a re-normalization of quasi particles.

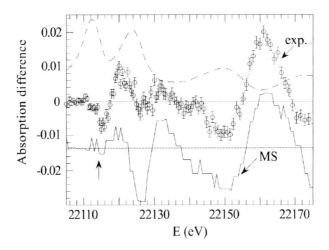

FIGURE 4. Absorption differences revealing effect of Ce in La$_{1-x}$Ce$_x$Ru$_2$. The differences are made between the measured spectrum of La$_{0.5}$Ce$_{0.5}$Ru$_2$ and LaRu$_2$ measured at 200 K (symbols) and calculated (solid line). The derivative of the absorption spectrum is shown as a reference (dashed line).

In summary, we have made temperature dependent Ru K-edge XANES measurements on for $LaRu_2$ system and studied effect of Ce doping on the local lattice and electron density of states. A real space multiple scattering approach has been used to identify the XANES features and a comparison of experimental and calculated spectra have been made. We have found that the temperature dependence of local density of states is mainly controlled by a compression of the $LaRu_2$ lattice by lowering the temperature. On the other hand, the effect Ce substitution in the $LaRu_2$ is two fold; the lattice shows an overall compression and suppression of density of states at the Fermi level. The suppression of the density of states does not seem to be related with variation of single particle density of states but may be renormalized due to quasi-particle scattering with magnetic excitations introduced by localized and itinerant Ce 4f electrons. The suppressed spectral density of states at the Fermi level could be one of the possible reasons to the decrease of the superconducting transition temperature in the $La_{0.5}Ce_{0.5}Ru_2$.

ACKNOWLEDGEMENTS

The authors would like to thank the ESRF staff for the experimental help. This research has been supported by the Ministero dell'Università e della Ricerca Scientifica (MURST) under the Programmi di Ricerca Scientifica di Rilevante Interesse Nazionale coordinated by R. Ferro, by Istituto Nazionale di Fisica della Materia (INFM) and by Progetto 5% Superconduttività del Consiglio Nazionale delle Ricerche (CNR).

REFERENCES

1. Belli, M., Scafati, A., Bianconi, A., Mobilio, S., Palladino, L., Reale, A., and Burattini, E., *Solid State Commun.* **35**, 355 (1980).
2. see e.g. a review by Bianconi, A., in *X Ray Absorption: Principle, Applications Techniques of EXAFS, SEXAFS and XANES*, edited by Prinz, R., and Koningsberger, D., J. Wiley and Sons, New York, 1988.
3. Durham, P.J., in *X-ray Absorption: Principles, Applications, Techniques of EXAFS, SEXAFS, XANES*, edited by Prinz, R., and Koningsberger, D., J. Wiley and Sons, New York, 1988.
4. Bianconi, A., Garcia, J., Benfatto, M., *Topics in Current Chemistry* **145**, edited by E. Mandelkow Springer Verlag, Berlin, 1988, pp. 29; Bianconi, A., Garcia, J., Marcelli, A., Benfatto, M., Natoli C. R. and Davoli, I., *Journal de Physique (Paris)* **46**, Colloque C9, 101 (1985); Benfatto, M., Natoli, C. R., Bianconi, A., Garcia, J., Marcelli, A., Fanfoni, M., Davoli I., *Phys. Rev. B* **34**, 5774 (1986).
5. Shelton, R.N., Lawson, A. C., and Baberschke, K., *Solid Stat. Comm.* **24**, 465 (1977).
6. Chudinov, S., Brando, M., Marcelli, A., and Battisti, M., *Physica B* **244**, 154 (1998).
7. Rettori, C., Davidov, D., Chaikin, P., and Orbach, R., *Phys. Rev. Lett.* **30**, 437 (1973).
8. Asokamani, R., Subramoniam, G., Mathi Jaya, S., and Pauline, S., *Phys. Rev. B* **44**, 2283 (1991).
9. Suzuki, T., Goshima, H., Sakita, S., Fujita, T., Hedo, M., Inada, Y., Yamamoto, E., Haga, Y., and Onaki, Y., *J. Phys. Soc. Jpn.* **65**, 2753 (1996).

10. Nakama, T., Hedo, M., Maekawa, T., Higa, M., Resel, R., Sugawara, H., Settai, R., Onuki, Y., and Yagasaki, K., *J. Phys. Soc. Jpn.* **64**, 1471 (1995); Nakama, T., Uwatoko, Y., Kohama, T., Bukov, A.T., Mori, N., Yoshida, H., Abe, S., Kaneko, T., and Yagasaki, K., *Rev. High Pressure Sci Technol.* **7**, 632 (1998).

11. Wu, Z.Y., Saini, N.L., Agrestini, S., Di Castro, D., Bianconi, A., Marcelli, A., Battisti, M., Gozzi, D., and Balducci, G., *J. Phys. : Condens. Matter* **12**, 6971 (2000).

12. Battisti, M., Marcelli, A., Chudinov, S., Bozukov, L., Piquer, C., and Chaboy, J., *J. Magn. Magn. Mat.* **196-197**, 888 (1999).

13. Huxley, A.D., Dalmas de Rèotier, P., Yaouanc, A., Caplan, D., Couach, M., Lejay, P., Gubbens, P.C.M., and Mulders, A.M., *Phys. Rev. B* **54**, R9666-9669 (1996); Huxley, A.D., Boucherle, J.X., Bonnet, M., Bourdarot, F., Schustler, I., Caplan, D., Lelievre, E., Bernhoeft, N., Lejay, P., and Gillon, B., *J. Phys.: Condens. Matter* **9**, 4185 (1997).

Lanthanum Dilution on the C15 Laves Phase of CeRu$_2$: An X-Ray Absorption Spectroscopy and Thermodynamic Investigation

M. Battisti[a], A. Marcelli[a], Z. Y. Wu[a,b], A. Bianconi[c], and N. L. Saini[c]

[a]Laboratori Nazionali di Frascati, INFN, P.O. Box 13, 00044 Frascati, Italy
[b]Beijing Synchrotron Radiation Facility, Institute of High Energy Physics, Chinese Academy of Sciences, P.O. Box 918, Beijing 100039, P.R. China
[c]Dipartimento di Fisica, Universita' di Roma "La Sapienza", and Unita' INFM, P.le Aldo Moro 2, 00185 Roma

Abstract. The CeRu$_2$ and the LaRu$_2$ compounds represent valuable, still not understood, exceptions to the magnetic behaviors of standard BCS superconductor materials. To understand the anomalous behavior of these superconductors we investigated the solid dilution La$_{1-x}$Ce$_x$Ru$_2$ by both X-ray absorption spectroscopy and thermodynamic experiments. Among the large class of REM$_2$ type C15 Friauf-Laves phase compounds, this series offers a good opportunity to investigate the correlation between structural and electronic modifications associated to the interstitial doping. Moreover the understanding of the structural, electronic and magnetic properties of intermetallic rare earth (RE) and the $4d$ transition metal (M) compounds addresses a large amount of research triggered by technological applications. These experiments have been performed to compare the results obtained by a local spectroscopy with thermodynamic data. Comparative and theoretical analysis demonstrate the large influence of the doping on the electronic and structural behavior of these systems.

INTRODUCTION

The CeRu$_2$ compound is a clean type II superconductor with a critical temperature of 5.8 K. It's usually considered a "typical" f-electron superconductor that however exhibits unusual magnetic and superconducting properties.

The close proximity of the Ce atoms differentiates the Ce Laves phase compounds from the majority of the other Ce alloys; the proximity favours the delocalization of the Ce f-electrons respect to the formation of localized magnetic moments. Actually, the cerium-based Laves phase compounds behave somewhat differently from other rare-earth based materials because the $4f$ electrons can never be regarded as entirely localized at the rare-earth site, and consequently their properties cannot be predicted either, by assuming a trivalent Ce ion.

To clarify the interplay between magnetism and superconductivity in these intermetallic compounds we studied the solid solution La$_{1-x}$Ce$_x$Ru$_2$. Attention was devoted to the role of the valence-fluctuation due to the Ru $4d$-Ce $4f$ hybridization [1]

CP554, *Physics in Local Lattice Distortions*, edited by H. Oyanagi and A. Bianconi
© 2001 American Institute of Physics 1-56396-984-X/01/$18.00

and to the existing relationship between superconductivity and hybridization phenomena.

In these materials the observed anomalies were initially ascribed to the promotion of Ce electrons to the conduction band [2]. Most probably the dilution in solid compounds brings to the f-electron delocalization by an "hybridization process" [3].

In the last years different high-energy spectroscopical studies have been performed and they clearly demonstrated the existence of an f-electron component of the density of states, near the Fermi surface [4]. They also show that the $4f$ electrons become more itinerant as the hybridization strength of the Ce $4f$ states increases.

More recently theoretical and experimental data suggest that the hybridization in CeRu$_2$ is large enough to provide a mixed valence state and to take the CeRu$_2$ far from both Kondo and heavy-fermion regime [5].

Both the CeRu$_2$ and the LaRu$_2$ are intermetallic compounds with a C15 cubic Laves phase structure (MgCu$_2$ type) and a space group Fd3m. The symmetry is the same also for the whole La$_{1-x}$Ce$_x$Ru$_2$ series; in these compounds the Ru atoms form a three-dimensional open network of corner-sharing tetrahedra interpenetrated by a diamond structure of Ce or La atoms [Fig. 1].

The end member of the dilution is the rare earth compound LaRu$_2$, one of the most appropriate reference to study the role of f-electron both in structural and electronic transitions.

Indeed, a pressure applied either by external apparatus or by chemical factor has a direct influence on the hybridization between $4f$ and d electrons. Because the dilution generates a chemical pressure in the CeRu$_2$ crystal lattice by the substitution of Ce ions with ions (La) of different ionic radii, our study on the La$_{1-x}$Ce$_x$Ru$_2$ system provides a useful tool to understand both the mixed valence behavior and the anomalous magnetic properties. To understand the role of the local density of states and the lattice structure we measured high-resolution X-ray absorption near edge structure (XANES) at the L edge of La and Ce, and at the K edge of Ru.

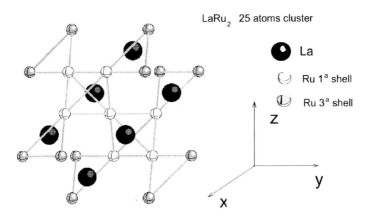

LaRu$_2$ 25 atoms cluster

● La
◔ Ru 1^a shell
◑ Ru 3^a shell

FIGURE 1. The crystallographic structure of the cubic C15 Laves phases LaRu$_2$ showing a cluster of 25 atoms around the Ru atom.

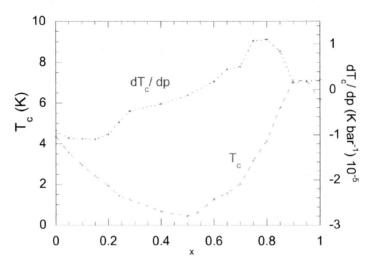

FIGURE 2: T_c and T_c/dp versus Ce concentration in the $La_{1-x}Ce_xRu_2$ series.

From thermodynamic measurements we estimated the superconducting temperature (T_c) at different concentration (x) in our solid solution, and we noticed a different trend when compared with the similar $La_{1-x}RE_xRu_2$ compounds (where RE = rare earth). In fact in RE based solid solutions the T_c drops to zero when x increases (x ~ 0.2). On the contrary, as we show in Fig. 2 in our solid solution the T_c has a minimum at x ~ 0.5, but there is no appearance of ferromagnetic properties [5].

In addition, because for Ru substitution the T_c drops immediately to zero, an important role in the superconductivity has to be associated to Ru atoms that form linear chains in the crystal plane [Fig. 1].

To better understand the role of the doping on the states at the Fermi level, that drives the BCS superconductors behaviors, we investigated the XANES Ru K-edge and the La (and Ce) L-edges. The K-edge transitions have the same p-symmetry (of the final state) of the L_1-edge transitions and in the framework of a one-electron theory the theoretical description is the same for both. However, due to the different energy, the resolution between K and L_1 edges may be significantly different.

EXPERIMENTAL AND THEORETICAL DATA

Ru K-edge absorption experiments were performed on characterized powders samples of $La_{1-x}Ce_xRu_2$. A detailed description of both sample preparation and characterization, and experimental set-up used can be found in [5].

Ru K-edge experiments in transmission mode were performed in air at the beamline 4-3 of the Stanford Synchrotron Radiaton Laboratories (SSRL) using a Si(220) double crystal monochromator. Fluorescence measurements at the same beamline were performed in air using a Lytle's detector and Soller slits.

La and Ce L-edges absorption data were collected in transmission mode, in air, at Grenoble, at the European Synchrotron Radiation Facility (ESRF) using a double crystal monochromator equipped with Si(311) crystals.

Multiple scattering (MS) calculations, based on the one-electron full MS theory, were made using the CONTINUUM code. The calculated Ru K-edge spectra were convoluted with a Lorenzian function with a full width $\Gamma_h \sim 5.5$ eV to account for the Ru $1s$ core hole lifetime and with a Gaussian function to simulate the experimental resolution [6]. For the calculated La L_1-edge spectra we taken $\Gamma_h \sim 4$ eV to account for the different La $2s$ core hole lifetime.

Thermodynamic experiments were performed both at the Camerino University (Italy) and at Zaragoza University (Spain), measuring heat capacity (> 1.2 K), resistivity (> 1.8 K) and susceptibility (>1.8 K).

DISCUSSION

The XANES spectroscopy probes the local and partial empty states. Although system dependent. XANES features produce information associated to the multiple scattering (MS) between the absorber atom and the neighbouring atoms within a cluster of atoms extended of the order of a few Å (5-10 Å) in the real space.

In our case we were able to simulate the experimental spectra [Fig. 3] using a cluster of 57 atoms (5 shells) around the absorbers (La and Ru). From these multiple scattering calculations, in the case of the Ru K-edge we associate the pre-peak a (Fig. 3, left) to the mixed unoccupied Ru p-d states [5-6], because due to the interatomic hybridization, the Ru p-d states are mixed also with the La (and Ce) d-f orbitals. The b peak is assigned to the main $1s$-$4p$ atomic transition in the multiple scattering regime, while the c peak is sensitive to the single scattering events between the Ru absorber and the neighbouring 6 Ru atoms.

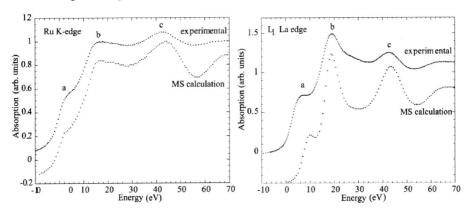

FIGURE 3. Experimental and calculated Ru K-edge and La L_1-edge in the LaRu$_2$ compound.

In the similar way, at the La L_1-edge (Fig. 3, right) the peak a is ascribed to the mixed unoccupied La p-d-states while the b peak is associated to the $2s$-$5p$ like atomic transition. The b peak is not a white-line (as in the L_3 edge) because there is no localized empty p DOS in the final state. Finally the c peak reflects the single scattering events between the La absorber and the neighbouring 12 Ru atoms.

As we can see in Fig. 3 there are no significant differences between the calculated and the experimental spectra. However, at the La L_1-edge in the energy range between the a and the b peak, the discrepancy between experiment and theory can be ascribed to the practical difficulty to remove from the experimental data the residual long-range oscillations due to the La L_2 edge [5].

To conclude, the non-linear behaviour of T_c vs. x [Fig. 2], a still not understood behavior of these systems, is associated for the first time to the intraatomic and interatomic hybridization effects. These represent a powerful mechanism because influence the properties of the empty DOS (density of states) above the Fermi level.

In particular, in the $La_{1-x}Ce_xRu_2$ series the main role in the low temperature superconductivity regime seems to be ascribed to the Ru linear chains and not to the $4f$ Ce electrons as suggested previously [4].

Curves in Fig. 2 indicate that in these metallic low T_c systems an "optimal doping" may be also achieved and as a consequence the superconductivity mechanism can be the same proposed recently for high T_c oxides, where charge-ordering phenomena associated with "stripes" have been observed and discussed [7].

ACKNOWLEDGMENTS

We would like to thank S. Tchoudinov of Camerino University and J. Chaboy of Zaragoza University for stimulating discussions. We are grateful to the Department of Physics of Camerino University, the Department of "Fisica de la Materia Condensada" of Zaragoza University, the SSRL and ESRF laboratories for the support during experiments.

REFERENCES

1. Joseph R.R., *Phys. Rev. B* **6**, 3286 (1972)
2. Crabtree G.B., Dunlap, B.D., and Koelling, D.D., *Physica B* **135**, 38-40 (1985)
3. Battisti M., Marcelli, A., Chudinov, S., Bozukov, L., Piquer, C., and Chaboy, J., *J. Mag. Mag. Mat.* **196-197**, 888-890 (1999)
4. Yanase A., *J. Phys. F:Met. Phys.* **16**, 1501-1504 (1986)
5. Battisti M., Thesis, University of Rome "La Sapienza" (2000)
6. Wu Z., Saini, N.L., Agrestini, S., Di Castro, D., Bianconi, A., Battisti, M., Marcelli, A., Gozzi, D., and Balducci, G., *J. Phys. Condens. Matter* **12**, 1-8 (2000)
7. Bianconi A., Valletta, A., Perali, A., and Saini, N.L., *Solid State Commun.* **102**, 369-374 (1997)

Mapping of Deformation Potentials

A. V. Tkach

Institute for Metal Physics,
18, Kovalevskaya st., Ekaterinburg 620219, Russia

Abstract. We discuss possible ways to get maps of deformation potential (DP) at Fermi Surfaces (FS). Also, we present the first attempt to reconstruct – from experimental data – the spatial DP-distribution over a FS-sheet of tungsten. All the patterns reveal considerable variability of deformation potential on the FS-sheet.

GENERAL

Deformation potential Λ_{ij} (DP) takes into account variations of the electron energy E due to a lattice distortion $u_{ij} = du_i / dr_j$. This concept plays an important part while describing ultrasonic and other phenomena in metals. A difficulty consists in the fact that DP is a local characteristic of a given state at the Fermi Surface (FS); each point of the FS must be assigned by the corresponding local DP value:

$$\Lambda_{ij} = \frac{\partial(E - E_F)}{\partial u_{ij}} \tag{1}$$

where E_F is the Fermi energy.

The task of experimental studies of Fermi Surfaces (called fermiology) cannot be considered as a completely resolved one, until the problem of deformation potential is resolved. But such maps of local DP-values, as a rule, could not be restored on the basis of experimental information. This is why the task of DP-mapping on FSs now seems to become a question of the hour in the experimental fermiology.

Specificity of the contemporary state in fermiology (well defined geometric parameters of FSs and detailed picture of Fermi–velocities v_F) gives us an opportunity to offer two ways, at least, [1] to trace the DP-maps on the basis of experimental data. Both of them are based on a simple interconnection between the *normal* FS-displacement $[\Delta k_F]_n$ under the given lattice distortion u_{ij} – and corresponding DP-value Λ_{ij}:

$$[\Delta \mathbf{k}_F]_n = \left\{ -n_j \cdot k_i - \frac{\Lambda_{ij}(\mathbf{k}_F)}{\hbar v_F(\mathbf{k}_F)} \right\} \cdot u_{ij} \tag{2}$$

where $\mathbf{n} \equiv \mathbf{v}_F / v_F$.

CP554, *Physics in Local Lattice Distortions*, edited by H. Oyanagi and A. Bianconi
© 2001 American Institute of Physics 1-56396-984-X/01/$18.00

1. *Parametrization.* Now some physically justified methods are developed to calculate the strain dependences of FSs. Corresponding models contain a number of parameters as a rule. Such a method can be taken as a parametrization scheme. Thus, its parameters have to be evaluated from the experimental data on the effective cross-sections of the given FS under the strain. As a result, contours of the strained FS must be revealed − on the basis of experimental information. Finally, Eq.(2) gives the way to get a solution of the task.

2. *Inversion.* It is believed that an inversion scheme can be applied to the case of distorted lattice (by the direct analogy with that for effective dGvA-areas in the unstrained state, [2]). It must reveal the normal displacements of FS under a distortion − from the experimental information. After that, the map of Λ_{ij} $(\mathbf{k}_F)$ can be contoured − on the totally model-free basis.

It is worth of mention that relationship Eq. (2) can be simplified by the transition to the so called "accompanying" reference frame (for example, [1]). In such a case, first term in the braces vanishes.

RECONSTRUCTION OF DP ON THE N-ELLIPSOID OF TUNGSTEN

Now we utilize the parametrization approach to reconstruct DP-maps over the "N-ellipsoid" of the tungsten FS. The main requirement for the procedure: all the known experimental data on the deformational dependence of effective areas S (over 40 experimentally measured values [3] for different types of deformation) have to be reproduced exactly.

Taking into account small linear dimensions of the FS-sheet discussed, one can accept the simple model of dispersion:

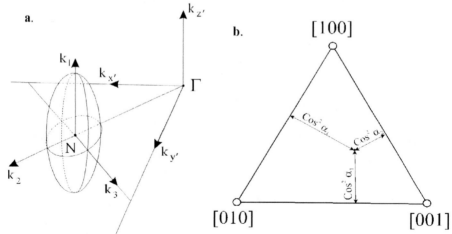

FIGURE 1. a. N-ellipsoid on the FS of tungsten and b. Orientation triangle

$$E^*(k) = \sum_{i \leq j} B_{ij} k_i k_j , \qquad (3)$$

which is supposed to be written in the "accompanying" reference frame. We introduce an additional limitation. It is based on the fact that two of the main N-ellipsoid semiaxes a_2 and a_3 in tungsten are close one to another and the both are far different of the third one. This is why we accept:

$$\frac{\partial a_2}{\partial u_{ii}} = \frac{\partial a_3}{\partial u_{ii}} \qquad (4)$$

for any longitudinal distortion u_{ii}.

From the symmetry considerations, one can conclude that off-diagonal components vanish in initial (*i.e.* unstrained) state of the lattice and also – under the longitudinal deformations along the main axes of ellipsoid. On the other hand, only the off-diagonal component B_{ij} can vary under a shear distortion u_{ii}, $i \neq j$. All the diagonal components are changeable under longitudinal deformations only.

The model proved to be fully consonant with the experimental information published [4, 3, 5] (effective FS-areas, cyclotron masses and also – area variations under the dilation $\Omega = \sum u_{ii}/3$ and under a tetragonal shear). To illustrate the consonancy, we compare (see Table 1) the experimentally obtained dilational variations of effective areas with those predicted by the model. Here we obtain the variations of inverse squares of the main ellipsoid areas $\partial(s_i^{-2})/\partial\Omega$, to be 66850, 41680, and 29820 $(a.u.)^{-2}$, correspondingly. Transformation from the laboratory reference frame to the "accompanying" one was accounted for by the special correction of [1].

Then, directly from the experimental data, we reveal number of model parameters (Fermi energy and diagonal components of **B** in initial, unstrained lattice, and also the parameters responsible for variations of the FS under different types of lattice deformation). Finally, local values of corresponding deformation potentials are reconstructed and mapped.

TABLE 1. Dilational variations of the inverse squared areas of the N-ellipsoid.

$h'k'l'$	S, a.u.	$\partial(S^{-2})/\partial\Omega, (10^3 \cdot a.u.^{-2})$			
		exp. value	ref.	weighted mean	model
		71 ± 7	[6]		
		69 ± 7	[7]		
$0'0'1'$	0.0159	56 ± 9	[8]	67 ± 4	66.9
		66 ± 13	[9]		
		58 ± 13	[9]		
$0'1'1'$	0.0184	53 ± 5	[7]	53 ± 5	51.3
$1'1'1'$	0.0188	49 ± 7	[6]	49 ± 7	49.4
$1'0'0'$	0.0228	31 ± 10	[9]	35 ± 4	35.8
		36 ± 5	[9]		
$\bar{1}'1'1'$	0.0205	42 ± 6	[6]	42 ± 6	41.5

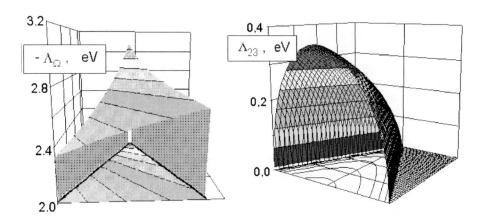

FIGURE 2. Maps of deformation potentials Λ_Ω and Λ_{23} at N-ellipsoid.

All the maps are drawn over the "orientation triangle". An equilateral triangle of the unit height is taken. Each point within the triangle represents a direction in the main octant of **k**-space (and the corresponding state at the N-ellipsoid surface). The distance from the point to the 'i-arm' of the triangle is taken as $\cos^2(\alpha_i)$, where α_i is a direction cosine. As to all other octants of N-ellipsoid, the corresponding DP-values can be easily derived from those of the main octant by simple symmetry relations.

The available experimental information gives us an opportunity to map, for the first time, the following deformation potentials (in the framework of the model validity): $\Lambda_{11}, \Lambda_{22}, \Lambda_{33}, \Lambda_{23}, \Lambda_\Omega$. The two later items in the list are determined quite regardless of the additional assumption Eq. (4). It will be observed that all the maps reveal considerable variability of deformation potential on the relatively small FS-sheet.

REFERENCES

1. Tkach, A. V., *Sol. St. Commun.* **107**, 407 (1998).
2. Mueller, F. M., *Phys. Rev.* **148**, 636 (1966).
3. Joss, W., Griessen, R., and Fawsett, E., in Landolt-Börnstein numerical data and functional relationships in science and technology. New Series. **Group III, 13(b)**. Springer, Berlin (1983).
4. Girvan, R. F., Gold, A.V., and Phillips, R.A., *J. Phys. Chem. Solids, 1968,* **29**, 1485 (1968).
5. Tkach, A. V. and Rinkevich, A. B., *Fiz. Met. Metallogr. (rus)*, **84**, 79 (1977).
6. Schirber, J. E., *Phys.Lett.* **35A**, 194 (1971).
7. Svechkarev, I. V. and Pluzhnikov, V. P., *Phys. Stat. Solidi (b)*, **5**, 315 (1973).
8. Stanley, D. J., Perz, J. M., and Au, H. P., *Can. J. Phys.*, **54**, 1234 (1976).
9. Stanley, D. J., Perz, J. M., and Lee, M. J. G., *Can. J. Phys.* **55**, 344 (1977).

Charge Ordering and Optical Conductivity of MMX Chains

M. Kuwabara[*] and K. Yonemitsu[* †]

[*]*Dept. of Theoretical Studies, Institute for Molecular Science, Okazaki 444-8585, Japan*
[†]*Dept. of Functional Molecular Science, Graduate University for Advanced Studies, Okazaki 444-8585, Japan*

Abstract. We study the optical conductivity of the halogen-bridged binuclear metal complexes, using a one-dimensional dimerized 3/4-filled-band model and the Lanczos method. The spectra are quite sensitive to the charge ordering pattern and the long-range Coulomb interaction.

INTRODUCTION

In quasi-one-dimensional systems such as conducting polymers and spin-Peierls systems, local lattice distortions (solitons, polarons, etc.) have been of great interest due to their relevance to conductivity, magnetism, and optical properties [1,2]. The halogen-bridged binuclear metal complexes (MMX chains) are linear chains consisting of dimer units of transition-metal (M) ions bridged by halogen (X) ions. These materials are interesting because they show a variety of charge ordering states accompanied with lattice modulations caused by strong electron-lattice coupling and electron-electron interaction; an averaged-valence (AV) state, a charge-density-wave (CDW) state, a charge-polarization (CP) state, and an alternate-charge-polarization (ACP) state, which are schematically shown in Fig. 1 [3,4]. Possibility of the existence of solitons is a natural consequence of such rich doubly-degenerate states.

Solitons may be found in $Pt_2(dta)_4I$ (dta = CH_3CS_2), which is in the ACP state below 80K [4]. The ACP state is analogous to a spin-Peierls state because the electrons on the M^{3+} sites on both sides of an X site form a singlet pair. However, no spin gap has so far been observed at low temperatures [4]. Solitons in the ACP state possibly account for the small but appreciable magnetic susceptibility. In the other compounds, R_4 [$Pt_2(pop)_4X$] (pop = $P_2O_5H_2$), the AV, CDW and CP states are realized by choosing the counter ions (R = K, $CH_3(CH_2)_7NH_2$, Na, Li) and the halogen ions (X = I, Br) [5-9]. Theoretical studies for the origins of these ordering states have been performed with the extended Hückel calculation [10,11], in the mean field approximation [12-14], and by the exact-diagonalization method [15,16].

In this paper, we investigate the optical conductivity of the MMX chains. The character of excited states in these systems reflects the charge ordering pattern and the strengths of Coulomb interactions.

CP554, *Physics in Local Lattice Distortions*, edited by H. Oyanagi and A. Bianconi
© 2001 American Institute of Physics 1-56396-984-X/01/$18.00

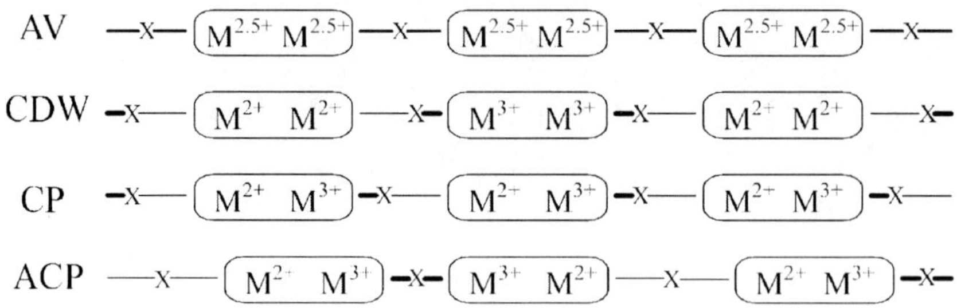

AV —x—$\boxed{M^{2.5+}\ M^{2.5+}}$—x—$\boxed{M^{2.5+}\ M^{2.5+}}$—x—$\boxed{M^{2.5+}\ M^{2.5+}}$—x—

CDW —x—$\boxed{M^{2+}\ M^{2+}}$—x—$\boxed{M^{3+}\ M^{3+}}$—x—$\boxed{M^{2+}\ M^{2+}}$—x—

CP —x—$\boxed{M^{2+}\ M^{3+}}$—x—$\boxed{M^{2+}\ M^{3+}}$—x—$\boxed{M^{2+}\ M^{3+}}$—x—

ACP —x—$\boxed{M^{2+}\ M^{3+}}$—x—$\boxed{M^{3+}\ M^{2+}}$—x—$\boxed{M^{2+}\ M^{3+}}$—x—

FIGURE 1. Schematic structures of the MMX chains.

MODEL AND METHOD

We use a one-dimensional dimerized 3/4-filled-band model for the M d_{z^2} orbitals on $2N$ sites;

$$H = -\sum_{i,\sigma}^{N} t_i (c_{a,i,\sigma}^{\dagger} c_{b,i,\sigma} + \text{h.c.}) - \sum_{i,\sigma} t_{i,i+1} (c_{b,i,\sigma}^{\dagger} c_{a,i+1,\sigma} + \text{h.c.})$$
$$- \beta \sum_i (y_{a,i} n_{a,i} + y_{b,i} n_{b,i}) + U \sum_i (n_{a,i,\uparrow} n_{a,i,\downarrow} + n_{b,i,\uparrow} n_{b,i,\downarrow}) \quad (1)$$
$$+ \sum_i (V_{\text{MN}} n_{a,i} n_{b,i} + V_{\text{MXM}} n_{b,i} n_{a,i+1}),$$

where $c_{a,i,\sigma}^{\dagger}$ ($c_{b,i,\sigma}^{\dagger}$) creates an electron with spin σ at M site a (b) in the i-th dimer, $n_{a,i,\sigma} = c_{a,i,\sigma}^{\dagger} c_{a,i,\sigma}$ and $n_{a,i} = \sum_{\sigma} n_{a,i,\sigma}$. The intradimer transfer integral is denoted by $t_i = t_{\text{MM}}$, and the interdimer transfer integral through the X p_z orbital by $t_{i,i+1} = t_{\text{MXM}} - \alpha(y_{b,i} + y_{a,i+1})$, where $y_{a,i}$ ($y_{b,i}$) is the bond length, relative to that in the undistorted phase, between the M site a (b) in the i-th unit cell and its neighboring X site. The strengths of the site diagonal and off-diagonal electron-lattice couplings are denoted by β and α, respectively. U is the on-site repulsion. V_{MM} and V_{MXM} are the intradimer and interdimer nearest-neighbor repulsion, respectively.

We exactly diagonalize the 12-site cluster with the periodic boundary condition. The real part of the optical conductivity is expressed, with the use of the current-current correlation function $\chi_{jj}(\omega)$, by

$$\sigma_1(\omega) = D\delta(\omega) + \frac{e^2}{\omega} \text{Im} \chi_{jj}(\omega), \quad (2)$$

$$\chi_{jj}(\omega) = -\frac{1}{2N} \langle \varphi_0 | j \frac{1}{E_0 - H + \omega + i\eta} j | \varphi_0 \rangle, \quad (3)$$

where $|\varphi_0\rangle$ is the ground state, E_0 is its energy, j is the paramagnetic current-density operator, $2N$ is the number of sites, and η is a small positive number. The current-

current correlation function can be obtained as a continued fraction by using the Lanczos method [17].

RESULTS AND DISCUSSIONS

We focus on the AV, CDW and CP states, which are observed in the pop compounds. Figure 2 shows a typical example of the optical conductivity in these states, where the lattice distortions are fixed to be $y_{a,i} = y_{b,i} = 0$ in the AV state; $y_{a,i} = y_{b,i} = (-1)^i y_0$ in the CDW state; $y_{a,i} = -y_{b,i} = y_0$ in the CP state. A single peak is seen in the AV and CDW states, while two peaks appear in the CP state in the present one-band model. The AV state is equivalent to a simple dimer system with a 3/4-filled (quarter-filled in the hole picture) band, where the site-energy modulation is absent. The single peak is ascribed to the dimerization gap because the system can be mapped to an effectively half-filled spinless fermion model due to strong on-site U [18]. The difference in the number of peaks between the CDW and CP states is understood in the atomic (strong-coupling) limit ($t_{MM} = t_{MXM} = \alpha = 0$). In Fig. 3, the electronic configurations of the ground states are schematically illustrated for (a) CDW and (d) CP. Two kinds of charge excitations are present; an intradimer charge excitation [Fig. 3 (b) and (e)] and an interdimer one [Fig. 3 (c) and (f)]. In the CDW state, the

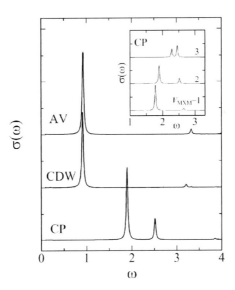

FIGURE 2. Optical conductivity in the AV, CDW and CP states with the parameters $t_{MXM} = 1$, $t_{MXM} = 0.8$, $\alpha = 0.2$, $\beta = 4$, $y_0 = 0.1$, $U = 6$, $V_{MM} = 3$, $V_{MXM} = 1.5$, $\eta = 0.02$. Inset: Variation of the optical conductivity with V_{MXM} in the CP state.

467

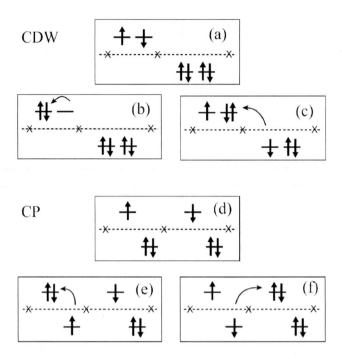

FIGURE 3. Schematic ground states and charge-excited states by the intradimer and interdimer processes in the CDW and CP phases.

excitation energy is estimated in the atomic limit to be $E_{MM}^{CDW} = U - V_{MM}$ for the intradimer process and $E_{MXM}^{CDW} = -V_{MM} + 2\beta y_0$ for the interdimer process.

The intradimer charge excitation needs the larger energy due to strong on-site U. Therefore, in the CDW state, only a single peak assigned to the interdimer charge excitation is observed in this range. In the CP state, the two peaks are ascribed to the (low-energy) intradimer and (high-energy) interdimer charge excitations, whose energies are $E_{MM}^{CP} = V_{MXM} + 2\beta y_0$ and $E_{MXM}^{CP} = V_{MM} + 2\beta y_0$, respectively, in the atomic limit. The intensities of these two peaks depend sensitively on the nearest-neighbor repulsions, V_{MM} and V_{MXM} (inset in Fig. 2). When V_{MXM} is substantially weaker than V_{MM}, only a single peak may be observed because the oscillator strength of the interdimer charge excitation is much weaker.

In the experiments [8,19], a single peak is observed in $Li_4[Pt_2(pop)_4I]$ (AV state), $Na_4[Pt_2(pop)_4I]$ (CDW state) and $K_4[Pt_2(pop)_4Br]$ (CDW state), which are consistent with our results. In $\{CH_3(CH_2)_7NH_2\}_4[Pt_2(pop)_4I]$ possibly inferred to be a CP state [19], two peaks are not apparently observed yet, suggesting the smallness of the interdimer repulsion.

In this study, we investigate the optical conductivity of the MMX chains within the 12-site system. In order to see the effects of the local structures such as solitons and polarons, we will perform and report the RPA calculations based on the unrestricted Hartree-Fock state.

ACKNOWLEDGMENTS

The authors thank K. Kanoda, H. Kitagawa, and H. Okamoto for fruitful discussions. This work was supported by a Grand-in-Aid for Scientific Research (C) and for Scientific Research on Priority Area "Metal-Assembled Complexes" from the Ministry of Education, Science, Sports and Culture, Japan.

REFERENCES

1. Heeger, A. J., Kivelson, S., Schrieffer, J. R., and Su, W. P., *Rev. Mod. Phys.* **60** 781 (1988).

2. Bray, J. W., Interrance, L. V., Jacobs, I. S., and Bonner, J. C., *in Extended Linear Chain Compounds*, ed. J. S. Miller (Plenum Press, New york and London, 1983) Vol. III.

3. Kurmoo, M., and Clark, R. J. H., *Inorg. Chem.* **24** 4420 (1985).

4. Kitawagta, H., Onodera, N., Sonoyama, T., Yamamoto, M., Fukawa, M., Mitani, T., Seto, M., and Maeda, Y.,*J. Am. Chem. Soc.* **121** 10068 (1999).

5. Yamashita, M., and Toriumi, K., *Inorg. Chim. Acta.* **178** 143 (1990).

6. Kimura, N., Ohki, H., Ikeda, R., and Yamashita, M., *Chem. Phys. Lett.*, **220**, 40 (1994).

7. Mitani, T., Wada, Y., Yamashita, M., Toriumi, K., Kobayashi, A., and Kobayashi, H., *Synth. Met.* **64** 291 (1994).

8. Wada, Y., Furuta, T., Yamashita, M., and Toriumi, K., *Synth. Met.* **70** 1195 (1995).

9. Yamashita, M., Miya, S., Kawashima, T., Manabe, T., Sonoyama, T., H. Kitagawa, T. Mitani, H. Okamoto, and R. Ikeda, *J. Am. Chem. Soc.* **121** 2321 (1999).

10. Whangbo, M., and Canadell, E., *Inorg. Chem* **25** 1726 (1986).

11. Borshch, S. A., Prassides, K., Robert, V., and Solonenko, A. O., *J. Chem. Phys.* **109** 4562 (1998).

12. Yamamoto, S., *Phys. Lett. A* **258** 183 (1999); **261** 125 (E) (1999).

13. Yamamoto, S., *J. Phys. Soc. Jpn.* **69** 13 (2000).

14. Kuwabara, M., and Yonemitsu, K., *Physica B* **284** 1545 (2000)

15. Kuwabara, M., and Yonemitsu, K., *Mol. Cryst. Liq. Cryst.* **343** 47 (2000).

16. Kuwabara, M., and Yonemitsu, K., to *appear in J. Phys. Chem. Solids*.

17. Dagotto, E., *Rev. Mod. Phys.* **66** 763 (1994); and references therein.

18. Favand, J., and Mila, F., *Phys. Rev. B* **54** 10425 (1996).

19. Okamoto, H., private communication.

A Combined EXAFS and Powder Diffraction Study of Local and Averaged Structure in Complex Oxides

N. Binsted[a], M. Stange[b], H. Fjellvåg[b] and M. T.Weller[b]

[a]*Department of Chemistry, University of Southampton, Southampton SO17 1BJ, England,*
[b]*Department of Chemistry, University of Oslo, N-0315 Oslo, Norway*

Abstract. Combined EXAFS/powder neutron diffraction analysis provides an excellent means of detecting local lattice distortions in metal oxide structures. Progress in modelling such distortions in perovskites is reported using the examples of $La_{6.4}Ca_{1.6}Cu_7CoO_{20}$ and $LaGa_{.5}Co_{.5}O_3$. Studies of some other oxide structures, $La_{6.4}Ca_{1.6}Cu_6Ni_2O_{20}$ and $La_8Ni_4Co_4O_{20}$, are also briefly reported.

INTRODUCTION

Combined EXAFS/Powder Neutron Diffraction (PND) analysis [1] provides the ideal way of utilising the power of the two techniques in defining the structure of materials on the scale of tens or hundreds of Ångstroms required for understanding their electronic and magnetic properties. Often EXAFS alone cannot even resolve nearest neighbour distances, as in many materials of interest a metal may occupy several distorted sites, and splitting of the oxygen shells cannot be reliably determined due to correlations with the Debye-Waller (DW) terms. In most perovskite derivatives, scattering paths with lengths between 2.5×2 Å and 4.0×2 Å make a large contribution to the spectrum and contain most of the useful structural information. However, the number of distances and angles required to describe such paths leads to an underdetermined refinement.

Powder neutron diffraction is the definitive technique for structure determination where the local and long-range structures correspond, but in many materials this is not the case. In a recent study of $Gd_2Ba_2CaCu_2Ti_3O_{14}$, in which the five metal EXAFS spectra and the PND profile were refined simultaneously [2], none of the sites were accurately described by a long-range structural model, and anomalies were large for the Ca and Ti sites. In such cases, analysis using both techniques is required in order to develop and refine structural models. The modelling of structures which involve local lattice distortions, in a way which allows both PND and EXAFS data to be refined, remains a challenge. In a previous study of the framework silicate gallobicchulite, $Ca_8Ga_8Si_4O_{24}(OH)_8$ [3], a single set of positional parameters was used to model a structure with ordered domains, using two different space groups (with partial PND occupancies for the higher-symmetry long-range space group). This

CP554, *Physics in Local Lattice Distortions*, edited by H. Oyanagi and A. Bianconi
© 2001 American Institute of Physics 1-56396-984-X/01/$18.00

approach has been found to be very useful where cation positions are locally ordered, and is used in the $La_{6.4}Ca_{1.6}Cu_7CoO_{20}$ example below. If some or all of the cation positions are randomly occupied this approach is less valid. In such cases, the use of a reverse Monte-Carlo approach [4] is appropriate but the resulting model, which may include thousands of atoms, cannot easily be applied to electronic structure calculations. Quasi-random periodic models may then provide a useful approximation [5]. The difficulties involved in developing models of random structures is apparent from the case of $LaGa_{.5}Co_{.5}O_3$ discussed below.

EXPERIMENTAL

Fully oxidised materials were prepared from acid solutions of oxides, acetates or metals, and then annealed in oxygen for 48 hours at 1298 K. Oxygen deficient materials were subsequently reduced using zirconium metal with a sample temperature of 673 K.

Fixed-wavelength PND measurements were made at the JEEPII reactor, Kjeller, Norway, using wavelengths from 1.5554 to 1.5563 Å. EXAFS spectra were recorded at SRS stations 8.1 and 9.2, and at the Swiss-Norwegian beamline, ESRF. Data were measured at room temperature unless stated otherwise.

ANALYSIS

EXAFS data were background-subtracted using the author's program PAXAS. Combined refinements were performed using the program P [1]. The Rehr and Albers approximation [6] was used for multiple scattering paths. The simplest possible treatment of EXAFS DW terms was used which assigned a parameter $A = 2\sigma^2$ Å^2 to 'blocks' of single-scattering distances (described by a range of distances (R) and atomic numbers). High-R blocks were pre-assigned a value, or set to a multiple of the shorter distance parameters, which were refined. Constraints were used to ensure that parameters for distances X-Y at edge X must equal parameters for Y-X at edge Y. Parameters for the legs of multiple scattering paths were interpolated from the single-scattering values by the program. Where multiple edges were available, this resulted in about three refined terms per spectrum.

Atomic positional parameters, cell parameters and peak-profile parameters were entered as for a typical structure refinement using PND data. PND isotropic thermal factors were refined independently from those used to fit the EXAFS spectra. While this is clearly not ideal, the descriptions used in the two analysis techniques cannot be easily correlated. An EXAFS energy zero EF was required for each spectrum. Additional positional terms were required to describe the local lattice distortions. The amplitude factor AFAC was set to 1 throughout. The inadequacies in fit resulting from these approximations are not sufficient to mask a general agreement between the local and long-range structure, where this exists.

RESULTS AND DISCUSSION

$LaGa_5Co_5O_3$

This material and the two end-members of the series $LaGa_{1-x}Co_{1-x}O_3$, $LaCoO_3$ and $LaGaO_3$ were investigated mainly at room temperature though some spectra were obtained at 78K. $LaCoO_3$ and the intermediate compound, $LaGa_5Co_5O_3$, are rhombohedrally distorted perovskites with space group $R\bar{3}c$. $LaGaO_3$ is orthorhombic, space group *Pbnm*. B-cation (B = Co, Ga) disordering was indicated by both EXAFS and PND data. Data from the end members were easily fitted using a simple structural model. Fig. 1 shows the fit to the Co K- and La L_{III}- edges of $LaCoO_3$ using positions derived from the PND data alone. Scattering paths to 15 Å are included. It is clear that the local structure showed no significant departure from the long-range structure, and that there was excellent agreement between the two techniques. The DW factors for the shortest B-O distances were A = .007 (1) $Å^2$ and those for the shortest B-B distances were A = .010 (1) $Å^2$ for both end members.

By contrast $LaGa_5Co_5O_3$ (Fig. 2) showed significant local departures from the averaged structure. B-O distances appeared similar to those in the end-members (1.93 Å Co-O, 1.96-1.97 Å Ga-O), rather than the single B-O distance of 1.945 Å given by the disordered model, and the amplitude of the 'second peak' at the Ga edge could only be approximated using a much smaller DW factor for Ga-B than for Co-B distances at the Co K-edge. The data suggests that the Ga-O-Ga bond angles are larger than those for Co-O-Co, with oxygen moving towards the Ga-Ga axis, but this is incompatible with Ga-O distances being larger than those for Co-O. The discrepancies were found to be as large for data collected at 78K, so anharmonic thermal effects are not responsible. There was no evidence of local ordering, formation of distinct layers or of off-site displacement of B cations. An attempt to model the local lattice distortions by neighbour-dependent displacements of the oxygen position did not significantly improve the fit, if the necessary constraints R_{B-O} = R_{O-B} and Ga-Ô-Co = Co-Ô-Ga were applied. It is clear from the EXAFS data and analysis so far that a model involving different local structures for Co and Ga is required. Further investigation of this problem is in progress, with an RMC approach offering the most likely solution.

$La_{6.4}Ca_{1.6}Cu_7CoO_{20}$

For this material the ordered oxygen vacancies give rise to an 8-8-20 superstructure with B cations occupying 4, 5 and 6 coordinated sites. The data showed the cations to be ordered with Co in half the octahedral sites. The averaged *P4/mbm* structure did not give a good fit to the EXAFS, especially at the Co K-edge, and the refined B-B DW terms were too high (> .02 $Å^2$). Both the combined refinement, and a PND-only refinement suggested a series of local distortions, which could, at least in part, be explained by the larger size of the Cu atoms in a lattice whose z-axis is determined by the smaller Co^{3+} ion (Co-Ô-Co=180°). The distortions determined from the combined refinement are summarised in Table 1, and fits to both PND and EXAFS data using this model (EXAFS refined in P1, PND in *P4/mbm*) are shown in Fig. 3.

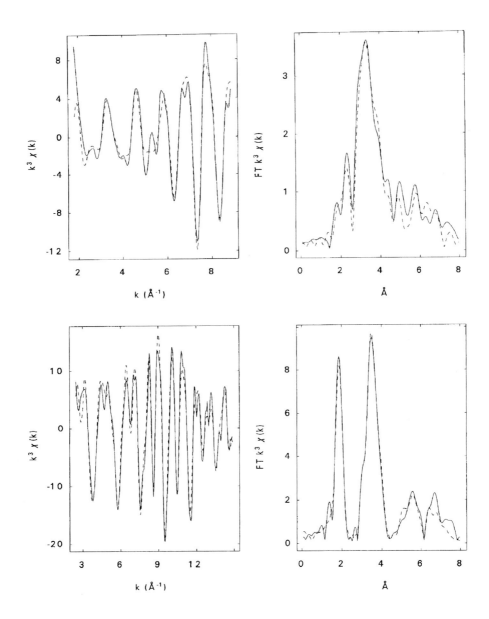

FIGURE 1. Fit to the La L_{III}–edge EXAFS (above) and Co K-edge EXAFS (below) of $LaCoO_3$ using positions derived from PND alone.

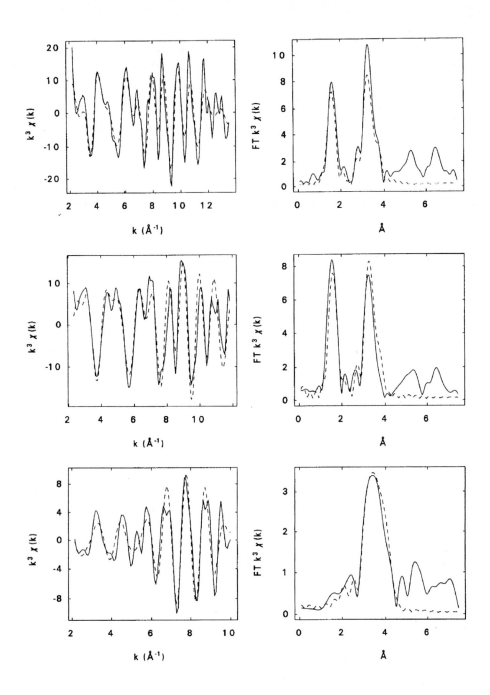

FIGURE 2. Fit to the Ga K-edge EXAFS (above), Co K-edge EXAFS (middle) and La L$_{III}$-edge EXAFS (below) of LaGa$_{.5}$Co$_{.5}$O$_3$ from a combined EXAFS/PND refinement using a $R\bar{3}c$ model without local lattice distortions.

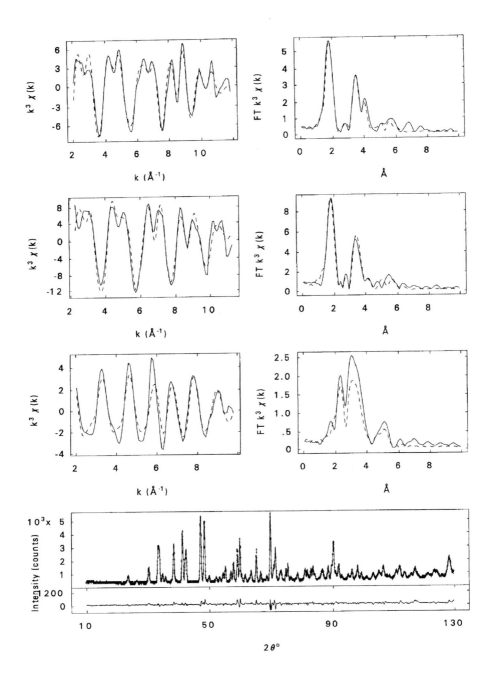

FIGURE 3. Fit to the Cu K-edge EXAFS (above), Co K-edge EXAFS (middle) and La L_{III}-edge (below), and to the PND profile of $La_{6.4}Ca_{1.6}Cu_7CoO_{20}$, using a *P4/mbm* model with local lattice distortions (EXAFS refined in P1, PND in *P4/mbm*).

The model used, based on a doubled c-axis, still has some drawbacks. The coordinates of the two O9 positions favoured by a PND-only refinement differ from that of the combined refinement. Clearly the modelling of the distortion of this site is not yet perfect. The fit is not really improved when the cluster size is increased from 5 to 6.2 Å, and the B-B DW terms, although improved, are still rather high (.018 Å^2). These two factors are most probably concerned with local ordering of the distortions associated with the pseudo Jahn-Teller distorted 4-fold Cu site. It is expected that an improved model will resolve these discrepancies. The fit to the PND data is better in *P4/mbm* (with statistical occupation of sites involved in lattice distortions), than either *P1* or *P4/m*. This suggests that ordering of the local lattice distortions is not coherent on the scale of the partical size of the material. It has been plausibly argued, that copper in this material is all in an average Cu^{2-3+} oxidation state. Our results suggest otherwise. The bond valence sum [7] for the 4- and 5-fold sites are 2.11 and 2.19 respectively for Cu^{2+} while for the 6-fold site it is 3.05 for Cu^{3+}, and the site is not Jahn-Teller distorted. However, stoichiometry determines that at least some octahedral Cu must be present as Cu^{2+}, and in practice a small percentage of a third type of octahedron is possible, although beyond the resolution of the technique.

Other 8-8-20 Structures

La$_{6.4}$Ca$_{1.6}$Cu$_6$Ni$_2$O$_{20}$ is similar to La$_{6.4}$Ca$_{1.6}$Cu$_7$CoO$_{20}$ in having Cu^{2+} in 4- and 5-fold sites. Ni^{3+} occupies the octahedral sites, which are undistorted. All PND peaks can be indexed in *P4/mbm*, but *P4/m* is required for the local structure. A good fit (R$_{exafs}$ [1] = 24%) to the EXAFS spectra can be obtained but only if Cu-Ni, Cu-Cu and Ni-Ni DW terms are significantly different. This implies similar distortion of the Cu-square planar units to that seen in La$_{6.4}$Ca$_{1.6}$Cu$_7$CoO$_{20}$ involving off-centre Cu atoms.

La$_8$Ni$_4$Co$_4$O$_{20}$, by contrast, has Ni^{2+} and Co^{2+} randomly distributed on 4-, 5- and 6-fold sites. As with LaGa$_{.5}$Co$_{.5}$O$_3$, there is an apparent size difference between the Co and Ni sites. This is to a large extent resolved if the metal coordinate for the 5-fold site differs depending on whether it is occupied by Co or Ni.

TABLE 1. *P4/mbm* coordinates for La$_{6.4}$Ca$_{1.6}$Cu$_7$CoO$_{20}$ and local lattice distortions. x, x+.5 > x, y implies displacement from a special position in *P4/mbm* to a more general position x, y. O1 moves off site only when bonded to Cu. The coordinates of O9 differ depending on whether it is bonded to Co or Cu.

Label	x, y, z	Wyckoff no.	Coord.	Local distortion
La/Ca	.2600, .4690, .5	8k	10	no change
Cu/Co	.0, .0, .0	2a	6	no change
Cu3	.5, .0, .0	2d	4	off site in (110)
Cu4	.2212, .7212, .0	4g	5	x, x+.5 > x, y
O1	.0, .0, .5	2b		offsite in (100) for Cu octahedra
O2	.5, .0, .5	2c		offsite in (nm0)
O7	.2188, .7188, .5	4h		x, x+.5 > x, y
O8	.3763, .8763, ..	4g		no change
O9	.1596, .0916, .0	8i		two positions

CONCLUSIONS

Combined EXAFS/PND analysis is necessary to resolve and describe local lattice distortions in powdered samples of defect and mixed cation site perovskites. Information is obtained on the local order associated with the differing B-cation locations, leading to an improved overall model of the structure. The improvements in the fit to the PND data alone could not justify the more complex model were it not for the significant simultaneous improvements to the EXAFS fit.

ACKNOWLEDGMENTS

This work has received financial support from the EPSRC, UK and the Research Council of Norway. The assistance of the Swiss-Norwegian project team, ESRF, and of Lorrie Murphy, and Fred Mosselmans, Daresbury Laboratory, is gratefully acknowledged.

REFERENCES

1. Binsted, N., Pack, M., Weller, M. T., and Evans, J., *J. Am. Chem. Soc.*, **118**, 10200 (1996).
2. Weller, M. T., Pack, M. J. and Binsted, N., *Angew. Chem. Int'. Ed.,* **37**, 1094 (1996).
3. Binsted, N., Dann, S. E., Pack, M., and Weller, M. T.,.*Acta Cryst.*, **B54**, 558 (1998).
4. McGreevy, R. L., *Nucl. Instrum. Methods A,* **354**, 1 (1995).
5. Wei, S-H, Ferreira, L. G., Bernard, J. E. and Zunger, A., *Phys. Rev B* **42**, 9622 (1991).
6. Rehr, J. J. and Albers, R. C., *Phys. Rev. B* **41**, 813 (1990).
7. Altermatt, D. and Brown, I. D., *Acta Cryst.*, **41**, 240 (1985)

Electron State and Local Structure around Sm-atom in $Sm_2Fe_{17}N_x$

H. Kasatani[*], M. Ohmura[*], K. Kobayashi[*], K. Yagi[¶], and H. Terauchi[¶]

[*]*Department of Materials Science, Shizuoka Institute of Science and Technology, 2200-2 Toyosawa, Fukuroi, Shizuoka 437-8555, JAPAN*
[¶]*Advanced Research Center of Science & School of Science, Kwansei –Gakuin University, Sanda, Hyogo 669-1337, JAPAN*

Abstract. High energy XASF studies of Sm:K-edge in $Sm_2Fe_{17}N_x$ (x = 0.0, 0.5, 1.07, 1.55, 1.85, 2.04 and 3.0) were performed at room temperature on SPring-8 BL01B1. The environment of the electron state of Sm-atom was changed by nitriding. But, this change was discontinuous change for the value of concentration of N-atom. The interatomic distances between Sm-atom and Fe-, N-atoms, roughly, increased with increasing the concentration of N-atom.

INTRODUCTION

The excellent magnetic properties of $Sm_2Fe_{17}N_3$ were discovered by Coey and Sun [1]. They found that the introducing three N-atoms to Sm_2Fe_{17} compound increased the Curie temperature, the saturation magnetization and the unaxial magnetic anisotropy. The Curie temperature increased dramatically from 398K (Sm_2Fe_{17}) to 752K ($Sm_2Fe_{17}N_3$). This Curie temperature is higher than $Nd_2Fe_{14}B$ (588K). The saturation magnetization (1.54T in $Sm_2Fe_{17}N_3$) was comparable $Nd_2Fe_{14}B$ (1.60T) and the uniaxial magnetic anisotropy (26T in $Sm_2Fe_{17}N_3$) was three times stronger than that of $Nd_2Fe_{14}B$ (7T) [2].

The crystal structure was rhombohedral Th_2Zn_{17}-type ($R\bar{3}m$) and crystal lattice of Sm_2Fe_{17} expanded more the about 6% to accommodate three N-atoms [2]. But, it is not established the interstitial site of N-atom because of two sites as 9e and 18g sites.

XAFS studies were carried out by using the L_3-edge of Sm-atom and K-edge of Fe-atom [2-5]. Sm-atom was predominantly trivalent in the both Sm_2Fe_{17} and their nitride. Furthermore, from the XANES spectra of Fe:K-edge, it was reported that the environment of the electron state of Fe-atom in Sm_2Fe_{17} nitride sample was intermediate state between its anionic state in Sm_2Fe_{17} and covalent state in α-Fe, and some charge transfer was occurred between N- and Fe-atoms, because of the small edge shift. From the XANES spectra of Sm:L_3-edge in the magnetically oriented powder samples, the small shift was reported in nitrides. On the other hand, the interstitial site of the N-atom in Sm_2Fe_{17} nitride compound was reported as 9e site from the interatomic distance between Sm- and N-atom.

Recently, we carried out the high energy XAFS studies of Sm:K-edge in $Sm_2Fe_{17}N_x$ (x = 0, ≤ 0.1, 0.5, 2.0 and 3.0) at room temperature [6]. We firstly observed the

CP554, *Physics in Local Lattice Distortions*, edited by H. Oyanagi and A. Bianconi

change of the structure at the just behind edge jump. The monotonous expansion of the interatomic distance between Sm- and Fe-atom was reported. But, the edge shift could not be checked. There was no measurement of $Sm_2Fe_{17}N_x$ with $x \approx 1$.

In this study, we carried out the re-examination of high energy XAFS study of Sm:K-edge in $Sm_2Fe_{17}N_x$ ($x = 0.0, 0.5, 1.07, 1.55, 1.85, 2.04$ and 3.0). The samples of $x = 1.07, 1.55$ and 1.85 are first measurement in this study, and other ones are same in previous study [6]. The aim of this paper is the systematic study in the relationship between the concentration of nitrogen and the change of the magnetic properties in $Sm_2Fe_{17}N_x$ compounds.

EXPERIMENTAL

Samples of $x = 1.07, 1.55$ and 1.85 were prepared as the same method in previous study [6]. From the preliminary x-ray powder diffraction measurements, it was obtained that a few % α-Fe was mixed in these samples. But, in high energy XAFS study of Sm:K-edge (46.834keV), it was possible to ignore this mixed α-Fe. XAFS measurement was carried out at BL01B1 of SPring-8. Measurement was transmission mode with Si(511) planes of a fixed-exit double crystal monochromator. In order to reject the higher harmonics, rhodium coated mirrors were used. The tablet sample with mixing boron nitride was prepared. The incident and transmitted x-ray intensities were monitored with ionization chambers with Ar(50%) + Kr(50%) and Kr(100%) gas, respectively. Using the UWXAFS3.0 program package carried out EXAFS analysis. The k-weighted EXAFS spectra were fit in the range of 3.5 - 14.0 Å^{-1} for k-space and 1.5 - 3.5 Å for r-space.

RESULTS AND DISCUSSION

Observed Sm:K-edge XAFS spectra of $Sm_2Fe_{17}N_x$ ($x = 0.0, 0.5, 1.07, 1.55, 1.85, 2.04$ and 3.0) are shown in Fig. 1. The origins of each data are shifted. The data with $x = 0.0, 0.5, 2.04$ and 3.0 are the almost same as previous study [6], but S/N is improved in this measurement. An inset in Fig.1 shows the expansion of the parts after the absorption edge jump, which is a square enclosed by dotted lines. It is evident that the structures at just behind absorption edge jump are changed by nitriding. In the case of $x = 0$, the peak is seen. As discussed in previous paper [6], this is the $1s \rightarrow 6p$ transition. In the case of $x = 3$, this peak is almost disappeared. In the cases of the intermediate concentration of N-atom; i.e., $x = 0.5, 1.07, 1.55, 1.85$ and 2.04, these parts are not peak like $x = 0$, but not completely disappeared like $x = 3$. It is difficult to identify whether the change of this part is dependent on the concentration of N-atom, or not. The normalization of all XAFS was done, and the expansion of the parts at just behind absorption edge jump is shown in Fig. 2. In the intermediate region of the concentration of N-atom, there is no systematic change. These patterns fall into four classes. The first pattern is the case with $x = 0$. The second one is the case with $x = 0.5$ and 2.04. The third one is the case with $x = 1.07, 1.55$ and 1.85. The fourth one is the case with $x = 3$. It is not understand why the patterns in intermediate concentration of N-atom fall two classes, and why the almost same pattern was observed in x = 0.5 and 2.04. The change of the environment of electron state of Sm-atom by absorption of N-atom is obtained so that it is expected the edge shift. The very small edge shift (about 0.6 eV shift to higher energy side in all nitrides compounds) was observed. However, this shift value is too small to compare with the energy resolution (about 4eV). The shift in Sm:K-edge is not clear at this moment.

Fourier transforms of XAFS spectra were shown in Fig. 3. The correction of phase-shift was not carried out, so the value of the horizontal axis is not absolute interatomic distance. It is thought that the first peak is Fe-atom. The position of this first peak shifts to the higher-R value with increasing the concentration of N-atom. This imply the expansion of average Sm-Fe interatomic distance. In the case of x = 3, the shoulder is obtained at the lower-R side of the first peak. It is thought that this shoulder corresponds to the N-atom. But, in the samples with intermediate concentration of N-atom, this shoulder is not seen.

The interatomic distance between Sm-atom and Fe-, N-atoms as the function of the concentration of N-atom are shown in Fig. 4. These interatomic distances, roughly, increase with increasing the concentration of N-atom. There are six kinds of interatomic distances between Sm-atom and Fe-atom. The differences between these interatomic distances are the region of 0.01 - 0.2 Å. But, in the process of fit, many parameters were restrained because of the large correlation. It is hardly to determine the precise model and to discuss in detail these six kinds of interatomic distances.

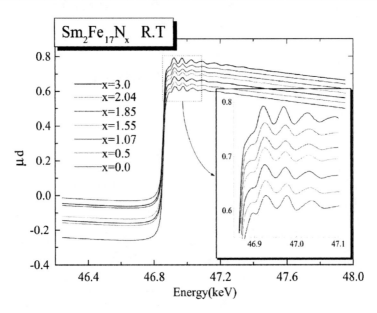

FIGURE 1. Observed Sm:K-edge XAFS spectra of Sm$_2$Fe$_{17}$N$_x$ (x = 0.0, 0.5, 1.07, 1.55, 1.85, 2.04 and 3.0), and inset is the expansion of the parts after the absorption edge jump.

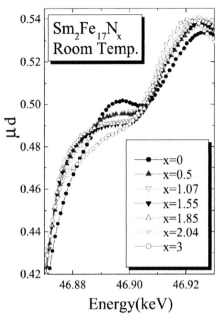

FIGURE 2. Expansion of the parts at just behind absorption edge jump from the normalized data.

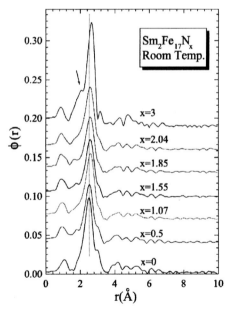

FIGURE 3. Fourier transforms of XAFS spectra with x = 0.0, 0.5, 1.07, 1.55, 1.85, 2.04 and 3.0.

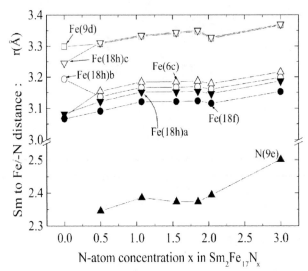

FIGURE 4. Interatomic distances between Sm-atom and Fe-, N-atoms as the function of the concentration of N-atom.

In summary, we carried out the high energy XAFS studies of Sm:K-edge in $Sm_2Fe_{17}N_x$ with x = 0.0, 0.5, 1.07, 1.55, 1.85, 2.04 and 3.0 at room temperature. (1) The environment of the electron state of Sm-atom is changed by nitriding. But, this change is discontinuous change for the value of concentration of N-atom. (2) The interatomic distances between Sm-atom and Fe-, N-atoms, roughly, increase with increasing the concentration of N-atom.

ACKNOWLEDGMENTS

This work was performed at SPring-8 with the approval of the Japan Synchrotron Radiation Research Institute (JASRI) (Proposal No. 1999A0300-CX-np).

REFERENCES

1. Coey, J.M.D., and Sun, H., *J. Magn. Magn. Mater.* **87**, L251-L254 (1990).
2. Fujii, H., and Sun, H., "Interstitially Modified Intermatallics of Rare Earth and 3D Elements," in *Handbook of Magnetic Materials, Vol.* **9**, edited by Buschow, K.H.J., Elsevier, 1995, pp. 303-404.
3. Coey, J.M.D., Lawler, J.F., Sun, H., and Allan, J.E.M., *J. Appl. Phys.* **69**, 3007-3010 (1991).
4. Capehart, T.W., Mishra, R.K., and Pinkerton, F.E., *Appl. Phys. Lett.* **58**, 1395-1397 (1991).
5. Capehart, T.W., Mishra, R.K., and Pinkerton, F.E., *J. Appl. Phys.* **75** 7018-7020 (1994).
6. Kasatani, H., Ohmura, M., Kobayashi, K., Nishihata, Y., Yagi, K., and Terauchi, H., *Proc. Int. Conf. SRMS-2 Jpn. J. Appl. Phys.* **38**, Suppl. 38-1, 433-435 (1999).

Local Structure of $(Ge_4Si_4)_5$ Monolayer Strained-layer Superlattice Probed by Fluorescence X-ray Absorption Fine Structure

S. Q. Wei[1,2], H. Oyanagi[1], K. Sakamoto[1], Y. Takeda[3] and T. P. Pearsall[4]

[1]Electrotechnical Laboratory, 1-1-4 Umezono, Tsukuba, Ibaraki 305-8568, Japan
[2]National Synchrotron Radiation Laboratory, University of Science & Technology of China, Hefei 230029, P.R.China
[3]Department of Materials Science and Engineering, Nagoya University, Nagoya 464-01, Japan
[4] CERF, CORNING, SA, 7-bis Avenue de Valvins 77210 Avon, France

Abstract. Local structure of $(Ge_4Si_4)_5$ monolayer strained-layer superlattice (MSLS) on Si(001) has been studied by fluorescence x-ray absorption fine structure (XAFS). The observed Ge-Ge and Ge-Si bond lengths, R_{Ge-Ge} (2.42 Å) and R_{Ge-Si} (2.38 Å), indicate that the the mismatch strain in $(Ge_4Si_4)_5$ MSLS is accomodated by both bond-compression and bond-bending in the (Ge_4) layer. The determined Si/Ge coordination number ratio for $(Ge_4Si_4)_5$ MSLS ($N_{Si}:N_{Ge} = 2.2:1.8$) deviate from that of an ideal interface model ($N_{Si}:N_{Ge} = 1:3$), which indicates a substantial interface mixing. A simple mechanism of intermixing via site-exchange and surface segregation is proposed.

INTRODUCTION

Artificially ordered GeSi superlattices possessing novel electrical and optical properties, have opened a doorway to band-structure engineering through heterostructures formed by the strained-layer coherent epitaxy of Si and Ge [1,2]. The (Ge4Si4)5 monolayer strained-layer superlattice (MSLS) grown on Si(001) substrate, has shown strong optical transitions (0.75, 1.25, 2.31eV) unique to the superlattice period, which are found neither in constituent crystals nor in the $Ge_{0.5}Si_{0.5}$ alloy [3].

In the early stage, optical transitions of Ge_nSi_n (n = 1, 6) MSLS and Ge_xSi_{1-x} alloys have been studied by electroreflectance spectroscopy [3-5]. The results demonstrated that both Ge_1Si_1 and Ge_6Si_6 grown on Si(001) substrates show the electroreflectance spectra similar to random $Ge_{0.5}Si_{0.5}$ alloy whereas those for Ge_2Si_2 and Ge_4Si_4 show significant difference from that of $Ge_{0.5}Si_{0.5}$ alloy. In particular, for $(Ge_4Si_4)_5$/Si(001), new optical transitions have been resolved at energies that can not be explained by a simple combination of those for crystalline Ge or Si. People and Jackson [6] have pointed out that the lowest-lying states in the conduction band are derived from the (100) valleys of Si. This result has been supported by the band calculations based on ideal superlattice model [7-9]. However, the magnitude of the calculated matrix elements for the new transitions are several orders lower than what is observed in experiment. On the other hand, the band calculations that take account of the

CP554, *Physics in Local Lattice Distortions*, edited by H. Oyanagi and A. Bianconi
© 2001 American Institute of Physics 1-56396-984-X/01/$18.00

deviations from an ideal superlattice result in optical transition matrix elements that are in better agreement with experimental results [10,11]. More recently, the surface sensitivity of XAFS experiment has been significantly improved by a grazing-incidence fluorescence excitation and a high efficiency x-ray detector. In fact, submonolayer sensitivity has been achieved [12,13].

In this work, we report the local structure around Ge atoms in $(Ge_4Si_4)_5/Si(001)$ MSLS using fluorescence XAFS. The results are discussed in relation to the strain accomodation and relaxation in the Ge layers, which have been a subject of numerous studies of Ge_xSi_{1-x} alloys [14-24]. Our results on MSLS show that the Ge layers are tetragonally deformed and the lattice matching is mostly achieved by bond bending (75%) with a minor contribution of bond compression (25%), and that the coordination geometry around Ge atoms indicates an appreciable amount of interface-mixing. A simple model is proposed to take into account both overlayer-substrate site-exchange.

EXPERIMENTAL

$(Ge_4Si_4)_5$ MSLS was prepared on Si(001) substrate in an ultra-high vacuum (UHV) molecular beam epitaxy (MBE) growth chamber with a base pressure of 2×10^{-11} Torr. The detail of MSLS sample preparation is reported elsewhere [5]. The layer thickness of the Ge_nSi_n MSLS was analyzed *in-situ* by reflection high-energy electron diffraction (RHEED). The fluorescence-yield spectra were recorded for $(Ge_4Si_4)_5$ MSLS and Ge_xSi_{1-x} alloys on Si(001), while transmission spectra were taken for

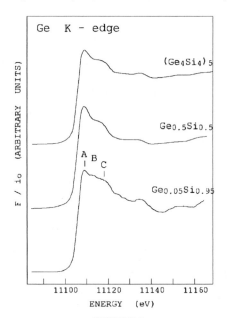

FIGURE 1.

crystalline c-Ge. All measurements were peformed at the BL-13B at the Photon Factory, National Laboratory for High Energy Physics (PF, KEK). The incidence angle for fluorescence-detected XAFS measurements was chosen so that the contribution of substrate Si is minimized. The electron beam energy was 2.5 GeV and the maximum stored current was 400 mA. XAFS data were collected using a fixed-exit double-crystal Si(111) monochromator. For each data point, the signal was integrated for ten seconds and ten scans were averaged. The energy window of the detector electronics for each channel was chosen to record only the Ge K_α peak.

RESULTS

The Ge K-edge x-ray near-edge absorption structure (XANES) are plotted in Fig. 1 for $(Ge_4Si_4)_5$ MSLS (top column), $Ge_{0.5}Si_{0.5}/Si(001)$ (middle) and $Ge_{0.05}Si_{0.95}$ / Si(001) (bottom). The sharp "white line" peak shows fine features A, B and C for Si-rich GeSi alloy ($Ge_{0.05}Si_{0.95}$). Feature A is enhanced while the peak is narrowed in width making it difficult to resolve B and C on going to higher Ge concentration. The low-k EXAFS region (11140~11160 eV) reflects the profile of back scattering amplitude of coordinated atom. The white line peak profile and the low-k EXAFS region for $(Ge_4Si_4)_5$ MSLS are similar to that of $Ge_{0.5}Si_{0.5}$, indicating that the average Ge composition of MSLS is close to that of $Ge_{0.5}Si_{0.5}$ although the intensity of feature A is suppressed.

The results of Fourier transform (FT) of EXAFS oscillations $\chi(k)$ multiplied by k, representing the radial distribution function (RDF) are shown in Fig. 2. The FT magnitude for c-Ge shows a characteristic features of a diamond-type tetrahedral structure (Td) upto the third nearest neighbors. These features are essentially the same as those for $Ge_{0.05}Si_{0.95}/Si(001)$ with a shorter interatomic distance. Compared with

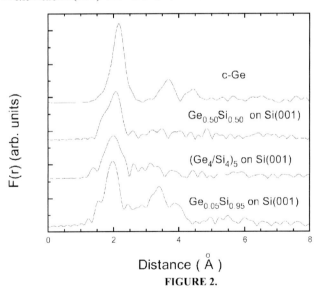

FIGURE 2.

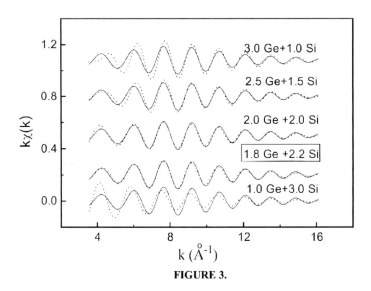

FIGURE 3.

the results for c-Ge, the magnitude of the prominent peak for $(Ge_4Si_4)_5$ MSLS decreased by about 50% shifting toward a smaller distance direction by 0.16 Å. The second and third nearest neighbor peaks in the RDF for $(Ge_4Si_4)_5$ MSLS are not well-resolved. Upon comparing the RDFs for $(Ge_4Si_4)_5$ MSLS and $Ge_{0.5}Si_{0.5}/Si(001)$ as shown in Fig. 2, one can notice that smearing is due to the interference between Ge-Ge and Ge-Si pairs. In fact, Kajiyama et al. [22] and Aldrich et al. [14] reported that no evident second and third nearest neighbor peaks appear for the relaxed crystalline $Ge_{0.61}Si_{0.39}$ and $Ge_{0.59}Si_{0.61}$ alloys.

The fitting results are summarized in Table 1. The first-shell EXAFS oscillations were calculated for several model structures for $(Ge_4Si_4)_5$ MSLS with various N_{Ge}/N_{Si} ratio and compared with the experimental curve in Fig. 3. In this simulation, the Ge-Ge and Ge-Si distances separately determined by a least-squares fit were used and only the ratio of Ge/Si coordination number is varied keeping the total coordination number (4). As can be seen, the best agreement is achieved for $N_{Ge}=1.8$ and $N_{Si}=2.2$.

Fig. 4a illustrates the schematic structures for $(Ge_4Si_4)_5$ MSLS with an ideal Ge/Si interface (left) and with an interface mixing (right). In the latter model, the Ge overlayer is assumed to have 0.5 ML site-exchange upon deposition and 1 ML

TABLE 1. Structural Parameters of Ge/Si samples obtained from XAFS data

sample	bond pair	R(Å)	N	σ(Å)	ΔE_0(eV)
$(Ge_4Si)_5$/Si(001)	Ge-Ge	2.42±0.01	1.8±0.2	0.062±0.005	6.5±1.0
	Ge-Si	2.38±0.01	2.2±0.2	0.068±0.005	4.5±1.0
$Ge_{0.5}Si_{0.5}$ alloy	Ge-Ge	2.41±0.01	2.0±0.2	0.058±0.005	7.0±1.0
	Ge-Si	2.38±0.01	2.0±0.2	0.050±0.005	3.8±1.0
$Ge_{0.05}Si_{0.95}$ alloy	Ge-Si	2.35±0.01	4.0±0.2	0.045±0.005	4.5±1.0
	Ge-Ge		~0		
C-Ge	Ge-Ge	2.45±0.01	4.0±0.2	0.054±0.005	8.0±1.0

segregation upon Si overlayer growth. Fig. 4b illustrates the schematic presentations of the interface local structure for the first Ge layer deposited on Si(001) with and without a site exchange effect. (a) shows the ideal interface while (b)-(d) are model interface structures where 1/2 ML Ge sites exchange with Si atoms. Note that the site exchange decreases the N_{Ge}/N_{Si} ratio depending on which sites are involved, although (c) and (d) give the same value.

DISCUSSION

For Ge overlayers grown on Si(001), a biaxial compression due to a lattice mismatch (4%) between Ge and Si causes a tetragonal deformation (elongation) of unit cell. Lattice distortion is achieved by both bond bending and bond shortening for the Ge-Ge and Ge-Si pairs. Previous model structures of $(Ge_4Si_4)_5$ MSLS, however, assumed that the Ge-Si interface is atomically sharp, and described the tetragonal deformation with a macroscopic elastic theory. This is based on the assumption that only the Ge layers are deformed along the c-axis, and ab plane lattice spacings (a=5.43 Å, b=5.43 Å) is kept constant. On the other hand, the c-axis lattice spacing 2c=11.24 Å expands by 0.38 Å compared with the 2c=10.86 Å of c-Si [3].

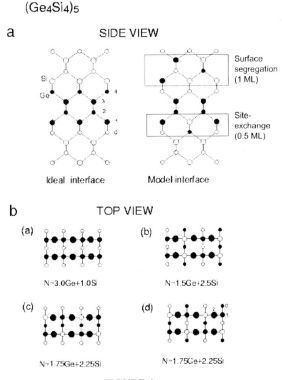

FIGURE 4

As can be seen in Table 1, we find that the bond lengths R_{Ge-Ge} (2.42 Å) and R_{Ge-Si} (2.38 Å) in $(Ge_4Si_4)_5$ MSLS are slightly shorter than those of c-Ge (R_{Ge-Ge}=2.45 Å) and the sum of covalent radii (R_{Ge-Si}=2.40 Å). Hitchcock et al. [25] and Aebi et al. [26] studied the local structures of the strained-layer $[(Si)_8(Ge)_2]_{100}$ and $[(Si)_9(Ge)_4]_{24}$ superlattices grown at 658 K. They reported R_{Ge-Ge} = 2.413 Å and R_{Ge-Si} = 2.388 Å for $[(Si)_8(Ge)_2]_{100}$ superlattice, R_{Ge-Ge} = 2.403 Å and R_{Ge-Si}=2.394 Å for $[(Si)_9(Ge)_4]_{24}$ superlattice. The bond lenghts in $[(Si)_8(Ge)_2]_{100}$ superlattice are in good agreement with that of $(Ge_4Si_4)_5$ MSLS in Table 1. More recently, we found that the bond length R_{Ge-Si} is 2.38 Å for the dilute Ge atoms doped into Si crystal. [27]. This shows that the bond length R_{Ge-Si} = 2.38 Å in $(Ge_4Si_4)_5$ MSLS takes the dilute limit value.

In summary, 1/4 of the mismatch strain in $(Ge_4Si_4)_5$ MSLS is accomodated by the Ge-Ge bond length variation while 3/4 is compensated by bond bending, in agreement with the bond length values of Ge_xSi_{1-x} alloys which are linearly dependent on composition [14-17,19-24]. Secondly, the observed R_{Ge-Ge} and R_{Ge-Si} values of MSLS rule out a possible model structure in which R_{Ge-Ge} is fully relaxed. Thus the energy level lowering due to the relaxed Ge-Ge bond length is not likely the case. Thirdly, the observed ratio of Ge and Si coordination number in $(Ge_4Si_4)_5$ MSLS reveals that the Ge-Si mixing occurs at the interface as reported earlier for Ge overlayers on Si(001) [28] although the cross section transmission microscopy shows that the (Ge_4) layers are clearly separated from (Si_4) layers [29]. There are two factors to be considered which contribute to the interface mixing: the site exchange during the Ge layer deposition on Si surface and the surface segregation during the Si overlayer growth on Ge layers. Our XAFS studies of Ge epitaxial overlayers on well-oriented Si(001) (Ge_n/Si(001), n < 7) have shown that ~ 1/2 ML of Ge atoms in the first deposited layer are replaced by the substrate Si atoms, relieving elastic strain in the second layer caused by a large atomic size mismatch between the adatom (Ge) and substrate atoms (Si) and bond-bending due to dimers [30]. In the present study, we assume that the Ge-Si interface mixing during the Ge growth on Si (001) is also about 1/2 Ge ML. Two models of Ge/Si site-exchange during the Ge overlayer growth, (b) and (c) or (d) are shown in Fig. 4b. In this figure, we consider four layers, i.e., the larger circles (3, 2) indicate the top and 2nd layer atoms while the small open circles (1, 0) indicate the 3rd and 4th layer atoms. As illustrated in the figure, the average Ge coordination number N_{Ge} decreases from 3.0 in (a) to 1.5-1.75 in (b)-(d) as the site exchange occurs. The nominal composition $Ge_{0.44}Si_{0.56}$ for model (c) or (d) is close to that of $(Ge_4Si_4)_5$ MSLS, i.e., $Ge_{0.45}Si_{0.55}$. Although whether the interface has an ordered structure or not is not experientally established yet, we believe that the interface mixing is roughly explained by the two contributions, i.e., Ge-Si site-exchange and Ge surface segregation.

Two factors influencing the optical transition are considered. First, the interface-mixing (chemical and structural disorder) would relax the k-conservation rule which may increase the transition matrix element. Second, the Ge_xSi_{1-x} alloy interface can be an intermediate "buffer" between Ge and Si layers helping the flat layer-by-layer growth. This would decrease the localized strain contributing to stabilize the MSLS, which may increase the gap energy. As a result normally inhibited direct transition might be allowed or the weak indirect transition might be enhanced on the contrary.

ACKNOWLEDGMENTS

We express our thanks Prof. M. Ikeda for useful discussion on the electronic structure, one of authors (S.Q. Wei) thanks the partial support by "100 people plan" and the "9•5" programs of Chinese Academy of Sciences.

REFERENCES

1. Bean, J.C., Feldman, L.C., Fiory, A.T., Nakahara, S., and Robinson, I.K., Vac, J., *Sci. Technol. A* **2**, 436–440 (1984).
2. Temkin, H., Pearsall, T.P., Bean, J.C., Logan, R.A., and Luryi, S., *Appl. Phys. Lett.* **48**, 963-965 (1986).
3. Pearsall, T.P., Bevk, J., Feldman, L.C., Bonar, J.M.., Mannaerts, J.P., and Ourmazd, A., *Phys. Rev. Lett.* **58**, 729-732 (1987).
4. Pearsall, T.P., Pollak, F.H., Bean, J.C., and Hull, R., *Phys Rev. B* **33**, 6821-6830 (1986).
5. Pearsall, T.P., Bevk, J., Bean, J.C., Bonar, J., Mannaerts, J.P., and Ourmazd, A., *Phys. Rev. B* **39**, 3741-3757 (1989).
6. People, R., and Jackson, S., *Phys. Rev. B* **36**, 1310-1313 (1987).
7. Satpathy, S., Martin, R.M., and Van de Walle, *Phys. Rev. B* **38**, 13237-13245 (1988).
8. Ciraci, C., and Batra, I.P., *Phys. Rev. Lett.* **58**, 2114-2117 (1987).
9. Hybertsen, M.S., and Schluter, M., *Phys. Rev. B* **36**, 9683-9693 (1987).
10. Wong, K.B., Jaros, M., Morrison, I., and Hagon, J.P., *Phys. Rev. Lett.* **60**, 2221-2224 (1988).
11. Morrison, I., Jaros, M., and Wong, K.B., *Phys. Rev. B* **37**, 9693-9707 (1987).
12. Oyanagi, H., Shioda, R., Kuwahara, Y., and Haga, K., *J. Synchrotron Rad.* 2, 99-105 (1995)
13. Murphy, L.M., Dobson, B.R., Neu, M., Ramsdale, C.A., Strange, R.W., and Hasnain, S.S., *J. Synchrotron Rad.* **2**, 64-69 (1995).
14. Aldrich, D.B., Nemanich, R.J., and Sayers, D.E., *Phys. Rev. B* **50**, 15026-15033 (1994).
15. Woicik, C., Miyano, K.E., King, C.A., Johnson, R.W., Pellegrino, J.G., Lee, T.-L., and Lu, Z.H., *Phys.Rev. B* **57**, 14592-14595 (1998).
16. Aubry, J.C., Tyliszczak, T., and Hitchcock, A.P., *Phys. Rev. B* **59**, 12872-12883 (1999).
17. Ridgway, M.C., Yu, K.M., Glover, C.J., Foran, G.J., Clerc, C., Hansen, J. L., and Nylandsted Larsen, A., *Phys. Rev. B* **60**, 10831-10836 (1999).
18. Ikeda, M., Terakura, K., and Oguchi, T., *Phys. Rev. B* **48**, 1571-1582 (1993).
19. Woicik, J.C., Bouldin, C.E., Bell, M.I., Cross, J.O., Tweet, D.J., Swanson, B.D., Zhang, T.M., Sorensen, L.B., King, C.A., Hoyt, J.L., Pianetta, P., and Gibbons, J.F., *Phys. Rev. B* **43**, 2419-2422 (1991).
20. Woicik, J.C., Bouldin, C.E., Miyano, K.E., King, C.A., *Phys. Rev. B* **55**, 15386-15389 (1997).
21. Matsuura, M., Tonnerre, J.M., and Cargill III, G.S., *Phys. Rev. B* **44**, 3842-3849 (1991).
22. Kajiyama, H., Muramatsu, S., Shimada, T., and Nishino, Y., *Phys. Rev. B* **45**, 14005-14010 (1992)
23. Mousseau, N., and Thorpe, M.F., *Phys. Rev. B* **46**, 15887-15893 (1992).
24. Laradji, M., and Landau, D.P., *Phys. Rev.* **B51**, 4894-4902 (1995)
25. Hitchcock, A.P., Tyliszczak, T., Aebi, P., Xiong, J.Z., Sham, T.K., Baines, K.M., Mueller, K.A., Feng, X.H., Chen, J.M., Yang, B.X., Lu, Z.H., Baribeau, J. M., and Jackman, T.E., *Surface Science*, **291**, 349-369 (1993).

26. Aebi, P., Tyliszczak, T., Hitchcock, A.P., Baines, K.M., Sham, T.K., Jackman, T.E., Baribeau, J. M., and Lockwood, D.J., *Phys. Rev. B* **45**, 13579-13590 (1992).

27. Oyanagi, H., Sakamoto, K., Shioda, R., and Sakamoto, T., *Jpn. J. Appl. Phys.* **33**, 3545-3552 (1994).

28. Wei, S.Q., Oyanagi, H., Kawanami, H., Sakamoto, K., Sakamto, T., Tamura, K., Saini, N.L., and Uosaki, K., *J. Appl.Phys.* **82**, 4810-4815 (1997).

29. Pearsall, T.P., *Appl. Phys. Lett.* **60**, 1712-1714 (1992).

30. Oyanagi, H., Sakamoto, K., Shioda, R., Kuwahara, Y., and Haga, K., *Phys. Rev. B* **52**, 5824-5829 (1995).

Local Structures of Nanocrystalline GaN Studied by X-ray Absorption Fine Structure

S. Q. Wei[1], Z. Li[1], X. Zhang[1], Y. Wang[1], K. Lu[2] and X. Chen[2]

[1] National Synchrotron Radiation Laboratory, University of Science and Technology of China, Hefei, 230029, P.R.China
[2] Institute of Physics, Chinese Academy of Sciences, Beijing, 100080, P.R.China

Abstract. X-ray absorption fine structure (XAFS) was used to investigate the local structures around Ga atoms in the hexagonal nanocrystalline and crystalline GaN under 78 and 300 K. For the first nearest neighbor coordination shell of Ga-N, the average bond length R (0.194 nm), coordination number N (4.0), thermal disorder σ_T (0.0052 nm) and static disorder σ_S (0.0007 nm) are nearly independent of the measured temperature and the crystalline state. This indicates that the Ga-N covalent bond is much stronger, and the 4 nitrogen atoms in first nearest neighbor around Ga atoms keep the tetrahedral structure Td. For the second nearest neighbor coordination shell of Ga-Ga, their bond lengths are about 0.318 nm. However, the σ_S (0.0057 nm) of nanocrystalline GaN is 0.0047 nm larger than that of crystalline GaN (0.001nm), and the σ_T of nanocrystalline is 0.0053 nm and 0.0085 nm at the temperature of 78 and 300 K, respectively. The result indicates that the difference of local structure around Ga atoms between nanocrystalline and crystalline GaN occurs mainly at the Ga-Ga second nearest-neighbor coordination shell. The reason is explained as the local lattice distortion and unsaturated surface atoms existing in nanocrystalline GaN.

INTRODUCTION

GaN based material with a wide band gap of 2 ~ 6 eV, has long been recognized as a promising candidate for semiconductor device application [1,2]. Recent improvements in material quality have made possible the utilization of GaN in fabrication of blue to ultraviolet light-emitting diodes, lasers and transistor [3-6]. In particular, nanocrystalline GaN shows size-tunable optical and electrical properties [7], the GaN multiple quantum-wall lasers are currently commercialized and its lifetime under continuous work operation is above 10000 hours [1], and the GaN based field effect transistors can operate under high-temperature of 673 K, high-power and high-frequency conditions [8]. So far, various opto- and micro electronic devices of GaN are already established or approaching the markets.

Despite exceptional progress, group III-nitride technology still suffers from the lack of a perfect substrate and therefore has to cope with strongly mismatched heteroepitaxial growth. GaN is characterized from high concentrations of point defects such as vacancies, nitrogen antisites and intersititials [9] as well as extended defects, *i.e.* dislocations and stacking faults, which have detrimental effects on the performance of lasers. Nevertheless, the relationship between information the effect

CP554, *Physics in Local Lattice Distortions*, edited by H. Oyanagi and A. Bianconi
© 2001 American Institute of Physics 1-56396-984-X/01/$18.00

of defects and the performance of device is still in debate for GaN material. In order to improve the performance of GaN-based semiconductor device, it is essential to study the local structure of nanocrystalline and crystalline GaN. In the present work, the microstructures of nanocrystalline and crystalline GaN were determined under the temperature of 78 and 298 K, using XAFS measurements at the Ga K-edge.

EXPERIMENTAL

Nanocrystalline GaN samples were prepared via decomposition of Ga-containing metalorganic compound at 873 K. Crystal GaN powder was prepared from 99.999 % pure gallium and high pure NH_3 gas. Small amount of gallium metal was loaded into a pure-quartz reaction boat which was placed in a horizontal tube furnace, and heated under an Ar flow rate of about 20 ml/min up to 1273 K, and then Ar gas was stopped and NH_3 was transferred at a flow rate of about 12 ml/min for 4 h. The heating temperature was controlled at 1300 ± 10 K.

The Ga K-edge XAFS spectra for nanocrystalline and crystal GaN were measured on the U7C beam-line of Hefei National Synchrotron Radiation Laboratory (NSRL) and Beijing Synchrotron Radiation Facility (BSRF). The storage ring of NSRL was operated at 0.8 GeV with a maximum current of 160 mA, and the hard x-ray beam from a three-pole superconductor wiggler with a magnetic field intensity of 6 Tesla was used. The fixed-exit Si(111) flat double crystals were used as a monochromator. The energy resolution was about 2 eV by the Cu foil $3d$ near K-edge feature. The XAFS data were collected in a transmission mode with ionization chambers filled with flowing the mixed gases of nitrogen and argon at the temperature of 78 and 300 K, using Keithley Model 6517 Electrometer to collect the electron charge directly. XAFS Data were analyzed by USTCXAFS1 software package compiled by Wan and Wei

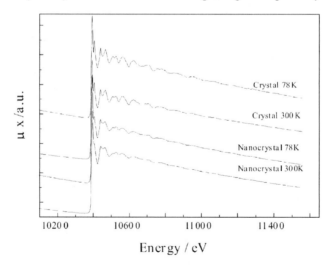

FIGURE 1. Ga K-edge absorption Spectra for GaN Samples

according to the standard procedures [10,11]. The backscattering amplitude and phase shift functions were obtained using the model hexagonal GaN (a = 0.318 nm, c = 0.5168 nm) constructed with the program FEFF7 [12].

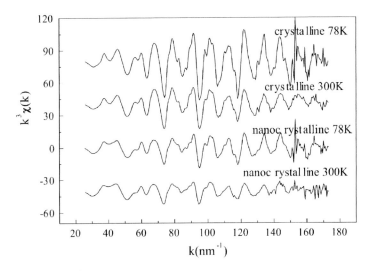

FIGURE 2. EXAFS oscillation function $k^3\chi(k)$ of GaN samples

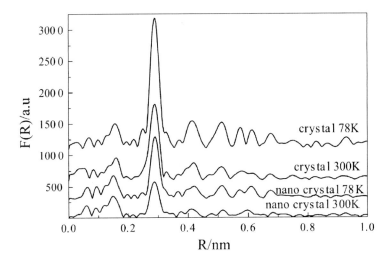

FIGURE 3. Radial distribution functions of GaN samples

RESULTS AND DISCUSSION

The Ga K-edge adsorption spectra of nanocrystalline and crystal GaN were demonstrated in Fig. 1. In the region of 300 eV above the Ga K-edge, it can be observed that there are eight strong oscillation peaks for crystalline GaN. The feature of Ga K-edge adsorption spectra of nanocrystalline GaN is similar to that of crystalline GaN, despite the oscillation intensity is decreased, especially, significantly decreases in the high-energy region. As compared with XAFS signals measured at 78 K, the oscillation magnitude of both nanocrystalline and crystalline GaN shrink dramatically at room temperature. The result suggests that the disorder degree of GaN is dependent on the temperature.

The EXAFS oscillation function $\chi(k)$ of GaN samples was obtained from the curves shown in Fig. 1, by removing background, μ_0 fitting and transformation from energy to wave vector. The detailed data analysis have been shown elsewhere [13,14]. The EXAFS functions of $k^3\chi(k)$ are illustrated in Fig. 2 for nanocrystalline and crystalline GaN. It clearly demonstrates that the oscillation magnitude of nanocrystalline GaN is smaller than that of crystal GaN. This result further reveals that there is difference in the local structure around Ga atom between nanocrystalline GaN and crystalline GaN.

Fig. 3 displays the radial distribution functions (RDF) of GaN samples, which were obtained from Fourier transformation of their $k^3\chi(k)$. There are two strong coordination peaks located at 0.16 nm and 0.29 nm, corresponding to the first and second nearest neighbor coordination shells of Ga-N and Ga-Ga, respectively. Moreover, many peaks in the region of 0.4 ~ 0.7 nm, which is due to higher coordination shells. For the two samples measured at different temperature, the intensity of the first nearest Ga-N coordination shell are similar while the intensity of the second nearest Ga-Ga coordination shell strongly depends on the temperature and crystal status. The magnitude intensities of Ga-Ga coordination shell are 576, 987,

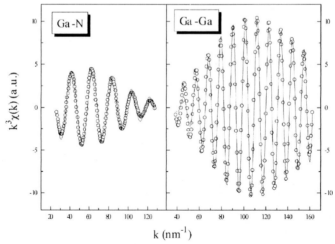

FIGURE 4. Fitting results of the experimental data (open circle) and the theoretical data (solid line) for nanocrystalline GaN (78K)

1210, and 2071 for nanocrystalline GaN (300 K), GaN (78 K), crystalline GaN (300 K) and GaN (78 K), respectively. This result suggests that the RDF intensity of nanocrystalline GaN is about 50 % as high as that of crystalline GaN. It exhibites that the local neighbor environment of Ga atom in nanocrystalline GaN sample is quite different from that of crystalline GaN.

In order to improve the precision for fitting the structural parameters around Ga atom in GaN sample, the peaks of Ga-N and Ga-Ga shells were isolated by a window function, and then an EXAFS oscillation signal of single shell was obtained by inverse Fourier transformation. The Debye-waller factor was separated into static disorder σ_S and thermal disorder σ_T for the benefit of considering the temperature effect on the disorder. Coordination distribution function $g(R)$ was assumed as the convolution of Gaussian function P_G and exponential function P_E [15].

$$g(R) = P_G \times P_E$$
$$P_E = (1/2\sigma_S)(R - R_0)^2 \exp[-(R - R_0)/\sigma_S] \qquad R \geq R_0 \qquad (1)$$
$$= 0 \qquad\qquad\qquad\qquad\qquad\qquad\qquad\quad R < R_0$$

where R_0 is the distance of central atom to the close packing atom, the average distance of $R_0 + \sigma_S$, EXAFS oscillation function should be expressed as following,

$$\chi(k) = \frac{N_j F(k) S_0^2(k)}{kR^2 \sqrt{1 + 4k^2 \sigma_S^2}} \times \exp(-2\sigma_i^2 k^2) \times \exp(-\frac{2R}{\lambda(k)})$$

$$(2)$$

$$\times \sin[2kR + \delta(k) + \arctan(2k\sigma_S)]$$

where N_j is the number of atoms in the jth shell, k is the wave vector of photoelectron, $F(k)$ is the magnitude of the backscattering amplitude of the jth-shell atoms, $S_0^2(k)$ is about 0.7 to 0.9 and is caused by many-body effects and dynamic relaxation, $\lambda(k)$ is the mean free path of electron, $\delta(k)$ is the phase shift function.

Using the equation (2), $F(k)$ and $\delta(k)$ function from FEFF7 [10] for fitting the structural parameters of GaN samples, the experimental data are well fitted with the theoretical curves as shown in Fig. 4. The fitting results were listed in Table 1.

TABLE 1. Structure parameters obtained from XAFS data of GAN samples

Sample	Pair	R / nm	N	σ_T / nm	σ_S / nm
Crystal, 78K	Ga-N	0.194	4.0	0.0047	0.0005
	Ga-Ga	0.318	12.0	0.0051	0.0008
Crystal, 300K	Ga-N	0.194	4.1	0.0050	0.0005
	Ga-Ga	0.318	12.1	0.0080	0.0010
Nanocrystal, 78K	Ga-N	0.194	4.2	0.0051	0.0007
	Ga-Ga	0.318	10.5	0.0053	0.0055
Nanocrystal 300K	Ga-N	0.193	4.3	0.0052	0.0007
	Ga-Ga	0.319	10.3	0.0085	0.0057

Crystalline GaN in the bulk form always thermodynamically stabilizes in the hexagonal (wurtzite) structure belonging to space group of *P63mc* [16]. The first and second nearest neighbors of Ga atom are 4 N atoms at 0.194 nm and 12 Ga atoms at 0.318 nm. Seen from Fig. 3 and Table 1, it can be found that the peak position and intensity of Ga-N coordination shell are almost the same for crystalline and nanocrystalline GaN samples. The bond length and coordination number are 0.194 nm and 4, which are identical with those of standard hexagonal GaN crystal. Moreover, the σ_T and σ_S of Ga-N coordination shell is nearly independent of temperature and crystal state. The results imply that the Ga-N covalent bond is much stronger, and the first nearest neighbor geometry is tetrahetral structure T_d. However, the structural difference between nanocrystalline and crystal GaN occurs in the second nearest Ga-Ga coordination shell. Although the Ga-Ga magnitude peaks shown in Fig. 3 keep at the same position, the intensity of Ga-Ga coordination shell decreases by half while the measured temperature is from 78 to 298 K or the state is from the crystalline to the nanocrystalline. The structural parameters in the Table 1 indicate that the σ_T of the second nearest Ga-Ga coordination shell is 0.005 nm at 78 K and 0.008 nm at 300 K for crystalline GaN. Our results are in good agreement with Katsikini *et al.*'s [17] results which reported that the disorder of the second nearest Ga-Ga coordination shell is 0.0052 nm and 0.0073 nm at 100 and 290 K, respectivly. In addition, there is nearly the same value for the σ_T of Ga-Ga shell between of nanocrystalline and crystalline GaN, their difference is below 0.0005 nm. Nevertheless, the static disorder σ_S (0.0055 nm) in the second nearest Ga-Ga shell of nanocrystalline GaN (78 K) is 0.0047 nm larger than that (0.0008 nm) of crystalline GaN (78 K). This reveals that there are local lattice distortions in nanocrystalline GaN.

CONCLUSIONS

The XAFS results indicated that the local structure difference between nanocrystalline and crystalline GaN samples, mainly occurs in the second nearest Ga-Ga neighbor shell. The σ_S (0.0055 nm) of Ga-Ga shell of nanocrystalline GaN (78 K) is 0.0047 nm larger than that (0.0008 nm) of crystalline GaN (78 K). We consider that the reason is local structure distortions and unsaturated surface atoms in nanocrystalline GaN. The Ga-N covalent bond in GaN samples is much stronger, whose thermal disorder σ_T and static disorder σ_S are nearly independent of the measured temperature and crystalline state. The interaction force in the second nearest Ga-Ga neighbor coordination shell is relatively weaker, and the temperature and the crystalline state can strongly influence the σ_T and σ_S.

ACKNOWLEDGMENTS

We would like to thank National Synchrotron Radiation Laboratory and Beijing Synchrotron Radiation Facility for giving us the beam time for XAFS measurement.

This work was support by "100 people plan" and the "9•5" programs of Chinese Academy of Sciences.

REFERENCES

1. Nakamura, S., Senoh, M., Iwasa, N., and Nagahama, S., *Jpn. J. Appl. Phys., Part 2* **34**, L797-799 (1995).
2. Mcmurran, J., Kouvetakis, J., and Smith, D.J., *Appl.Phys.Lett.* **74**, 883-885(1999).
3. Nakamura, S., *Meter. Res. Bull.* **22**(2), 29-35(1997).
4. Shur, M.S., and Khan, M.A., *Mater. Res. Bull.* **22**(2): 44-48(1997).
5. Cheng, G.S., Zhang, L.D., Chen, S.H., Li, Y., Li, L., Zhu, X.G., Zhu, Y., Fei, G.T., and Mao, Y.Q., *J. Mater. Research* **15**, 347-350 (2000).
6. Zhang, H.X., Ye, Z.Z., Zhao, B.H., *J. Appl.Phys.* **87**, 2830-2834 (2000).
7. Alivisatos, A.P., *Science* **271**, 933-937 (1996).
8. Yoshida, S., and Susuki, J., *J. Appl. Phys.* **85**, 7931-7936 (1999).
9. Neugebauer, J., and Van de Walle, C., *Phys. Rev. B* **50**, 8067-8071(1994).
10. Wan X.H., and Wei, S.Q., *USTCXAFS Software Package,* 1999.
11. Sayers, D.E., and Bunker, B.A., *X-ray Absorption, Principles, Applications, Techniques of EXAFS, SEXAFS and XANES*, edited by Koningsberger, D.C., and Prins, R., John Wiley and Sons, Inc., 1988, pp. 211.
12. Rehr, J.J., Zabinsky, S.I., and Albers, R.C., *Phys. Rev. Lett.* **69**, 3397-3400 (1992).
13. Wei, S.Q., Oyanagi, H., Kawanami, H., Sakamoto, K., Sakamto, T., Tamura, K., Saini, N.L., and Uosaki, K., *J. Appl. Phys.* **82**, 4810 (1997).
14. Wei, S.Q., Oyanagi, H., Sakamoto, K., Takeda, Y., and Pearsall, T.P., *Phys.Rev. B* **62**, 1883 (2000).
15. Wu, L.W., Wei, S.Q., Wang, B., and Liu, W.H., *J. Phys. CM.* **9**, 3521-3528 (1997).
16. Wells, F. W., *Structural Inorganic Chemistry*, Oxford University Press, 1975, pp. 1012.
17. Katsikini, M., Rossner, H., Fieber-Erdmann, M., Holub-Krappe, E., Moustakas, T.D., and Paloura, E.C., *J Synchrotron Rad.* **6**, 561-563 (1999).

Local Structure Analysis of GaAs:Zn by X-ray Fluorescence Holography Using Multi-element SSD

K. Hayashi[*], M. Matsui[*], Y. Awakura[*], T. Kaneyoshi[†],
H. Tanida[‡] and M. Ishii[‡]

[*]Department of Materials Science and Engineering, Kyoto University, Sakyo-ku, Kyoto, Japan
[†]Hyogo Prefectural Institute of Industrial Research, Suma-ku, Kobe, Japan
[‡]Japan Synchrotron Research Institute, SPring-8, Mikazuki-cho, Sayo-gun, Hyogo, Japan

Abstract. X-ray fluorescence holography is a technique which can record the amplitude and phase of X-rays scattered by atoms and directly provide three dimensional atomic image of an environment around the atoms, emitting the fluorescence. We used this method to image the local atomic environment of Zn atoms doped in a GaAs wafer using synchrotron radiation and a multi-element solid state detector. The first two neighbor atomic image on (001) plane were successfully obtained, revealing that the Zn atoms occupied substitutional site.

INTRODUCTION

The holographic method was first proposed and demonstrated by Gabor [1] in order to improve the spatial resolution of electron microscopy. Szöke pointed out that a photoexcited atom within a sample was an ideal source of coherent waves for atomic resolution holography [2], and proposed the concepts of an X-ray fluorescence holography (XFH) and an X-ray photoelectron holography (XPH). In last decade, XPH was applied for direct imaging of surfaces. However, it has been difficult to obtain clear atomic image due to phase shift of electron scattering and multiple scattering. Since the phase shift and multiple scattering of X-ray scattering are negligible in data analysis, XFH has recently been attracting attention.

In 1995, Tegze and Faigel measured the XFH for a strontium titanate ($SrTiO_3$) crystal [3]. Subsequently, several types of XFH methods were developed, such as multiple energy X-ray holography (MEXH) [4], γ-ray holography [5], bremsstrahlung x-ray holography [6] and mixed XFH [7]. Though the atomic images have been successfully obtained using these methods until now, most of the papers report demonstrations on the XFH for structural determinations of single crystals, whose atomic configuration were already known by the X-ray diffraction method. In 1998, we measured the hologram of Zn doped in GaAs wafer by the MEXH and successfully obtained an atomic image, suggesting that the Zn atoms occupied the substitutional site. But, the atomic image was blurry and distorted [8]. In this experiment, we improved MEXH experimental system and again measured the holograms of the Zn in

CP554, *Physics in Local Lattice Distortions*, edited by H. Oyanagi and A. Bianconi
© 2001 American Institute of Physics 1-56396-984-X/01/$18.00

GaAs at two different energies using synchrotron radiation in order to obtain high quality atomic image.

PRINCIPLE AND EXPERIMENTAL

The MEXH bases the idea of the optical reciprocity of the normal XFH and the application of known principles for X-ray standing waves. As shown in Fig. 1, the atoms emitting fluorescence serve as the detector of the interference field originating from the incident and scattered X-rays, which constitutes the reference and object beams, respectively. The holographic pattern can be obtained by detecting the fluorescence while varying the sample orientation relative to the incident beam. Since the incident X-rays can be of any energy above the absorption edge of the emitter, MEXH allows holograms to be recorded at an arbitrary energy.

The hologram measurement was carried out using synchrotron beam line BL10XU at SPring-8. The synchrotron radiation from an undulator was monochromatized by a Si (111) double-crystal monochromator. The GaAs:Zn wafer (001) was purchased from Furuuchi Chemical Co. (Tokyo Japan). The Zn concentration in the wafer was 1.0×10^{19} atoms cm^{-3}. The diameter and thickness of the sample were 50.0 and 0.25 mm, respectively. The incident X-ray energy was 9.7 and 10.0 keV, which was between the Zn and Ga K absorption edges, so as to avoid excitation of the Ga and As X-ray fluorescence. Fig. 2 shows a schematic illustration of the experimental setup. The sample was mounted on a two-axis (θ-ϕ) rotatable stage, where θ was the polar angle between the incident beam and the surface normal, and ϕ was the azimuthal angle between the [110] and the projection of the incident beam. Multi element SSD was placed parallel to the incident X-ray electric field, as shown in Fig. 2. The multi element SSD is suitable for the MEXH of the impurity, because it has a large detecting area and enables high countrate measurement. The intensity of Zn Kα X-ray fluorescence was measured as a function of the azimuthal angle, ϕ, and polar angle, θ, within the ranges of $0° \leq \phi \leq 360°$ and $26° \leq \theta \leq 60°$. The dwelling time for one pixel was 1 s with 2° steps in θ and 1° steps in ϕ. The total integrated intensity of the Zn Kα X-ray fluorescence at each pixel was about two hundred thousand counts.

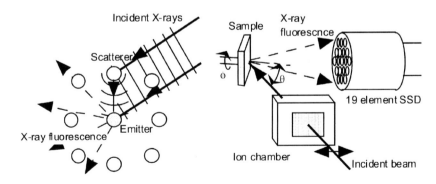

FIGURE 1. Priciple of MEXH.   **FIGURE 2.** Experimental setup for MEXH.

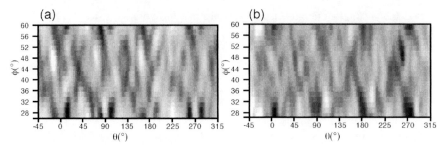

FIGURE 3. MEXH Holograms of Zn in GaAs (001). (a):9.7 keV. (b): 10.0 keV.

RESULTS AND DISCUSSION

Since the overall anisotropy of hologram was about 0.3 % in the total intensity of the fluorescence, the holographic undulation can hardly identified from the raw data due to their noise. Thus, the data were smoothed by band-pass filter. Fig. 3 (a) and (b) show the holographic patterns recorded at the incident energies of 9.7 and 10.0 keV, respectively, after smoothing. The holographic patterns in Fig. 2 (a) and (b) were different, while they show periodical patterns reflecting the symmetry of GaAs (001) plane. We reconstructed the atomic image from these holograms using the Barton algorithm [9], and successfully obtained the atomic images within the distance of 7 Å from Zn atoms, as shown in Fig. 4. Since we could not observe the atoms above 4 Å distant from the emitter in the previous paper [8], it was considered that the present hologram data was precise, compared to the previous one.

From Fig. 4, 110, $\bar{1}$10, 1$\bar{1}$0 and $\bar{1}\bar{1}$0 atoms are clearly seen at the distance of 4 Å from the emitter. The crystal structure of GaAs is a ZnS-like structure with a = 5.65 Å;

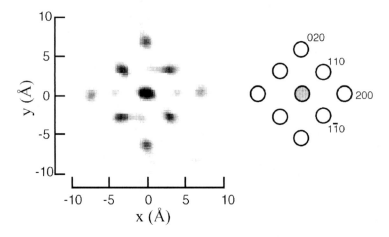

FIGURE 4. Reconstructed holographic image of (001) plane around Zn (left) with atomic configuration of GaAs.

that is, it consists of two face-centered-cubic cells. The Ga and As layers stack alternately along the c-axis; these two layers are separated by 1.41 Å. The atomic configuration of the Ga layer is the same as that of the As layer, and the nearest Ga-Ga or As-As distances are 4.00 Å. Thus, the Zn atoms are found to substituted for a Ga or As site. The possibility of As-site substitution may be negligible because of the charge neutrality. This result well agreed with EXAFS one [10]. 200, $\overline{2}00$, 020, $0\overline{2}0$ atoms are visible at distances of 7 Å, which is 25% larger than the distance (5.65 Å) expected from the GaAs crystal structure. It is considered that this displacement originate from the weakness of the holographic signal from 200, $\overline{2}00$, 020, $0\overline{2}0$ atoms. The intensity of holographic signal is inversely proportional to the distance from the emitter.

CONCLUSION

We measured the holograms of Zn in GaAs (001) using multi element SSD at third generation synchrotron facility, SPring-8. The reconstructed image of (001) plane shows the first two Ga neighbor atoms and revealed that most of the Zn atoms occupied the substitutional site, similarly to the previous result [8]. The limitation of the observable area was largely improved in this experiment. The XFH technique is applicable to determine the three-dimensional atomic arrangement around impurities in single crystal. In future, this method will become promising method for observation of the local lattice distortion around certain elements in semiconductors, superconductors and magnetic materials.

ACKNOWLEDGMENTS

This work was performed under the approval of the Spring-8 Program Advisory Committee (1999B0121-ND-np). Part of this work financially supported by a Grant-in-Aid for Scientific Research from the Ministry of Education, Science, Sports and Culture, Japan.

REFERENCES

1. Gabor, D., *Nature* **161**. 777-778 (1948).
2. Szöke, A., *Short Wavelength Coherent Radiation: Generation and Applications,* edited by. D. T. Attwood and J. Boker, AIP Conference Proceedings 147. New York: American Institute of Physics, 1986, pp.361-367.
3. Tegze, M. and Faigel, G., *Nature* **380**, 49-51 (1996).
4. Gog, T., Len, P. M., Materik, G., Bahr, D., Fadley, C. S. and Sanchez-Hanke, C., *Phys. Rev. Lett.* **76** 3132-3135 (1996).
5. Korecki, P., Korecki, J. and Slezak, T., *Phys. Rev. Lett.* **79**, 3518-3521 (1997).
6. Bompare, S. G., Peterson, T. W. and Sorensen, L. B., *Phys. Rev. Lett.* **83**, 2741-2744 (1999).
7. Hayashi, K. *J.pn. J. Appl. Phys.*, to be published.

8. Hayashi, K., Yamamoto, T., Kawai, J., Suzuki, M., Goto, S., Hayakawa, S., Sakurai, K. and Gohshi, Y., *Anal. Sci.* **14** 987-990 (1998).
9. Barton, J. J., *Phys. Rev. Lett.* **67,** 3106-3109 (1991).
10. Kitano, T., Matsumoto, Y. and Matsui, J., *Appls. Phys. Lett.* **53,** 1390-1392 (1988).

Nanoscale Heterogeneity in Crystalline Materials

S. D. Conradson, F. J. Espinosa, A. Henderson and P. M. Villella

*Materials Science and Technology Division, Los Alamos National Laboratory,
Los Alamos, NM, 87545, USA*

Abstract. Nanoscale heterogeneity in crystals is described, a phenomenon that creates a third class of materials that is not on the sequence between the standard dichotomy of perfectly ordered crystals and glasses. Multiple conformations of atoms may arise spontaneously or by random or cooperative composition fluctuations in compounds whose chemical formula is not an integral ratio of the unit cell stoichiometry. Several examples are given, establishing the association between this heterogeneity and the interesting properties of complex materials.

INTRODUCTION

A structural property found in many complex materials, *i.e.*, those that display interesting correlated and transformation behaviors, is the presence of nanoscale phase separation or heterogeneity. Although multi-domain structures are accepted as a fundamental aspect of glasses [1], the putative rigidity and order of crystalline solids poses problems in understanding why and how different coexisting arrangements of atoms form. One origin is spontaneous charge localization and atom displacements, *e.g.*, charge density waves and stripes [2]. However, the distribution anomalies may also be caused by nanoscale composition fluctuations. To wit, in a crystal containing two or more elements that substitute for each other in the lattice; how do the atoms arrange themselves when the stoichiometry is not identical in every unit cell? This can be the outcome of chemical formulas that do not have integral ratios with the stoichiometry of the unit cell or because the atoms are distributed (semi-) randomly instead of as a rigidly ordered phase. The crystallographic answer to this question is Vegard's law, the size of the unit cell expands and contracts according to the population weighted average size of the substituting element pair. However, local structure measurements show that the sizes of the different atoms are largely conserved, implying a very much different ordering of the atoms (Fig. 1) [3]. The atoms are displaced off the lattice and the connectivity is maintained by flexing of the bond angles. The crystallographic unit cell becomes a useful construct that does not actually occur anywhere in the crystal, resulting from averaging over all of the actual conformations of the atoms.

While this second conceptual model of the atomic arrangements is more sophisticated it is nevertheless still an oversimplification. It assumes that the electronic structure and bonding of every atom of each element are fixed and do not

CP554, *Physics in Local Lattice Distortions*, edited by H. Oyanagi and A. Bianconi
2001 American Institute of Physics 1-56396-984-X

respond to their local environment. In fact, these atomic characteristics that determine the local connectivity and conformation - and, by extension, the electronic behavior and chemical reactivity - can be highly dependent on the types of nearest neighbors (nn). The next step is therefore to consider the ramifications of a randomly substituted lattice. There will be not only substantial disorder of the atom positions with respect to the lattice and the average structure of the crystal but also small (but connected) regions with local compositions differing from the average (Fig. 2). Furthermore, the attributes of an A-type atom surrounded by all B's may differ from one with all A neighbors, and one with equal amounts of adjacent A's and B's does not necessarily behave as an average of the two extremes. This dependence can have large consequences on the structure and properties of the material because randomly positioned substituting atoms give a distribution with nanometer scale composition fluctuations rather than an even one. Examination of a randomly ordered, two-dimensional AB compound shows that it does not resemble chemically ordered AB. Instead, using a 75% rate of accuracy in atom placement and relatively restrictive rules on domain sizes and shapes, it can be seen to contain all possible ordered structures from pure A to pure B.

The final step in constructing an accurate depiction of the arrangement of the atoms is to include cooperative effects that promote pair specific interactions between neighboring sets of atoms instead of average behavior. The domains in which random

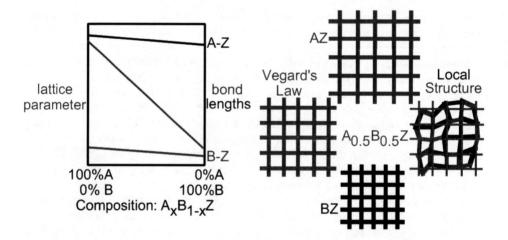

FIGURE 1. Because diffraction measurements are biased towards higher symmetry, periodic crystal lattices via their sensitivity to the long range average arrangement of the atoms, conventional crystallographic methods give linear changes in lattice parameter with composition in substitutional compounds (Vegard's law) and find highly ordered atom positions of intermediate size and average composition. In contrast, local structure measurements show that, on the atomic scale, atomic sizes and bond lengths are conserved and that, in such substitutional compounds, the resulting displacements of atoms result in the lattice reflecting average rather than actual positions and the unit cell is merely a construct that does not actually occur anywhere within the crystal.

or induced local fluctuations of the composition shift it away from the average may then attempt to reorganize into a configuration that is more stable with respect to that local combination of atoms. This will be countered by the elastic forces and interfacial energies from the surrounding lattice. When the coupling between the atoms is weak, *i.e.*, when their intrinsic tendency to organize into a particular conformation is small, the ordering and properties may be characterized by modest, purely local composition-driven fluctuations around the average order. In contrast, when the minima in the free energy surface are surrounded by steep changes in energy with conformational coordinate, then the local composition fluctuations may result in domains that have snapped into the stable structure despite only partially complete chemical ordering (Fig. 2). In this latter case the most accurate nanoscale description of the material is isolated fully ordered domains separated by fully dis- or differently ordered regions. In addition, random fluctuations in composition are not necessarily solely responsible for the final chemical ordering of the compound. Bond energies and relaxation in response to elastic strains may actively promote clustering during the formation of the material, resulting in even larger domains with non-average local compositions and structures.

The identification of the multiple conformations usually defies conventional crystallographic approaches, apparently because the sizes of the domains are at or below the diffraction limit and because the coherence of the isolated domains is poor.

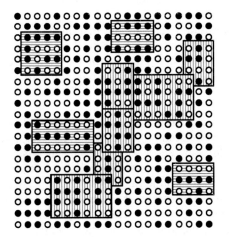 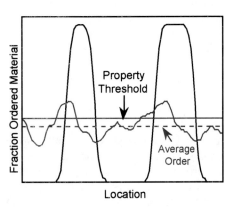

FIGURE 2. Left) A random distribution of AB in two dimensions demonstrates that all types of ordered structures occur, albeit within small domains that would be at or below the diffraction limit. Almost 40% of the atoms are involved in a layered ABABA structure (domains in boxes) using rectangular domains with a minimum 75% accuracy in atom placement and 3×4 atoms as the minimum size. Right) Even in a randomly ordered material, fluctuations from the average order could result in it displaying a property whose threshold is greater than the average order. Cooperative effects during synthesis or among the atoms within a composition fluctuation could produce a material consisting of highly ordered but isolated domains separated by totally disordered interfacial regions. Both of these behaviors have been observed.

505

Available data suggest a model on this scale where up to a fifth or even a quarter of the material occurs as small, isolated domains of a second stable atomic configuration that are embedded within the diffracting host lattice formed by the majority of the atoms [4,5]. It is also essential to ask how these materials will behave. Suppose a compound with a certain chemical formula displays a property when ordered in a particular way and that this property can occur at small domain sizes. A material containing the elements of the compound will then exhibit that property if the local fluctuations in composition result in domains sufficiently ordered so as to exceed the threshold where the property begins, even if the average order falls below the minimum value (Fig. 2). In addition, the interactions between these domains (as opposed to isolated mononuclear clusters) and the formation of loose networks may endow the materials with unexpected, unusual, and even unique properties.

SUPPORTING EXPERIMENTAL RESULTS

A system where these organizational principles are observed is cubic stabilized zirconia (CSZ). The high temperature cubic phase of ZrO_2 is stabilized by substitution of Zr with a variety of alio- and isovalent cations that range from smaller to larger than Zr^{4+}. The common theme in the stabilization mechanisms is the reduction of the steric strain around the very small, highly charged Zr^{4+} ion in the cubic geometry. This is accomplished by expansion of the Zr-O bond lengths and reduction of the O nn number from eight [6]. Calculations show that the placement of the oxygen vacancies is dependent on the charge and size of the substituting cations and that the distortions are extended with second and third shell atoms displaced from their average sites in the lattice by a few hundredths to a few tenths of an Å [7]. Element-specific local distortions are observed in multi-edge XAFS measurements on quaternary Y-Er-Ce/U CSZ candidate inert matrix fuel compounds. The conservation of the O shell characteristics around Zr and Er and expansion of their M-second nn cation distances to compensate for a larger ion (U) follow the pattern previously observed in binary CSZ [8]. However, steric strain also appears to be accommodated around Zr, Ce, and Y with U by a 0.5 - 0.7 Å splitting and additional subsplittings of the nn O shell, around the Er by distorting the second nn cation shell, or around U by both. The local distortions in response to substitution and stabilization are thus highly element specific. Because of the virtual identity of size and charge of Er and Y it is notable that they behave differently. In addition, the Y-O distribution displays the highly distorted pattern in the U-containing compound but a much smaller O distortion and wider second near neighbor distribution with Ce. That the details of the distortion around the Y (but not Zr or Er) depend upon whether the compound is made with Ce or U demonstrates that not only are the local environments element specific but also that the extended structure beyond the nn is susceptible to cooperative, pair specific interactions.

Collective behavior also occurs in, $Ca_{0.8}Pu_{0.2}HfTi_{1.8}Al_{0.2}O_7$, the synroc mineral zirconolite with 20% substitution of Ca by Pu [9]. Based on this substitution a small number of Pu-Pu pairs would be expected at the Ca-Ca distances. This is not observd experimentally. Because the large number of overlapping shells make it difficult to unambiguously identify Pu nn by XAFS, anomalous PDF measurements at the Pu L$_{III}$ edge with their high sensitivity to Pu-Pu pairs were used instead. Apart from the amplitude differences caused by small errors in peak intensity, the differential PDFs show only the one significant variance near 5.5 Å, which is a Ca-Zr and not a Ca-Ca distance. The magnitude of this difference is also unexpected, it corresponds to a very large fraction of the total Pu. The Pu is evidently not randomly distributed in the host but preferentially occurs in domains with high local Pu concentrations, most likely formed by favored interactions of the local strain fields around the Pu. XAFS measurements demonstrate that it does not form PuO$_2$ inclusions but is found as a locally ordered structure that contains many of the expected features of the zirconolite. This suggests two possible models. Half of the Pu atoms could reside in Zr sites that are next nn to the Pu atoms on the Ca sites in nanoscale Pu-enriched domains. Alternatively, the Pu could all reside on Ca sites in such domains, with the local

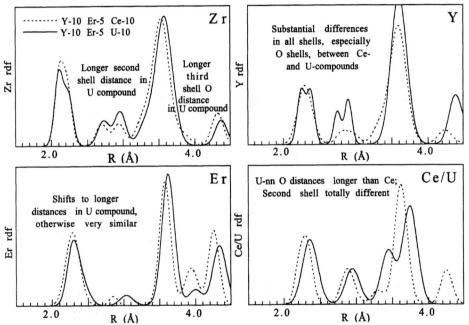

FIGURE 3. Radial distribution functions (rdf) around indicated elements in the quaternary 10 mol-% Y, 5 mol-% Er, 10 mol-% Ce/U CSZ compounds, showing element specific changes in local environments resulting from substituting U for Ce. Shells below 3.2 Å are always O. The M-O nn shell is highly split for M = Zr, Ce/U, and Y with U, but not for Er or Y with Ce. The second nn cation shells beyond 3.2 Å are the composite of all types and is split around Er, U, and broader around Y with Ce. The larger size of U(IV) relative to Ce(IV) in the CSZ lattice (*cf.* nn O positions in Ce/U plot) results in an expansion of the second and third nn for Zr and Er, whose rdfs are otherwise very similar in both the Ce and U compounds. The substantial differences in the Y environment, including splitting of the O shell at 2.3 Å and increased separation of the more distant nn O set, and in Ce/U environment including the splitting of the second shell cations, imply strong interaction between the Y and U sites that results in preferential Y-U second nn aggregation.

structure ordered but sufficiently distorted to form a non-diffracting network extending through the host. Whichever of these two is correct, these data show both the presence of pair specific interactions and local distortions and the consequence as non-diffracting, locally ordered, nanoscale heterogeneity in the arrangement of the atoms.

Magnetic alloys are a class of compounds where the direct coupling between the local structure and short range order and the magnetic properties can be elucidated. A correlation between perpendicular magnetic anisotropy and locally ordered PtCo layers has been observed in Pt_3Co [4]. Another example is NiMn. The chemically ordered, tetragonally distorted structure is antiferromagnetic whereas the disordered fcc structure is not [10]. However, significant exchange coupling occurs prior to the observation of chemical ordering in diffraction patterns. Standard analysis of the Mn XAFS is inconclusive; relatively poor fits to the data suggest significant disorder. However, comparisons of the experimental phase shifts from the combined third and fourth shell contributions with those from calculated spectra ranging from fully fcc to fully tetragonal are consistent with a very high degree of local order as the origin of the observed exchange coupling. This results implies a structure consisting of small, highly ordered, poorly coherent, tetragonal domains embedded in highly disordered material that contributes only minimally to the XAFS spectrum. The structure averaged over the diffraction limit is fcc. Such a structure suggests that, instead of the purely random arrangement described previously, cooperative effects among small groups of atoms promote rapid chemical ordering and the formation of the tetragonal

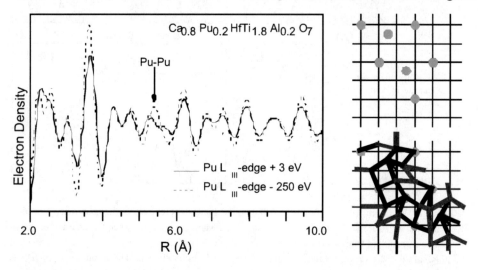

FIGURE 4. Pu anomalous PDF measurements (left) reveal a large number of Pu-Pu pairs only at around 5.5 Å. Because there are no Ca-Ca separations of this length in the crystal, this result implies that the local strain fields around the Pu atoms act collectively to either 1) make Pu pairs on adjacent Zr sites around 5.5 Å away the most stable configuration (as in upper right), or 2) create a highly distorted, Pu-rich, pseudo one-dimensional network in which nn or next nn Ca sites are displaced from their normal sites in the lattice to this distance (as in lower right).

structure at many nucleation sites. Consolidation into larger, diffracting domains is hindered by different orientations and misregistry of these domains at their interfaces, which also causes these interfaces to have poor local as well as long range order.

Although the most extensive experimental data as well as recent modeling results appertain to δ-Pu alloys our most interesting new results are from XAFS measurements of the layered manganite, $La_{1.2}Sr_{1.8}Mn_2O_7$. Polarized spectra from single crystals promote detailed characterization of the apical O distribution, with spectra from each of the cations available for identification of any element specific behavior. A change in the Mn-O distribution consistent with charge localization has been observed. However, an even more interesting corollary to this result follows from examination of the Sr and La K XAFS spectra (Fig. 6). Although both the Sr-O and La-O distributions display a modulation at T_c, only the Sr-O distribution is different above and below, with a 0.08 Å expansion of the distance in the insulating phase. A longer Sr-O distance implies localization of the itinerant holes in the MnO_2 planes, that no change in the La-O distance occurs implies that this localization is specific to the Mn sites with neighboring Sr ions. Although it is not unreasonable to expect that the static charge distribution in the rock salt layer would be mirrored by (or at least affect) the dynamic distribution in the conducting planes, the "inert" rock salt layer with its fixed charge cations has never been considered as a possible contributor

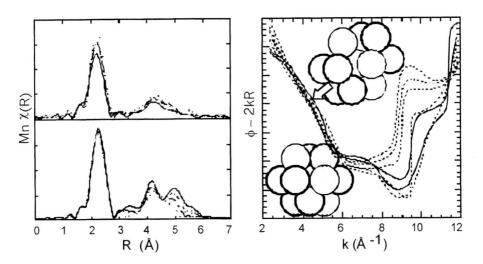

FIGURE 5. Annealing sputtered NiMn films for longer times slowly increases the amplitude of $\chi(R)$ (upper right) without ever duplicating the calculated spectra of the disordered fcc, layered tetragonal, or mixtures of these structures (lower right). However, the phase shifts of the calculated spectra (left, dahsed lines) backtransformed over the third and fourth shell region displays trends in the beat including a displacement from k = 8.5 to 10.5 Å^{-1} on going from fcc to completely tetragonal that is duplicated by the same phase shift from the experimental data (left, solid lines). The degree of ordering derived from this analysis is in good agreement with the measured exchange coupling, demonstrating that the short range order turns the coupling on before tetragonal splitting is observed by diffraction.

to the properties of these compounds. These data show that the interaction between the two components of the structure is actually quite strong. A corollary consideration is even more intriguing. Because crystallographic measurement have shown no evidence for ordering within the rock salt layer it is likely to have a random distribution of the two ions. Referring back to Fig. 2, the presence within the material of nanoscale domains enriched in both Sr and La would be expected. The different local compositions and charge distributions they possess could promote their separation and rearrangement into a heterogeneous type of structure. The structure within and interactions between these domains could be an important factor in determining the properties of these and similar materials.

CONCLUSIONS

Nanoscale heterogeneous crystalline solids, a third class of materials that is not part of the standard dichotomy of perfectly ordered crystals and glasses, have been described and experimental evidence identifying examples presented. In addition to being coupled to the interesting properties of a number of complex systems, this conceptual model itself motivates further investigation, particularly in regards to their energetics. Nanoscale heterogeneity emulates a phase transition. The very wide temperature range over which multiple conformations are observed is consistent with a strong, scale-dependent entropic contribution, which is in turn consistent with the intrinsic chemical disorder ascribed to them. The significance of entropy in the context of the double minimum potentials that determine the two conformations and its coupling to the length scale of the associated atom displacements may be essential.

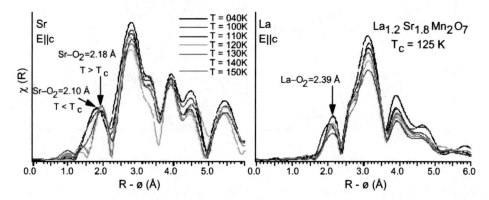

FIGURE 6. T Polarized, multiple element, temperature dependent XAFS measurements on a layered manganite compound reveal this intriguing behavior in the low and high temperature metallic and insulating phases. Although both the Sr and La show signs of disordering in the c direction at T_c, the avearage Sr-O2 distance is significantly shorter in the metallic phase whereas the average La-O2 distance never changes. This implies that the reordering of the apical Mn-O pair between the metallic and insulating phases is specific to the Mn-O-Sr combination.

ACKNOWLEDGMENTS

The authors would like to thank Claude Degueldre, Lou Vance, Bruce Begg, Robin Farrow, Mike Toney, Y. Tokura, and T. Kimura for the compounds used in these experiments. All experimental data were obtained at the Stanford Synchrotron Radiation Laboratory, which is supported by the U.S. DOE Office of Basic Energy Research. Direct support for this work came from DOE OBES and DP.

REFERENCES

1. Greaves, G. N., Smith, W., Giulotto, E., Pantos, E., *J. Non-Cryst. Solids* **222**, 13–24 (1997).

2. Bianconi, A., Saini, N. L., Lanzara, A., Missori, M., Rossetti, T., Oyanagi, H., Yamaguchi, H., Oka, K., and Ito, T., *Phys. Rev. Lett.* **76**, 3412 (1996).

3. Boyce, J. B. and Mikkelsen, J. K. Jr., in *EXAFS and Near Edge Structure III,* edited by Hodgson, K.O., Hedman, B., and Penner-Hahn, J.E., New York, Springer Verlag, p. 426 (1984).

4. Tyson, T. A., Conradson, S. D., Farrow, R. F. C., and Jones, B. A., *Phys. Rev.* **B 54**, R3702–R3705 (1996).

5. Conradson, S. D., *Appl. Spectr.* **52**, 252A–279A (1998).

6. Ho, S.-M., *Mat. Sci. Eng.* **54**, 23–29 (1982).

7. Stapper, G., Bernasconi, M., Nicoloso, N., and Parrinello, M., *Phys. Rev.* **B59**, 797–810 (1999).

8. Li, P., Chen, I.-W., and Penner-Hahn, J. E., *J. Am. Ceram. Soc.* **77**, 1289–1295 (1994).

9. Begg, B. D., Vance, E. R., and Conradson, S. D., *J. Alloys and Compounds* **271-273**, 221 (1998).

10. Farrow, R. F. C., Weller, D., Marks, R. F., Toney, M. F., Horn, S., Harp, G. R., and Cebollada, A., *Appl. Phys. Lett.* **69**,1166 (1996).

Strain Induced Synthesis of High Temperature Superconducting Thin Films

M. Mukaida, S. Makino, K. Chiba, M. Kusunoki, and S. Ohshima

Yamagata University, 4-3-16 Jonan, Yonezawa Yamagata 992-8510 Japan

Abstract. Microstructures of $YBa_2Cu_3O_{7-\delta}$ / $BaSnO_3$ superlattice are analyzed by Transmission electron microscopy. $YBa_2Cu_3O_{7-\delta}$ / $BaSnO_3$ superlattices are fabricated by a pulsed laser deposition technique with in-situ rotation of target materials. Typical growth conditions of substrate temperature and oxygen pressure are 710°C and 400 mTorr, respectively. A few nm thick $BaSnO_3$ insulating layers and $YBa_2Cu_3O_{7-\delta}$ superconducting layers are sequentially deposited with total thickness of around 150 nm. Crystallographic characteristics of these superlattices are evaluated by x-ray diffractometry (XRD) and transmission electron microscopy (TEM). Several layers of $YBa_2Cu_3O_{7-\delta}$ and $BaSnO_3$ are clearly observed by TEM. The observed thickness of one layer of $YBa_2Cu_3O_{7-\delta}$ and that of $BaSnO_3$ films are 12 nm and 6.5 nm, respectively. From a high-resolution TEM observation, the strain induced by lattice length difference between $YBa_2Cu_3O_{7-\delta}$ and $BaSnO_3$ are released by introducing misfit dislocations in the $YBa_2Cu_3O_{7-\delta}$ films not in the $BaSnO_3$ films. A new insulating layer is required with small lattice mismatch to $YBa_2Cu_3O_{7-\delta}$ for avoiding introduction of misfit dislocations.

INTRODUCTION

Since the discovery of high temperature oxide superconductors, many materials were synthesized to get higher T_C superconductors. One of the methods for synthesizing higher T_C superconductors was substitution of lanthanide elements with different ion radius [1]. In the $YBa_2Cu_3O_{7-\delta}$ system, Y atom was replaced by Nd, Sm, Ho and so on.

$NdBa_2Cu_3O_{7-\delta}$ has the highest T_C in the 1-2-3 system. However, Nd atoms easily substitute with Ba atoms resulting in low T_C s. In order to avoid this substitution, high processing temperature above 800°C is required. The above 800°C processes critically give damages to thin film growth apparatuses. In this paper, we try to introduce lattice distortions by using superlattice structures. Microstructures of $YBa_2Cu_3O_{7-\delta}$ / $BaSnO_3$ superlattice are analyzed by Transmission electron microscopy.

EXPERIMENT

Thin $YBa_2Cu_3O_{7-\delta}$ and $BaSnO_3$ films were deposited by an ArF excimer laser on (100) $SrTiO_3$ or (100) MgO substrates. Detailed experimental procedures have already been reported elsewhere [2]. Deposition of $BaSnO_3$ was also carried out *in-situ* in the same chamber using a sintered stoichiometric $BaSnO_3$ target. The

CP554, *Physics in Local Lattice Distortions*, edited by H. Oyanagi and A. Bianconi
© 2001 American Institute of Physics 1-56396-984-X/01/$18.00

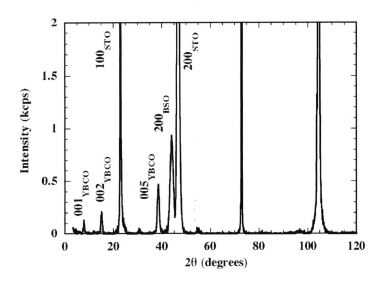

FIGURE 1. A typical x-ray diffraction pattern of a YBa$_2$Cu$_3$O$_{7-\delta}$ / BaSnO$_3$ superlattice

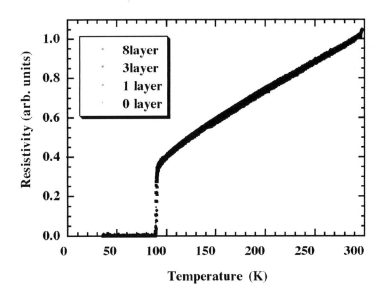

FIGURE 2. A typical ρ-T curves of superlattices

temperature was the same as that for the deposition of YBa$_2$Cu$_3$O$_{7-\delta}$. The targets were rotationally changed at a periodic growth time, which controls thickness of layers. The total thickness of the YBa$_2$Cu$_3$O$_{7-\delta}$ / BaSnO$_3$ superlattice is fixed for different number of BaSnO$_3$ layers.

Crystallinity was mainly characterized by x-ray diffractometry (XRD). c-axis length was calculated by using a NR-function. A transmission electron microscope (TEM) was used to investigate the interfacial crystalline structure. A T_C was estimated by a conventional four probe method. Surface morphology was observed by atomic force microscopy (AFM).

RESULTS AND DISCUSSIONS

Fig. 1 shows a typical x-ray diffraction pattern of a YBa$_2$Cu$_3$O$_{7-\delta}$ / BaSnO$_3$ superlattice. All peaks are assigned to be 00l peaks of YBa$_2$Cu$_3$O$_{7-\delta}$, h00 peaks of BaSnO$_3$ and h00 peaks of a SrTiO$_3$ substrate. The peaks are a little broad comparing to single layer of YBa$_2$Cu$_3$O$_{7-\delta}$.

Temperature dependence of resistivity of YBa$_2$Cu$_3$O$_{7-\delta}$ / BaSnO$_3$ superlattices with BaSnO$_3$ layers was shown in Fig. 2. In the figure, resistivities are normalized by its resistivity at 290K [ρ(290)] in order to emphasize the differences. However the characteristics of the ρ-T curves are almost the same. They showed a metallic behavior above the transition temperature.

Next we investigated the c-axis length of YBa$_2$Cu$_3$O$_{7-\delta}$ / BaSnO$_3$ superlattices with different number of BaSnO$_3$ layers, however the total thickness of the superlattices are the same. The c-axis length showed a strong BaSnO$_3$ layer dependence. As the number of BaSnO$_3$ layer increases, in other words, thickness of the YBa$_2$Cu$_3$O$_{7-\delta}$ layer decreases, the c-axis length becomes long. Usually, the c-axis length is strongly correlated with oxygen deficiency. YBa$_2$Cu$_3$O$_{7-\delta}$ films with long c-axis length show

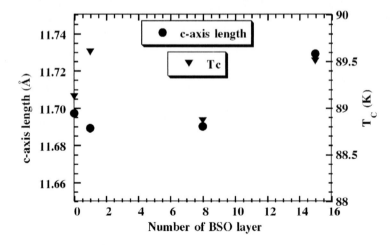

FIGURE 3. *C*-axis length of a YBa$_2$Cu$_3$O$_{7-\delta}$ film as a function of number of BaSnO$_3$ layer.

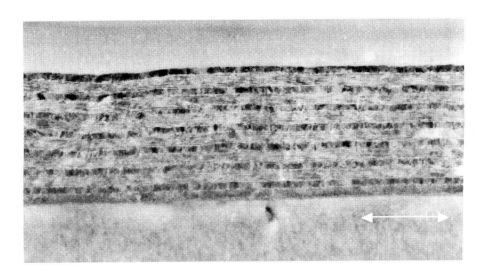

FIGURE 4. A cross-sectional TEM image of a 7-layered YBa$_2$Cu$_3$O$_{7-\delta}$ / BaSnO$_3$ superlattice.

FIGURE 5. A high magnification cross-sectional TEM image of the superlattice with a 3 nm-thick BaSnO$_3$ layer

low T_Cs. However the T_C of the YBa$_2$Cu$_3$O$_{7-\delta}$ / BaSnO$_3$ superlattice with 15 BaSnO$_3$ layers was still as high as 89.5K as shown in Fig. 3. All of the superlattices showed T_Cs between around 89K and 90K.

In order to estimate crystal structure of the YBa$_2$Cu$_3$O$_{7-\delta}$ / BaSnO$_3$ superlattice, we observed a transmission electron microscopy of a super lattice. Fig. 4 shows a low magnification cross sectional TEM image of a YBa$_2$Cu$_3$O$_{7-\delta}$ / BaSnO$_3$ superlattice grown on a SrTiO$_3$ substrate. In the figure, several layers of YBa$_2$Cu$_3$O$_{7-\delta}$ and BaSnO$_3$ layers are clearly observed. Layers with very fine lines correspond to CuO$_2$

planes are $YBa_2Cu_3O_{7-\delta}$ layers. Those without fine lines are $BaSnO_3$ layers. Thickness of $YBa_2Cu_3O_{7-\delta}$ and $BaSnO_3$ films are 12 nm and 6.5 nm, respectively. Fig. 5 shows a high magnification TEM image of a cross-section of a super lattice at the interface between $YBa_2Cu_3O_{7-\delta}$ and $BaSnO_3$. The TEM image are focused on a $BaSnO_3$ layer with the thickness of ~ 3 nm. A $BaSnO_3$ layer about 3 nm thick covers well the surface steps of the lower $YBa_2Cu_3O_{7-\delta}$ layer, and noticeably the top surface of $BaSnO_3$ reflects well the surface roughness made of steps of the lower $YBa_2Cu_3O_{7-\delta}$ film, even though some spirals exist. Careful observation of the lattice continuity at interfaces revealed that periodical discontinuity of lattice planes was predominantly in the $YBa_2Cu_3O_{7-\delta}$ layer adjacent to the $BaSnO_3$ layer. Lattice discontinuity occurs in only one third of the c-axis unit of $YBa_2Cu_3O_{7-\delta}$ adjacent to $BaSnO_3$ and discontinuity looks like partial dislocations or half-dislocations. These discontinuity release the lattice distortion induced by $BaSnO_3$ layers.

SUMMARY

Microstructures of $YBa_2Cu_3O_{7-\delta}$ / $BaSnO_3$ superlattice grown by pulsed laser deposition with an *in-situ* target rotating technique were estimated. A few nm $BaSnO_3$ insulating layers and $YBa_2Cu_3O_{7-\delta}$ superconducting layers were sequentially deposited with total thickness of around 150 nm. Crystallographic characteristics of these superlattices were evaluated by x-ray diffractometry (XRD) and transmission electron microscopy (TEM). Several layers of $YBa_2Cu_3O_{7-\delta}$ and $BaSnO_3$ are clearly observed by cross sectional TEM. The observed thickness of one layer of $YBa_2Cu_3O_{7-\delta}$ and $BaSnO_3$ films are 12 nm and 6.5 nm, respectively. From a high-resolution TEM observation, the strain induced by lattice length difference between $YBa_2Cu_3O_{7-\delta}$ and $BaSnO_3$ are released by introducing misfit dislocations in the $YBa_2Cu_3O_{7-\delta}$ films not in the $BaSnO_3$ films. A new insulating layer is required with small lattice mismatch to $YBa_2Cu_3O_{7-\delta}$ for avoiding introduction of misfit dislocations.

ACKNOWLEDGMENTS

The author acknowledges Dr. Miyazawa of Ushio Research Institute for his encouragement. They also thank Mr. Aizawa, Mr. Okai and Mr. Ehata of Yamagata University. This program is partially supported by SUZUKI FOUNDATION.

REFERENCES

1. Tsurumi, S., Iwata, T., Tajima, Y., and Hikita, M., *Jpn. J. Appl. Phys.* **26**, L1009-1012 (1987).
2. Mukaida, M., and Miyazawa, S., *Jpn. J. Appl. Phys.*, **32**, 4521-4528 (1993).

Superconducting Glass State in Binary Alloy $Ti_{0.5}V_{0.5}$ Single Crystal

K. Tsutsumi[1], S. Takayanagi[2] and T. Hirano[3]

[1]Department of Physics, Kanazawa University, Kanazawa 920-1192, Japan
[2]Physics Department, Hokkaido University of Education Sapporo, Sapporo 002-8502, Japan
[3]National Research Institute for Metals, Tsukuba, Ibaragi 305-0047, Japan

Abstract. We report the results of the specific heat measurements at low-temperatures of the binary alloy $Ti_{0.5}V_{0.5}$ single crystal which undergoes the superconducting transition at 7.20 K. The single crystal of $Ti_{0.5}V_{0.5}$ is not only one of the strong-coupling superconductors but also a dirty type-II superconductor above about 2 K. However, this material exhibits anomalous behaviors in the specific heat below about 2 K. We attribute this anomalous behavior to the transition from the superconducting state into a superconducting glass state around about 2 K.

INTRODUCTION

The study of superconductivity in the binary alloys has been mainly concentrated to the polycrystal because the synthesis of the single crystals is very difficult. The low-temperature specific heat of $Ti_{0.5}V_{0.5}$ was studied on polycrystalline specimens about 40 years ago [1]. This study showed that the superconducting transition of $Ti_{0.5}V_{0.5}$ is bulky. But there is no work on the single crystal. The present study was performed in order to obtain the information on the superconductivity characteristics of a binary alloy $Ti_{0.5}V_{0.5}$ single crystal at low-temperatures by measuring the specific heat.

We already reported the results of the specific heat measurement in comparison with the BCS theory and then those of polycrystals. We clarified the strong-coupling effect of the superconductivity and discussed in comparison with the theory derived by Sheahen and Gubser [2].

EXPERIMENTS

The single crystals of the 3d transition-metal binary alloy $Ti_{0.5}V_{0.5}$ were grown by the floating zone method with the infrared heating. It was confirmed that the as-grown crystals are single crystals by the transmission and back X-ray Laue photograph method and then the lattice constant was determined as a = 0.334 nm. The diameter of the single crystal employed for the present study is 3.8 mm, the length 5.0 mm and the weight 289 mg, respectively. In order to measure the specific heat, we employed an adiabatic method, which is described in ref. [3], in the temperature range from 0.5 to

CP554, *Physics in Local Lattice Distortions*, edited by H. Oyanagi and A. Bianconi
© 2001 American Institute of Physics 1-56396-984-X/01/$18.00

20 K. The temperature dependence of electrical resistance was measured by using an ordinary dc four-probe method from 4.2 to 300 K.

RESULTS AND DISCUSSION

The temperature dependence of the specific heat is shown in Fig. 1. The superconducting transition temperature T_c was determined from the midpoint of the step-like structure shown in Fig. 1. This is consistent with that obtained in the temperature dependence measurement of the electrical resistance. The temperature dependence of the electronic specific heat C_{el} in the superconducting regime was obtained as a function of the inverse of the reduced temperature $t \equiv T/T_c$ by the subtruction of the lattice contribution. In the temperature range $1 < t^{-1} < 5$, C_{el} could be approximated by

$$C_{el}/\gamma T_c = 14.6\exp(-1.69t^{-1})$$

where γ is the electronic specific heat coefficient in the normal metallic state. However, in the lower-temperature range which is $t^{-1} > 5$, C_{el} exhibits a temperature independent behavior, namely, a plateau. This plateau can never been explained by the BCS theory. We show this anomalous behavior of C_{el} devided by γT_c as a function of the inverse of the reduced temperature $t \equiv T/T_c$ in Fig. 2.

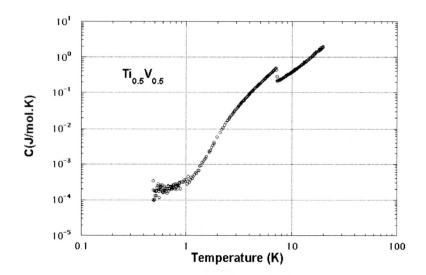

FIGURE 1. Temperature dependence of the specific heat of $Ti_{0.5}V_{0.5}$

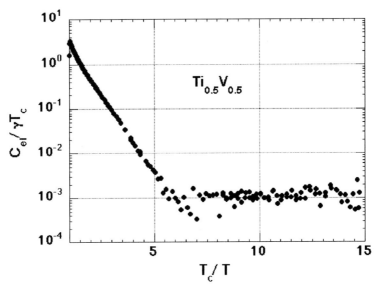

FIGURE 2. Electronic specific heat C_{el} divided by γT_c in the superconducting regime as a function of the inverse of the reduced temperature $t \equiv T/T_c$. Note that a clear plateau is seen in the low-temperature range which is $t^{-1} > 5$.

We calculated the temperature dependence of the entropy S based on the data of the specific heat. This result is shown in Fig. 3. We clearly observed a gradual change of the entropy S between 1 and 2K besides the sharp change at 7.20 K which corresponds to the superconducting transition temperature. This gradual change suggests a glass transition rather than a phase transition.

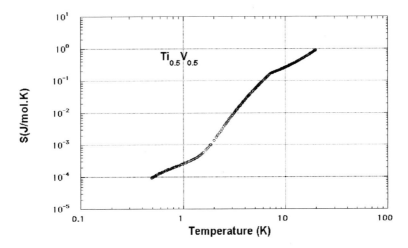

FIGURE 3. Temperature dependence of the entropy S calculated based on the specific heat data.

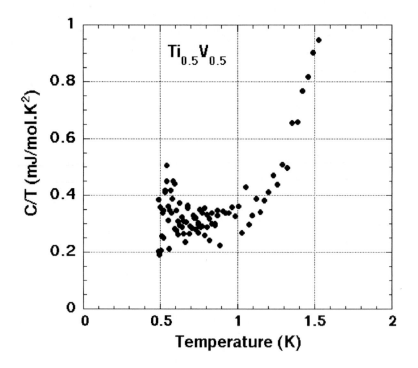

FIGURE 4. Temperature dependence of C/T in the temperature range from 0.5 to 1.6K.

We calculated the temperature dependence of C/T in the temperature range from 0.5 to 1.6 K. The result is shown in Fig. 4. The C/T shows a minimum around 0.9K and then the upturn of C/T is clearly observed below 0.8K. This corresponds to the low-energy excitations between metastable states which are characteristic of the glass state.

CONCLUSION

We insist that we might be able to describe the state below about 2K in the binary alloy $Ti_{0.5}V_{0.5}$ single crystal as a superconducting glass state by measuring the low-temperature specific heat.

REFERENCES

1. Cheng, C.H., Gupta, K.P., van Reuth, E.C., and Beck, Paul A., *Phys. Rev.* **126**, 2030-2033 (1962).
2. Tsutsumi, K., Takayanagi, S., and Hirano, T., *Physica B* **237-238**, 607-608 (1997).
3. Takayanagi, S., Wada, N., Onuki, Y., and Komatsubara, T., *J. Phys. Soc. Jpn.* **60**, 975-976 (1991).

Local Disorder in Mixed Crystals
as Viewed by XRPD

V. Sh. Machavariani[†1], Yu. Rosenberg[‡], A. Voronel[†], S. Garber[†],
A. Rubstein[†], A. I. Frenkel[*2] and E. A. Stern[§3]

[†]R.&B. Sackler School of Physics and Astronomy, Tel-Aviv University, 69978, Israel
[‡]Wolfson Center for Materials Research, Tel-Aviv University, 69978, Israel
[*]Materials Research Laboratory, University of Illinois at Urbana-Champaign
[§]Physics Department, Box 351560, University of Washington, Seattle, WA 98195-1560, USA

Abstract. A correlation between precise X-ray powder diffraction patterns and atomic size mismatch in disordered mixed crytalls (alloys and ionic crystals) is observed. The anisotropy of the elastic moduli has been taken into account for evaluation of the strain energy density of the mixed crystals revealed in XRPD measurements.

INTRODUCTION

Our extended study of mixed ionic salts with atomic size mismatch RbBr-KBr, RbCl-RbBr, and AgCl-AgBr [1] revealed strong deviations of the local structure obtained by X-ray absorption fine structure (XAFS) technique from the average structure obtained by X-ray powder diffraction (XRPD). The equilibrium atomic positions have been found by the XAFS analysis to be shifted from the periodic lattice sites, ascertained by diffraction. Our further XAFS investigation of the disordered metallic alloy Au_xCu_{1-x} [2] also reveals considerable deviations of the interatomic distances from averaged ones measured by XRPD. In our present careful XRPD investigation we are trying to find the characteristic features of diffraction patterns of crystals with disorder.

Earlier [3] the diffuse scattering of the ordered Cu_3Au alloy were extensively studied. In this paper the alloys with random distribution of the atoms within the crystal lattice are investigated. Typical diffuse intensities [4] are several orders of magnitude below Bragg peak intensities. Therefore, an investigation of the diffuse scattering demands very high signal-to-noise ratio. In the present work we use the width of the Bragg peak itself. This experiment is easier because of large peaks' intensities. Another feature of this work is that we use the XPSD measurement which is easier to do and it can be applied for the wider range of materials than the usual (single crystal) X-ray

[1] This work was supported by The Aaron Gutwirth Foundation, Allied Investments Ltd. (Israel).
[2] This work was supported by DOE Grant No. DE-FG02-96ER45439 through the Materials Research Laboratory at the University of Illinois at Urbana-Champaign (AIF).
[3] This work was supported by DOE Grant No. DE-FG03-98ER45681 through the University of Washington, Seattle (AIF and EAS).

diffraction. It is especially important for the thin film growth monitoring and for the investigation of the properties of cold-rolled steels. The line widths as a function of diffraction angle have been carefully measured for different concentrations of Au-Cu, Au-Ag, RbBr-RbCl and AgBr-AgCl. A definite correlation has been found between the built in atomic size disorder and the XRPD line broadening in all the investigated mixed crystals.

EXPERIMENTAL

The Au-Cu alloys were quenched to avoid phase separation. The details of samples' preparation were presented in Ref. [1,2]. Homogeneity of the alloys and mixtures was verified by XRPD and no trace of phase separation or superstructure were observed for all the concentrations. The compositions of the mixed samples were established by energy dispersive spectroscopy with a scanning electron microscope. All specimens were found to be homogeneous within the accuracy of 1 %.

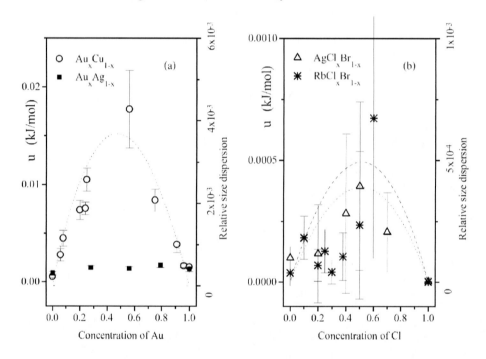

FIGURE 1. Part a: The density of the deformation energy u for Au_xCu_{1-x} and Au_xAg_{1-x} metallic alloys. Dotted (Au-Cu) and dashed (Au-Ag) curves show the relative size dispersion (right scale). Part b: The density of the deformation energy u for $AgCl_xBr_{1-x}$ and $RbCl_xBr_{1-x}$ ionic crystals. Dotted (AgCl-AgBr) and dashed (RbCl-RbBr) curves show the relative size dispersion (right scale).

XRPD data were collected in the range 30-120° 2Θ range with CuK$_\alpha$-radiation on Θ : Θ powder diffractometer "Scintag" equipped with a liquid nitrogen cooled Ge solid state detector. Peak positions and widths of Bragg reflections were determined by self-consistent profile fitting technique with Pearson VII function [5]. Contributions of K$_{\alpha 2}$ radiation were subtracted from the total profiles; the obtained results correspond to only the K$_{\alpha 1}$component of the K$_\alpha$-doublet. Lattice constant computation was carried out by a reciprocal lattice parameters refinement.

DISCUSSION

The Williamson-Hall [6] approach allows us to separate two different causes for a line broadening: (i) a finite grain size broadening and (ii) a stress-induced broadening. The usual interpretation scheme [6] assumes that the deformation is uniform in all crystallographic directions. This assumption is doubtful for a crystalline material. It is more physical to consider an anisotropic magnitude of deformation. It was shown earlier [7] that an assumption of uniform deformation pressure gives more adequate picture. However, it may be more reasonable to expect the real parameter of deformation to be a density of a deformation energy u. We compared [8] all the three approaches and found, that the density of the deformation energy better describes the experimental data on mixed alloys.

Fig. 1 presents the calculated density of the deformation energy for the metallic (part a) and non-metallic (part b) samples. One can see from both figures that the maximal deformation energy density roughly corresponds to the 50-50 concentration range. In substitutionaly disordered crystal each site may be occupied either by atom of size d_1 or by that of size d_2 with probabilities $(1-x)$ and x respectively. The relative size dispersion γ can be written as [9]:

$$\gamma = \frac{\left\langle (\Delta d)^2 \right\rangle}{\left\langle d \right\rangle^2} = x(1-x)\frac{(d_1 - d_2)^2}{\left\langle d \right\rangle^2}$$

Indeed, this dispersion has a maximum at 50-50 concentration independent of the disparity of size but its maximal value is greater for the greater disparity. In the case of ionic crystal a distance between anions is a corresponding relevant parameter. Fig. 1 demonstrates the correlation between the density of deformation energy and the size dispersion. In spite of the fact that the relative size dispersion (and, simultaneously, the deformation energy) within the ionic crystals is about an order of magnitude less than in the metallic alloys and is within the uncertainty of our measurements, the results do not contradict a definite correlation between u and the size dispersion.

Another aspect of the results presented here to be noted is that the Au-Cu alloys were rapidly quenched to avoid the thermal equilibrium ordering and phase separation while the Au-Ag alloy and the mixed salts were cooled more slowly, since their disordered phases are thermally stable. Thus, one expects that the quenched alloy would have higher strains than the others, as observed. The microstrains are an inherent feature of the non-thermal equilibrium quenched state and any attempt to anneal them would cause the alloy to change phase. These microstrains are an inherent feature of the disordered alloy at the lower temperatures and their large magnitude is related to the large relative atomic size disparity in the disordered Au-Cu alloys. The energy associated with the microstrains is the important factor in making the disordered phase of Au-Cu alloys thermally unstable at lower temperatures.

CONCLUSIONS

Density of strain energy u for mixed crystals Au-Cu, Au-Ag, AgCl-AgBr and RbCl-RbBr was extracted from our experimental XRPD data. This strain energy density depends on concentration the same way as the relative atomic size dispersion. A definite correlation exists between the line widths of the XRPD pattern and the relative atomic size disparity of the disordered mixed crystals Au-Cu, Au-Ag, AgCl-AgBr and RbCl-RbBr.

REFERENCES

1. Frenkel, A.I., Stern, E.A., Voronel, A., Qian, M., and Newville, M., *Phys. Rev. Lett.* **71**, 3485 (1993); Frenkel, A.I., Stern, E.A., Voronel, A., Qian, M., and Newville, M., *Phys. Rev. B* **49**, 11662 (1994); Frenkel, A.I., Stern, E.A., Voronel, A., and Heald,S., *Solid State Commun.* **99**, 67 (1996); Frenkel, A.I., Voronel, A., Katzir, A., Newville, M., and Stern, E.A., *Physica B* **208-209**, 334 (1995).
2. Frenkel, A.I., Stern, E.A., Rubshtein, A., Voronel, A., and Rosenberg, Yu., *J. Phys. IV* **7**, C2-1005 (1997); Frenkel, A.I., Machavariani, V.Sh., Rubshtein, A., Rosenberg, Yu., Voronel, A., and Stern, E.A., *Phys. Rev. B* **62**, 9364 (2000).
3. Cowley, J.M., *J. Appl. Phys.* **21**, 24 (1950); Warren, B.E., Averbach, B.L., and Roberts, B.W., *J. Appl. Phys.* **22**, 1493 (1951).
4. Welberry, T.R., and Butler, B.D., *Chem. Review* **95**, 2369 (1995).
5. Langford, J., and Louër D., *Rep. Prog. Phys.* **59**, 131 (1996).
6. Williamson, G.K., and Hall, W.H., *Acta Metallurgica.* **1**, 22 (1953).
7. Reimann, K., Würschum R., *J. Appl. Phys.* **81**, 7186 (1997).
8. Rosenberg, Yu., Machavariani, V.Sh., Voronel ,A., Garber, S., Rubstein, A., Frenkel, A.I., and Stern, E.A., *J. Phys.: Condense. Matt.*, **12**, 8081 (2000).
9. Voronel, A., Rabinovich, S., Kisliuk, A., Steinberg, V., and Sverbilova, T., *Phys. Rev. Lett.* **60**, 2402 (1988).

Collapsed-core Atoms as a Source of Disorder within Transition Metals

V. Sh. Machavariani[1] and A. Voronel[2]

R. & B. Sackler School of Physics and Astronomy, Tel-Aviv University, 69978, Israel

Abstract. A considerable heat capacity anomaly of metals with unfilled d-states at high temperature is explained using Core Collapse Model (CCM). The XPS measurement of Pt atoms embedded in Cu matrix has been performed at different concentrations. The analogous measurement of Au atoms has been used as a reference. A difference in Pt levels' behavior and their dramatic narrowing with diluting have been interpreted in favour of the CCM.

INTRODUCTION

In 1956 J. Friedel [1] explained an anomalous heat capacity behavior of uranium as a result of $7s \rightarrow 6d$ transition in a part of atoms resulting from thermal fluctuations. Such a behavior is not a particular property of U (and Th [1]) but is typical for atoms [2,3], with unfilled d level which is distinctive enough from the energetically nearest s state (in particular transition metals but actually also lantanoides, actinoides, Cs, Ba). The heat capacity of these metals considerably deviates from the linear behavior common for "regular" metals as Cu, Ag, Au, Al, etc., and two- three times exceeds the usual "Dulong-Petit" law [4] value.

Such a transition, indeed, had been observed in Cs and Ba for $6s \rightarrow 5d$ levels under high pressure [5]. From these experiments we knew that the Cs and Ba atoms may shrink considerably (~ 15 % of their volume) changing their quantum state with absorbing the energy ΔE (so-named isostructural phase transition [6]). Then the Core Collapse Model (CCM) ([7]) was generalized also for thermodynamic behavior of Ce and Ce-Th alloy [8.9]. (Although in the case of Ce the matter is complicated by involving $4f$ level). The CCM gives a plausible way to explain peculiar diagrams of states of Cs, Ce and many others [10]. Thus these metals behave actually as binary alloys with considerable size disparity. The concentration of shrunk atoms (and corresponding level of disorder) is a function of thermodynamic state of the matter. The CCM might be a cause of a shortening of La-Al bonds' length observed in La-Al metallic glass [11] and two humps in its heat capacity temperature dependence [12] as well.

---

[1] This work was supported by The Aaron Gutwirth Foundation, Allied Investments Ltd. (Israel).
[2] Authors are deeply grateful to M. Karpovskii for the samples' preparation.

CP554, *Physics in Local Lattice Distortions*, edited by H. Oyanagi and A. Bianconi
© 2001 American Institute of Physics 1-56396-984-X/01/$18.00

X-RAY PHOTOELECTRON SPECTROSCOPY

The high resolution X-ray photoelectron spectroscopy (XPS)measurement gives an interesting opportunity to detect the atom, which underwent the core collapse transition. To make this problem concrete let us consider a behavior of Pt atoms embedded into a copper lattice in comparison with the analogous Au atoms' behavior. The electronic configuration of Pt is $[Kr]4d^{10}5s^25p^64f^{14}5d^96s^1$, where [Kr] is the Kr-structured core. It is important here that the $5d$ level of Pt is not saturated and $6s \rightarrow 5d$ transition is possible. The distinction between the $5d$ and $6d$ levels has a trend to be preserved for many transition metals within the crystalline lattice [13]. Moreover, it is known from Refs. [5,8] and [9] that such a transition may be induced by implementing pressure or by alloying. Since the size of the collapsed atom is smaller (the $5d$ wave function has the smaller radius than the $6s$ one) this transition reduces the inner pressure in the alloy.

The Au atom follows immediately after Pt in the Periodic Table and its electronic structure can be distinguished from the Pt one only by a presence of one additional electron making its $5d$ level fully saturated. As it was mentioned above the Au heat capacity behavior is "regular". The size of both Pt and Au atoms (2.78 Å for Pt and 2.88 Å for Au) considerably exceeds the interatomic space within the Cu matrix (2.56 Å). Therefore, the both atoms should be compressed within the Cu surrounding. The atomic size of Au is even greater than the Pt diameter, thus the effect of compression is expected to be even more pronounced. This effect has to be higher for the lower concentrations since in a dilute limit the greater atom is surrounded by the smaller ones only.

Our XPS measurement gives the positions of the Pt and Au $4d_{3/2}$, $4d_{5/2}$, $4f_{5/2}$ and $4f_{7/2}$ levels relative to the Fermi energy. The levels' positions of both atoms are expected to be shifted since they are implanted into the Cu matrix.

The results for the Au-Cu alloy exhibit a homogeneous shift of all the checked energy levels as a result of the compression within the Cu lattice. In the contrast to Au the spectra for Pt-Cu alloys exhibit a sharply different picture for Pt atoms: the shifts are drastically different for different levels and the difference in the shift depends on the concentration. The increase of this difference in dilute alloys agrees with the CCM hypothesis. Fig. 1a illustrates this difference between the Au ("regular") and Pt ("anomalous") metal behavior.

It is important to note that these experimental data are consistent with the CCM. Really, for the implanted Au atom (without collapse) one may expect the constant shift originated from the smooth redistribution of the electron density in the alloy. This redistribution is approximately constant on the scale of the radius of the wave functions of the deep inner levels. However, in the case of $6s \rightarrow 5d$ transition the shift may be different for different levels since the core collapse changes the conditions for deeper electronic states in a different way. Indeed, the $5d$ wave function has both a smaller radius [5,6,8] and a different symmetry (angular dependence and spin) than the $6s$ one. Thus one may expect that the effective potential for each deeper level changes in a different way.

To check up our hypothesis of the core collapse presence the widths of both Pt and Au $4d_{3/2}$, $4d_{5/2}$, $4f_{5/2}$ and $4f_{7/2}$ levels in Pt_xCu_{1-x} and Au_xCu_{1-x} alloys as functions of concentration have been measured. The results are presented in Fig. 1b. The widths of the Au levels become a little bit wider in the dilute limit (because of a strong dependence of the compression on the local surrounding). However, the widths of the Pt levels are unexpectedly narrowing for the smaller concentrations. It might be understood if one assumes that Pt in the alloy is a mixture of two states (regular and collapsed one). Therefore, the measured XPS spectra is a superposition of two signals from these states. The closer we are to the dilute limit the higher part of implanted Pt atoms is transformed into the collapsed state. As a result the observed peak width becomes narrower.

CONCLUSIONS

Pt atoms (with unfilled d-state) have been implanted into the Cu matrix to initiate the core collapse. The Au atoms (with filled d-state) similarly implanted into the Cu matrix have been used as a reference. The XPS of the deeper ($4f$ and $4d$) levels of both Pt and Au have been measured as a function of concentration. The different behavior of the Pt and Au levels' positions as well as the dramatic narrowing of the $4d$ and $4f$ Pt levels in the dilute alloys have been observed in the contrast to the widening of the $4d$ and $4f$ Au levels as it has been expected from the Core Collapse model.

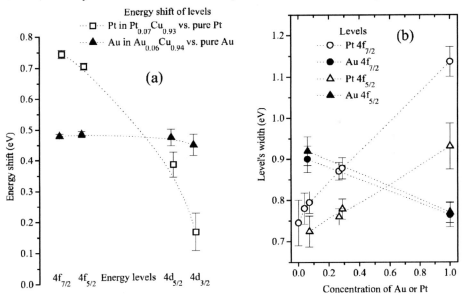

FIGURE 1. Part a: Energy shift ΔE of Pt levels in $Pt_{0.07}Cu_{0.93}$ alloy from the values for pure Pt (squares) and Au levels in $Au_{0.06}Cu_{0.94}$ alloy from the values for pure Au (triangles). Part b: Widths γ of Pt and Au $4f_{7/2}$ and $4f_{5/2}$ levels in Cu matrix versus concentration. Dashed curves are added to lead the eye.

REFERENCES

1. Friedel, J., *J. Phys. Chem. Solids* **1**, 175 (1956); Wallace D.C., *Phys. Rev* **120**, 84 (1960).
2. Cezairliyan, A., Morse, M.S., Berman, H.A., and Beckett, C.W., *Journal of Research of the National Bureau of Standarts* (*A. Physics and Chemistry*). **74A**, No. 1, 65 (1970); Cezairliyan, A., McClure, J.L., ibid, **75A**, No. 4, 283 (1971); Cezairliyan, A., McClure, J.L., and Beckett, C.W., ibid., **75A**, No. 1, 1 (1971); Cezairliyan, A., ibid., **75A**, No. 6, 565 (1971); Cezairliyan, A., Righini, F., and McClure, J.L., ibid., **78A**, No. 2, 143 (1974).
3. *JANAF Thermochemical Tables*, 2nd edition, 1971 (National Bureau of Standarts, Washington D.C., 1971.
4. Dulong, P.L., and Petit, A.T., *Ann. Chim.* [2] **7**, 113, 225, 337 (1819).
5. Sternheimer, R., *Phys. Rev.* **78**, 235 (1950); Hall, H.T., Merrill, L., Barnett, J.D., *Nature* **146**, 1297 (1964).
6. See, for instance, in *Valence Instabilities and Related Narrow Band Phenomena*, edited by Parks, R.D. Plenum, NY, 1977.
7. Kincaid, J.M., Stell, G., and Hall, C.K., *J. Chem. Phys.* **65**, 2161 (1976); ibid., **65**, 2172 (1976); ibid., **67**, 420 (1977).
8. Beecroft, R.I., Swenson, S.A., *J. Phys. Chem. Solids* **15**, 234 (1960); Wilkinson, M.K., Child, H.R., McHargue, C.J., Koehler, W.C., Wollan, E.O., *Phys. Rev.* **122**, 1409 (1961); Davis, B.L., Adams, L.H., *J. Phys. Chem. Solids* **25**, 379 (1964); Gschneidner, K.A., Elliott, R.O., and McDonald, R.R., *J. Phys. Chem. Solids* **23**, 555, 1191, 1201 (1962); McHargue, C.J., and Yakel, H.L., *Acta Metall.* **8**, 637 (1960).
9. Lawrence, J.M., Croft, M.C., and Parks, R.D., *Phys. Rev. Lett.* **35**, 289 (1975); Bates, L.F., Newmann, M.M., *Proc. Phys. Soc. London.* **72**, 345 (1958).
10. Jayaraman, A., Newton, and R.C., McDonough, J.M., *Phys. Rev.* **159**, 527 (1967).
11. Frenkel, A., Stern, E., Voronel, A., Rubstein, A., Ben-Ezra, Y., and Fleurov, V., *Phys. Rev. B* **54**, 884, (1996); Frenkel, A., Rubstein, A., Voronel, A., and Stern, E.A., *Physica B* **208 & 209**, 398, (1995); Manov, V., Rubshtein, A., Voronel, A., Popel, and P., Vereshagin, A., *Mater. Sci. Eng.* **A179-A180**, 91, (1994).
12. Ben-Ezra, Y., and Fleurov, V., *J. Phys.: Condenc. Matt.* **11**, 135, (1999).
13. *Electronic Density of States*, edited by Bennett, L.H., National Bureau of Standards Special Publication 323, Washington, D.C., December 1971, pp. 222, pp.217.

A

Abe, S., 350
Adachi, T., 202
Akoshima, M., 202
Alam, S., 246
Anderson, M., 384
Aoyama, M., 202
Arai, M., 191
Arao, M., 442
Asai, Y., 350
Asamitsu, A., 417
Asokan, K., 323
Awakura, Y., 498
Azuma, Y., 296

B

Babushkina, N. A., 391
Balatsky, A. V., 182
Barilo, S., 291
Battisti, M., 456
Belova, L. M., 391
Bennington, S. M., 191
Bianconi, A., 124, 141, 449, 456
Bianconi, G., 124
Binsted, N., 470
Bishop, A. R., 182
Bridges, F., 384
Bruls, G., 291

C

Cao, D., 384
Capone, M., 371, 395
Castro Neto, A. H., 209
Chen, X., 491
Chiba, K., 512
Chiou, J. W., 323
Chistotina, E. A., 391
Choy, J.-H., 217
Conradson, S. D., 503

D

Dalela, B., 217
Dalela, S., 217
Di Castro, D., 124
Diop, D., 304
Dogan, F., 154

E

Egami, T., 38
Emura, S., 309
Endoh, Y., 191, 422
Escribe-Filippini, C., 291
Espinosa, F. J., 336, 503
Eto, T., 379

F

Feinberg, D., 262, 371, 395
Fjellvåg, H., 470
Fratini, S., 371
Frenkel, A. I., 521
Fujishiro, H., 433

G

Garber, S., 521
García, J., 357
Garg, K. B., 217
Girshberg, Y., 278
Golovashkin, A. I., 300
Gontchar, L. E., 403
Gorbenko, O. Y., 391
Goto, T., 417
Grilli, M., 371, 395
Gudenko, S. V., 300
Gusakov, V., 119, 291

H

Hamada, T., 3
Harigaya, K., 257

Haskel, D., 154
Hasselmann, N., 209
Hayamizu, N., 84
Hayashi, K., 498
Hazama, H., 417
Hemberger, J., 408
Henderson, A., 503
Hervieu, M., 379
Higemoto, W., 202
Hirano, T., 517
Hirota, K., 422
Honda, F., 379

Kotliar, G., 115
Koyama, Y., 442
Kristoffel, N., 98
Kubota, M., 422
Kubozono, Y., 309
Kugel, K. I., 391
Kusmartsev, F. V., 133, 262
Kusunoki, M., 512
Kuwabara, M., 465
Kuwahara, H., 437
Kuznetsov, A. V., 269

I

Ignatov, A. Y., 169
Ikebe, M., 433
Ikeda, S., 304
Inoue, J., 399
Inoue, Y., 442
Ishii, M., 498
Ito, Y., 309

L

Lee, J. F., 217
Li, Z., 491
Lin, I. N., 323
Loidl, A., 408
Lu, K., 491
Lüthi, B., 291

J

Jain, D. C., 217
Jan, J. C., 323

M

Machavariani, V. S., 521, 525
Maignan, A., 379
Makino, S., 512
Marcelli, A., 456
Martin, C., 379
Martin, I., 182
Matsui, M., 498
Menushenkov, A. P., 269
Mishima, A., 242
Miyanaga, T., 304
Miyazaki, M., 222
Mizouchi, H., 328
Morais Smith, C., 209
Mori, S., 428
Moritomi, Y., 422
Moritomo, Y., 442
Motome, Y., 115
Mukaida, M., 512
Mustre de León, J., 336
Mutaguchi, K., 309
Myasnikov, E. N., 315
Myasnikova, A. E., 315

K

Kagan, M. Y., 269
Kageyama, H., 344
Kamimura, H., 3
Kanazawa, I., 253
Kaneyoshi, T., 498
Kasatani, H., 478
Katsufuji, T., 428
Kaul, A. R., 391
Kawamata, T., 202
Kawamoto, T., 350
Kida, N., 366
Klementev, K. V., 269
Klopov, M., 98
Kobayashi, K., 478
Koike, S., 222, 232
Koike, Y., 202
Kon, H., 304

Weller, M. T., 470
Wichert, D., 291
Wolf, B., 291
Wu, Z. Y., 449, 456

Yamamoto, N., 428
Yanagisawa, T., 222, 232, 246
Yonemitsu, K., 465
Yoshizawa, H., 422

Y

Yacoby, Y., 278
Yagi, K., 296, 478
Yakubovsky, A. Y., 300
Yamaji, K., 222, 232

Z

Zaharov, A. Y., 413
Zhang, X., 491
Zherlitsyn, S., 291
Zoli, M., 92